Farm Management in a Changing World: Improving Smallholder Livelihoods in Low-income Countries

Farm Management in a Changing World: Improving Smallholder Livelihoods in Low-income Countries

David Kahan

Former Chief Technical Advisor and Senior, Agro-enterprise Development and Extension Specialist, FAO, Rome, Italy (retired)

Steven Worth

Retired Professor of Agricultural Extension and Former Dean and Head of the School of Agriculture, Earth and Environmental Sciences at the University of KwaZulu-Natal, South Africa

CABI

CABI is a trading name of CAB International

CABI
Nosworthy Way
Wallingford
Oxfordshire OX10 8DE
UK

Tel: +44 (0)1491 832111
E-mail: info@cabi.org
Website: www.cabi.org

CABI
200 Portland Street
Boston
MA 02114
USA

T: +1 (617)682-9015
E-mail: cabi-nao@cabi.org

A catalogue record for this book is available from the British Library, London, UK.

ISBN-13: 9781836990512 (hardback)
9781836990529 (ePDF)
9781836990536 (ePub)

DOI: 10.1079/9781836990536.0000

Commissioning Editor: David Hemming
Editorial Assistant: Theresa Regueira
Production Editor: Helen Elliott

Typeset by Exeter Premedia Services Pvt Ltd, Chennai, India
Printed and bound by CPI Group (UK) Ltd, Croydon, CR0 4YY

Contents

Preface

There are more than 570 million family farms worldwide (Lowder, S. K., *et al.*, 2021). Whatever your social or economic background, if you are reading this book, you must have an interest in the future of agriculture, food production and the fate of smallholder farmers – because that is what this book is about.

Food is essential for all human life. Other agricultural products like fibre and fuels also form a critical part of our existence. While you are likely to have a better understanding, we would guess that when most people think about agriculture, they think about what is referred to as industrial agriculture – large scale intensive production of crops and animals. And they would not be wrong. But they would also not be right.

According to the UN Environment Programme (UNEP, 2020), 'industrial agriculture produces mainly commodity crops which are used in a wide variety of inexpensive, calorie-dense and widely available foods'.[1] The main commodity crops are rice, maize and wheat. Together they account for 60% of all dietary energy globally. Industrial agriculture is massive and pervasive. It is somewhat obvious because people living in urban areas rely on industrial agriculture for their main source of nutritional calories.

What is less known is that, despite the enormity of industrial agriculture, smallholder farmers play a vital role in the agricultural sector economy of low-income countries, serving as the backbone of food production, rural livelihoods and poverty reduction.

Of the more than 570 million family farms worldwide (Lowder, *et al.*, 2021), smallholder farms, of less than 2 hectares, represent more than 80% of all farms globally. While they occupy only 12% of the world's farmland, they account for a substantial portion of agricultural production and between 30-32% of the world's food supply (Ritchie, 2021).

Unfortunately, while smallholder farmers make such a massive contribution, they get very little benefits in return. They are the most vulnerable players in the agri-food system and, more often than not, are the most adversely affected by the many rapid changes occurring globally.

The future of these smallholder farmers (and their farms) is at risk. And that is the main driver for this book on farm management for smallholder (family) farmers. We hope to contribute to changing that dynamic – and we hope you will join us in this mission.

What is causing the risk to smallholder farmers and their farms? There are many factors. A few key ones are outlined here.

The Evolving Landscape of Agriculture Demands

Over the last several decades, the world has become more connected – communication and transactions take place seemingly at the speed of light. Agriculture has been and continues to be affected by climate change, shifting dynamics in global trade, liberalization of markets and changes in demand for agricultural products driven by population trends and the massive changes that are occurring in technology, including production, business management and communications technologies. Agriculture – farming – is no longer just about producing and selling products. The social, economic and natural environments in which agriculture operates is ever-changing, growing more and more complex with each passing year. This has implications for smallholder farmers – particularly those in low-income countries – who must find ways to rise to the challenges that demand they not only continue to meet home consumption requirements, but also to participate in the wider agri-food system.

In this evolving landscape of agriculture and the modern dynamics of the world agri-food system, it has become increasingly evident that farm management, once a prominent discipline in developed countries, has lost some of its allure. In low-income countries, these very same changes in agriculture highlight the fact that farm management is more critical than ever and is indispensable for protecting and enhancing rural livelihoods. Farmers must make sound farm management decisions to navigate these complexities effectively and ensure that their operations are sustainable and profitable.

Climate Change as a Risk for Smallholder Farmers

Climate change poses a significant risk to smallholder farmers, exacerbating the challenges they already face. Unpredictable weather patterns, increased frequency of extreme weather events, and shifting growing seasons can lead to crop failures, reduced yields, and increased pest and disease pressures. These impacts threaten the livelihoods of smallholder farmers who are often the least equipped to adapt to such changes. Addressing climate change requires concerted efforts to build resilience through sustainable farming practices, access to climate-smart technologies, and robust support systems. It is crucial for smallholder farmers to make sound management decisions that consider both immediate and long-term environmental impacts to sustain their agricultural practices and secure their futures.

The Greying of Smallholder Farming and Young People's Disinterest in Agriculture

Currently, we are encountering a pressing concern: the farming population is gradually aging and young people are turning away from agriculture. Without new entrants into farming, there will be no future for smallholder farming. While industrial agriculture may appear to be a straightforward solution for scaling up production, it poses significant risks to socio-economic systems, livelihoods, human health and the natural environment. Ways must be found to attract young people into the system and great potential lies in innovation.

Indeed, the future of farming depends on innovation. Innovation, particularly through digital technologies, offers the opportunity to transform agriculture into a profitable enterprise, potentially rekindling interest among the youth. By embracing novel approaches, technologies, strategies and management systems, we have the opportunity to make farming an attractive and financially rewarding venture, ensuring its sustainability for future generations. But it will happen only if there is real commitment to attracting the youth into this space.

The Unique Situation of Smallholder Farmers

To effect the changes needed to address the challenges posed by the evolving agricultural landscape and the greying of smallholder farming, constant communication and engagement with these farmers is required.

However, reaching smallholder farmers in low-income countries presents a unique set of challenges. The situation of smallholder farmers in the Global South is very different to that of industrial agriculture. Firstly, their farms are family farms. Their farming activities and associated finances are closely intertwined with other household activities and financial matters. Secondly, these farm households are increasingly dependent on a broad range of livelihood strategies – some as income-generating opportunities and others as coping strategies. Farming is just one of those income sources. A third factor is that often smallholders (farmer and family) take care of everything related to their farms and, thus, spend only a portion of their working time in production, with the rest divided between processing, transporting and selling farm produce. They may also have further demands on their time, such as daily responsibilities or, even, additional employment. Finally, smallholder farms are often scattered in more remote areas with limited infrastructure.

These differences require an understanding of farm management from a broader farm-household-livelihood perspective, and the application of both traditional and new decision-making methods to address this wider system. By exploring these diverse avenues and understanding the specific needs and circumstances of smallholder farmers, we can reshape the landscape of farm management in low-income countries and create pathways towards improved livelihoods. By pursuing this, we can help change the dynamic where smallholder farmers account for such a large proportion of agricultural production in the world, but see little benefit from it other than survival, to a situation where their efforts not only maintain their families but provide opportunities for advancement.

Traditional extension approaches are not designed to achieve this. The key to achieving this lies in harnessing the power of advisory services and information and communication technologies to achieve effective outreach and scaling with a specific focus on farm (business) management. The premise of this book is that there is a critical need for further developing, expanding and implementing farm management extension and advice with the express purpose of empowering these farmers to make decisions that will enable them to navigate the turbulent waters of the changing agricultural landscape, and that will aid them in responding to market changes, including gaining access to more profitable and sustainable markets. To effect such a change will also require at least three things: (i) policy commitment to fostering the adoption of farm management as a focus for extension and advice; (ii) developing extension workers' skills in the discipline of farm management; and (iii) promoting the creation and deployment of farm management specialists to augment the frontline extension efforts.

Important Realities to Acknowledge

We do not see a dichotomous choice between supporting smallholder and large-scale farming. Both have their place in the agri-food system, and both have a distinct and valuable contribution to make to create rural wealth, ensure sustainable food production, strengthen livelihoods and create vibrant economies.

While we believe in the potential of smallholder farming, we acknowledge the reality that, for very small-scale subsistence farms, transforming into viable businesses that sustainably support livelihoods may not be feasible. In such cases, it may be necessary to provide exit strategies that help such farmers find alternative livelihood opportunities outside of farming.

The farm management methods and tools proposed in this book to engage small family farmers require careful consideration, as they differ significantly from the approaches commonly found in the Western farm management discipline, which assume that farmers are already operating their farms as businesses.

A Journey of Exploration

In this book, we embark on a journey to explore the intersection of farm management, strengthened livelihoods, and sustainable agricultural practices. By blending traditional wisdom with innovative techniques, we aim to equip farmers, extension workers, specialists and stakeholders with the knowledge and tools to make informed decisions and optimize the farm enterprises of smallholder family farms. Our focus extends beyond profit margins, encompassing social, environmental and economic sustainability. We understand that successful farm management is not just about maximizing yields, or even maximizing incomes, but also about creating resilient and thriving rural communities.

In this book, the authors draw on their expertise and experience, as well as the insights of experts, practitioners and researchers globally. This book seeks to bridge the gap between theory and practice, offering practical insights, case studies and actionable strategies. The authors' intention is to ignite a renewed enthusiasm for farm management in the context of low-income countries, where its significance remains paramount.

The book is organized into four interlinking parts. Part 1 (Chapters 1–3) sets the scene by exploring the small family farm context. Part 2 (Chapters 4–8) explores the concept of farm management. Part 3 (Chapters 9–14) discusses management techniques and tools that support decision making in the context of small family farms. Finally, Part 4 (Chapters 15–17) considers the role of farm management advisory services in supporting smallholder farming. The book also provides references and a generous list of additional readings for consideration.

Each part of the book can be read as a standalone section, allowing readers to explore topics of particular interest without needing to read the book from cover to cover. This flexible structure is designed to meet the diverse needs of readers, whether they are students, practitioners or policymakers.

Readers are invited to envision the future of smallholder farming – where 'farming as a business' drives better livelihoods and greater opportunities. And they are invited to determine how they will be part of this much needed transformation to ensure that smallholder farming can advance and thrive well into the future.

Terminology Used in this Book

In this book, we use several terms interchangeably to reflect the diverse ways small-scale farming systems are described in literature and practice. Terms such as **smallholder farmer**, **small family farm**, and **small-scale family farm** are used interchangeably, recognizing that they broadly refer to the same category of farms characterized by their limited size, family-based management, and strong reliance on family labour. The book is deliberately pitched at the level of the **small farmer**, addressing their unique needs and circumstances.

Similarly, we use the terms **farm management** and **farm business management** interchangeably throughout the book. While **farm business management** is often considered a subset of the broader **farm management** discipline, it is particularly relevant in the context of viewing farming as a business. This emphasis on farming as a business aligns with our focus on supporting small family farms in becoming more market-oriented and economically sustainable.

The book also refers to **agribusiness development and management**, which we define as the off-farm provision of goods and services by stakeholders along the value chain. While the focus of this book remains on **small family farm business management**, references to **agribusiness** are made specifically in the context of private or public-private enterprises that interact with farmers – whether through the purchase of their products or the supply of inputs and services necessary for modern agriculture. These interactions, particularly those related to off-farm value chain stakeholders, are acknowledged as essential but are not the primary focus of this book.

In discussing the systems that provide advisory support for small family farm development, we also use the terms **agricultural advisory services** and **extension services** interchangeably, though they have distinct connotations. Agricultural advisory services refer to a pluralistic system that includes public, private and NGO-based providers. In contrast, the term extension services is often used in its historical context to describe the public sector's role in bridging the gap between research institutions and farmers, primarily through the dissemination of technologies and practices via farmer education. Extension, fundamentally, is an informal educational process aimed at offering rural populations advice and information to address their challenges.

Over time, the concept of extension has expanded to incorporate innovations from private sector initiatives, market-driven solutions, adaptations of indigenous technologies, and farmer-to-farmer learning. Today, a wide range of public, private and NGO-based advisory service providers are involved in delivering farm business management advice alongside traditional extension roles.

The individuals who provide these services to farmers are commonly referred to as extension agents, officers, or workers and, more broadly, as agricultural advisors. These terms will also be used interchangeably throughout this book. Finally, the term **farm management specialist** refers to the formal position of a subject matter specialist situated within a public sector extension service. However, we note that many of the specialist's functions are also found in private sector consultants working with farmers. This is an advantage to the promotion of farming as a business as it creates opportunities for collaboration.

Recognizing the evolving agricultural landscape, we emphasize the need for advisory and extension systems to adapt to changing demands. Whilst the **small family farm** remains central, extension services must also account for the perspectives of consumers and the agricultural businesses that support modern farming. Moreover, we highlight the importance of attracting rural youth to this broad agricultural ecosystem, encouraging them to initiate enterprises that foster meaningful self-employment and income generation both in production and along agricultural value chains. These efforts are essential to ensure that the development, evolution and sustainability of small family farming continues to be propelled forward in the face of changing demographic and economic realities.

The authors request that you join them, as they explore the possibilities for strengthening the position of smallholder family farms in the global agri-food system, and the contribution you can make to this.

Why this Book will be Relevant to Readers

This book is designed to serve a diverse audience, as outlined here, each playing a critical role in the agricultural sector and smallholder farming development.

- Higher education students reading agricultural development, agricultural extension, rural development, rural sociology, and farm management will find that this book gives them a non-traditional perspective on farm management, particularly for smallholder farmers in low-income countries. The book provides theoretical insights, practical applications and case studies, serving as a valuable academic resource.
- Government policymakers working in agricultural and/or rural development, or related fields, can use this book to expand their understanding of the challenges faced by smallholder farmers and to help them design policies that foster resilience, sustainability and inclusivity. This book

offers guidance for creating strategies on climate adaptation, women in development, youth engagement, and rural development in general.

- Extension workers and agricultural advisors already working on the frontline with smallholder farmers, can use this book to serve as a training and reference manual to enhance advisory services. It provides insights on integrating traditional and innovative farm management approaches to support smallholder farmers in complex agricultural environments.
- Farm management specialists will be able to use this book to refine their skills in providing targeted advice on farm business development, climate-smart practices and sustainability strategies. The book equips specialists to navigate both traditional extension roles and modern challenges such as market integration and value chain participation.
- Agricultural economists can use this book to help them analyse the economic implications of smallholder farming within the broader agri-food systems. It offers data, trends, and strategies for evaluating farm-level productivity, profitability and resilience.
- Rural planners can use this book as a guide to help them when designing community-level interventions that enhance infrastructure, access to markets, and the integration of smallholder farms into local and national economies. This book highlights the intersection of farm management and rural livelihoods.
- NGOs and development practitioners will find this book a valuable resource for designing and implementing programmes that improve smallholder livelihoods, strengthen climate resilience and promote the involvement of women and youth in agriculture. It serves as a practical guide for field-level application and scaling of successful strategies.
- Private sector stakeholders, such as agribusinesses, input suppliers and financial institutions, can use this book to better understand the needs of smallholder farmers. It can also inform their plans when developing partnerships or services catering to smallholder farmers.

By addressing the unique needs of these audiences, this book aims to foster innovation, resilience and sustainability within smallholder farming systems, ensuring that their essential contributions to food security and rural development endure.

Endnote

[1]Commodity crops are crops grown, typically in large volume and at high intensity, specifically for the purpose of sale to the commodities market.

References

Lowder, S. K., Sánchez, M. V. and Bertini, R. (2021) Which farms feed the world and has farmland become more concentrated? *World Development*, 142 DOI: 10.1016/j.worlddev.2021.105455

Ritchie, H. (2021) Smallholders produce one-third of the world's food, less than half of what many headlines claim. OurWorldInData.org. Available at: https://ourworldindata.org/smallholder-food-production (accessed 12 March 2025).

UNEP (2020) 10 Things you Should Know about Industrial Farming. Available at: https://www.unep.org/news-and-stories/story/10-things-you-should-know-about-industrial-farming/ (accessed 3 July 2025).

About the Authors

David Kahan

David Kahan is an agricultural economist and agricultural extension expert with more than 40 years of experience in international agricultural development. His work has spanned across Africa and Asia, where he has collaborated with leading development organizations including the FAO, IFAD, the World Bank, and the European Union. For 25 years, he served as a senior officer in agro-enterprise development and agricultural extension at FAO headquarters in Rome, Italy. Following his retirement in 2013, David continued his career at CIMMYT (a CGIAR research centre) as principal scientist in agribusiness development. Since then, he has worked as an independent consultant, advising governments, development partners and programme managers in over 70 countries. His advisory work has focused on strengthening farm management, agricultural extension systems, food security and sustainable agriculture. David has authored numerous influential publications, which are widely used in the field. He holds a PhD in agricultural economics and rural development from the University of Reading, UK.

Steven Worth

For 20 years Steven Worth worked in agricultural extension and development as a field practitioner and manager in South Africa. He then spent a further 20 years as a professor, researcher, Director of the African Centre for Food Security, and as Dean and Head of the school of agriculture, earth and environmental sciences at the University of KwaZulu-Natal (UKZN), from which he recently retired. He continues serving the university as a curriculum developer and manager of international projects in education. He brings to this book a wide-ranging expertise augmented by years of working as a private consultant with numerous public, private and NGO agencies, and agricultural educational institutions, creating policies, plans and programmes on capacity development, wealth creation, rural development, and, especially, agricultural extension and education. Steven holds a Bachelor of Science in agricultural business, a Masters in agricultural development, and a PhD in agricultural extension and rural resource management from UKZN, South Africa.

Glossary

Please note, definitions reflect the context of this book and may vary from more general definitions.

Agro-processing	The transformation of raw agricultural products, such as crops and livestock, into consumable food, beverages or industrial products.
Agribusiness	Commercial farming and the industries related to agricultural production, including supply chains and processing.
Agroecology	An approach to agriculture that views agricultural areas as ecosystems and promotes sustainable and biodiverse farming practices.
Artificial Intelligence (AI)	Computer systems capable of performing tasks that typically require human intelligence, such as decision making, pattern recognition and learning.
Benchmarking	Comparing farm performance indicators with standard references or industry best practices to identify areas for improvement.
Big data	Extremely large data sets that can be analysed computationally to reveal patterns, trends and associations in agriculture.
Biodiversity	The variety of plant and animal life in a particular habitat, crucial for resilient and productive farming systems.
Biosecurity	Measures taken to prevent the introduction and spread of harmful organisms to crops and livestock.
Blockchain	A secure digital ledger system that can enhance transparency and traceability in agriculture and food supply chains.
Budgeting	The process of creating a plan to spend money, which helps farmers allocate resources efficiently.
Business plan	A written document that outlines a farm's operational and financial objectives and how they will be achieved.
Climate change	Long-term shifts in temperatures and weather patterns, largely caused by human activities.

Climate change adaptation Strategies and actions taken to adjust to the effects of climate change, such as selecting drought-resistant crops or adjusting planting times.

Commodities Agricultural products such as grains, coffee or livestock that can be traded in bulk.

Comparative advantage The ability of a farm or country to produce a good or service at a lower opportunity cost than others, forming the basis for mutually beneficial trade.

Contract farming A system in which farmers enter into agreements with buyers before production begins, often involving pre-agreed pricing.

Cost-benefit analysis A method for comparing the costs and benefits of a decision or investment to determine its feasibility or profitability.

Data analytics The process of examining agricultural data to draw conclusions and inform decisions.

Decision Support Systems (DSS) Tools that combine data and models to support farm management decisions.

Depreciation The amount of a capital asset's value that has been used up as a result of wear and tear, age, obsolescence and usage.

Digital extension The delivery of advisory services using digital tools such as mobile phones or apps.

Digital literacy The ability to use digital tools and platforms effectively and safely.

Digital marketing (e-marketing) The use of internet and mobile platforms to promote and sell agricultural products.

Digital technologies Electronic tools, systems, devices and resources that generate, store or process data.

Diminishing returns The principle stating that, as more of a variable input is used, the added benefit eventually decreases.

Diversification The practice of engaging in a variety of agricultural activities, including multiple crop types, livestock enterprises, and strategies such as vertical and horizontal integration, to reduce risk and improve resilience.

Drones Unmanned aerial vehicles used for tasks such as mapping, monitoring crops or spraying inputs.

E-extension Technology-enabled agricultural advisory services, often delivered through the internet or mobile devices.

Economic efficiency Achieving the highest value of output at the lowest possible cost.

Economic viability The ability of a farm to be financially sustainable over time.

Economies of scale The cost advantages that a farm business experiences as it increases production. As output grows, the average cost per unit of production typically decreases, making the operation more efficient.

Entrepreneurship The act of starting and managing a business, including taking financial risks to do so.

Equi-marginal principle The rule that resources should be allocated where they yield equal marginal returns across all enterprises.

Extension services Educational services that provide practical information and training to farmers.

Extension worker	A trained professional who provides on-the-ground advisory services, technical guidance and education to farmers to promote farmer learning and improved use of agricultural practices.
Factor-factor relationship	The relationship between two or more inputs (factors of production) used in the production process, while keeping output fixed.
Factor-product relationship	The relationship between the quantity of a single input (factor) used in production and the resulting level of output (product), assuming all other inputs are fixed.
Farm budgeting	Planning income and expenditures for a farm enterprise over a specified period.
Farm management	The process of making and implementing decisions about the use of resources on a farm.
Farm management software	Digital tools used to track inputs, yields and finances and improve planning and reporting.
Farm management specialist	An expert who advises farmers or institutions on efficient use of farm resources, strategic planning and decision making.
Farm planning	Strategic allocation of resources, timing and operations on a farm.
Financial literacy	Understanding how to manage money, including budgeting, saving, investing and borrowing.
Fixed costs	Costs that do not change with the level of production, such as rent or salaries.
Food security	The condition in which all people, at all times, have physical and economic access to sufficient, safe and nutritious food.
Gender equity	Fairness in treatment for women and men, according to their respective needs and interests.
Geospatial Information Systems (GIS)	Technologies that collect, store and analyse location-based data for land use planning.
Global South	In agricultural and development contexts, the Global South refers to low- and middle-income countries, primarily in Africa, Asia, Latin America and Oceania, that face structural challenges in development but are also sites of innovation and critical focus for sustainable and equitable agricultural growth.
Good Agricultural Practice (GAP)	Principles and standards that ensure safe, sustainable and high-quality agricultural production while protecting the environment and promoting worker and animal welfare.
Gross income	The total revenue a farm earns from all sources before deducting any expenses. It includes the value of products consumed and/or stored, but not sold.
Gross margins	The difference between gross income from the sale of a product and the variable costs of producing it.
Herd health	Management practices to maintain the health and productivity of livestock.
Horizontal integration	When a farm or agribusiness expands its operations at the same stage of the value chain, such as by acquiring or merging with similar enterprises.
Information and Communication Technologies (ICT)	Tools used to handle telecommunications, broadcast media, intelligent building management systems, and audiovisual processing and transmission systems.

Income statement	A financial report showing income, expenses and profit or loss over a period.
Indigenous knowledge	Traditional understanding and practices developed by local communities over generations to manage crops, livestock, soils and natural resources in ways that are adapted to their specific environments and cultural contexts.
Information management	The process of collecting, storing, and using data and information effectively to support farm operations.
Inputs	Resources such as seeds, fertilizer, labour and capital used in agricultural production.
Input supply	The provision and distribution of essential agricultural inputs such as seeds, fertilizers and equipment.
Integrated Pest Management (IPM)	A strategy that combines different pest control methods to reduce reliance on chemical pesticides.
Internet of Things (IoT)	A system of interconnected devices that collect and transmit data, such as soil sensors or automated irrigation systems.
Inventory management	Tracking and controlling stock of inputs and outputs.
Labour	Human effort used in the production and value-adding processes, including planting, weeding, harvesting, animal care and processing.
Labour productivity	The amount of output per unit of labour input.
Land tenure	Legal or customary rights individuals or groups hold to land.
Livelihood assets	The resources and capabilities – such as human, social, physical, natural and financial capital – that people draw on to pursue sustainable livelihoods.
Livelihoods	The capabilities, assets and activities that people use to make a living.
Local markets	Markets within close geographical proximity, often important for smallholder sales.
Logistics	The coordination of moving, storing and delivering agricultural products, inputs and equipment, etc.
Manure management	The handling, storage and disposal of animal waste to improve productivity and environmental health.
Marginal analysis	The study of the costs and benefits of using one more unit of a resource.
Marginal cost	The additional cost incurred from producing one more unit of output.
Marginal returns	The additional output or profit resulting from the use of one more unit of input.
Marketing	Activities involved in promoting and selling agricultural products, including market research, advertising and distribution.
Marketing channels	The routes through which agricultural products move from farm to consumer.
Market access	The ability of farmers to reach and sell in profitable markets.
Mechanization	The use of machinery in farming operations.
Mobile weather apps	Digital applications that provide real-time weather forecasts and alerts to help farmers plan agricultural activities.

Monitoring and Evaluation (M&E)	Systems used to track progress and assess the effectiveness of farming interventions.
Net returns	Profit remaining after all production costs have been deducted from revenue. Also known as gross margin.
Nutrient management	The application of fertilizers and organic matter to meet crop needs while minimizing environmental harm.
On-farm sensors	Devices placed in fields or livestock areas to collect real-time data on environmental conditions.
Opportunity cost	The value of the next best alternative foregone when a decision is made.
Organic farming	An agricultural approach that relies on natural processes and avoids synthetic inputs.
Output	The total quantity of agricultural products produced.
Participatory learning	Learning processes that involve active farmer engagement and peer-to-peer knowledge sharing.
Pedagogy	The method and practice of teaching.
Precision agriculture	Technology-enabled approach to optimize field-level management regarding crop farming.
Processing	Converting raw agricultural materials into finished goods or ingredients.
Product-product relationships	The interactions between different enterprises on a farm, such as how increasing one affects the output or profitability of another.
Production plan	A schedule outlining the inputs and steps required to produce agricultural goods.
Profitability	The financial gain from a farm operation after all expenses have been subtracted.
Quality control	Processes to ensure agricultural products meet certain standards.
Record-keeping	The systematic documentation of farm activities and finances.
Regenerative agriculture	Farming practices that restore soil health, biodiversity and ecosystem services.
Relative price	The price of one good or service in terms of another, used to compare opportunity costs and trade-offs.
Remote sensing	The use of satellite or drone imagery to monitor farm conditions and environmental changes.
Resilience	The ability of a farm system to withstand and recover from shocks and stresses.
Resource allocation	The process of assigning resources – time, labour, capital – to various farming activities.
Risk management	Strategies to minimize potential losses in farming, including insurance, diversification and planning.
Rotational grazing	Moving livestock between pastures to allow forage plants to recover.
Seasonal planning	Planning farm activities around climatic and market seasons.
Small family farm	A family-run farm that relies mainly on family labour and is often of modest size.

Smallholder farm	A farm typically under 2 ha, operated and managed by a household.
Smart irrigation	Automated irrigation systems that use data from sensors and weather forecasts to optimize water use.
Soil health	The continued capacity of soil to function as a vital living ecosystem.
Stakeholder engagement	Involving all actors – farmers, government, NGOs, private sector – in decision-making processes.
Storage systems	Facilities and technologies used to store crops and inputs safely.
Strategic planning	Long-term planning that aligns farm goals with available resources and environmental factors.
Subsidy	Financial assistance provided by the government to support farming.
Substitution	The process by which farmers replace one good or input with another in response to changes in relative prices or availability.
Supply chain	The entire network involved in producing and delivering agricultural goods to consumers.
Sustainability	Meeting current farming needs without compromising the ability of future generations to meet theirs.
Sustainable livelihoods	Approaches that promote lasting means of living by enhancing access to key livelihood assets and reducing vulnerability.
Total costs	The sum of variable and fixed costs, providing a comprehensive view of the financial investment required.
Technical efficiency	Producing the maximum possible output from given inputs.
Technical training	Instruction focused on improving specific farming skills and knowledge.
Time value of money	The principle that money today is worth more than the same amount in the future, because it can be invested to earn returns.
Traceability	The ability to track agricultural products through the supply chain.
Value of production	Value of production is the total monetary worth of all goods and services produced by a farm over a given period.
Value chain	The full range of activities to bring an agricultural product from conception to end use.
Variable costs	Costs that change depending on the level of production, such as seeds or fertilizer.
Vertical integration	When a farm or agribusiness expands its operations into different stages of the same production path, such as from production to processing and retail. It also includes pre-production activities such as broiler farmers setting up their own hatchery.
Water management	Strategies for the efficient and sustainable use of water resources in agriculture.
Weather forecasting	The use of data and models to predict weather conditions that affect farming operations.
Working capital	Working capital refers to the funds or resources available - or required - to cover a farm's day-to-day operational expenses, such as seeds, fertilizers, labour and transport, between planting and harvest. Working capital refers to the readily available funds used to cover day-to-day operational expenses in farming, such as seeds, fertilizers, labour and transport between planting and harvest.

Part 1: The Small Family Farm Context

This part of the book provides a comprehensive overview of the small family farm as a unique and integral component of rural economies. It explores the defining characteristics of small family farms, including their reliance on family labour, limited resources and vulnerability to external factors. The section highlights the challenges faced by small-scale farmers, such as limited market access, the impacts of climate change and evolving government policies. It underscores the importance of developing tailored support strategies to address these challenges and improve the sustainability and productivity of small family farms.

1 The Changing Farming Environment

In this chapter, the authors explore how agriculture is changing under the influence of globalization, market changes, population shifts, climate change and new technology. As the world becomes more connected, small family farmers in low-income countries face the challenge of meeting home consumption requirements, market demands and climate change. The authors look at how global trade and market openness affect traditional farming, the rise in demand for certain agricultural products due to demographic trends, and the critical role of technology in farming today. This chapter also examines changes in the way farmers get advice and learn new methods, highlighting the need for new knowledge and skills to succeed.

1.1 Context

The primary context for this book is the small family farm, often referred to as a smallholder farm – described in detail in Chapter 2. The small family farm was chosen as the focus of this book because of the tremendous contribution it makes to economic stability, livelihood and food security. It was also chosen because it is an important subset of the larger frameworks of 'family farming' (farms owned and operated by a family) and 'small family farming' (a farm operating on a relatively small amount of land with a small labour force). Unfortunately, small family farms are frequently overlooked in discussions on agricultural development and its future. This book aims to help correct this.

Although the global changes in the agricultural landscape described in this chapter affect all farms, they have a particularly significant impact on small family farms.

A second important context is understanding the concept of the global agri-food system. According to the FAO:

> Agri-food systems comprise the entire range of actors and interlinked activities that add value in agricultural production and related off-farm activities such as food storage, aggregation, post-harvest handling, transportation, processing, distribution, marketing, disposal and consumption. (FAO, 2023)

This is important because the small family farm is very much a part of these systems. They may appear to be small players, but their role is critical.

A third context is agricultural extension. This is discussed extensively in Part 4. However, there is frequent reference to terms such as 'extension', 'agricultural extension' and '(agricultural) advisory services'. These all essentially describe a similar process – that of working with farmers to help them be more successful. As will be discussed briefly at the end of the chapter, the premise of this book is that extension services, supported by specially trained farm business management advisors (Chapter 16), will be the primary vehicle through which small family farmers are to be assisted in adapting to the many serious changes taking place in agriculture globally.

1.2 Changes in Farming as a Result of Globalization and Market Liberalization

Over the last four decades, economic liberalization policies have reduced the role of the state in managing economies. This has been especially true in low-income countries, where government support for farmers has fallen

 Farm Management in a Changing World: Improving Smallholder Livelihoods in Low-income Countries (Kahan and Worth)
DOI: 10.1079/9781836990536.0001

significantly (Pingali and Feder, 2017). As a result, farmers now carry more responsibility for their decisions and their consequences. This shift has accelerated the move towards market-led development. Globalization has further sped up this transition. It has opened new opportunities to access global export markets. The move to free markets was expected to improve efficiency. It was based on the belief that market forces could better manage production, pricing and distribution. However, this has not always worked out as expected – especially for small family farmers.

In addition to long-term changes, recent global crises and geopolitical tensions have created new challenges for small family farmers. Input costs for fertilizers, seeds and fuel have increased, straining already tight budgets and reducing productivity. Rising shipping and logistics costs, along with trade barriers, restrict market access and weaken farmers' income stability and competitiveness. Large industrial farms are better able to handle these risks; small farms often struggle to stay profitable and grow.

Despite these difficulties, greater market integration offers new opportunities. Better transport, fewer trade barriers and digital tools allow small farmers to reach more consumers. But this also brings more competition – from both local and international producers. Farmers must change their practices and business models to stay competitive.

At the same time, consumers are becoming more aware of social sustainability issues. These include child and forced labour, poor working conditions and health and environmental concerns. There is growing pressure – especially in developed countries – for farms to meet higher social, health and environmental standards. This puts extra costs and demands on small farms.

To face these growing challenges, small family farmers need strong support. Agricultural advisory services are key; they help farmers deal with changing economic conditions and meet new standards. Access to finance, markets and advice supports farmers in adopting ethical and market-focused practices. Building resilience is essential. This means diversifying crops, using sustainable practices and working together in cooperatives or networks. These steps help reduce risks and improve competitiveness; profitability and sustainability are both critical to long-term success. Advisory services will be needed as farmers continue to adapt.

Changing demographics and globalization are also shaping what consumers want. People now seek diverse, high quality and sustainably produced food. Cultural exposure and increased connectivity are changing food preferences, which creates new chances and challenges for small family farms.

Consumers are asking for more variety; they want fruits, vegetables, specialty items and organic products, and also rate products with added value. These trends reflect lifestyle changes and a growing awareness of health, environmental and social justice concerns.

For small family farmers in low-income countries, responding to these changes is vital. They can do this by expanding what they produce and using market-driven approaches. This opens opportunities to enter niche markets, such as organic, fair trade and local products, which can bring higher prices. But entering these markets requires change: farmers must adopt new standards, gain certifications and improve marketing and financial practices. They will need support to make these changes.

Agricultural advisory services play an essential role. They can help farmers with technical advice, farm management and decision making. They also help farmers connect to markets and understand how to meet quality and certification requirements. Beyond this, advisory services can help farmers gain more control over value chains. By becoming owners or shareholders, farmers can benefit more from value addition.

Small family farms located in peri-urban areas or close to urban centres have a unique advantage. Being close to cities lowers transport costs and allows delivery of fresh produce. Urban consumers often prefer local and freshly harvested food but competing in urban markets is not easy. Larger producers and imports are strong competitors. Small family farmers cannot do this alone; they need partnerships with other small family farmers, perhaps in the form of a cooperative or association. They will also need partnerships with wholesalers, retailers and food service providers. Such partnerships can help small family farmers access urban markets and negotiate better prices.

Clearly, the world has changed for small family farmers in low-income countries; they need to adapt to survive and thrive.

1.3 Climate Change and Farming

Climate change is real, and it poses a significant threat to small family farming. Increasing variability in weather patterns, rising temperatures and extreme weather events affect agricultural productivity and rural livelihoods, as well as the transportation and communication needed to run the farm business. Small family farmers in low-income countries are particularly vulnerable to the impacts of climate change due to their reliance on rain-fed agriculture, limited access to resources and lack of adaptive capacity. See Box 1.1. for a brief discussion on the importance of rain-fed farming. Addressing the challenges posed by climate change requires concerted efforts to enhance the resilience of small family farming systems, and to support farmers in adapting to changing climatic conditions.

The impacts of climate change on small family farming are multifaceted. They affect every aspect of agricultural production and rural livelihoods, as well as non-production aspects of farm management. The following are some of the key challenges faced by small family farmers as a result of climate change:

- **Altered rainfall patterns:** Changes in rainfall patterns, including prolonged droughts and intense rainfall events, pose significant challenges to small family farmers. Drought conditions can lead to crop failures, reduced water availability for irrigation and increased risk of land degradation. Conversely, heavy rainfall and flooding can cause soil erosion, damage crops and lead to loss of agricultural inputs. Small family farmers often lack the resources and infrastructure to mitigate the impacts of these extreme weather events, making them especially vulnerable to such events.
- **Increased temperatures:** Rising temperatures can have direct and indirect effects on agricultural productivity. Higher temperatures can exceed the optimal growing conditions for many crops, leading to reduced yields and increased susceptibility to heat stress. Livestock production can also be affected, with heat stress reducing fertility, milk production and overall animal health. Small family farmers may struggle to adapt to these changing conditions without access to heat-tolerant crop varieties, improved animal husbandry practices and cooling technologies.
- **Pests and diseases:** Climate change can alter the distribution and life cycle of pests and diseases, leading to increased outbreaks and the emergence of new threats. Warmer temperatures and changing moisture conditions can create favourable environments for pests and pathogens, resulting in increased crop losses and reduced productivity. Small family farmers often lack access to effective pest management strategies, making it difficult to control outbreaks and protect their crops.
- **Land degradation:** The increased frequency of extreme weather events, such as droughts and floods, can accelerate soil erosion and degradation. Loss of soil fertility

Box 1.1. Importance of rain-fed agriculture. From World Bank, 2010; Chakrabarti (IFAD), 2016; FAO, 2017a; FAO, 2017b.

Over 80% of the world's agricultural land is rain-fed, with this being particularly pronounced in low-income countries where smallholder farmers operate. Rain-fed agriculture accounts for 95% of farmed land in sub-Saharan Africa and 60% in South Asia, where irrigation infrastructure is limited. This heavy reliance on rain-fed systems leaves smallholder farmers highly vulnerable to climate variability and droughts.

and structure can reduce the productive capacity of agricultural land, making it more difficult for small family farmers to sustain their livelihoods. Soil conservation practices, such as cover cropping, reduced tillage and agroforestry, can help mitigate soil degradation, but implementing these practices requires knowledge, resources and support. See Box 1.2. for additional insight about land degradation in the context of small family farming.

- **Damage to transportation and communication infrastructure:** We are increasingly witnessing the impact of extreme weather events – particularly storms and floods – on transportation and communication infrastructure. Access to external markets depends heavily on this infrastructure. When it is damaged or disrupted, farmers lose access to vital market information and are unable to transport their products to market.

Collectively, climate change significantly threatens small family farmers, their communities, food security and local economies in low-income countries. The livelihoods of these farmers are intrinsically tied to the land and the climate. This makes them highly susceptible to the vagaries of weather and changes in climate. To mitigate these impacts, there is an urgent need for concerted efforts to enhance the adaptive capacities of small family farmers. This requires that farmers not only simply adjust but proactively innovate and reconfigure their agricultural practices to withstand the unpredictability of climate change.

Empowering small family farmers to adapt to these challenges is not only critical for their survival but also for global food security and the preservation of ecosystems. Although industrial agriculture produces most of the marketed products, small- and medium-scale family farmers, in general, are responsible for 80% of the total value of food produced, of which 32% is produced by small family farms (Ritchie, 2021). This is a major contribution to global food security. But it is somewhat 'invisible' because these farmers and their families normally consume their own production. Hidden as it may be, when their crops fail, these small family farming families and communities face food challenges, leading to a substantial impact on global food security.

Agricultural advisory services and climate-smart agriculture programmes play a

Box 1.2. Land degradation and small family farming. From IFPRI, 2020; FAO, 2021; UNCCD, 2021; CGIAR, 2024.

It is estimated that over 40% of global agricultural land is degraded, and smallholders – who cultivate around 80% of the world's farmland – are disproportionately affected. In sub-Saharan Africa, for example, 65% of agricultural land is affected by degradation, leading to a reduction in productivity by up to 50%. In South Asia, more than 30% of the land is degraded, and the impact on smallholder farming systems is pronounced due to the overuse of land and lack of access to sustainable land management techniques. In Latin America and the Caribbean, an estimated 20% of agricultural land is degraded, with severe degradation affecting around 14% of the region's total land area. The resulting decline in yields increases vulnerability, food insecurity and poverty among land-dependent smallholder farmers. The annual cost of land degradation globally is estimated to be about 10% of global GDP, with a significant share of this burden falling on smallholder farmers in low-income regions.

In many regions, smallholder farmers operate on marginal lands that are highly prone to soil erosion, nutrient depletion and desertification. In sub-Saharan Africa alone, soil erosion removes an average of 30–40 tons of topsoil per ha annually, severely impacting crop yields. This soil degradation contributes to an estimated 20% decrease in productivity for rain-fed agriculture, where smallholders dominate. In India, smallholders operating on degraded lands see productivity losses ranging from 15–30% compared to areas with better managed soils. This leads to food insecurity and increasing poverty rates, as these farmers are heavily dependent on their land for their livelihoods.

vital role in providing farmers with the knowledge and tools needed to respond to climate challenges. By supporting small family farmers in their efforts to mitigate and adapt to climate change, all of us working in these spaces can help ensure the sustainability of agricultural production and the livelihoods of rural communities in the face of an uncertain climate future. The challenges of climate change from a management perspective are further elaborated on in Chapter 7.

1.4 Technological Change

The 21st century is all about technological advancements on every front and in every aspect of life, economy, business and society, including within agriculture. Applied correctly and with wisdom, technological advancements have the potential to transform small family farming in low-income countries. They offer new opportunities to enhance productivity, sustainability, profitability and resilience. Innovations such as mobile technology, precision farming and biotechnology can help solve many challenges faced by small family farmers. However, leveraging these technologies also presents significant challenges, including issues of access, appropriateness, affordability, adoption and discernment.

The advent of digital tools and mobile communications has revolutionized access to timely and accurate information for farmers. Mobile applications, SMS-based services, and online platforms provide real-time updates on weather, market prices and agricultural best practices, empowering farmers to make informed decisions about production and marketing. Digital tools – including mobile and online services – enable access to finance, marketing and transaction support. Despite these advantages, challenges such as digital literacy and the ability to analyse data remain significant barriers to adoption.

Precision agriculture and advanced biotechnologies offer further opportunities for small family farmers. Technologies such as GIS, drones and remote sensing optimize input use, reduce costs and minimize environmental impacts, while innovations like genetically modified crops promise higher yields and resilience against climate stressors. However, the high costs of equipment, training, and concerns about the environmental impact of GM crops limit their accessibility and widespread adoption. Similarly, renewable energy sources, such as solar and wind power, have the potential to reduce dependence on fossil fuels and mitigate climate change, though their initial costs and lack of rural infrastructure remain obstacles.

The integration of digital tools into farm management, such as software for tracking activities, sales and expenses, has enabled farmers to optimize the business aspects of farming. Online banking, marketing and app-based services provide solutions across the entire agricultural cycle, from preparation to sale. These tools empower farmers to streamline operations and make data-driven decisions, improving the performance of their farm businesses. Yet, the digital divide continues to hinder the adoption of such technologies, particularly in underserved rural areas with low internet penetration.

Bridging the gap requires a collective effort from governments, NGOs and private sector initiatives to improve digital infrastructure, enhance training programmes and ensure that information systems are inclusive and relevant. Investments in capacity-building, market access and infrastructure are crucial to addressing barriers such as high costs, limited knowledge and resistance to change. By equipping small family farmers with the tools, information and resources they need, these technological advancements can unlock their potential and position them as competitive players in the global agricultural market – this is explored in greater detail in Chapter 10.

1.5 Innovation and Innovation Systems

Innovation is essential for the growth and resilience of small family farms. It does not happen in isolation but is the result of collaboration among a wide range of actors. Farmers, researchers, advisory service providers, agribusinesses and policymakers all play interconnected roles in generating, sharing and applying new ideas and practices. When these actors work together effectively, they create an environment that supports the development and adoption of new technologies, better farm management practices, and strategies to cope with market and environmental pressures. Understanding how these collaborative processes work is key to helping small family farmers adapt, stay competitive and succeed in a rapidly changing agricultural landscape.

1.5.1 Innovation at the level of the small family farm

Just as it is for any business trying to keep pace with a rapidly changing world, innovation is critical for the future of small family farming in low-income countries. Innovations in agricultural practices and business management practices, including market strategies and financial management, can improve the resilience, productivity and profitability of small family farming systems. However, fostering innovation requires a supportive environment with access to information, financial resources, markets, training and accompaniment.

To support innovation, agricultural advisory services play a vital role. They can provide farmers with the tools to identify, and the knowledge and skills needed to adopt (and adapt), new technologies and practices that are relevant to their circumstances and aims. Governments, research institutions, private businesses and NGOs must collaborate to create and distribute practical innovations for small family farmers.

Innovations in farming practices, coupled with access to new technologies, can lead to improved crop yields, enhanced nutritional quality, and increased farm incomes. Furthermore, innovation can contribute to sustainable resource management, reducing the ecological footprint of agriculture and ensuring the long-term health of the land.

Innovations in business management practices – including the use of online services – can improve overall efficiency of the farm and each of its enterprises. Such innovations also facilitate and strengthen the decision making capacity of farmers.

For innovation to be truly transformative, it must be inclusive, considering the diverse needs and capabilities of small family farmers. This requires a supportive ecosystem that includes policy support, investment in rural infrastructure, research and development tailored to the needs of small family farmers, and programmes that foster farmer education and capacity building.

Part of forming a supportive environment for innovation involves building capacity for farmers to innovate. Rather than being reactive to potential and experienced challenges, or passive adopters of externally developed technologies, farmers will benefit from a well-developed capacity to innovate and solve problems. It will increase their resilience and reduce their risk in relation to changing environmental and market conditions.

Small family farms will not be able to increase their productivity and sustainability unless they are prepared to innovate and are supported in doing so. Given the critical importance of family farming to food security, natural resource management and poverty reduction, promoting innovation in family farming should be a priority for politicians and policymakers. The broad participation and involvement of farmers – including small family farmers, women and disadvantaged or marginalized groups – will be essential.

1.5.2 Innovation as a system of knowledge creation

Innovation in agriculture is often thought of as an isolated process of research and development, often carried out by a government department of agriculture, universities or private sector industry. Most certainly, such innovation does take place, but over the last several decades there has been a move towards formalizing innovation systems. This is part of the emergence of the growing trend towards pluralistic agricultural advisory services, which, in brief, is the provision of services to farmers by multiple agencies operating with some level of coordination. This will be discussed briefly in Section 1.6 of this chapter and in more detail in Part 4.

One key aspect of the shift has been the development of Agricultural Knowledge and Innovation Systems (AKIS), which have emerged both as a conceptual framework and as an operational practice. As illustrated in Fig. 1.1, AKIS involves multiple actors – including farmers, research institutions, educational

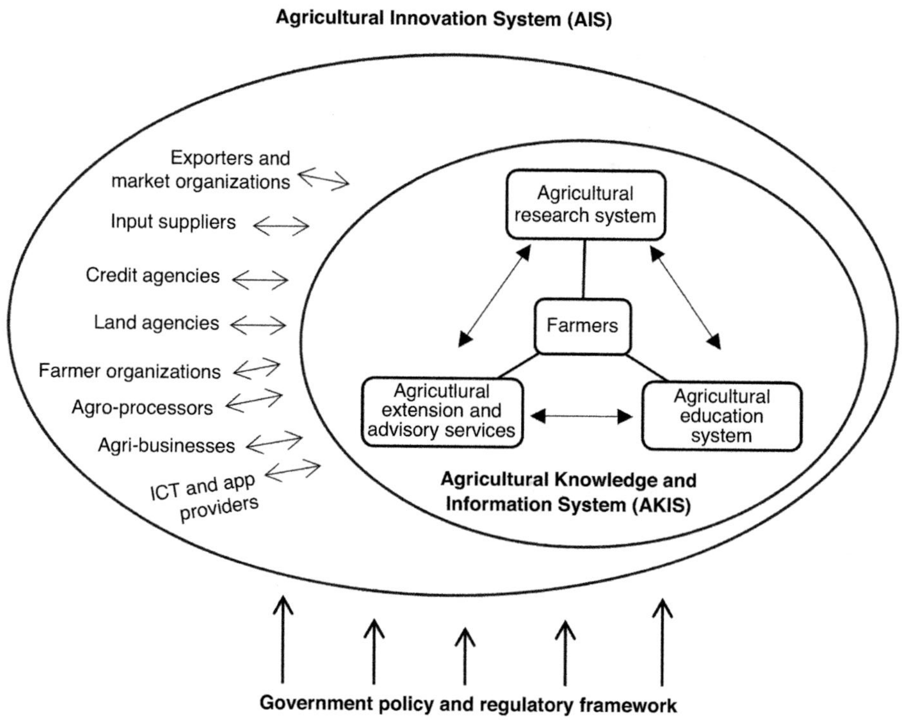

Fig. 1.1. Depiction of an agricultural innovation system. Adapted from World Bank, 2012, used under license CC-BY-3.0.

bodies and advisory services – working together to develop, share and apply agricultural knowledge and innovations. The aim is to coordinate the processes of knowledge generation and dissemination, particularly the development and adoption of new technologies and practices.

Over the years, AKIS – already a pluralistic system – has become part of the broader concept of Agricultural Innovation Systems (AIS). AIS extends beyond the traditional focus on technology transfer by public research and advisory bodies. It responds more comprehensively to production challenges, evolving market demands, and risk factors. AIS acknowledges that innovation is not driven solely by scientific or technological advances. It also emerges from the actions of private sector actors, such as agribusinesses and farmers, which adapt to shifting market conditions (Larsen *et al.*, 2009). These innovations may include changes in farm management, new business models, and strategic partnerships across the agri-food system.

In this wider context, innovation is facilitated not only through the transfer of technology but through the creation of an enabling environment that supports change. This environment is shaped by policy frameworks and institutional arrangements at both national and local levels. These, in turn, influence how services are delivered and how effectively public and private actors can collaborate. Such frameworks are especially important for integrating small family farmers and other economically disadvantaged groups into value chains, helping ensure that the benefits of innovation are broadly shared.

As shown in Fig. 1.1, AIS expands the boundaries of AKIS to encompass a wide range of agricultural and management services, along with additional stakeholders involved in knowledge creation and dissemination. This expanded system is instrumental in supporting small family farmers by improving their productivity, resilience, market access and managerial skills. Services go beyond traditional extension to include access to credit, inputs, business and organizational development, and post-harvest and market integration support.

As noted previously, AIS operates within a pluralistic advisory landscape where advice is sourced from a diverse array of providers from the public, private and NGO sectors and farmer organizations, each responding to the needs of a broad spectrum of value chain actors beyond just farmers. This necessitates a flexible and responsive approach to farm management, acknowledging the diverse requirements of stakeholders, including input suppliers, traders, processors and policymakers.

Effective coordination among diverse stakeholders is essential. It depends on strong networks and linkage mechanisms – both horizontally across institutions and vertically through different levels of public administration and value chains. Such coordination fosters innovation by facilitating communication and collaboration. The ultimate goal of the AIS is to turn new ideas into marketable products, services or processes, all forms of innovation.

1.6 Agricultural Services System

In addition to the AIS, the agricultural services system provides essential support to farmers in every stage of production and beyond. In the context of small family farms in low-income countries, an efficient agricultural services system ensures that farmers can access the resources, knowledge and markets they need to succeed.

As illustrated in Fig. 1.2, the agricultural system comprises a wide range of interconnected components, each contributing to the sector's overall effectiveness. Beginning with input supply – such as seeds, fertilizers and tools – farmers depend on services that help maximize yields and reduce production costs. Animal health services are also essential, providing advisory support and protecting livestock, which are key sources of income and nutrition.

Post-harvest services, including processing, transportation and storage, ensure that products reach markets in good condition and align with consumer expectations. Equally important are marketing and business management services, which support farmers in understanding market trends, negotiating better prices and sustaining profitability. Finally, food safety and quality assurance play a vital role in meeting consumer demands and enabling access to high-value markets.

Moreover, farmers need access to finance for investment in modern technology; and ICT-based information systems can bridge knowledge gaps by providing timely weather forecasts, market trends and crop management advice. Farmers are at the centre of this system; they are the ones who need both to navigate and integrate the wide array of services and to draw upon them in accordance with their particular needs and circumstances. Given the complexity of the system, many small family farmers will, undoubtedly, need support from agricultural advisory services to make sense of the system and to make appropriate choices. To the extent that it is possible, advisory services should foster the integration of these interconnected services. In this way the agricultural services system not only enhances productivity but also builds resilience and ensures sustainable livelihoods for small family farmers.

1.7 Agricultural Advisory Services Support in these Times of Change

As stated at the onset of this book, the future of small family farmers (and their farms) in low-income countries is at risk. And as mentioned several times in the previous sections, agricultural advisory services are crucial role-players in helping these farmers face and address this risk. But to do this, agricultural advisory services will also have to make some serious changes – particularly with regard to their capacity to address farm (business) management issues.

Fortunately, agricultural extension, as a discipline, practice and service, is familiar with change. Agricultural extension has transformed dramatically over the past three decades, driven in part by broader governance and economic shifts. It has moved away from the traditional linear model of technology transfer – where innovations flow directly from research institutions to extension agents, and then to farmers. This model has evolved into a more dynamic, interactive approach that recognizes the complexity of agricultural innovation

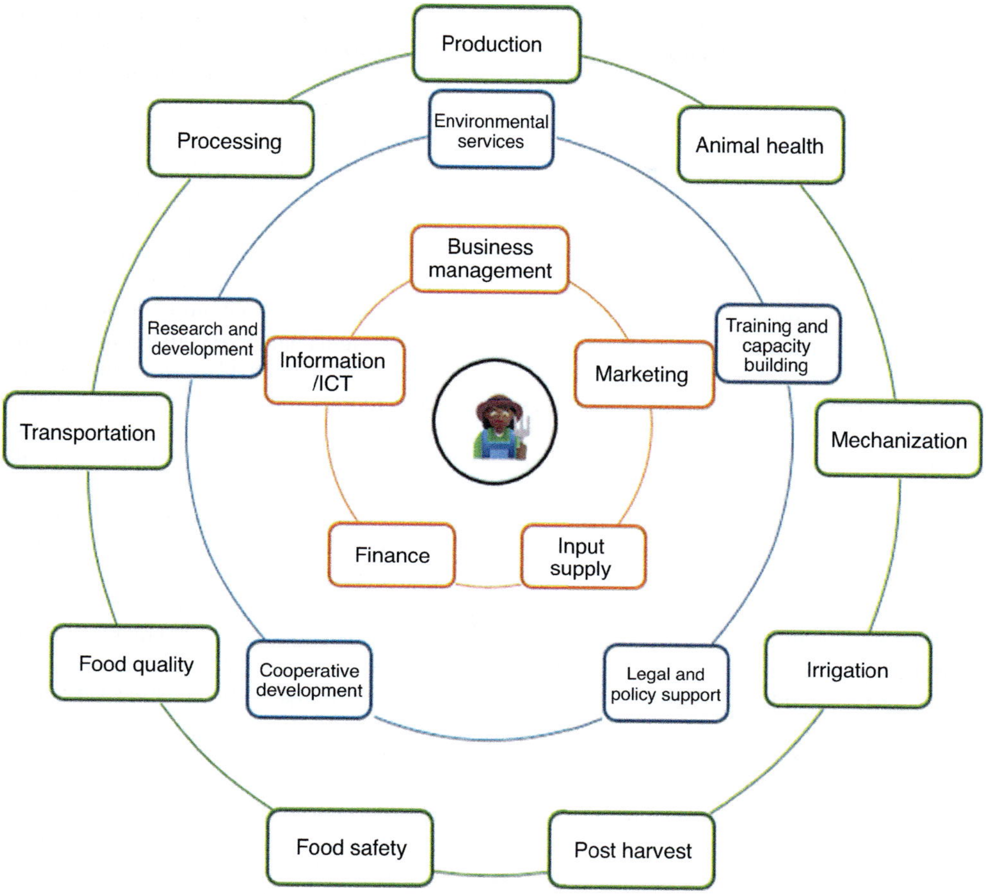

Fig. 1.2. The agricultural services system.

and the diverse sources of knowledge and influence. Even its name has changed; increasingly extension services are taking on broader names, such as 'extension and rural advisory services', or even simply 'rural advisory services'. With the recent increase in urban agriculture, the name may yet change again. As a reminder, for the purposes of this book, we will call these forms of advisory support 'agricultural advisory services' and, in specific contexts, the more traditional 'agricultural extension'.

These changes to extension have come about partly due to a wider change taking place in governance over the last 30 years. A key aspect of this has been the deliberate trend globally to reduce the role of the State in managing the national economy and to increase private sector involvement. These changes have led to reductions in public spending and a shift in thinking about public agencies being the sole provider of rural services. As a result, many support services, once purely government-run, have been partly or wholly privatized. The growing role of the private sector and civil society organizations and their increased engagement in service provision has broadened the institutional landscape of service systems in rural areas. The wider range of actors and service providers include public sector research-extension agencies, private sector companies, non-governmental organizations, and farmer associations and cooperatives. This diversity reflects the complexity of agricultural innovation and the need for comprehensive support for small family farms. This paradigm shift reflects an understanding that change and innovation in agriculture occur within a web of interactions among various stakeholders.

There is another aspect that is not usually associated with pluralistic extension and advisory services. That is the fact that – as we have mentioned before – farming is much more than primary production. In fact, farming, at whatever level, is quite complex with many moving parts. Agricultural advisory services need to be able to address all these complexities. Thus, advisory services need a plurality of disciplines.

In short, agricultural advisory services are critical to the survival and eventual thriving of small family farmers and their farms. These services must be able to respond to the ever increasing complexities that both characterize and challenge small family farming. Agricultural advisory services must collaborate within these pluralistic systems to meet the diverse needs of different farmers and their farming contexts. The 'service' itself must be offered through multiple agencies that employ various methods and provide support and advice in several areas of expertise. This book focuses on one such expertise: farm management. (Agricultural advisory services is explored in more detail in Part 4.)

1.8 Implications of Agricultural Changes on Small Family Decision Making

The dynamic changes occurring in agriculture due to globalization, market liberalization and climate change necessitate a paradigm shift in how small family farmers in low-income countries make decisions. The impact of agricultural changes on small farms highlights the need for better farm management, business skills, market access, finance and technology adoption. As stated by IFAD, 'there is a need and an opportunity to enhance the role that smallholders play in food production and natural resource stewardship' (IFAD, 2013, p. 6). These impacts are outlined here as follows:

- **Access to market information and integration:** The globalization of agricultural markets and the shift towards market-oriented farming requires small family farmers to have an intimate understanding of market dynamics. This involves not just knowledge of local markets but an awareness of global market trends, consumer preferences and quality standards. Farmers need to develop competencies in market analysis to identify potential market opportunities and challenges. Furthermore, with the advent of digital platforms offering real-time market data, farmers must become adept at using these technologies to access timely market information. However, challenges such as digital literacy, connectivity issues and the reliability of market information pose significant barriers to effective market integration.
- **Climate-smart agricultural practices:** As previously noted, climate change introduces a layer of complexity to small family farming, with increased variability in weather patterns and the occurrence of extreme weather events. To mitigate these impacts and ensure the sustainability of their farming systems, small family farmers must adopt climate-smart agricultural practices. This requires not only access to climate information and early warning systems but also knowledge and skills in implementing adaptive and mitigative farming techniques. However, the challenge lies in translating complex climate data into actionable insights that farmers can apply in their context. Moreover, the financial and technical resources required to implement climate-smart practices remain out of reach for many small family farmers.
- **Leveraging new technologies:** To remain competitive in a rapidly changing agricultural environment, small family farmers must embrace new technologies. This includes precision agriculture tools, biotechnologies, renewable energy sources and digital platforms for market access and advisory services. The adoption of these technologies can significantly improve productivity, efficiency and market responsiveness. Nevertheless, challenges such as high upfront costs, technical complexity,

and lack of access to technology and support services pose significant barriers to technology adoption.

- **Access to markets and financial sources:** Access to markets and financial sources is critical for the sustainability and profitability of small family farming. Farmers need to develop networks and partnerships with buyers, processors and retailers to secure stable market outlets. Additionally, access to credit, insurance and investment is essential for financing farm improvements, managing risks, and seizing new market opportunities. However, small family farmers often face difficulties in accessing formal financial services due to collateral requirements, high interest rates and limited financial literacy.
- **Development of business and entrepreneurial skills:** The evolving agricultural landscape necessitates a shift from traditional farming to a more business-oriented approach. Small family farmers must develop skills in financial management, market negotiation and strategic planning to navigate the risks and opportunities presented by market and climate changes. Entrepreneurial skills are crucial for identifying innovative opportunities, diversifying income sources and enhancing farm profitability. However, limited access to business training, advisory services and mentorship programmes hinders the development of these skills among small family farmers.
- **Ensuring sustainable profitability:** The sustainability and profitability of small family farming depend on effective business management. Farmers must possess the skills to analyse financial performance, manage costs and optimize returns on investment. This involves a continuous process of planning, monitoring and adjusting farm operations in response to market and climate conditions. However, the lack of business management training and advisory support limits farmers' ability to implement sound business practices.
- **Considering issues of social responsibility and justice:** Sustainability and profitability must be pursued alongside social responsibility and justice. This includes environmental sustainability. While profits are essential, profits should not be made at the expense of the natural environment or, equally if not more importantly, at the expense of human beings. Small family farmers must earn their profits by honest and noble means. This can be challenging in a world that increasingly prizes material wealth and gives less consideration to how that wealth is created. The decisions made by small family farmers must be weighed against such questions as: is it environmentally sustainable? Is it socially just? Is it fair to others who might be affected by those decisions?

1.9 Conclusion

The dynamics discussed underscore a new era of market-oriented agriculture, where independent small family farmers navigate an economy governed by market rules. Economic liberalization, globalization and the technological revolution have produced opportunities for these farmers with increased market access and the potential for enhanced productivity and profitability. However, notwithstanding these opportunities, there are also risks. The intensification of competition, both locally and internationally, challenges small family farmers to innovate and adapt to remain viable. Moreover, the shift in consumer preferences towards more diverse and high-value agricultural products demands a reorientation of farming systems towards market demands. Farmers are now compelled to navigate the complexities of a globalized market landscape without the safety net of government assurance, making market-oriented approaches and strategies essential for competitive survival and eventual thriving. The evolving global agricultural landscape thus highlights a critical need for farmers to adapt their practices, develop new skills, and embrace technological advancements to meet the demands of a changing world.

Understanding the opportunities and risks of globalization and market liberalization is essential for farmers. For small family farmers in low-income countries to thrive in this new era, a concerted effort towards adopting market-oriented approaches, understanding global consumer preferences, and navigating the challenges of increased competition is indispensable. The implications of the changes in agriculture for small family farmers underscore the need for a comprehensive approach to skill and competency development amongst both extension and advisory service providers and farmers. Meeting these needs requires coordinated action from governments, NGOs, the private sector and farmer organizations, to provide targeted support and training. By equipping small family farmers with the necessary skills, knowledge and resources, they can navigate the challenges and opportunities of the changing agricultural landscape more confidently and competently, ensuring their resilience and long-term sustainability. Much of this will centre on developing capacity around farm (business) management.

References

CGIAR (2024) Sustainable land use systems: Advancing Land Degradation Neutrality through Local Action and Global Integration. CGIAR. Available at: https://www.cgiar.org/news-events/news/sustainable-land-use-systems-advancing-land-degradation-neutrality-through-local-action-and-global-integration/ (accessed 15 March 2025).

Chakrabarti, S. (2016) *The drylands advantage: Protecting the environment, empowering people*, IFAD Advantage Series 304739. International Fund for Agricultural Development (IFAD). DOI: 10.22004/ag.econ.304739. (accessed 15 March 2025).

FAO (2017a) The future of food and agriculture: Trends and challenges. Innovations in agricultural extension: Lessons from around the world. In: *Proceedings of the International Conference on Agricultural Extension*, Bangkok, Thailand. Available at: https://openknowledge.fao.org/handle/20.500.14283/i6583e (accessed 15 March 2025).

FAO (2017b) Water for Sustainable Food and Agriculture; World Bank, 2020, Agricultural Irrigation and Water Management in Developing Countries: Trends and Challenges. FAO. Available at: https://openknowledge.fao.org/handle/20.500.14283/i7959en (accessed 15 March 2025).

FAO (2021) The state of the world's land and water resources for food and agriculture. FAO, Rome, Italy. Available at: https://openknowledge.fao.org/handle/20.500.14283/cb9910en (accessed 15 March 2025).

FAO (2023) *The Status of Women in Agrifood Systems*. FAO, Rome, Italy. DOI: 10.4060/cc5343en. (accessed 11 December 2024).

IFAD (2013) *Smallholders, Food Security and the Environment*. IFAD & UNEP, Rome.

Larsen, K., Kim, R. and Theus, F. (eds) (2009) *Agribusiness and Innovation Systems In Africa*. World Bank. Available at: https://openknowledge.worldbank.org/entities/publication/8804e270-6f8c-5ffe-ba71-9be1487aacf3 (accessed 27 January 2025).

Pingali, P. and Feder, G. (eds) (2017) *Agriculture and Rural Development in a Globalizing World: Challenges and Opportunities*. Routledge. Available at: https://www.routledge.com/Agriculture-and-Rural-Development-in-a-Globalizing-World-Challenges-and-Opportunities (accessed 26 January 2025).

Ritchie, H. (2021) *Smallholders produce one-third of the world's food, less than half of what many headlines claim*. OurWorldInData.org. Available at: https://ourworldindata.org/smallholder-food-production (accessed 12 March 2025).

UNCCD (2021) Global Land Outlook, second edition: Land Restoration for Recovery and Resilience. UNCCD, Bonn, Germany. Available at: https://www.unccd.int/resources/global-land-outlook/overview (accessed 15 March 2025).

World Bank (2010) *Improving water management in rainfed agriculture: Issues and options in water-constrained production systems (English)*. World Bank Group, Washington, D.C. Available at: http://documents.worldbank.org/curated/en/608111468156864510/Improving-water-management-in-rainfed-agriculture-issues-and-options-in-water-constrained-production-systems (accessed 15 March 2025).

World Bank (2012) Agricultural Innovation Systems: an investment sourcebook. World Bank, Washington, D.C. Available at: https://openknowledge.worldbank.org/entities/publication/cb73650c-7ef4-555c-af9b-6ca602b29266 (accessed 25 January 2025).

2 The Importance of Small Family Farming to Sustainable Agricultural Development

In the previous chapter we explored the rapid and extensive changes that affect the global agricultural landscape at all levels throughout the agri-food system. We noted that these changes present both opportunities and threats to small family farmers. A primary concern is that globalization and rapid technological advances, particularly in information and communication, might unintentionally exclude small family farms from the agricultural system. This would have a devastating effect, not only on these farmers but also on global food security. Thus, this chapter explores the importance of small family farming, and its multifaceted role in fostering sustainable agricultural development.

Explored in this chapter is the vast diversity of smallholder farming systems; each adapted to its specific environmental and socio-economic context. This diversity is a testament to the innovative spirit of small family farmers and their invaluable contribution to natural resource management, conservation of biodiversity and adaptation for climate resilience. Through this chapter, the aim is to shed light on the indispensable role of small family farming and advocate for greater support and recognition, which is needed to motivate and interest these farmers to farm for profit and long-term sustainability. The mindset of small family farmers – and of those who support them – must shift to unlock the potential to operate their farms as businesses.

2.1 The Global Significance of Small Family Farming

Small family farming plays a vital role in the agricultural sector economy of low-income countries. They serve as the backbone of food production, rural livelihoods and poverty reduction. But what constitutes a small family farm?

It is estimated that there are 570 million family farms in the world. Some 74% are in Asia, including China (35%) and India (24%). Around 9% of these farms are in sub-Saharan Africa; and 4% are found in high-income countries. Of the world's 570 million family farms, 84% are smallholdings operating on less than 2 ha (Lowder *et al.*, 2016). This means that there are approximately 480 million small family farms around the world. If one expands the size of the farm to include 5 ha or less, the number of farms greatly increases.

There is much debate around the definition of a small family farm. There is also much debate around their contribution to global food production. Part of the challenge is that different metrics are used. For example, some studies measure contribution to food production using kilocalories. Other studies use the value of the production – sometimes within a specified range of food items such as staples. This book has generally adopted the terms 'small family farm' or 'smallholder farm' and has further adopted the value of food production as the metric for measuring their contribution to global food security. The main point to be made is that regardless of

DOI: 10.1079/9781836990536.0002

the precise number of farms globally, the criteria used to define a 'small family farm' or the metric used to measure their contribution, small family farms are substantial in number and their contribution to food security is considerable and significant.

More importantly, these families and their farms are under threat. In the majority of low-income countries, the average size of farms is decreasing, making viability increasingly difficult to achieve. These farms are also 'greying', that is, the average age of small family farmers (currently about 50 years old) is increasing (Barrett, 2008)[1]. And, most significantly, they are under threat from the impacts of the climate crisis, and they run the risk of being left behind in the rapid changes occurring to the global agricultural landscape.

Small family farms make a substantial contribution to agriculture. Their potential for profitability adds pressure on agricultural resources. This highlights the need for innovative solutions to ensure economic viability and sustainability.

2.2 Defining a Small Family Farm

What constitutes a small family farm? In truth, the answer varies. Size is often relative. In many parts of Latin America, 10 ha is smaller than the national average, whereas in Bengal, India, 10 ha would be above the national average (Nagayets, 2005). To provide clarity, we adopt three frameworks to define what constitutes a 'small family farm'.

One framework is the definition of small family farmers according to the Food and Agriculture Organisation (FAO). The FAO defines small family farmers as those who operate small plots of land, typically less than 2 ha, and that rely predominantly on family labour (FAO, 2014a)[2]. This a relevant definition for this book because, in low-income countries more than 72% of farms are smaller than 1 ha and a further 12% are 1–2 ha. Still, size is relative, and the actual size of the small family farms globally is often determined by context.

The second framework, economic viability, means that farms at least generate enough income to cover their production costs. In many regions, many of these small family farms are or may potentially be economically viable depending on their location, their ability to access markets for their produce and capacity to innovate. At the same time, there are many small family farms which, for a range of reasons, may never have the potential for economic viability. It is accepted that for such farms, the ideas presented in this book are less likely to be appropriate. For such farms other policy approaches – such as safety nets and even exit strategies – may be more relevant.

The third framework is drawn from a livelihood's perspective. While there are many family-run businesses, the family farm is somewhat unique because of the high level of integration of the farming activities within the household (family) activities. As much as it may be desirable to do so, it is often difficult to separate the farming business and its related activities from the 'non-farming' activities of the family. The farm is often the main livelihood strategy for a small farm household. While the farm is ostensibly a discrete business enterprise, in practice it often dominates the choices around other livelihood strategies which may include non-farm business activities, off-farm wage earning and remittances from family members working away from home.

Operating within this three-fold framework, we can establish a basic description of a small family farm in low-income countries:

It is a family-owned and operated agricultural production enterprise operating on a relatively small hectarage of land (probably 2–5 ha). The farming activities are carried out mostly using family labour. The farm operation is a principal livelihood strategy among other household livelihood strategies. The activities on the farm are largely integrated into other household activities and often play a dominant role in the household. The labour requirements of the farm are often given priority over other household activities, including off-farm wage earning.

The premise of this book is that efforts to strengthen small family farms should focus on farms that fit this description but with the additional criteria that they have the potential to advance beyond production for home consumption and become economically viable and profitable, given the right support.

2.3 The Global Contribution of Small Family Farming

The FAO report on the state of food and agriculture (FAO, 2014a) argues that small family farmers in low-income countries are essential to the future of agriculture, since they account for a significant portion of agricultural production in many low-income countries. However, as pointed out in the previous chapter, these farmers face limited resources, knowledge and skills gaps and unpredictable environmental conditions.

These farms, small and constrained as they are, are a testament to the enduring spirit of the farmers and their families in the face of modern challenges.

While farms of less than 2 ha account for around 500 million farmers, representing more than 80% of all farms globally, they occupy only 12% of the world's farmland and produce approximately 32% of the world's food (Ritchie, 2021). This highlights the significant role that small family farmers play in feeding local populations, particularly in regions where agriculture is the primary source of livelihood (see Box 2.1.). However, despite their crucial contribution to global food production, many small family farms are themselves poor and food insecure, struggling to sustain their livelihoods.

As alluded to previously, another factor to explore when considering the issue of viability and profitability is the trends towards changes in farm size. Many studies note that in most parts of the developing world, due to the way inheritance works, farm sizes are shrinking in many low-income countries. Some studies argue the opposite, that there is a move towards consolidating small farms into 'larger' ones (Lowder *et al.*, 2016). General observations show that as countries develop and become richer, the average farm size tends to increase. It is, therefore, important to understand the diversity of farming systems worldwide in order to grasp global trends among small family farms. This topic is discussed further in Chapter 3.

Whatever the trend, it is well established that small-scale family farms often achieve higher yields per hectare than larger farms for staple crops, due to more intensive management and labour. This highlights their potential to contribute significantly to national food production. However, to shift beyond the goal of national food security there is a need to overcome the challenges of lower labour productivity and to promote innovation.

Innovation, as discussed previously, encompasses the development and application of new technologies, the adoption of existing practices on a wider scale, access to markets and the adoption of better production and business management practices. For many small family farmers, diversifying household income through off-farm employment or innovative agricultural practices is essential for sustainability and decent living conditions. Thus, focusing public and private support on small-scale family farms with innovation potential could significantly boost the productivity, sustainability and economic viability of smallholder farming in low-income countries.

It must also be stressed, as stated earlier, that the importance of small family farming extends beyond mere agricultural production; it is integral to the management of global natural resources, environmental sustainability and the battle against rural poverty. Small family farms make numerous contributions to local, regional and global agri-food systems and their corresponding economies. The capacity of these farmers to innovate and adapt to changing conditions is crucial for the future of global agriculture. This,

Box 2.1. Agriculture's centrality to rural development and poverty alleviation. From Hazell *et al.*, 2007.

Agriculture is likely to be central to rural development and rural poverty alleviation. Farming has high potential to create jobs, to increase returns to the assets that the poor possess – their labour and in some cases their land – and to push down the price of food staples, which is crucial when so many of the poor are net buyers of food. Historically, few countries have industrialized successfully without prior development of their agriculture. Recent comparisons made across countries show that increases in agricultural productivity are closely related to poverty reduction. In most rural areas, moreover, there are few alternatives to farming as a large-scale source of jobs.

again, highlights the need for supportive policies, targeted investments and inclusive agricultural innovation systems that recognize the central role of small family farmers in achieving sustainable agricultural development.

2.3.1 Food production

As we noted earlier, farms smaller than 2 ha, despite occupying only about 12% of the world's agricultural land, play an outsized role in ensuring both global and national food security. Small family farms are crucial, contributing approximately 35% of the world's food supply. This contribution is remarkable given their limited access to land and resources, underscoring the importance of these farms in sustaining the diets of millions of people worldwide.

Small family farms are particularly vital in the production of staple foods such as rice, maize and wheat, which are crucial elements of national food security. In countries like India and Bangladesh, these farms form the bedrock of agricultural production. For instance, in Bangladesh, small family farms contribute significantly to the country's rice production, which is the staple food for the majority of the population. Despite their limited size, intensive farming practices, including multiple cropping seasons, enable these farms to achieve high yields that are essential for both local consumption and national food security.

In sub-Saharan Africa, small family farms are the backbone of the agricultural sector, producing most of the food consumed locally. In Ethiopia, small family farms are responsible for producing key staples like teff, maize and sorghum. These locally produced foods are vital for national food security, as they reduce dependence on food imports and help buffer against the volatility of global food prices. This local production is especially important in regions vulnerable to external shocks, such as price fluctuations or supply chain disruptions, which can have severe consequences for food availability and affordability.

Overall, the significant role of small family farms in food production highlights their contribution not only to local and national food security but also to global food systems. Their ability to grow large amounts of food on small plots helps keep food supplies stable. This is important in both developing and developed regions, highlighting the need for support and investment in these vital farming systems.

2.3.2 Diversity of production

Small family farms are often more diversified than larger commercial operations, growing a variety of crops and raising different types of livestock. This diversity not only helps manage risks associated with farming, such as crop failure or market volatility, but also contributes to a more stable and varied food supply. In countries like Vietnam, small family farms are known for their diverse production systems, where farmers grow rice alongside other crops such as vegetables, fruits and spices, while also raising poultry and fish. This diversification is crucial for maintaining a balanced diet for local populations and ensuring a steady supply of various food products to both rural and urban markets. In Guatemala, small family farmers often cultivate a mix of maize, beans and coffee, which not only supports household food security but also provides a source of income through local markets and exports. These systems not only enhance the availability and accessibility of a wide range of nutritious crops but also promote resilience against market fluctuations and environmental challenges.

2.3.3 Resilience and sustainability

Small family farmers are deeply connected to their environment, often drawing on generations of traditional practices that are inherently sustainable. These farmers maintain a symbiotic relationship with their land and natural resources, employing techniques that minimize the use of external inputs while maximizing resource efficiency. This close connection to the land compels small family farmers to adopt practices that enhance the long-term sustainability of food production, which is crucial given their limited access to resources.

Sustainable practices such as crop rotation, intercropping and organic farming are commonly employed to preserve the environment and maintain soil fertility. For instance, in the mountainous regions of Nepal, terraced farming is widely practiced on small plots. This method not only prevents soil erosion but also conserves water, making agriculture more sustainable and viable over the long term. Terraced farming effectively transforms steep slopes into productive agricultural land, allowing for the efficient use of water resources and reducing the risk of landslides.

In regions like sub-Saharan Africa, small family farmers often integrate agroforestry into their farming systems, combining crops with trees and shrubs. This practice helps maintain soil fertility, provides additional sources of food and income, and enhances biodiversity. Agroforestry is particularly beneficial in areas with degraded soils, as the presence of trees helps to restore nutrients to the soil, improve water retention and protect crops from extreme weather conditions. By integrating these sustainable practices into their farming systems, small family farmers contribute to the resilience of food systems, particularly in the face of climate change and other environmental challenges.

Small family farmers also act as caretakers of their land. They use traditional farming methods that help manage natural resources sustainably. Techniques like natural pest control, crop rotation and terrace building protect soil fertility, prevent erosion and support ecosystem services. This careful approach to resource use is key to sustainable farming. Supporting these practices can help fight environmental degradation, protect biodiversity and strengthen ecosystems. These methods also help farmers deal with climate change. Farmers who use them are better prepared for extreme weather like droughts and floods. By protecting biodiversity and using eco-friendly techniques, small family farms build resilience in local and global food systems. This helps ensure continued food security despite growing environmental challenges.

2.3.4 Rural employment and livelihoods

The World Bank estimates that more than 75% of the world's poorest people live in rural areas, with most depending on agriculture for their livelihoods (World Bank, 2024). Small family farms are a major source of rural employment, especially in regions where agriculture drives the local economy. In countries like Nigeria and Uganda, these farms provide the main source of income for millions of rural households. Small farms not only feed the household but also generate income by selling surplus produce in local markets. This income is vital for rural families to meet basic needs and improve their quality of life. The employment created by small family farms strengthens food security by ensuring people can produce and afford food.

In places like Indonesia, where small family farms dominate, agriculture provides jobs for a large share of the population. This helps reduce rural poverty and supports more stable and resilient rural livelihoods. Strengthening small family farms is, therefore, essential for promoting employment, reducing poverty and improving rural living conditions.

2.3.5 Poverty alleviation

Small family farming not only provides these communities with a means of subsistence but also generates income, creates employment opportunities and fosters overall economic development. This helps reduce poverty and strengthens rural communities, contributing to sustainable and inclusive growth. Programmes that boost the productivity and profitability of small farms can directly reduce poverty, which is closely tied to food insecurity. In Rwanda, for example, government efforts to supply improved seeds, fertilizers and training, have led to higher crop yields and increased household incomes. These gains improve food security and help small farmers earn more from their land. In India, microfinance programmes have enabled small farmers to invest in their farms. This has raised productivity and incomes, improving food security for both families and communities. Overall, small family farms contribute to food security in many ways. They produce much of the world's food, support local economies and use sustainable practices. It is vital to address challenges like limited access to resources and markets, in order to boost productivity, increase resilience and enhance national food security.

2.4 Characteristics of Small Family Farming Households

As this book focuses on small family farming, it is useful to examine the characteristics of small family farm households. Some of the key characteristics are given below.

Research highlights the widespread poverty among small farm households, with a significant proportion living below national poverty lines in all surveyed countries (FAO, 2014a). As Ritchie (2021) notes, 'most of the world's farms are small family farms, and they are some of the poorest people in the world'.

While there is much variation among small scale farming families, they tend to prioritize food for the household, with only surpluses being sold.

The concept of land and labour productivity on smaller farms is a critical area of study. Small family farmers utilize their limited resources productively. Smaller farms often achieve higher crop yields per hectare for selected crops (e.g. staple crops). They achieve higher overall land productivity than larger farms. However, they face lower labour productivity. Lower labour productivity underscores the challenges smaller farms encounter, including lower household incomes and higher poverty rates due to the excessive, often unpaid, use of family labour and limited access to markets and modern farming technologies.

Small family farms typically operate with limited resources and assets. The livelihoods framework categorizes household assets as: human assets (e.g. knowledge, skills, health), social assets (e.g. networks, trust arrangements); natural resources (e.g. land, water); financial assets (e.g. sources of income, savings); and physical assets (e.g. tools, equipment, structures, infrastructure, productive livestock) (UNDP, 2017). Small family households are often constrained in the majority of these types of livelihood household assets. Education and training are often very limited. Personal health is often low due to nutritional challenges. The land available to the family is usually quite small and access to irrigation is often limited. Tools and equipment are also limited by their limited financial assets. However, typically, small family farming families have well-developed social and trust networks such as community banks, extended family and community structures that provide safety nets.

Diversifying income sources is a strategy employed by farming households to manage risks and make the best use of available labour. Small family farms, in particular, tend to rely more on off-farm income due to insufficient incomes generated from their small plots. Conversely, larger farms often find farming to be the main source of revenue, highlighting a correlation between farm size and the share of income derived from agricultural activities.

The adoption of modern farming technology (e.g. mechanization, genetic material, etc.) is markedly low across both large and small farms in low-income countries, but substantially lower on smaller farms. This lack of mechanization and limited use of improved seeds and other genetic material suggests significant potential for boosting agricultural productivity through greater adoption of existing technologies and farming processes, as well as the potential for developing new technologies specifically for smaller farms.

Similarly, smaller farms tend to use inputs such as seeds and fertilizer more intensively, striving to maximize output from their limited land resources.

Because their livelihood assets are often so limited, many small farm households often have multiple livelihood strategies to ensure the wellbeing of the family. They also tend to prefer livelihood strategies that are stable and offer security over livelihood strategies that are perceived as risky.

Smaller family farm households typically sell a smaller share of their agricultural production compared to larger farmers. Further, whatever productivity challenges smallholder farmers face, entry into markets remains a pivotal factor in their overall livelihood strategy choices. Even if productivity issues are addressed successfully, market entry remains a challenge.

2.4.1 Land productivity on small family farms

The farm household characteristic of high land productivity and low labour productivity discussed above warrants further discussion as a characteristic of the farm itself.

In this instance, an illustration will be useful. Fig. 2.1 provides insights into selected crop yields by farm size, particularly focusing on the smallest farm categories (1st and 2nd quartile). Yields demonstrate notable differences in productivity levels across various countries and crops, highlighting both the strengths and challenges faced by small family farmers. We will look at each of these more closely, as they again demonstrate the diversity and variability around the world. More importantly, they underscore the need to consider local contexts when designing support programmes aimed at improving farmer decision making and enhancing the viability and profitability of small family farms.

Small family farms, particularly those in the smallest size categories (1st quartile), demonstrate a diverse range of productivity across different regions and crops, highlighting both the potential and challenges inherent in small-scale agriculture. For instance, in Vietnam, the smallest rice farms achieve an impressive yield of 9.7 t/ha, a level of productivity that remains consistent even as farm size increases slightly. This suggests that effective farming practices and favourable growing conditions contribute significantly to the efficiency of small rice farms in the country.

Conversely, in Tanzania, the smallest maize farms produce 3.1 t/ha, but this yield drops dramatically to 0.73 t/ha in slightly larger farms. This significant decrease indicates difficulties in maintaining productivity with increased farm size, possibly due to limited access to resources or differences in farming techniques. Similarly, in Kenya, maize yields are relatively low, with the smallest farms producing only 1.10 t/ha, a figure that declines further as farm size increases, underscoring consistent challenges in maize production for small farm families.

Nepal presents a contrasting scenario where the smallest rice farms achieve exceptionally high yields of 16 t/ha. However, this productivity plummets to 3.9 t/ha in the 2nd quartile, indicating that while the smallest farms can be highly productive, they face significant barriers that hinder sustained high yields as farm size increases. In Ethiopia, the yield for teff, a staple crop, is also low, with the smallest farms producing 1 t/ha and slightly larger farms yielding only marginally more. This suggests that increasing farm size does not significantly impact teff productivity, pointing to some of the other constraints that affect small family farms broadly.

In Bolivia, the situation is somewhat different, with maize yields increasing from 6.6 t/ha on the smallest farms to 7.7 t/ha on slightly larger farms. This upward trend indicates that small increases in farm size can lead to higher

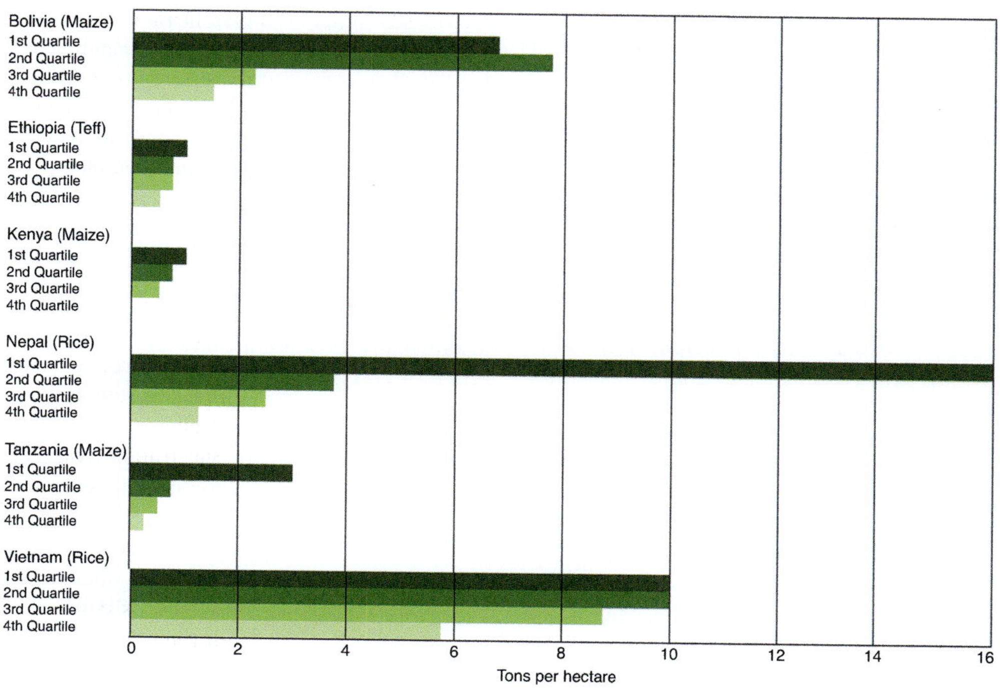

Fig. 2.1. Selected crop yields, by farm size (Source: FAO, 2014c)

productivity, possibly due to better resource allocation or the benefits of economies of scale.

These variations in productivity among small family farms across different regions illustrate the complex dynamics of small-scale agriculture. While some small farms, particularly in Vietnam and Nepal, achieve high productivity, others, like those in Tanzania and Kenya, face significant efficiency challenges. The disparities in yields highlight the need for targeted interventions to support small family farmers by improving access to markets, modern technologies and better resource management practices. Such support can enhance the productivity of small family farms, contributing significantly to rural development and food security.

2.4.2 Labour productivity on small family farms

Labour productivity refers to the output per unit of labour. It varies significantly across farms of different sizes, particularly when comparing small family farms to larger agricultural enterprises. While small family farms, as we have previously seen, often achieve high land productivity due to the intensive use of their limited resources, they typically exhibit lower labour productivity. This inverse relationship between land efficiency and labour productivity is a crucial consideration in the context of small family agriculture, particularly in regions where small-scale farming is prevalent.

For example, in many parts of sub-Saharan Africa, small family farms, which are often less than 2 ha in size, rely heavily on manual labour, usually provided by family members. In countries like Kenya and Ethiopia, where small family farms dominate, labour productivity is often low due to the limited use of mechanization and modern farming techniques. In Ethiopia, small family farmers primarily cultivate staple crops such as teff and maize using traditional methods, which, while maximizing land use, result in low output per worker. This low labour productivity is further exacerbated by the challenging climatic conditions and the health status of the rural workforce, which is often compromised by poor nutrition and limited access to healthcare.

In Asia, the situation is similar, though with some regional variations. In Nepal, for instance, small family rice farmers achieve high yields on small plots of land, but the labour involved is intensive and mostly manual. The country's mountainous terrain limits the use of machinery, leading to lower labour productivity. Conversely, in Vietnam, small family rice farms have managed to achieve relatively higher labour productivity through the adoption of effective farming practices and favourable growing conditions, yet these gains are still modest when compared to larger, mechanized farms in the region.

In the Near East, small family farms also exhibit low labour productivity, particularly in rain-fed agricultural areas. In Tunisia, small family farms, often less than 2 ha, are labour-intensive and rely heavily on traditional farming methods. The labour productivity on these farms is low, particularly in the cultivation of cereals and olives, which dominate the region's agricultural output. The harsh climatic conditions, combined with limited access to modern agricultural technologies, further reduce the efficiency of labour on these small farms.

These examples illustrate the complex factors influencing labour productivity on small family farms across these different regions. Climate, health, cultural practices, crop types and the seasonality of agricultural activities all play a significant role in determining labour productivity. For agricultural advisory programmes aimed at building the capacity of small family farmers, it is essential to consider these factors and the inherent differences in productivity and profitability related to farm size (Gollin, 2018).

Customized interventions that meet the specific needs of small family farms can help close the labour productivity gap between small and large farms. This, in turn, can improve the sustainability and profitability of smallholder agriculture. These interventions may include improving access to modern farming technologies; enhancing worker health and nutrition; and strengthening farm management for more efficient practices.

2.5 Diversity of Small Farms

As noted before, the concept of 'small' varies widely. For example, in countries like Bangladesh where there are nearly 15 million farms, the maximum farm size is 2 ha while the average farm size is 0.4 ha. On the other end of the spectrum, Nicaragua has just 320,000 farms; the maximum farm size is 282 ha and the average farm size is 9.5 ha (FAO, 2014b).

Some countries have many very small farms, reflecting high population densities and intensive farming methods. Other countries have fewer small farms of a larger size. This diversity is important when making plans to support small family farmers in improving their farm management. While the principles, and even the tools, employed may be the same, their application should take into account the context of the specific country, or country region. Understanding these variations is crucial for developing targeted agricultural policies and interventions that address the unique challenges and opportunities in each context.

This analysis highlights the significant role these small family farms play in agricultural production relative to the amount of land they operate. Overall, the FAO study illustrated that the 75% smallest family farms are not only vital in terms of land use but are also incredibly productive, often contributing a disproportionately high share of agricultural output relative to the land they manage. This, again, highlights the need for continued support and investment in small family farms to sustain and enhance their productivity and contribution to national food security (see Box 2.2.).

In the next chapter (Chapter 3), we explore the diversity of small family farms in more detail, within varied farming systems.

Drawing insights from the preceding discussions and the reports reviewed, we can delineate common categories of farmers, each with unique challenges and opportunities. They can be classified broadly into three kinds: (i) small subsistence family farms; (ii) small market-oriented family farms; and (iii) small commercial family farms.

2.5.1 Small subsistence family farms

Small subsistence (or near-subsistence) family farms primarily produce for their own consumption, with little or no surplus for the market. These farms operate in less-favoured environments with poor market access, focusing on meeting the food needs of the household. They typically manage very small plots of land, rely heavily on family labour, and face significant constraints in accessing resources and markets. Production is often unstable, leading to frequent food insecurity. The emphasis on home consumption can limit crop choices to those that are not necessarily best suited to the natural resource base, further compounding their challenges. Some farms in this category do produce a modest surplus of food staples for local or national markets. These households make joint decisions about production and consumption, with farm and household activities closely intertwined.

2.5.2 Small market-oriented family farms

Small market-oriented family farms have the potential to transition into fully commercial operations when provided with appropriate incentives and reliable market access. These smallholders adapt their farming systems to meet market demand and are often seen as emerging commercial producers. However, their progress is frequently constrained by limited liquidity, high levels of risk and elevated transaction costs. With improved access to markets and targeted support, these farms can become more commercially viable. They constitute a significant share of the agricultural sector in many low-income countries and hold substantial potential to contribute to food security and poverty reduction. By improving production techniques, expanding market opportunities and strengthening support services, many of these farmers can build their asset base, adopt environmentally sustainable practices and transition towards profitable, market-oriented farming. Such a transformation can improve their livelihoods, bolster national food security and stimulate broader economic development. Advisory support in farm management is essential to enhance their capacity and skills, paving the way for higher levels of commercialization.

2.5.3 Small commercial family farms

Commercial farmers focus on market production and aim to maximize profits by meeting market demands. They operate in environments with strong infrastructure, efficient financial systems and secure property rights. These farmers use

Box 2.2. Regional variations of farm size.

Sub-Saharan Africa: In sub-Saharan Africa, a significant portion of farms are under 1 ha, reflecting the small-scale nature of agriculture in the region. For instance, in Ethiopia, approximately 80% of farms are less than 1 ha in size, with the average farm size being around 0.9 ha. Similarly, in Kenya, around 70–75% of farms fall into this category, with typical small family farms often around 0.5 ha. These small-scale farms are essential for the livelihoods of millions of people in these countries, where land is often passed down through generations and gradually subdivided, leading to smaller plots.

South Asia: In South Asia, the situation is somewhat similar, with a large percentage of farms being less than 1 ha due to high population density and extensive land fragmentation over time. In India, for example, about 47% of farms are less than 1 ha. This reflects significant land division, with small family farms (less than 2 ha) making up about 86% of all farms in the country. In Bangladesh, more than 90% of farms are under 1 ha, with many small family farmers working on plots smaller than 0.5 ha. Nepal also has a high prevalence of small farms, with approximately 60% of farms being less than 1 ha, influenced by the country's mountainous terrain and dense population.

East Asia: In East Asia, small family farming is also prevalent, particularly in countries like China and the Philippines. In China, about 70% of farms are less than 1 ha, a result of historical land redistribution and the country's tradition of small-scale farming. In the Philippines, around 40% of farms are less than 1 ha, with these farms often being fragmented and used for both subsistence and small-scale commercial agriculture.

Latin America and the Caribbean: The average farm size tends to be larger, but there are still many small farms, especially in countries like Haiti and Peru. In Haiti, 70–80% of farms are less than 1 ha, with many farmers practicing subsistence agriculture. In Peru, approximately 20–30% of farms are under 1 ha, with smaller plots more common in the Andean regions, where the terrain and farming practices differ significantly from the coastal areas.

North Africa and the Middle East: Small family farming is also widespread. In Egypt, around 70% of farms are less than 1 ha, a legacy of historical land reforms and the high population density along the Nile River. In Morocco, approximately 50–60% of farms are less than 1 ha, particularly in densely populated rural areas. These small farms are critical for sustaining rural communities in these regions, where agriculture remains a key component of the economy and social structure.

advanced technologies and best practices to improve production and profits. Their success also depends on access to reliable transport, effective financial services and legal protections for their investments. By integrating deeply into market systems and specializing in certain commodities, they drive innovation. They also make significant contributions to food security and economic development.

2.5.4 Other sub-categories of family farms

Of course, these three categories are not fixed. While it may be convenient to categorize small family farmers in this manner, the reality is likely to be more nuanced. For example, subsistence farmers may in fact actively seek market surpluses when they have them. Their desire may be to move away from producing solely for home consumption, but external circumstances may not permit this. Some in this category may find that the contribution of their farm to their livelihood no longer outweighs the cost of farming – whether in financial, practical or other terms. Such farmers may be open to seeking a strategy to exit farming altogether.

Within the broader typology of farms, there are distinct sub-categories based on their farming focus, market orientation and ecological practices. These sub-categories include small family farmers operating specialized farms, integrated, mixed farming systems and farmers who are running their farm according to agroecological principles.

Small specialized family farms are typically profit-oriented, focusing on high-value crops, organic farming or niche markets aiming for higher profitability through targeted production and marketing strategies. By targeting specific market segments, these farmers can achieve higher financial returns and sustainability. Their production and marketing strategies are customized to meet the demands of global or niche markets, making them key players in driving economic growth and contributing to food security.

Small integrated family farms can be found in diverse farming systems that combine crop production, livestock and agroforestry to create synergistic benefits. These farmers focus on resource efficiency and resilience by carefully integrating their various enterprises. While they are profit-driven, their approach is centred on minimizing waste and promoting a circular economy. The integration of different agricultural activities not only enhances productivity but also supports sustainable practices and biodiversity conservation.

Small agroecological family farmers are deeply committed to ecological principles, prioritizing organic practices, biodiversity and sustainable resource management over purely profit-driven goals. These farms play a crucial role in environmental conservation while also maintaining agricultural productivity and resilience. For example, emerging regenerative farmers focus on practices that restore soil health and contribute to environmental sustainability. While profitability is a consideration, their primary focus is on upholding sustainable farming practices, which they are unwilling to compromise for short-term gains.

Each of these sub-categories plays a vital role in the agricultural landscape, contributing to economic growth, food security and environmental sustainability in their own unique ways.

As useful as these categories are, it is important to focus on the reality of the specific farmer or group of farmers with which one is engaged. It is really the mental orientation and attitude of the farmer that is the focal point. We will explore this further in Parts 3 and 4 of this book. For now, it is sufficient to embrace the idea that while it may be useful to categorize farmers, each farmer and each farm family is unique. They have their particular reasons for being farmers and for what they produce and why and how they produce it. And while the intent of this book is to help foster the capacity of these farming families to engage more rigorously in market-oriented farming – that they may survive and thrive in the fast-changing agricultural landscape – their individuality must be acknowledged and respected.

A final word: Every farmer is, of course, constrained by available resources—such as land size, knowledge, skills and market opportunities. Farm management support can only be effective if the farmer is oriented towards, or at least open to, adopting a more business-minded approach to farming.

2.6 The Critical Role of Women in Small Family Farming

To fully understand small family farms, it is essential to examine the role of women as their participation is crucial to agriculture. Despite their significant contributions, women's roles in smallholder farming are often overlooked. Recognizing their impact and addressing the challenges they face is key to advancing modern agriculture in the 21st century.

Women play an indispensable role in small-scale agriculture. They contribute to food production, household income and global food security. Women make up 43% of the global agricultural workforce. Their participation is even higher in sub-Saharan Africa and Asia, where smallholders provide up to 80% of food. In sub-Saharan Africa, women make up 62% of the agricultural workforce and produce nearly 70% of the region's food. However, they own less than 20% of the land (FAO, 2023). These figures show both their essential role and the gender disparities that limit their productivity.

More female-headed farms are emerging. This is due to male outmigration, widowhood or cultural norms that make women primary decision makers. In sub-Saharan Africa, female-headed households range from 12.9% in Niger to 39.3% in Namibia (Buehren, 2023). These women often face economic challenges. They have fewer household members to share labour and they earn lower incomes and have limited access to land, credit and extension services.

Despite their contributions, women face many barriers. Legal and cultural norms often stop them from inheriting or buying land. This makes it hard to get credit or invest in modern farming. Labour shortages are another issue. Female-headed households often lack enough working-age members to handle tasks like planting and harvesting. Many women grow surplus food but lack direct market access. They do not have price information or transport. As a result, they rely on intermediaries who reduce their profits. Women also receive less training and fewer opportunities in technology, finance and value chains compared to men.

Empowering women in agriculture needs targeted support. They must be included in agricultural development programmes, especially extension services. The FAO says that if women had equal access to resources, their yields could grow by 20–30%. This could reduce global hunger by 12–17%. Women are also more likely to reinvest in their families. This improves child health, education and household resilience (FAO, 2023). To unlock their full potential, we must address gender gaps in land, credit and markets. This will support smallholder farming, promote sustainability and boost rural economic development (see Box 2.3.).

2.7 Challenges and Opportunities for Small Family Farmers

Small family farmers, in general, face many challenges that threaten their productivity and sustainability. These challenges range from limited access to essential resources like land, water and financial services, to a lack of knowledge regarding modern agricultural techniques, market trends, climate-smart practices and business practices. Unpredictable weather, pests and diseases complicate farm management and make effective decisions more difficult. Similarly, lack of adequate knowledge and skills in farm management also limit the farmer's ability to make decisions.

Small family farms are in a vulnerable position due to climate change. They rely heavily on rain-fed agriculture and often lack access to adaptive technologies. However, they have strong potential to lead in climate change mitigation and adaptation. By using practices like conservation agriculture, agroforestry and efficient water use, they can reduce greenhouse gas emissions, boost carbon storage and strengthen resilience to climate shifts. Their efforts in sustainable farming are crucial to global climate action.

Despite these challenges, many small family farms have great potential to boost productivity and integrate into modern markets. Their deep local knowledge and adaptability, combined with modern innovations, can transform farming practices. This can make agriculture more productive, resilient and sustainable.

Small family farming also brings important socio-economic benefits. Income from agriculture can drive local economic growth, boost rural markets and create jobs, both within and beyond the farming sector.

The communal nature of small family farming encourages collaboration, knowledge sharing and social cohesion, which strengthens community resilience. Small farmers also help to preserve biodiversity, cultural heritage and traditional farming practices. In doing so, they support both cultural identity and biodiversity conservation.

Sustainable productivity growth in small family farming can occur in three fundamental ways: (i) developing, adapting and applying new technologies and practices; (ii) increasing

Box 2.3. Insights on female-headed households.

Data on female-headed households (FHHs) in low-income countries outside Africa is limited and varies by region. However, some insights can be drawn from available sources:

- *Global Perspective*: Globally, an estimated 41% of female-headed households live below the locally defined poverty line.
- *Regional Variations*: The proportion of women in the agricultural labour force varies significantly by region, ranging from 20% in Latin America to 50% in Eastern Asia.
- *Impact of Migration and Conflict*: In regions affected by male outmigration, conflict or health crises, women often become primary household heads. For instance, in Nepal, male labour migration has led to an increase in female-headed households, with women taking on greater agricultural responsibilities.

These insights indicate that while specific data on FHHs outside Africa is scarce, women in low-income countries often assume headship roles due to various socio-economic factors, facing challenges related to poverty and limited access to resources.

and accelerating the adoption and application of existing technologies and practices; and (iii) helping less productive farmers or farm household members transition out of agriculture into other activities (Nossal and Gooday, 2009). These pathways are not mutually exclusive and often work in tandem, reinforcing each other. Innovation in family farming is central to the first two pathways and can be promoted through various instruments, such as research and development, advisory services, training, and access to finance and markets. The gradual exit of less productive farmers and household members is facilitated by overall economic development, particularly rural development, which provides alternative income and employment opportunities. This transition contributes to raising labour productivity in agriculture and improving rural incomes. This latter point is worthy of deeper exploration; however, a detailed discussion of relevant policy instruments is beyond the scope of this book.

Addressing the unique challenges faced by small family farmers requires customized support that acknowledges their diverse needs and potential. Supporting small family farmers with the capacity to innovate and adapt is not only an investment in individual livelihoods but also a strategic move towards achieving global food security and environmental sustainability. Cross-cutting support, such as enhancing access to information through digital platforms, promoting cooperatives for better market access and bargaining power, and facilitating access to microfinance can benefit all categories of small family farmers.

Differentiating support for diverse categories of small family farming acknowledges the unique challenges and opportunities each faces. Possible options are given in the table below. By customizing interventions, stakeholders can effectively enhance small family farm productivity, resilience and contributions to food security and sustainable development. However, for innovation to occur and for farming to be productive, profitable and sustainable, attention should be focused on those farmers that possess the potential to make farming work for them as a business.

Table 2.1 provides an example of the kinds of challenges faced by the different categories of farmers. It also gives some ideas about how support can be customized to address those challenges. The point is: support should be relevant to the situation of the farmer. Whether this categorization or a more detailed framework is used (see Chapter 16, this book), what is important to bear in mind is the farmer's reality on the ground. Approached in this way, there is less risk of providing support that is irrelevant, or even detrimental or counterproductive.

2.8 Customizing Policy and Support for Small Family Farming Innovation and Market Integration

Given the diversity of small family farms in terms of size, market access and other characteristics, generalized policy prescriptions are often ineffective. Public support for innovation must consider the specific structure of small family farming in each country and setting. Integrating small family farmers into markets is essential for fostering innovation and increasing productivity. Market participation and technology adoption are closely connected (Barrett, 2008). New technologies enable farmers to produce for the market, while market opportunities encourage greater production and innovation. Thus, markets significantly influence the technologies and practices adopted by farmers.

For small family farms to transition to market-oriented farming, they must not only adopt technical innovations but also manage their farms as businesses. This involves making decisions on what to produce, where and how to sell, financing investments, product differentiation, organizing production and collaborating with neighbours for collective action. Developing these business and decision-making skills requires support from farm business management advisory services. However, as previously noted, the transition towards more commercial farming is more challenging, particularly due to barriers such as limited access to productive assets, financing and technologies – challenges that are especially acute for women farmers. High transaction costs in remote areas further hinder market engagement. In addition to advisory support, addressing these barriers requires public investment in market infrastructure and institutions, as well as the development of strong farmers' organizations to reduce transaction costs through economies of scale.

To promote innovation in small family farming, public and private investments should be directed at those small family farms with the

Table 2.1. Challenges and customized support for small family farmers by category of farmer

Small family category	Challenges	Examples of customized support
Subsistence	• Access to resources, including land and water. • Often operate under the constant threat of food insecurity. • Limited market access and lack of knowledge about sustainable practices exacerbate their vulnerability to environmental shocks.	Advisory services contribute to wider support regarding: • Providing a multifaceted support system that includes safety net programs and alternative employment opportunities. • Ensuring access to adequate food to combat immediate food insecurity. • Establishing safety net programmes. • Creating pathways for off-farm employment and alternative livelihoods to diversify income sources and reduce dependency on farming alone.
Market-oriented	• Difficulties in accessing markets, adhering to quality standards and coping with price fluctuations. • Growth stifled by limited financial services and technological adoption.	• Organizing smallholder farmers into farmer organizations and cooperatives. • Strengthening market linkages, offering market information and enhancing business skills. • Financial management training and support in adopting efficient technologies to elevate their market competitiveness and income potential.
Commercial	• Obstacles to increasing the size of the business. • Challenges of scaling-up 'good practices' requiring assistance in business development, risk management and integrating into modern value chains. • Accessing finance for expansion and optimizing large-scale operations.	• Supporting business planning, accessing finance and risk management strategies. • Facilitate linkages with supply chains and improving operational efficiency.
Specialized	• Market demand variability. • Struggle to maintain product quality and accessing specialized knowledge. • Customizing their operations to specific market demands requires precise market intelligence and innovation in product development.	• Focused market research, product development assistance and quality management training to help these farms align with market demands and enhance their product value.
Intensive	• Distinct support to capitalize on their potential for high productivity and environmental sustainability. • Assistance in optimizing resource use and labour.	• Training in precision agriculture and sustainable intensification practices to increase productivity while reducing environmental impacts.
Agroecological	• Required to comply with certification requirements.	• Guidance on organic certification, ecological pest management and diversification strategies to enhance their sustainability and market appeal.

potential to innovate and operate commercially. This includes small family farms with market orientation potential, as well as semi-commercial and commercial small family farmers. While innovation and increased commercialization reinforce each other, not all small family farmers have the desire or capacity for innovation or commercial production. Some may achieve higher incomes and improved livelihoods through non-farm activities. Supporting innovation and market integration, while also enabling diversification into non-farm activities, can be mutually beneficial.

Most small-scale farms have the potential to become commercially viable and environmentally sustainable enterprises but currently lack the capacity or incentives to innovate. Public efforts must focus on providing a bundle of support services – research, advisory services, financial services and market infrastructure – not typically offered by the private sector. Applied agricultural research customized to the needs of small family farms, along with support for producer organizations, can significantly promote innovation. Given the critical role of small family farming in food security, natural resource management and poverty reduction, promoting innovation must be a priority for policymakers. Broad participation of farmers, especially including women, youth and marginalized groups, will be essential in this effort, and will contribute to generational sustainability.

The essence of this discussion is this: small family farmers have great potential to transform their farms into thriving, sustainable businesses. A key element for success in such a transition is specialized, customized advice and support through dedicated agricultural advisory services, to help farmers adopt innovations that will improve their productivity and business efficiency. However, targeted public and private investments in innovation and market integration for small family farms are also required both at the level of policy and practice. This mutually reinforcing approach will not only enhance individual livelihoods but also contribute to broader goals of food security, environmental sustainability and economic development.

2.9 Conclusion

Small family farming is indispensable to sustainable agricultural development, food security and poverty reduction. Due to their diversity in size and market access, generalized agricultural advice and policies often fail small family farms. Consequently, support for innovation must consider the specific structure of small family farming in each country and setting.

Integrating small family farms into markets is essential for fostering innovation and increasing productivity. Market participation and technology adoption are closely linked, as technologies enable farmers to produce a marketable surplus, and market opportunities incentivize farmers to produce more, add value and innovate. Thus, markets significantly influence the technologies and practices adopted by farmers. For small family farms to transition to market-oriented farming, they must not only adopt technical innovations but also manage their farms as businesses. This involves making decisions on what to produce, where and how to sell, financing investments, product differentiation, organizing production and collaborating with neighbours for collective action. Developing these business and decision-making skills requires support in farm management from advisory services.

To support innovation, both public and private investments should focus on small family farms with potential for commercial growth. This includes those with market opportunities and a clear pathway towards more commercial production. However, not all small family farmers are suited for modern and profitable farming. For some, better livelihoods may come from engaging in non-farm activities. Encouraging innovation and market integration, while also supporting diversification into non-farm income sources, can be complementary.

Many market-oriented small-scale family farms have the potential to become commercially viable and environmentally sustainable. However, they often lack the capacity, resources or incentives to innovate. Public support is crucial to bridging these gaps. This includes targeted investments in research, advisory services and market infrastructure – areas often overlooked by the private sector. Applied research and advisory services tailored to the specific needs of small family farms, along with support for producer organizations, can play a key role in fostering innovation. Given the vital contribution of small family farms to food security, natural resource management and poverty reduction, promoting innovation should be a top policy priority. This effort must be underpinned by an inclusive capacity-building approach and the provision of appropriate incentives. Engaging farmers – especially women and the youth – will be essential to ensuring the success and sustainability of these initiatives.

Endnotes

[1] There are some indicators in many sub-Saharan African countries, the age may be dropping because the rapid increase in populations is increasing the number of youths working on small family farms.
[2] Worldwide, farms of less than 2 ha account for 84% of all farms but control only 12% of all agricultural land (FAO, 2014a).

References

Barrett, C. (2008) Smallholder market participation: Concepts and evidence from Eastern and Southern Africa. *Food Policy* 33, 299–317. DOI: 10.1016/j.foodpol.2007.10.005.

Buehren, N. (2023) Gender & Agriculture in Sub-Saharan Africa: Review of Constraints and Effective Interventions, Gender and Innovation Lab https://openknowledge.worldbank.org/server/api/core/bitstreams/19966f50-f68b-42c9-9088-9098dfaca729/content (Accessed 15 November, 2025)

FAO (2014a) *The state of food and agriculture: Innovation in family farming*. FAO, Rome, Italy. Available at: https://www.fao.org/family-farming/detail/en/c/273649/ (accessed 15 March 2025).

FAO (2014b) *FAOSTAT. Food and Agriculture Data*. FAO, Rome, Italy. Available at: http://faostat.fao.org/site/291/default.aspx (accessed 7 March 2025).

FAO (2014c) *Small family farms Data Portrait*. FAO, Rome, Italy. Available at: https://www.fao.org/fileadmin/user_upload/smallholders_dataportrait/docs/Data_portrait_variables_description_new2.pdf (accessed 15 March 2025).

FAO (2023).The Status of Women in Agrifood Systems. FAO, Rome, Italy. Available at: https://doi.org/10.4060/cc5343en (accessed 11 Decemberl 2024)

Gollin, D. (2018) *Farm Size and Productivity: Lessons from Recent Literature*. IFAD RESEARCH SERIES 34. IFAD, Rome, Italy. Available at: https://ssrn.com/abstract=3321659 (accessed 7 March 2025).

Hazell, P., Poulton, C., Wiggins, S. and Dorward, A. (2007) *The future of small farms for poverty reduction and growth*. 2020 vision discussion papers 42. IFPRI, Washington, D.C. Available at: https://ideas.repec.org/p/fpr/2020dp/42.html (accessed 15 March 2025).

Lowder, S.K., Skoet, J. and Raney, T. (2016) The number, size, and distribution of farms, small family farms, and family farms worldwide. *World Development* 87, 16–29. Available at: https://www.sciencedirect.com/science/article/pii/S0305750X15002703 (accessed 12 March 2025).

Nagayets, O. (2005) Small farms: Current status and key trends. In: *The future of small farms: Proceedings of a research workshop, Wye, UK*, International Food Policy Research Institute, June 26–29, 2005. Available at: https://www.scirp.org/reference/referencespapers?referenceid=2325854 (accessed 12 March 2025).

Nossal, K. and Gooday, P. (2009) Raising productivity growth in Australian agriculture. Issues Insights 09.7, Canberra. Available at: https://www.researchgate.net/publication/239926982_Raising_Productivity_Growth_in_Australian_Agriculture (accessed 15 March 2025).

Ritchie, H. (2021) Smallholders produce one-third of the world's food, less than half of what many headlines claim. OurWorldInData.Org. Available at: https://ourworldindata.org/smallholder-food-production (accessed 12 March 2025).

UNDP (2017) Guidance Note: Application of the Sustainable Livelihoods Framework in Development Projects. United Nations Development Programme, Panama. Available at https://www.undp.org/latin-america/publications/guidance-note-application-sustainable-livelihoods-frameworkdevelopment-projects (accessed 27 January 2025)

World Bank. 2024. Poverty, Prosperity, and Planet Report 2024: Pathways Out of the Polycrisis. World Bank,Washington, DC, USA. DOI: 10.1596/978-1-4648-2123-3. (accessed 15 November 2025)

3 Understanding the Farming System Context for Small Family Farm Management

In the first two chapters of this book, we explored the nature, role and importance of smallholder family farming and how it is affected by a rapidly changing agricultural landscape. This has helped highlight both the urgency and complexity of the issues smallholders face. It also shows their potential to adapt and remain a key part of the global agri-food system in the future.

In this and the following chapter, we will explore yet another aspect of the complexity of smallholder family farms. In this chapter, we will learn about the systems within which smallholder family farms operate. In Chapter 4, we will learn about the individual farm as a system. Yes, they are similar topics, but their difference will become clear as you read on.

Let's start with the basics: What is a system? A simple definition of a system is a group of connected parts that work together as a whole to achieve a purpose. A system is influenced by its surroundings and shaped by its boundaries, structure and function.

> A system is a collection of elements connected together to form a purposive whole with properties that differ from those of its component parts. (Armson, 2011, p. 134)

Systems are everywhere; we use them every day. In the context of agriculture, the concept of systems is used frequently. We speak of the agri-food system, of farming systems, and also of the farm itself as a system. It is critical to understand smallholder family farming from the perspective of the system in which the farm operates (this chapter) and from the perspective of the farm (Chapter 4).

In this chapter, we will come to appreciate the diversity among smallholder family farms within their specific farming systems. To truly improve smallholder productivity, food security, livelihoods and incomes, support strategies that strengthen smallholder family farming must be factored into the farming systems in which they operate. This holistic approach is necessary for ensuring that development initiatives are precisely aligned with the unique opportunities and challenges each farmer faces.

As previously implied, the socio-economic characteristics of smallholder family farmers, as described in Chapter 2, provide one part of the picture. They do not reveal the full range of challenges and opportunities encountered. A comprehensive approach to agricultural development must also incorporate an understanding of the agro-ecological and physical conditions that form the 'system' within which the farm operates. These elements of the farming system – including soil types, water availability, climate patterns, infrastructure and access to markets – play a critical role in determining the success of practices and interventions aimed at promoting farm commercialization and improving smallholder farmers' market integration.

3.1 What is a Farming System?

A farming system includes all agricultural activities, resources and inputs involved in growing crops and raising livestock in a specific area. It includes the diverse interactions between

DOI: 10.1079/9781836990536.0003

farming practices, the environment and the socio-economic conditions under which these activities occur. Farming systems are defined based on various factors including climate, landscape, crop and animal types, and available technology; and socio-economic aspects such as market access, labour availability and cultural practices. A farming system includes all activities, resources and inputs used in crop cultivation and livestock rearing in a specific location. It covers the interactions between farming practices, the environment and socio-economic factors. These systems are not static but change over time in response to environmental, technological and market changes and, in the case of smallholder family farms, changes within the family household system. See Box 3.1 for more about farming systems and the farming systems approach.

Farming systems vary widely across the world. They reflect the adaptation of agricultural practices to the local environment and the needs of the farming community. At the heart of any farming system is the farm household goal set by the farm family. For market-oriented family farming this goal often involves maximizing profitability while ensuring long-term sustainability. This requires decisions to be taken regarding the strategic mix of crops and livestock enterprises, the use of available resources, and management practices that enable the farm to be run as a business. This also calls for minimizing adverse environmental, market and other impacts. Some farm systems may focus on monoculture, where a single crop is cultivated extensively. Others may adopt polyculture or mixed farming practices, integrating various crops and livestock to enhance diversity and resilience. The choice of farming system affects soil health, water use and biodiversity, as well as the economic and social well-being of the farming community, in particular, the farming family itself.

Understanding the systems in which small family farms operate is key to creating effective policies and practices that support food security, sustainability and economic stability. Recognizing the unique features of different systems helps policymakers, researchers and advisors design targeted solutions. These can include better crop varieties, improved land management, and policies that improve market access and fairness. Ultimately, the goal is to balance profitability, income growth, human nutrition and environmental health and sustainability within a framework of ethical practice.

3.2 Factors Influencing Farming Systems

The study of farming systems and their future growth requires a consistent framework across different regions and systems. Five key factors shape farming systems: (i) natural resources and climate; (ii) science and technology; (iii) trade changes; (iv) policy and institutions; and (v) information and education (Dixon *et al.*, 2011).

3.2.1 Population and population densities

Population growth and changing population density have played a major role in driving agricultural innovation. Today's settled farming systems have developed over time in response to these shifts. One key driver was the growing need to produce more food. Another was the rise of exchange between populations, marking the early beginnings of markets. As cities expanded, more farmers settled nearby, leading to smaller and more fragmented farms. At the same time, growing demand from non-farm activities like mining and manufacturing increased pressure on land. As we will see, choices in farming systems have largely been shaped by access to fertile soil, reliable water and favourable climate.

3.2.2 Natural resources and climate

The mix of natural resources, climate and population sets the foundation for farming systems. As the population grows, the area under cultivation increases, causing conflicts over land and water use. When high-quality land becomes scarce, farming intensifies. This pressure on land leads to biodiversity loss and strains between development and conservation efforts. The variety of crops, especially major grains like wheat, maize and rice, has dropped significantly, mirroring a similar trend

Box 3.1. Farming systems and the Farming Systems Approach. From Dixon and Ameda, 2023.

The term 'farming systems' is used firstly as a description of the patterns of farm and landscape use and production; and secondly as a research and development approach or process of multidisciplinary diagnosis, analysis, testing and scaling for the improvement of farming systems. Throughout history seven major drivers shaped the systematic evolution from hunter gatherer systems to modern farming and food systems, including climate, population density, natural resources, energy and technologies. Over the past five decades, the farming systems approach has been applied by agricultural research scientists, extension agents, planners, policy makers, geographers and other development professionals, to characterize and analyse types of farming and identify and scale technological and institutional innovations that support small and large farms. The farming systems approach itself is characterized by multidisciplinary diagnosis, analysis, innovation testing, targeting and scaling of farming system interventions using participatory and experimentation/learning methods. The farming systems approach has added substantial value to agricultural research and scaling, and merits renewed investment to tackle the current complex climate, poverty, food security and rural development challenges.

in livestock diversity. The rapid increase in population has halved the per capita availability of farmland since the 1960s. Despite this, the potential for agricultural expansion is limited, mainly to Latin America and sub-Saharan Africa. Although climate change now is a major threat to smallholder family farming, huge strides were made over hundreds of years to develop technologies that reduced the role of climate as a determinant of the farming system.

3.2.3 Science and technology

Science and technology has become increasingly important in shaping farming systems. There are two key aspects to this influence. The first aspect is formal agricultural research, which has produced many useful technologies. Much of this research has focused on increasing crop yields, especially for cereals. Since 1961, most production growth has come from these yield gains. However, many smallholders in less fertile areas have not benefited as the technologies are often too expensive or unsuitable for their conditions. Research has also been heavily focused on cereal crops, with little attention to non-cereal crops or sustainable farming methods. The second aspect is local science and technology development. This is often grouped with 'indigenous knowledge' but stands on its own. Farmers have long adapted and improved their practices to boost productivity. These local innovations have shaped farming system choices over time. More recently, there is greater acknowledgement of the symbiosis between these two aspects of science and technology.

In the past, the primary focus was on private and large-scale farming investments, which have advanced cash crop research. This shifted to a focus on smaller crop and livestock farms, which began to receive more attention. A great deal of emphasis has been put on creating technologies that are coherent with the smallholder farmer's system and requirements. More recently, because of the need to increase farm income and access markets, the research focus is shifting towards high-value crops and value chain development. Concurrently, there is also a shift towards improving overall system productivity through better resource management and targeting technologies to benefit women and poorer farmers more directly. Current research also looks at better management of mixed farm enterprises, the sustainable use of natural resources and the targeting of support to vulnerable groups. Importantly, there is more emphasis on public–private partnerships and responding to the needs of farmers. There is also greater respect for blending local knowledge with modern science. Undoubtedly, science and technology will continue to shape farming systems. The priority is to ensure smallholder farmers can access and have the ability to use these resources effectively. Ideally, farmers should be active partners in developing these technologies.

3.2.4 Trade liberalization and market development

Of the many processes included under the term globalization, a primary focus has been economic reform and trade liberalization. These have had the greatest impact on smallholder family farms – bringing initial challenges but also potential opportunities. Historically, a major change occurred with the Structural Adjustment Programmes (SAPs), promoted by the World Bank and International Monetary Fund in the 1980s in response to economic crises in developing countries. The economies of these countries had become distorted by heavy government control, and GDP growth often lagged behind population growth. The SAP reforms led to trade liberalization, reduced subsidies and changes to exchange rate policies in many developing countries. The reforms ended state marketing monopolies; reduced the role of parastatals in input supply, marketing and processing; removed subsidies and price controls; lifted trade restrictions; and encouraged private sector participation. The establishment of the World Trade Organisation in 1995 further accelerated trade liberalization. Markets expanded, and production patterns shifted in response to new global demands. However, these changes often hurt smallholder farmers in low-income countries. Many faced reduced government support and falling prices for key crops, leading to increased poverty in the 1980s and early 1990s.

Over time, developing countries were expected to grow more of the products in which they have a competitive advantage. But progress has been slow, partly due to ongoing protection of agriculture in wealthy countries. In response, 'fair trade' movements have emerged to support small-scale producers and promote a more equitable global trade system (UNCTAD, 2019).

Globalization has also reshaped market demand. As we saw in Chapter 1, urbanization, rising incomes, better communication and changing cultural preferences are all influencing what consumers want. For smallholder farmers to benefit, they must adapt to these trends. Advances in production, processing and transport technologies allow farmers to reach new markets with new or improved products.

Like science and technology, trade liberalization requires small family farmers to rethink their farming systems. To benefit, they must adopt approaches that align with changing markets and consumer demands.

3.2.5 Policies, institutions and public goods

Policy has played a central role in shaping farming systems. It can create opportunities, such as trade liberalization but can also limit options when there is poor investment in rural development. A supportive policy environment is, therefore, essential for the growth and adaptability of farming systems. In recent decades, the structural adjustment policies previously mentioned, shifted many countries away from food self-sufficiency towards open markets. This brought a focus on service efficiency and transferred many government roles to the private sector and local institutions. Policymakers increasingly restructured institutions, decentralised services and reduced public investment. While these shifts aimed to improve responsiveness and encourage local participation, they also created challenges. See Box 3.2. for additional insight into international policy and smallholder farming.

The private sector, however, has often been slow to fill the gap left by reduced public services, especially in key areas such as finance, research, extension, education and infrastructure. Smaller farmers and female-headed households have been particularly affected. Despite these issues, progress has been made in strengthening local institutions, with many countries promoting decentralization and community-level decision making. However, this has also led to tensions between central and local governments over development priorities, budget allocations and oversight responsibilities.

Access to and control of natural resources – particularly land and water – is another growing policy concern, especially for smallholder farmers. As populations rise and marginal lands degrade, demand for equitable resource access for poor and more vulnerable groups is increasing. Without effective policies on land rights, water use and market reform, governments risk social unrest and increased vulnerability for smallholder farmers.

Box 3.2. International policy and smallholder farming.

Fan and Rue (2020) argue that smallholders are central to achieving the Sustainable Development Goals, noting that '… smallholders have a unique role to play in the new global development agenda—the SDGs which world leaders agreed upon in September 2015. Smallholder agriculture, especially if well-integrated into a diversified rural economy and agri-food value chains, can contribute even more to inclusive growth and employment generation. Even very poor smallholder farmers can be empowered to manage resources sustainably, and benefit from goals around education, peace and gender equality. Assistance, through measures such as safety nets and support through off-farm employment to diversify livelihoods, can also help develop rural communities and interrupt cycles of poverty, hunger and undernutrition. In addition to promoting more inclusive patterns of growth, this support can also cushion the short-term impact of transitioning to non-farm activities.'
(Source: Fan and Rue, 2020)

Currently, smallholder agriculture faces multiple policy challenges that require urgent and innovative solutions. Beyond food security, land degradation and climate change, critical issues include inclusive market access, gender equity and youth employment. Addressing these areas is vital to building resilient, equitable farming systems and improving the livelihoods of millions of small-scale farmers.

3.2.6 Information and human capital

The evolution of farming systems is also deeply influenced by the availability and utilization of information and human capital. As agriculture transitions towards more intensive and market-oriented practices, the demand for knowledge, skills and training has grown significantly. Farmers require access to reliable and timely information on climate, markets, technology and institutional frameworks to adapt their practices and seize emerging opportunities. At the same time, the ability to utilize this information effectively depends on the level of education and capacity within farming communities. These elements are fundamental to fostering innovation, improving productivity and enabling smallholders to compete in increasingly complex agricultural markets.

Over time, the role of information and human capital in farming has evolved considerably. In earlier periods, the focus was primarily on expanding basic literacy and primary education to rural populations, recognizing that education is a key driver of development. Today, the information revolution offers unprecedented opportunities to deliver tailored and actionable insights to farmers. Advances in digital technologies have enabled the distribution of large amounts of data to remote areas, potentially transforming traditional farming.

The importance of human capital lies in its ability to bridge these gaps. Education and training equip farmers with the skills to assess, interpret and apply information to their specific contexts. Over the years, there has been growing recognition for the need to address these challenges inclusively, particularly for marginalized groups such as women and youth, whose participation is essential to the sustainability of farming systems. Efforts to expand education and capacity-building initiatives are increasingly focusing on these groups, ensuring they can access and benefit from the expanding information base.

3.3 Historical Context: Small Family Farm Evolution

Farming systems evolve in response to both biophysical and socio-economic factors. Biophysical conditions, such as climate, soil and water availability define the range of possible systems; while socio-economic factors, such as population pressure, labour availability and market access determine what farming practices are adopted. A common pattern of system evolution can often be observed depending on the context.

Smallholder farming systems have evolved significantly over time in response to these changing biophysical, social and economic conditions. Initially, these systems were largely subsistence-based, with small family farms relying on traditional knowledge to manage crops and livestock. Practices such as seed selection, intercropping and soil conservation were well adapted to local climates, soil types and water availability. Farming was deeply embedded in local ecosystems, with minimal reliance on external inputs.

Over time, economic and demographic pressures have driven transitions in farming systems. For example, a system that begins with simple tools such as hand hoes may evolve as market opportunities arise. Farmers may introduce cattle for draught power, which reduces reliance on manual labour and allows for expansion of cultivated land where available. As population increases and land becomes scarce, the focus typically shifts from area expansion to intensification. This stage may be characterized by the adoption of higher-yielding crop varieties, the use of fertilizers and investments in irrigation.

Market-driven transitions have also played a pivotal role. The introduction of cash crops, such as cotton, coffee and sugarcane, often promoted by colonial administrations, marked a major shift. These crops required different cultivation methods and contributed to the spread of monoculture. While they enhanced income and market integration, they also brought increased vulnerability to pests, diseases and price fluctuations, as well as social impacts, particularly notably increased food insecurity and displacement of women.

As regional markets developed, smallholders began producing surplus crops for sale, integrating further into the cash economy. This transition enabled access to technologies such as improved seeds, fertilizers and pesticides, increasing yields and supporting population growth. However, increased dependency on external inputs also led to environmental issues, including declining soil fertility, water contamination and biodiversity loss.

With further development, farming systems move towards mechanization, boosting efficiency and enabling larger-scale operations. Market-oriented systems may evolve into more commercial, export-driven enterprises, often characterized by land consolidation, intensive input use and high production levels. In some cases, intensive mixed farming systems emerge, integrating crops and livestock in technically advanced ways. Across all these stages, access to infrastructure, such as roads, storage and processing facilities; and the availability of technical and market information remain critical drivers of system transformation.

Amid these changes, many farmers have blended modern technologies with traditional practices. Organic manure, local seed varieties and water-saving techniques continue to complement the use of external inputs. This hybrid approach has enabled smallholders to adapt to diverse conditions, especially in areas where access to modern technologies is limited.

Social dynamics have also significantly shaped system evolution. Population growth and urbanization have intensified land use, led to farm fragmentation, and encouraged a shift to less labour-intensive crops. Gender inequality has limited women's access to land, finance and advisory services, constraining innovation and diversification. Insecure land tenure has discouraged long-term investments in sustainable practices, while the erosion of communal land systems has weakened collective natural resource management.

More recently, global shocks such as the Covid-19 pandemic have exposed vulnerabilities in supply chains and labour markets, prompting many farmers to shift back towards systems that focus on food security. Climate change has added further complexity, with shifting rainfall patterns, rising temperatures and increasingly frequent extreme weather events, forcing adaptation.

The capacity of farmers to respond to these challenges depends not only on environmental conditions but also on access to advisory services, climate information, drought-resistant seeds and irrigation infrastructure. In this context, the value of traditional knowledge has been reaffirmed, especially in managing risk, sustaining productivity and maintaining resilience under increasingly unpredictable conditions.

In summary, the evolution of smallholder farming systems reflects a dynamic interplay of biophysical, social and economic forces. While environmental factors such, as climate and soil have long shaped agricultural practices; shifts in markets, policies, technologies and social structures have significantly transformed how small farms operate.

3.4 Regional Farm Systems Variations

The characteristics of small family farm systems vary significantly across regions, shaped by a combination of the factors previously described. These regional differences influence not only the types of farming systems that are prevalent but also the strategies that small family farms use to manage their resources and sustain their livelihoods.

In sub-Saharan Africa, for instance, small family farms are often characterized by low input, rain-fed farming systems. These systems are adapted to the region's diverse agro-ecological zones, ranging from the humid tropics to arid and semi-arid areas. In many parts of Africa, small family farms rely on traditional knowledge and practices to manage soil fertility, conserve water and cope with climate variability. However, these farms also face significant challenges, including limited access to markets, inputs and technology, as well as vulnerability to climate change and land degradation.

In South Asia, small family farms are commonly found within densely populated rural areas, where landholdings are typically small and fragmented. Here, both rain-fed and irrigated farming systems are prevalent, with a strong emphasis on rice and wheat cultivation in regions like the Indo-Gangetic Plains. Mixed farming is also common, with livestock playing a crucial role in household nutrition and income. The Green Revolution has had a profound impact on farming systems in South Asia, leading to increased productivity but also raising concerns about environmental sustainability, equity and the overuse of chemical inputs.

In Latin America, the diversity of ecosystems, from tropical rainforests to high-altitude Andean regions, has given rise to a wide range of farming systems. Small family farms in this region often engage in both subsistence and market-oriented production, with crops like maize, beans, coffee and vegetables being prominent. In many areas, indigenous knowledge and practices play a key role in managing diverse cropping systems and maintaining soil fertility. However, the expansion of large-scale commercial agriculture, particularly in countries like Brazil and Argentina, has put pressure on small family farms, leading to issues such as land concentration, deforestation and socio-economic inequality.

In East Asia, particularly in countries like China and Vietnam, small family farms are integrated into highly organized and intensive farming systems. Irrigated farming is widespread, particularly in the production of rice, which is a staple crop in the region. These farms often benefit from strong government support, access to markets, and the adoption of modern technologies. However, rapid industrialization and urbanization are transforming rural landscapes, leading to challenges such as labour shortages, land fragmentation and environmental degradation.

3.5 Categorization of Farming Systems

The broad categorization of farming systems is based on natural resources, livelihood patterns, production intensity and regional political and geographical differences. In a seminal work by Dixon *et al.* (2011), some 72 farming systems were identified globally. These include core systems of irrigated, rain-fed and mixed farming, as well as multitudes of sub-systems each with unique attributes and challenges. These systems are delineated based on an array of critical factors such as water usage, the integration of livestock, the degree of mechanization, the cultivation strategies employed and their proximity to urban centres. Since some of the socio-economic aspects of the smallholder family farm were described in Chapter 2, focus here will be on physical and agronomic aspects of the system within which they operate. A nuanced understanding of these diverse systems is pivotal to formulate strategies that strengthen smallholder family farms and help farmers navigate the challenges discussed in Chapter 1.

3.5.1 Rain-fed farming systems

Rain-fed farming systems, relying entirely on rainfall, form the backbone of global agriculture. They rely solely on natural precipitation for watering crops. This traditional system predominates in regions blessed with ample annual rainfall. Rain-fed systems can be found under humid, highland and dry/cold conditions.

However, the system is fraught with challenges, such as susceptibility to drought, soil degradation and acute water shortages. Despite these hurdles, rain-fed agriculture underpins a large share of food production worldwide. Its heavy reliance on the vagaries of climate, however, renders it particularly susceptible to the impacts of climate change. This vulnerability underscores the urgent need for the development and implementation of adaptive strategies aimed at bolstering the resilience and sustainability of rain-fed systems, such as improved soil management practices, drought-resistant crop varieties and innovative water conservation techniques.

The examples of rain-fed systems given in Box 3.3 illustrate the global prevalence of rain-fed farming and highlight the challenges faced by small family farms in regions where rainfall is unpredictable, and water resources are limited. The success of rain-fed farming systems is closely tied to the ability of farmers to manage soil moisture, adapt to changing climatic conditions and diversify their cropping patterns to mitigate risks.

3.5.2 Irrigated farming systems

By using artificial methods to supply water to crops, irrigated farming systems not only reduce the risks associated with rain-fed systems, they significantly boost agricultural productivity and enable the cultivation of multiple crops within a single year. These systems are especially vital in regions and countries where natural rainfall is insufficient to support reliable agricultural production. When situated in lowland areas with dependable water infrastructure, irrigated farming systems are among the most productive and commercially oriented smallholder systems. Wetland rice systems, commonly found in South-east Asia, also offer considerable potential. Irrigated systems are indispensable in arid and semi-arid areas where rainfall is either scarce or highly unpredictable.

Although more productive than rain-fed systems, irrigated agriculture requires substantial investment in infrastructure and resource management. Major challenges include limited water availability, the complexity of irrigation

Box 3.3. Regional and country examples: rain-fed farming systems.

Sub-Saharan Africa: In the Ethiopian highlands, rain-fed agriculture is the mainstay of small family farms. Crops such as teff, maize and sorghum are commonly grown, relying heavily on seasonal rains. The region's dependence on unreliable rainfall makes it vulnerable to droughts, which can lead to food insecurity. In the Sahel region, countries like Burkina Faso, Niger and Mali in the Sahelian zone practice rain-fed farming, with millet, sorghum and cowpea as staple crops. The area is characterized by low and variable rainfall, making farming highly susceptible to climatic fluctuations.

South Asia: In the central and eastern regions of India, including parts of Madhya Pradesh, Odisha, and Jharkhand, rain-fed farming is common. Farmers in these areas grow crops like pulses, oilseeds and coarse cereals, relying on the monsoon rains. The erratic nature of the monsoon poses challenges for stable crop production. In Bangladesh, specifically the Barind Tract and other upland areas, rain-fed farming is practiced with crops like rice and wheat. The region faces challenges due to irregular rainfall patterns and drought risks.

Latin America: In the semi-arid region of North-east Brazil known as the Sertão, rain-fed farming is widely practiced. Farmers grow crops like maize, beans and cassava, relying on the seasonal rains. The region frequently experiences droughts, making agriculture challenging. Similarly, in the central highlands of Mexico, rain-fed farming dominates, with maize being the primary crop. The region's agriculture depends on summer rains, which can be highly variable and lead to significant fluctuations in crop yields.

Middle East and North Africa: In the rain-fed agricultural zones of the Atlas Mountains and the surrounding plains of Morocco, small family farms grow wheat, barley and olives. Rainfall is unpredictable and often inadequate, making these areas prone to droughts and water scarcity. Before the conflict, the Syrian steppe (Badia) region relied heavily on rain-fed farming for crops like barley and wheat. The variability in rainfall and the frequency of droughts has historically made agriculture in this area challenging.

management – including soil maintenance, salinity control and management and maintenance of irrigation infrastructure and equipment – and the environmental impacts of excessive water extraction. Addressing these challenges calls for the adoption of water-efficient irrigation technologies and sustainable water and soil management practices. See Box 3.4 for geographic examples of irrigated farming systems.

These examples illustrate the reliance of many countries around the world on irrigated farming systems in regions where natural rainfall is insufficient to sustain agricultural production. While these systems have enabled significant increases in productivity, they also highlight the importance of managing water resources sustainably to ensure the long-term viability of irrigated agriculture.

3.5.3 Livestock-based farming systems

As the term suggests, these systems focus on livestock production. Livestock-based systems range from extensive pastoral farming to intensive dairy and poultry farming. They offer a diversified income stream through the production of meat, milk, eggs and/or other animal products such as fibre (e.g. wool). Within the focus, a livestock-based farming system may have many variations. They may be grassland/grazing-based, as is often the case with cattle, sheep and goats; landless, as is often the case with poultry and rabbits; or they may be part of a mixed farming system. Among the principal challenges faced is the efficient management of feed resources, mitigation of greenhouse gas emissions, safeguarding of animal health and welfare, and compliance with market and government health regulations.

The examples in Box 3.5 demonstrate the diversity and importance of livestock-based farming systems across different regions of the Global South. These systems are integral to the livelihoods of millions of small family farmers, providing essential food, income and ecological benefits. However, they also face significant challenges, including the need for sustainable

Box 3.4. Country variations: irrigated farming systems.

Examples of irrigated farming systems can be found in:

- **India (Indo-Gangetic Plains):** The Indo-Gangetic Plains, encompassing states like Punjab, Haryana and Uttar Pradesh, offer a prime example of irrigated farming systems. This region, often referred to as the 'breadbasket of India', relies heavily on extensive canal irrigation networks and tube wells to grow water-intensive crops like rice and wheat. The green revolution significantly boosted agricultural productivity in this region, largely due to the expansion of irrigation infrastructure. However, the over-extraction of groundwater has led to declining water tables, posing a significant challenge to the sustainability of irrigated agriculture.
- **Egypt (Nile delta):** Egypt's agriculture is almost entirely dependent on irrigation from the Nile River. The Nile delta and valley is among the most intensively irrigated areas in the world, supporting the cultivation of crops such as cotton, rice and wheat. The Aswan High Dam plays a crucial role in regulating the flow of the Nile, allowing for year-round irrigation. However, the challenges of water scarcity, salinization and the environmental impact of damming the river are ongoing concerns for the sustainability of this irrigated farming system.
- **China (North China Plain):** The North China Plain, encompassing parts of Hebei, Henan and Shandong provinces, is a major agricultural area that relies heavily on irrigation to grow crops such as wheat, maize and vegetables. Irrigation in this region primarily comes from both surface water (rivers and reservoirs) and groundwater. However, the overuse of groundwater has led to a serious depletion of aquifers, prompting the Chinese government to promote water-saving technologies and sustainable water management practices.
- **Pakistan (Indus River basin):** The Indus River basin supports one of the world's largest contiguous irrigation systems, covering much of Punjab and Sindh provinces. This system is essential for the cultivation of staple crops such as rice, wheat and sugarcane. However, the region faces significant challenges, including waterlogging, salinization and the inefficient use of water resources, which have led to efforts to promote more sustainable irrigation practices.

resource management, improved animal health and welfare, and strategies to mitigate environmental impacts such as greenhouse gas emissions. Addressing these challenges is crucial for ensuring the long-term sustainability and productivity of livestock-based farming systems.

3.5.4 Peri-urban and urban farming systems

Peri-urban farming systems located in, or on the outskirts of, urban centres are essential for supplying cities with fresh produce, dairy and meat. These systems leverage their proximity to urban markets to provide a continuous supply of food, often reducing the need for long-distance transportation and ensuring fresher products reach consumers. Peri-urban farming faces unique challenges such as competition for land from urban expansion, pollution and limited farming space. Despite these challenges, peri-urban agriculture plays a vital role in ensuring urban food security and can contribute significantly to sustainable urban planning and the development of green spaces. To thrive in the context of rapid urbanization, these farming systems must adopt innovative approaches to policy, land use planning and agricultural practices (FAO *et al.*, 2022).

Urban farming systems, often referred to as 'urban agriculture', include the cultivation, processing and distribution of food and non-food agricultural products within an urban area. This is similar to peri-urban farming and often shares similar resource bases and markets. Recently, however, there has been growing interest in intensive urban agricultural production (production within the urban area), which employs a variety of technologies including hydroponics, rooftop farming, aquaponics and container farming. While urban agriculture may not directly form part of the small family

Box 3.5. Regional and country variations: livestock-based farming.

Examples of livestock-based farming systems across the Global South include:

- **Sub-Saharan Africa:** In Kenya's central highlands, smallholder dairy farming provides a successful example of intensive livestock production. Farmers keep improved dairy breeds, such as Friesians, in zero-grazing units and feed them with cultivated fodder. Supported by cooperatives for milk collection and marketing, this system boosts household income, nutrition and employment in rural communities.
- **South and South-east Asia:** Livestock systems in this region are closely integrated with crop farming. In Gujarat, India, dairy cooperatives like Amul help smallholders market milk from cattle and buffalo, while animals also provide manure and draught power. In Bangladesh, chickens and ducks are raised alongside rice cultivation, improving soil fertility and food security. In Indonesia's Bali, Balinese cattle are integrated with rice and corn farming. In Vietnam's Red River delta, pigs complement rice and vegetable farming, with manure recycled as fertilizer. These integrated systems enhance resource efficiency, incomes and resilience.
- **Middle East and North Africa (MENA):** Livestock farming supports rural livelihoods in challenging climates. In Morocco's Atlas Mountains, Berber farmers raise sheep and goats alongside growing cereals, using manure for soil fertility. In Egypt's Western Desert, sheep, goats and camels support the oasis farming of dates and olives, where efficient water use is critical. In Jordan and Yemen, traditional pastoralism and terrace farming integrate livestock with crops like sorghum. Across MENA, livestock systems face shared challenges like water scarcity and desertification but offer opportunities for innovation in resource management and crop–livestock integration.
- **Latin America:** Livestock-based farming systems are adapted to a range of environments, from high-altitude regions to semi-arid plains. In the Altiplano region of Bolivia, mixed farming systems combine the cultivation of potatoes, quinoa and barley with the rearing of llamas and alpacas. The integration of livestock with crop farming helps sustain livelihoods in this challenging environment. In Paraguay, the semi-arid Gran Chaco region is dominated by cattle ranching. Livestock are raised extensively on natural pastures, often integrated with the cultivation of soybeans and other crops. This system supports diversified income streams but faces challenges such as deforestation, land degradation and water management.

farming system in rural areas, it has been identified as an emerging farming system with great potential to diversify and strengthen local food systems, create employment opportunities, and encourage technological innovation that could benefit rural farmers.

The examples in Box 3.6, from Nairobi, Mumbai, São Paulo and Accra illustrate the diverse challenges and opportunities associated with peri-urban and urban farming systems in rapidly urbanizing areas. By adopting innovative agricultural practices and supportive policy frameworks, these cities can ensure that these farming systems continue to play a vital role in providing fresh food, enhancing urban sustainability, and contributing to the overall well-being of their populations.

3.5.5 Mixed farming systems

Mixed farming systems integrate crop and livestock production, and, when combined with agroforestry, may also incorporate forestry practices. They leverage the synergistic relationships between these agricultural components to maximize efficiency in resource utilization and mitigate risks. They also offer the advantage of diversified income sources and overall sustainability because they often use fewer external inputs. If they are well managed, these systems are versatile and can adapt to a wide range of environmental and socio-economic settings.

Mixed farming systems are a model for sustainable agricultural production. By introducing innovative practices in crop-livestock integration, effective nutrient recycling and sustainable farm management, mixed farming systems can significantly enhance resilience and productivity. However, mixed farming systems do require more work and more specialized skills, more complex disease and pest management, and more complex marketing strategies.

The case studies in Box 3.7. highlight the adaptability and effectiveness of mixed farming and agroforestry systems across diverse environmental and socio-economic contexts. By integrating crop and livestock production, mixed farming systems optimize resource use, enhance

Box 3.6. Examples of peri-urban and urban farming.

Examples of peri-urban and urban farming systems in areas undergoing urban expansion:

- **Nairobi, Kenya:** Peri-urban farming in Nairobi is vital to the city's food system. As urban expansion increases demand for fresh produce, small-scale family-run farms on the outskirts grow vegetables like kale, spinach, tomatoes and potatoes, and raise dairy cattle and poultry. Proximity to urban markets ensures fast sales and a steady food supply. However, these farmers face land competition, water scarcity and pollution. Innovative practices such as vertical farming, rainwater harvesting and organic farming methods are helping to boost productivity. Supportive policies – regarding, for example, zoning laws and access to resources – are essential to sustain urban agriculture (Nairobi City County, 2015).
- **Mumbai, India:** Mumbai's surrounding peri-urban areas supply vegetables, fruits and flowers to city markets. Farmers benefit from high demand and better prices, but rapid urban growth has reduced farmland and increased pollution risks. In response, some farmers are adopting hydroponics and other space-efficient methods. Integrating urban agriculture into city planning is key to preserving these systems amid urbanization (Vazhacharickal *et al.*, 2013).
- **São Paulo, Brazil:** Peri-urban farming in São Paulo supports food security and sustainable development. These farms produce vegetables, fruits, herbs, dairy and poultry for local consumption, reducing reliance on long supply chains. Yet, urban sprawl threatens farmland and causes soil degradation and water pollution. Some farmers now use agroforestry, organic methods and biofertilizers. City policies that protect agricultural zones and promote green belts are crucial (Borgan Project, 2024).
- **Accra, Ghana:** In Accra, urban agriculture is practiced using simple techniques like hand-dug gardens, composting and basic irrigation. Farmers grow vegetables on small plots, improving food security and reducing reliance on imports. The Urban Agriculture Network in Ghana trains farmers in sustainable methods, supporting economic resilience (Nchanji, 2017).

nutrient recycling, and strengthen resilience to environmental and economic shocks. Similarly, agroforestry systems improve sustainability, livelihoods and environmental conservation by combining trees with agricultural crops. Both approaches underscore the need for innovative practices to ensure sustainability and address challenges such as land tenure, market access and knowledge dissemination, which are critical for their long-term success and scalability in modern agriculture.

Each of these farming systems encapsulates a unique set of opportunities and challenges that necessitate customized management and support strategies. By acknowledging the distinct needs, challenges and potential inherent in each system, agricultural stakeholders are better positioned to design and implement interventions that elevate productivity, advance environmental sustainability and deliver socio-economic advantages. Strategies could encompass a broad spectrum of initiatives, ranging from enhancing water use efficiency within irrigated schemes and advocating for climate-smart practices in rain-fed agriculture, to promoting sustainable livestock management techniques and safeguarding peri-urban agricultural lands from urban encroachment. They may also include a fresh look at organizational management, marketing and communication systems. A profound comprehension of these diverse farming systems paves the way for the formulation of comprehensive agricultural policies and practices that underpin global food security and champion sustainable development initiatives – all within the ambit of smallholder family farming.

3.6 Interactions Between Components: Crops, Livestock and Environment

Farming systems are formed of interconnected components that influence one another in complex ways. For small family farms, understanding and managing these interactions, particularly those between crops, livestock and the environment, is essential for optimizing productivity and ensuring long-term sustainability.

Box 3.7. Examples of mixed farming systems.

Mixed farming systems offer sustainable solutions by integrating crops and livestock or combining agriculture with forestry and aquaculture.

In southern Mali, West Africa, the Bambara people successfully combine maize, millet and groundnut cultivation with cattle, sheep and goat rearing. Livestock provide manure to fertilize fields, draft power for ploughing, and income from milk and meat, creating a resilient system through nutrient recycling and resource optimization.

In north-east Thailand, smallholders practice rice–fish farming by raising fish in flooded rice paddies. The fish control pests and fertilize the fields in a natural way, while ducks or chickens introduced after harvest manage weeds and provide eggs and meat. This system boosts food security, diversifies income and reduces reliance on chemical inputs.

In Ethiopia's central highlands, farmers integrate cereals like wheat and barley with livestock. Manure maintains soil fertility, while cereal residues feed animals, especially during dry periods. Livestock also provide draft power for ploughing terraced fields. Improved forage management and rotational grazing further enhance sustainability.

The Taungya system in Myanmar combines crop cultivation with reforestation. Communities grow maize, beans and vegetables alongside young trees, later shifting their focus to forest management as the trees mature. This supports livelihoods, biodiversity, soil conservation and water retention. However, its success depends on secure land tenure and collaboration between farmers and government.

In the Peruvian Amazon, the Inga alley cropping system restores degraded land using nitrogen-fixing Inga trees planted between food crops like maize and beans. The system enriches soil, suppresses weeds and provides shade, timber and firewood. While effective, challenges such as high costs, limited training and weak market access persist, requiring targeted support and incentives.

As we have seen, mixed farming systems are common among smallholders, where crops and livestock are integrated to maximize resource efficiency. Livestock provide manure to fertilize crops, while crop residues are used as animal feed. This cyclical relationship supports soil fertility, reduces dependency on external inputs and increases farm self-sufficiency. These interactions are heavily shaped by environmental conditions such as climate, soil quality and water availability. For example, farms in water-scarce areas may adopt drought-tolerant crop varieties or livestock breeds better suited to dry conditions, while soil type determines crop selection and influences management practices.

The interplay between these components is dynamic and requires careful management to balance productivity with sustainability. Small family farms often develop intricate knowledge systems to manage these interactions, drawing on experience and traditional practices that have been passed down through generations. However, these farms are also increasingly integrating modern technologies and practices to enhance the efficiency and resilience of their farming systems.

A key aspect of this interaction is the role of ecosystem services – natural processes that support and enhance agricultural productivity. Services such as pollination, soil fertility, pest control, water regulation and climate moderation are critical to the sustainability of small family farms. When managed well, these services reduce reliance on chemical inputs, enhance resilience to environmental shocks and support long-term viability.

Soil fertility and nutrient cycling are foundational services. Healthy soils, rich in organic matter, retain moisture, supply nutrients and resist erosion. Practices like composting, cover cropping and reduced tillage help maintain soil health, improve yields and reduce the need for synthetic fertilizers.

Water regulation and purification are also vital. Natural ecosystems such as wetlands and forests help regulate water flows and improve water quality. By protecting and restoring these natural ecosystems on or near their farms, small family farmers can enhance water availability and quality, which is crucial for sustaining agricultural production.

Pest control is another ecosystem service provided by natural predators, such as birds, insects and bats, which help keep pest populations in check. Small family farms that promote habitat diversity, such as through agroforestry or inter-cropping, can support these natural predators and reduce the need for chemical pesticides. This helps maintain a healthier environment and reduces the costs associated with pest management.

To meet rising demands for food and other agricultural products without degrading the environment, small family farms must adopt strategies of sustainable intensification. This means increasing productivity from existing land while conserving natural resources and ecosystem functions. It requires an integrated approach, balancing the need to produce more with the imperative of maintaining environmental health.

Efficient resource use is central to this approach. Technologies such as drip irrigation allow farmers to apply water directly to plant roots, reducing waste and improving yields. This enhances both productivity and sustainability by conserving water while supporting healthy crop growth.

In summary, managing the interactions between crops, livestock and the environment is fundamental to the success of small family farming systems. By maintaining ecosystem services, adopting sustainable practices and improving resource efficiency, these farms can achieve resilient and productive systems that meet both current and future needs.

3.7 The Need for a Farming Systems Perspective

The farming system perspective recognizes that small family farms operate within complex and dynamic environments where various elements – such as crops, livestock, natural resources and human factors – interact in ways that impact the farm's overall productivity and sustainability. This perspective views farms not as isolated units but as integral parts of broader ecological, economic and social systems.

This broader understanding is essential for several reasons. Firstly, as highlighted in Chapter 2, small family farms are inherently diverse in their operations, often involving a mix of crops, livestock, forestry and natural resource management. These components do not function in isolation but are deeply interconnected, influencing each other in ways that affect the farm's output and long-term viability. A farming system perspective enables a holistic understanding of these interactions, providing insights into how different elements can be optimized to improve family farm performance.

Secondly, small family farms vary widely in their resources, challenges and opportunities, meaning that development strategies and, in particular, agricultural advisory support must be customized to fit their specific contexts. By examining the specific farming systems within which these farms operate, it becomes possible to tailor development interventions that address the unique needs and conditions of different types of small family farms, whether they are subsistence-oriented, market-integrated or operating under a mix of these models.

A critical aspect of market-oriented small family farm development, as we described in Chapter 2, is enhancing their integration into market systems. Moving from subsistence farming towards more business-oriented models requires linking farms to markets in ways that are sustainable and profitable. A farming system perspective facilitates this by identifying the most effective methods to connect these farms with markets, considering factors such as supply chains, value addition, access to finance and the adoption of appropriate technologies. This approach not only helps small family farms become more competitive but also ensures that they can sustain their market presence in the long term.

In addition, the farming system perspective is invaluable for promoting sustainability and resilience within small family farms. The interconnectedness of the various components of a farming system means that changes or shocks in one area – such as climate variability, market fluctuations or policy shifts – can have a ripple effect throughout the entire system. Understanding these dynamics allows for the development of practices and strategies that enhance the farm's capacity to absorb and adapt to such changes, thereby ensuring its continued viability and contribution to the broader agricultural landscape.

Finally, the insights gained from a farming system perspective are critical for informing policy and institutional frameworks that support small family farm development. Policies attuned to the complexities and specificities of different farming systems are more likely to be effective in promoting sustainable resource use, increasing productivity and improving market access for small family farms. By aligning policy interventions with the realities of various farming systems, governments and development agencies can create more supportive environments that enable small family farms to prosper.

As diverse and complex as farming systems are, they have one thing in common: if they are to survive and thrive in the ever-changing agricultural landscape, they need to embrace the market and practice modern farm management.

3.8 Development Trajectories of Smallholder Farming Systems

Smallholder farming systems follow different development trajectories based on their agroecological characteristics, socio-economic context and market access. Across all systems, strategies for poverty reduction and livelihood improvement generally involve combinations of intensification, diversification, increased farm size, value addition, off-farm income and, in some cases, exiting from agriculture. The relevance and effectiveness of each strategy, or combination of strategies, depend heavily on the nature of the specific farming system, which are outlined here.

- **Irrigated farming systems:** Due to their relatively favourable growing conditions, irrigated systems show strong potential for poverty reduction through farm improvement, with intensification playing a central role. Productivity gains are achievable through improved water management, better soil fertility and crop management practices. Diversification, often into high-value or export-oriented cash crops, also plays a major role, particularly where marketing channels are well established.

  In addition, increased farm size, where possible, contributes to scaling up operations, especially for market-oriented households. However, the potential for land expansion is often limited due to population density and land tenure constraints. Off-farm income, including seasonal and local migration, remains a valuable strategy whereby income is supplemented, and farmers are able to reinvest, contributing to a poverty reduction of 20% (Dixon *et al.*, 2011).

Wetland rice systems, common in South-east Asia, also have favourable conditions for intensification and diversification, though their development trajectories vary slightly. In countries such as Cambodia and Myanmar, intensification – through improved rice varieties, fertilizer use and water control – is important but less dominant than in irrigated systems. Diversification into fish, poultry, vegetables and dual-purpose rice–fish systems is key to increasing income and nutrition.

Value addition through the local processing of agricultural produce (e.g. rice milling, fish smoking and fruit drying) plays an increasingly important role in these systems. A significant share of poverty reduction, however, will come from off-farm income, reflecting the strength of rural economies and the presence of large non-farm labour markets – particularly in areas like China, where nearly half of farmers are already engaged in off-farm activities.

- **Rain-fed farming systems:** Rain-fed farming can be sub-divided into humid systems and highland/cold systems. These systems are more reliant on multiple strategies to drive development due to variable rainfall and limited infrastructure. Diversification plays a particularly important role, accounting for over a quarter of poverty reduction (Dixon *et al.*, 2011). Off-farm income also plays a substantial role helping to reduce risk and smooth seasonal income. Because of the agroecological potential of these areas, intensification strategies (e.g. improved seed varieties and soil fertility enhancement) also contribute meaningfully to growth and resilience.

  Highland and dry/cold rain-fed systems face the most constraints in terms of natural resources and market access. Consequently, opportunities for farm improvement are limited and contribute only modestly (by around 40%) to poverty reduction (Dillon and Hardaker, 1993). Instead, these systems rely more heavily on off-farm income and, notably, exit from agriculture, as key livelihood strategies. Migration, whether seasonal or permanent, provides an important source of remittances and income diversification. Where local conditions permit, diversification into high-value, low-bulk products such as spices, herbs or honey is increasingly important, especially in areas with improved road access or tourism-related markets.
- **Livestock farming systems:** These play a crucial role in food security, income generation, and nutrient cycling, especially in mixed and pastoral systems. The potential for change lies in improving animal health, feed quality and breeding, alongside better integration with crop systems. Market access, value addition and sustainable grazing practices also offer significant opportunities. Climate-smart livestock strategies can enhance resilience while reducing environmental impacts. With targeted advisory support, infrastructure investment and policy incentives, livestock systems can become more productive, sustainable and inclusive, particularly for smallholders in resource-constrained environments. Among pastoralists located in arid areas, the risks of drought are acute and in cases where the incidence of livestock mortality is high, individual livestock herders may be forced to find employment outside agriculture.
- **Mixed farming systems:** These systems also offer substantial scope for diversification and intensification. These systems are particularly resilient due to the interdependence between livestock and crops – manure enhances soil fertility, and crop residues serve as feed. Farm management practices that optimize nutrient recycling and labour allocation can significantly improve productivity. Opportunities for value addition also exist, such as processing milk into cheese or yoghurt, or crops into dried or packaged goods. These systems benefit from flexible strategies, with the potential for increasing farm size where land is available, and for supplementing income through off-farm work.
- **Peri-urban and urban farming systems:** These systems are uniquely positioned to benefit from proximity to markets. Their development pathways rely heavily on diversification into high-value, perishable

crops such as vegetables, poultry and dairy, which cater directly to urban demand. These systems also benefit from value addition and intensification, often in the form of controlled-environment agriculture (e.g. greenhouses and vertical farming). Off-farm income remains an important livelihood strategy, particularly for households engaged in part-time farming. Some may also seek to expand the size of their business rather than their land area, focusing on higher-value production and direct marketing channels. Overall, these systems offer strong potential for income growth but face land access, waste management and regulatory challenges.

A typology of the strategies for poverty reduction by farming system is given in Table 3.1.

In conclusion, while each farming system follows a distinct development trajectory shaped by its environmental and socio-economic context, most smallholder systems require a mix of strategies to sustainably improve livelihoods. The balance between intensification, diversification, value addition and off-farm income depends on the specific opportunities and constraints faced by small family farmers within each system.

3.9 Customizing Advisory Support for Smallholder Farming Systems

For sustainable agricultural development, it is essential to customize advisory support to reflect the diverse farming systems in which small family farms operate. Effective, context-specific support is critical to promoting sustainability in both production and management, and to unlocking the full potential of smallholder agriculture.

Customizing support begins with identifying the key drivers of productivity, sustainability and resilience within each farming system and developing targeted extension programmes to strengthen them. In rain-fed systems, advisory services should encourage water harvesting, the adoption of drought-tolerant crops and the construction of soil conservation measures, along with practical farm management support to help farmers plan and evaluate resource use. While many rain-fed farmers have the potential to intensify their farming system, some subsistence farmers may pursue off-farm work or, alternatively, exit from farming altogether. This underscores the need for flexible advisory approaches that respect farmer choices and offer alternative livelihood pathways.

In irrigated systems, support for farmers should include precision irrigation, on-farm water resource improvements and training in efficient water management. An example of this is discussed in Box 3.8.

Similarly, in mixed farming systems, advisory support should focus on optimizing crop–livestock integration, nutrient cycling and agroforestry, with attention to managing trade-offs between farm enterprises. In all these contexts, the goal is to help farmers make the best use of scarce resources. Box 3.9. provides an example of development strategies for mixed farming systems.

Livestock-based systems require tailored advisory services, including veterinary care, improved feed practices and guidance in animal husbandry. These services are essential to boosting productivity while maintaining sustainability.

In peri-urban systems (see Box 3.10.), farmers often grow high-value perishables and need help accessing markets; identifying suitable outlets; and improving logistics, storage and processing infrastructure. Their position near urban centres offers direct marketing opportunities that can be harnessed with the right assistance. In this context, support should be given to finding suitable market outlets for the sale of high value products.

Ultimately, social and cultural dimensions are just as important as technical factors, in customizing advisory support. Interventions must consider gender roles, traditional knowledge and community dynamics. Advisory support should be tailored to the specific farming system and the realities of each farmer. It must be inclusive, participatory and responsive to local contexts. There is no one-size-fits-all approach – only carefully designed strategies that build resilience, improve livelihoods and promote the long-term sustainability of small family farms. The support needed for advisory services is further discussed in Chapter 16.

Table 3.1. Key strategies for poverty reduction by farming system category.

Poverty reduction strategies	Smallholder irrigated schemes	Wetland rice systems	Rain-fed humid	Rain-fed highland	Livestock farming system	Mixed farming	Peri-urban, urban
Intensification	x	x	x		x		
Diversification	x	x	x	x	x	x	x
Increased farm/ herd size					x	x	x
Increase in value addition	x	x				x	x
Increase off-farm income	x	x	x	x	x	x	x
Exit from agriculture				x	x		

Box 3.8. Irrigated farming systems in Punjab, India. From Sasaki, 2017.

In the Punjab region of India, irrigation systems are crucial to the success of farming, particularly for rice and wheat production. However, over-reliance on irrigation, especially groundwater, has led to significant depletion and waterlogging issues. In response, agricultural extension services have adapted their support to promote water-efficient practices.

Programmes like 'The Punjab Agricultural University Extension System' have introduced practices such as:

- System of Rice Intensification (SRI), which reduces water usage by optimizing the spacing of rice plants and the timing of water application.
- Laser land levelling, which ensures that fields are more uniformly levelled to use water efficiently.
- Drip irrigation techniques, which minimize water wastage and are being increasingly promoted for high-value crops like fruits and vegetables.

Box 3.9. Mixed farming systems in Tanzania. From FAO, 2011.

In Tanzania, mixed farming systems, which integrate crop production with livestock management, are common, particularly in the central and southern regions. Extension services have adapted to promote integrated approaches that help farmers manage both crops and livestock in a sustainable way. One example is the integrated soil fertility management approach, which aims to improve soil health while also supporting better livestock management practices.

Extension officers provide training in:

- Diversification of crop varieties that can benefit livestock feed.
- Integrated pest management for crops and livestock.
- Rotational grazing and improved pasture management.

Box 3.10. Peri-urban farming in Accra, Ghana. From FAO *et al.*, 2022.

Peri-urban farming is an increasingly important part of food production in cities like Accra, Ghana. The extension services in urban and peri-urban areas have adapted by promoting efficient, space-saving farming practices that are suited to the smaller land areas typical of these regions.

For instance, extension agents in Accra have introduced:

- Vertical farming techniques for vegetable production, allowing farmers to make the best use of limited space.
- Hydroponics for growing leafy greens and other high-value crops in urban areas.
- Urban livestock production, such as poultry and small ruminants, which require less space and can be managed alongside crop production.
- Composting and waste management practices to recycle organic waste into valuable inputs like compost for urban farming.

Additionally, urban extension services in Ghana have helped farmers connect to local markets, providing them with training on value addition, food safety and packaging.

3.10 Conclusion

This chapter has highlighted the importance of adopting a farming systems perspective to understand and support small family farms. Farming systems are shaped by a mix of environmental, economic and social factors that affect what farmers grow, and how they use resources, connect to markets and deal with challenges and opportunities. From rain-fed and irrigated systems to livestock-based, mixed and peri-urban models, each presents specific challenges and opportunities that demand context-sensitive responses.

Given the wide diversity of farming systems – both across and within countries – a uniform approach is neither feasible nor effective. Broad policy prescriptions fall short when confronted with the heterogeneity of farm types and household circumstances. Effective support must, therefore, be grounded in a deep understanding of the specific system in which a farm operates. This includes aligning interventions – whether focused on intensification, diversification, value addition or income generation – with the household's goals, resource endowments and surrounding environment.

Transforming smallholder family farming into a sustainable and viable livelihood requires tailored advisory services, inclusive development strategies and responsive policy frameworks. These must recognize the value of both traditional knowledge and modern science and build the capacity of farmers to innovate and adapt. At the same time, not all smallholders can or will benefit equally from on-farm improvements; some will need support to pursue off-farm income opportunities or alternative livelihood pathways.

Ultimately, support for smallholders should be part of a wider rural development agenda – one that respects farm diversity, reflects real farming conditions and helps build smallholder resilience while improving food security and sustainability.

References

Armson, R. (2011) *Growing Wings on the Way: Systems Thinking for Messy Situations*. Triarchy Press Ltd, Bridport, England. Available at: www.triarchypress.net/growing-wings.html (accessed 17 March 2023).

Borgan Project (2024) *Cultivating Hope: Urban Farming in São Paulo*. Website blog. Available at: https://borgenproject.org/urban-farming-in-sao-paulo/?utm_source=chatgpt.com (accessed 26 January 2025).

Dillon, J.L. and Hardaker, J.B. (1993) *Farm Management Research for Small Farmer Development*, 2nd Edition. FAO, Rome.

Dixon, J., Gulliver, A., Gibbon, D. and Hall, M. (2011) *Farming systems and poverty: Improving farmers' livelihoods in a changing world (English)*. World Bank group, Washington, D.C. Available at: http://documents.worldbank.org/curated/en/126251468331211716 (accessed 18 March 2025).

Dixon, J., Li, L. and Amede, T. (2023) A century of farming systems. Part 1: Concepts and evolution. *Farming System* 1(3). https://www.sciencedirect.com/science/article/pii/S2949911923000576

Fan, S. and Rue, C. (2020) The role of smallholder farms in a changing world. In: Gomez, Y., Paloma, S., Riesgo, L. and Louhichi, K. (eds) *The Role of Smallholder Farms in Food and Nutrition Security*. pp. 13–28. Available at: https://link.springer.com/chapter/10.1007/978-3-030-42148-9_2 (accessed 25 April 2025).

FAO (2011) Sustainable Land Management in Practice: Guidelines and Best Practices for Sub-Saharan Africa. Available at: https://www.fao.org/sustainable-food-value-chains/library/details/en/c/266323

FAO, Rikolto, and RUAL (2022) *Urban and peri-urban agriculture sourcebook – From production to food systems*. FAO and Rikolto, Rome, Italy. DOI: 10.4060/cb9722en. (accessed 25 January 2025).

Nairobi City County (2015) Nairobi City County Urban and Peri-Urban Agriculture, Livestock and Fisheries Policy. Available at: https://repository.kippra.or.ke/items/be02deea-3be7-4488-802f-5be65102077b (accessed 18 March 2025).

Nchanji, E.B. (2017) Sustainable urban agriculture in Ghana: What governance system works? *Sustainable Urban Agriculture* 9–11. Available at: https://www.mdpi.com/journal/sustainability/special_issues/SUA (accessed 18 March 2025).

Sasaki, T. (2017) *Achieving sustainable cultivation of rice Volume 2: Cultivation, pest and disease management*, (First edition). CRC Press. DOI: 10.4324/9781351114196.

UNCTAD (2019) World needs fairer and more sustainable trade, not less trade. Available at: https://unctad.org/news/world-needs-fairer-and-more-sustainable-trade-not-less-trade (accessed 26 January 2025).

Vazhacharickal, P.J., Predotova, M., Chandrasekharam, D. and Buerkert, A. (2013) Urban and periurban agricultural production along railway tracks: A case study from the Mumbai metropolitan region. *Journal of Agriculture and Rural Development in the Tropics and Subtropics* 114(2), 145–157. Available at: https://www.jarts.info/index.php/jarts/article/view/2013112544608 (accessed 17 March 2025).

Additional Reading

Chambers, R., Pacey, A. and Thrupp, L.A. (1989) *Farmer First: Farmer Innovation and Agricultural Research* Intermediate Technology Publications, London.

Collinson, M. (2000) *A History of Farming Systems Research*. CAB International, Wallingford, UK.

Darnhofer, I., Gibbon, D. and Dedieu, B. (2012). *Farming Systems Research*. Springer Dordrecht Heidelberg, New York. DOI: 10.1007/978-94-007-4503-2

Dixon, J., Li, L. and Amede, T. (2023) A century of farming systems. Part 1: Concepts and evolution. *Farming System* 1(3). https://www.sciencedirect.com/science/article/pii/S2949911923000576

McConnell, D.J. and Dillon, J.L. (1997) *Farm Management for Asia: A Systems Approach*. FAO, Rome.

Norman, D. W., Worman, F. D., Siebert, J. D. and Modiakgotla, E. (1995) *The Farming Systems Approach to Development and Appropriate Technology Generation*. FAO, Rome.

Oxfam (2013) Power, Rights, and Inclusive Markets: Public policies that support small-scale agriculture. Available at: https://policy-practice.oxfam.org/resources/power-rights-and-inclusive-markets-public-policies-that-support-small-scale-agr-292876/ (accessed 26 January 2025).

Part 2: Farm Management

Part 2 explores the concept of farm management, with a focus on the decision making processes that influence farm productivity, profitability and sustainability. Key topics include the farm as a system, marketing, financial management, risk management and the integration of technology on small family farms. It highlights how effective farm management is essential for balancing productivity with sustainability, enabling farmers to optimize operations in the face of changing environmental and market conditions. Additionally, the section discusses the evolving agribusiness environment and underscores the importance of small farmers organizing themselves to access and participate in modern value chains.

4 The Farm as a System: Unpacking the Smallholder Context

As noted in Chapter 3, farming systems are complex. Firstly, this is due to their fundamentally different components, such as crops, livestock, people, tools and mechanization, which means a multi-disciplinary approach is needed to study them. Secondly, there are several ways in which components can be combined.

Building upon our exploration of the role of small family farming in low-income countries and the diverse farming systems in which they operate, this chapter looks deeper into the concept of a farm, especially within the smallholder context. This takes us from the broader farming systems landscape into the confines of the individual farm considered as itself a system. In this context, a farm is understood as a complex system that encompasses an asset base, serves as a decision-making unit and maintains linkages to both input and output markets. This definition enables us to view the smallholder farm through a micro-level lens, focusing on the farmer and their family as the core decision-making unit responsible for its management.

By treating the farm as a system, we can better understand how its different parts work together to keep it running and sustainable. This approach helps us see how smallholder farmers make decisions – such as which crops to grow or how to use resources – based on both family needs and outside influences. This chapter explores how farms are structured and how they function, with a focus on the key role of the farmer and their family in managing challenges, making the most of opportunities and working towards long-term growth and sustainability.

4.1 The Farm Household System

Farming systems can be viewed as a hierarchy of systems and sub-systems. In the previous chapter we looked at the farming system from a global perspective, going beyond national boundaries and within broad diverse agro-ecological contexts. This classification was an agro-ecologically based classification and typology of farm systems. The hierarchical classification of the farm system presented here is distinctly different. It is specifically oriented to an individual farm household.

The farm household systems are complex and reflect the multiple objectives of the farm family. These systems are dynamic and evolve in response to environmental and socio-economic changes. These systems also incorporate indigenous knowledge. Farm household systems can also be modified since farm families are generally receptive to change (see Box 4.1.).

To fully understand the family farm, it must be viewed within the context of the family operating it. While the farm can be analysed as an independent business and can be looked at separately, support programmes will be more effective when they are developed with an understanding of the interaction between the farm and the family systems (Gasson and Errington, 1993).

Further, farm household systems do not operate in a vacuum. They are part of, and influenced by, other systems and environments, namely the physical, socio-cultural and policy/institutional environments. However, the actual influence that an individual farm household system has on these systems and environment is often limited.

DOI: 10.1079/9781836990536.0004

Box 4.1. The farm household system.

The farm household system consists of the following basic sub-systems, which are closely linked and interactive:

1. **The household decision-making unit:** establishes goals for the system; controls the system; provides labour; and demands food and cash in fulfilment of set objectives.
2. **The farm and its crop and livestock activities:** provides employment, food and cash for the farm family.
3. **The off-farm component:** competes with farm activities for labour (or absorbs surplus labour); provides employment and income generating activities; becomes increasingly important for supplementing the well-being of farm families.

To find our way through this complexity and increase our understanding we must simplify. This is achieved by either concentrating on particular sub-systems, such as crop and livestock enterprises, or by simplifying the interrelationships by concentrating on a few critical ones.

A more complete understanding of the farm household system is described here, starting with the question: What is a farm?

4.2 What is a Farm?

A very simple definition of a farm describes an area of land with some level of infrastructure, used for growing crops and/or rearing animals. While this definition is accurate, it does not explain the true complexity of the farm.

A more complex definition of a farm would be that it is a basic unit of agricultural production involving various activities, such as crop cultivation and livestock rearing; and serving as a cohesive entity in which resources are allocated, decisions are made and livelihoods are sustained (see Box 4.2.).

A further complexity is that the concept of a farm extends beyond individual plots of land. It also includes the notion of common land holdings owned or rented by farmers, collectively, which adds an important dimension to the understanding of the farm as a system.

A farm can be functionally defined as an agricultural production unit in which land, labour, capital and entrepreneurship are combined to generate agricultural outputs. It comprises physical structures, such as fields, pastures, buildings and machinery, as well as intangible assets like knowledge, skills and social networks. Farms vary in size and scale, ranging from smallholder family farms to large commercial enterprises.

In addition to individually owned or occupied land, farms often include common land holdings that are collectively owned or rented by farmers. Common land refers to resources that are accessible to multiple farmers within a community. These resources may include grazing areas, forests, water bodies or other shared assets. Common land holdings play a crucial role in supporting the livelihoods of farmers, especially in areas where land tenure is complex or limited, which often is the case in low-income countries.

The concept of common land holdings highlights the interdependence and cooperation among farmers within a specific geographic area. It enables farmers to access additional resources, expand their production capacity and mitigate the risks associated with individual land use. Common land holdings often require collective decision-making processes and governance mechanisms to ensure equitable access and sustainable use. Thus, in this context, a farm may represent a collective space where some resources, such as land and water, are shared but the agricultural activities may be individual. This arrangement often requires special measures to ensure shared resources - especially water, land and vegetation - are used wisely and fairly, and are protected from overuse.

Understanding the diverse forms of land tenure, including common land holdings, is essential for comprehending the complexities of farming systems. It acknowledges the social, economic and cultural dimensions that shape agricultural practices and rural livelihoods. By recognizing the significance of common land holdings, we can develop more inclusive and

Box 4.2. Understanding the farm and its context.

Consider the following extract from Richard Harwood:

The analysis of farming systems properly begins with the identification of significant interactions: of people with plants, plants with animals, plants with other plants, and so on. Each interaction propagates others; the problem for the analyst is to sort out the various reaction products, define the significant ones, and match them with the farmer's goals. To do this, the analyst must order the variety and complexity that are characteristic of tropical farming systems. The aim of the process is to identify situations in which existing farm resources are inefficiently used. The process succeeds when it defines changes in the farming system that result in increased productivity.

To understand the farmer's agricultural system, the analyst must classify the various environmental factors to which the farmer responds and identify local farms where these factors are expressed in varying degrees. The basic factors to be analysed are soil and climate. Local crop and animal production data is also important. These factors must be viewed in terms of the farmer's own goals and priorities, which figure as importantly as objective physical and biological considerations in his decisions about how to farm. His need for food, his competing need for cash income, his status in the community, his desire for stability and security, his motivation to conserve energy and other resources - such subjectively perceived values are also factored into the farmer's agricultural equations.

"Only after he has gathered this information and understood its meaning for the farmer can the analyst plan appropriate changes in the farming system."

(Harwood, 1979. pp. 6–7)

context-specific approaches to farm management and agricultural development.

One construct of a farm is the farming homestead, a construct that is common in many parts of the world, and which may also vary considerably. Some may comprise a family dwelling surrounded by farming land, while others may have additional land that is not directly attached or even adjacent to the dwelling and surrounding land. Still others may have access to common land, such as forest and grazing, which is shared by other homesteads. Within all these permutations is the range of farming activities, including, for example, a small garden around the dwelling; more extensive cropping on lands around, adjacent to and some distance from the dwelling; trees and shrubs – both natural and planted; and a range of livestock, including chickens (or other fowl), large and small ruminants, and other types of animals. Often in such a farmstead setting, one or two enterprises, such as staple crops, may dominate the production, while animal production may represent a less formally managed element of the overall farming operations. Because farmsteads are often small, the farming families must employ these multiple strategies to generate the most income and benefits (e.g. food).

Family farming is very common in low-income countries. It often reflects a historical way of life that has been maintained as a cultural norm. However, such farmsteads are under pressure due to the phenomenon of fragmentation, which refers to the division of land into smaller pieces over time. The main contributors to fragmentation are high population density and the number of children per household standing to inherit land from their parents. Fragmentation often results in a farmer having multiple small pieces of land that are scattered. One important consequence of fragmentation is the inability to leave land fallow, which is important for soil regeneration. Fragmentation poses many such challenges to farm management.

As the chapter explores the farm as a system, it delves into the dynamics and interactions within different types of farms, including those with common land holdings. By understanding the unique characteristics of farms and their broader social and environmental contexts, effective strategies can be developed that enhance farm productivity, sustainability and livelihoods to benefit farmers and their communities.

4.3 The Farm Livelihoods System

In this section, the farm as a part of a livelihood system is explored. In its simplest expression, a livelihood is a means of supporting one's existence. It is the entirety of all the actions one takes to sustain oneself. When livelihoods are analysed, we find that they are really decisions one makes about how to use one's resources or assets to acquire desired outcomes. The sustainable livelihoods approach, developed in 1999, is now widely used for analysing livelihoods from a system-wide perspective. It helps to identify the factors affecting livelihoods and the relationships between these factors.

Livelihoods praxis identifies several key elements that make up livelihoods and influence their sustainability. These have been adapted to the context of small family farming, set out in Table 4.1.

The sustainable livelihoods approach assembles these factors into a framework, which have been adapted here for a small family farm (Fig. 4.1).

This framework highlights that the foundation of any livelihood is one's, or in this case the farming family's, assets. As shown in Fig. 4.1, these assets are grouped as physical, financial, human, social and natural capital. The framework indicates that livelihood assets are subject to various influences: the influencing context, the influencing policies and the influencing institutions. These affect the condition of the assets, access to the assets, and how they can be used. The assets are used to develop livelihood strategies to achieve the farm-family's livelihood goals.

This approach emphasizes the integration of human, technological and economic changes, institutional and organizational support, and the contextual factors influencing farm performance and sustainable livelihood goals.

The key components of the framework include the farm's asset base, which comprises the following resources and capital that contribute to its productive capacity and overall livelihood:

- **Physical capital assets:** physical capital assets encompass tangible resources that directly contribute to agricultural production. These assets include buildings, machinery, tools, irrigation systems and infrastructure. Efficient management of physical assets involves optimizing resource allocation, maintaining infrastructure, and investing in appropriate technologies to enhance productivity and reduce production costs.
- **Financial capital assets:** financial capital assets represent the monetary resources available to the farm. These assets include income, savings, investments, credit and other forms of financial resources. Adequate financial capital is essential for purchasing inputs, acquiring machinery, accessing markets and managing risks. Farmers need to assess their financial situation, develop sound budgeting and financial management practices, and explore options for accessing credit and other financial services to support their farm operations.
- **Human capital assets:** human capital assets refer to the knowledge, skills, expertise and labour of individuals involved in farm activities. Farmers' own expertise, and the skills of their family members or hired labour, play a critical role in farm productivity and management. Continuous learning, capacity building and skills development are essential for adapting to evolving agricultural practices and technologies. Investing in human capital assets through fostering proficiency, training, education and knowledge-sharing initiatives can improve farm efficiency and competitiveness.
- **Social capital assets:** social capital assets represent the networks, relationships and social interactions that farmers have with other individuals, groups and institutions. These connections provide access to information, markets, support services and collective action opportunities. Strong social capital can facilitate knowledge exchange, cooperation and collaborative decision making among farmers. Building and nurturing social capital through participation in farmer associations, cooperatives, community groups and extension services can enhance farm resilience and unlock valuable resources and opportunities.
- **Natural capital assets:** natural capital assets encompass the natural resources and ecological services that are essential for farming. These include land, fertile soil, water resources, biodiversity and ecosystem services. Land is a fundamental asset that forms the foundation of farming activities. Sustainable farm management requires the conservation and responsible use of natural capital assets. Practices such as

Table 4.1. Key elements of the sustainable livelihood framework.

Livelihoods element	Explanation
Livelihood assets	The capital assets that the farm family has to work with. These include human capital, natural capital, financial capital, physical capital and social capital
Influencing context	The 'environmental' factors that influence access to, and use of, livelihood assets. These include the natural environment (climate, soil conditions, etc.), the economic environment and the social environment, as well as new technologies
Influencing policies	The local, national, regional and international policies, laws and cultural practices that define the rights and responsibilities regarding access to and use of capital assets. These include agricultural policy, land use policy, water policy, traditional rules, societal norms, etc.
Influencing institutions	The local, national, regional and international government structures, markets and organizations that define the rights and responsibilities regarding access to and use of capital assets. These include regulatory bodies, civil society, commodity organizations, businesses, cooperatives, etc.
Livelihood strategies	The on- and off-farm strategies farm families implement to achieve their livelihood goals
Livelihood goals	The goals the family has set for itself. These include sufficient sustainable income, improved well-being, improved food security, education of child dependants, etc.

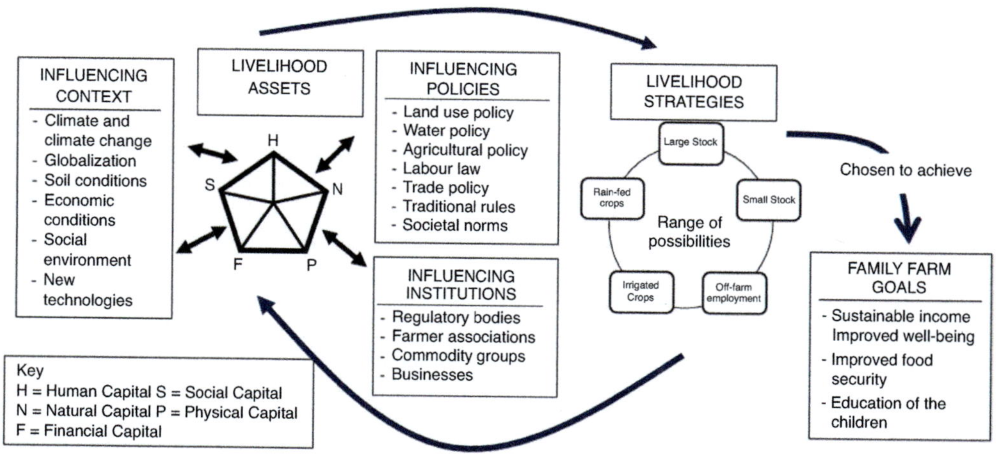

Fig. 4.1. A small family farm livelihood system.

soil conservation, water management, agroforestry and biodiversity conservation contribute to long-term farm viability, environmental sustainability and resilience in the face of climate change and other challenges.

The farm's asset base thus refers to the diverse resources and capital that contribute to its productive capacity and overall livelihood potential. Understanding the importance of each asset category is crucial in farm management decision making. Balancing and leveraging these assets effectively can help farmers optimize their productivity, profitability and sustainability. Farmers need to assess their asset base, identify areas for improvement or investment and develop strategies to enhance their asset portfolio. This may involve diversifying income sources, strengthening human and social capital, investing in appropriate technologies and adopting sustainable practices to ensure the long-term viability and resilience of the farm.

The influencing aspects of the framework warrant further discussion. The assets themselves exist in various quantities and qualities. The influencing context is primarily about quality. For example, a small family farm may have 2 ha of arable land, but the condition of the soil may be poor due to many years of inadequate conservation, or it may be some distance from the homestead. The family have some savings, but the rate of inflation is reducing its value. The family may have sufficient family labour and know that the younger children will soon be able to help, thereby increasing their human capital. However, their lack of training limits their ability to help manage the farm. They may have farm tools that are in good condition because they are well looked after. The farm may be in an area with historically good rainfall, but climate change has reduced the annual average rainfall. In each case, the assets available to the family farm will be in some current condition but either improving or deteriorating.

Similarly, policies also influence the farming family's assets. Land is a good example. In many low-income countries, there are traditional (and even legal) laws that govern access to land. In many countries, women have less access to land than men. The land is available, but it is not accessible. Access to credit is also subject to various regulations which may make it easier or harder to access. Many countries have policies limiting the amount of water that can be taken from primary sources such as rivers and dams. There may also be laws restricting the use of land for agricultural purposes, which is the case around riverbanks.

And institutions also influence assets. Institutions like banks, government structures, traditional authorities, farmers associations and commodity groups set out the processes and procedures for accessing various assets. Land allocation in traditional areas very often requires the approval of a traditional authority. Cooperatives and farmers' associations often act as agents to facilitate access to physical (e.g. inputs) and financial assets (e.g. income, savings, markets).

In sum, the mere existence of assets or capital is not sufficient when assessing assets for a livelihood. For assets to be useful, they must be adequate in quality and quantity, accessible and available for use by the farming family. Unfortunately, it is often the case in low-income countries that the poorer small family farmers have limited access to the assets needed to develop their farms.

Figure 4.1 shows that the process is highly dynamic. For example, the vulnerability context influences the assets, but the assets also influence the vulnerability context. This highlights the reality that, however small, livelihoods are the result of a complex process of many interacting elements. Livelihood strategies are the result of decisions made at the end of the process.

This framework focuses the need for effective farm management with robust linkages with public and private sector support services to ensure seamless operations. The performance of farm management practices is evaluated based on the effective, efficient and sustainable (environmental, financial and social) use of assets. By understanding and managing these diverse assets, farmers can make informed decisions to optimize productivity, profitability and sustainability. This comprehensive approach ensures the long-term viability and resilience of the farm, contributing to improved livelihoods for small family farmers.

The livelihoods framework provides a lens through which to view the status of asset categories and their significance in shaping farm outcomes. Fig. 4.2 applies the livelihoods framework to a small family farm in a simplified form, focusing on assets and strategies. This figure shows how a farm family combines their livelihood assets to produce farm products and to pursue off-farm activities to generate income which, either directly or indirectly, influences their assets. The agricultural products may be consumed or sold. Likewise, income can be consumed or invested. Investment simply means additions and/or improvement to the stock of capital. Some products are sold, and the proceeds may be used to buy consumer goods and items of capital, such as tools and machinery, or to hire labour and capital services, or even land. The farming environment consists of larger systems, of which the farm is only a part. Farming environments are particularly influenced by the ecological or natural and the socio-economic environment, as explored in Chapter 3.

4.4 The Farm and Its Enterprises

As discussed previously in relation to the farm as a system, a farm is not just a single entity engaged in a singular agricultural activity; it is a dynamic system that encompasses multiple enterprises. As shown in the following list, these enterprises are categorized as supplementary, complementary or competitive, depending on their roles and interactions within the farming system. Understanding the distinctions between these enterprises is crucial for effective farm management and the optimization of overall farm performance. These enterprise options are also on-farm livelihood strategies that the farm family can consider as they look at their resources, goals and the wider context – including policies and institutions – affecting their choices.

- **Supplementary enterprises:** these enterprises generate extra income and diversify the farm's earnings. They are often secondary to the main agricultural activity and serve as a means to maximize resource use and enhance overall profitability. Examples include agro-tourism, beekeeping, poultry production and agroforestry. They can capitalize on existing farm resources and infrastructure, adding value and generating

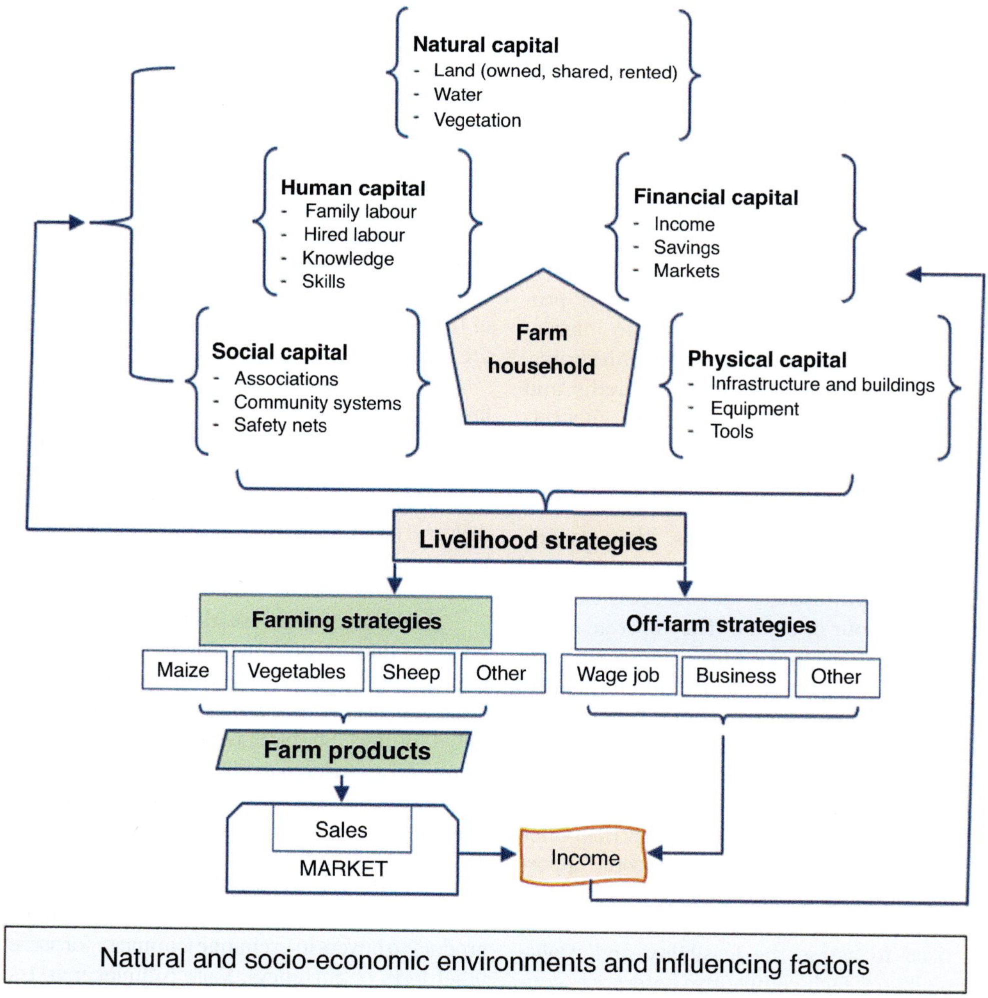

Fig. 4.2. A family farm livelihood system.

additional income without significant disruption to the primary agricultural operations.

Supplementary enterprises play a vital role in risk mitigation, especially in areas prone to climatic variability or market uncertainties. They provide a buffer against income fluctuations and contribute to the overall resilience of the farming system. By tapping into diverse income sources, farmers can reduce their dependence on a single enterprise and spread their risk across multiple activities.

- **Complementary enterprises:** these enterprises are closely linked to the primary agricultural activity, enhancing its productivity and value. They are mutually beneficial and create synergies within the farming system. They may involve integrating different crops, livestock or value-adding activities that complement and support one another. For example, a dairy farm may have a complementary enterprise of producing fodder crops to feed the livestock, creating a self-sufficient feed system.

  Complementary enterprises can optimize resource use, improve nutrient cycling, and increase overall farm productivity. They capitalize on the interdependencies between different enterprises, leveraging their combined strengths and minimizing potential conflicts. By integrating complementary activities, farmers can enhance the efficiency and sustainability of their farm operations.
- **Competitive enterprises:** these enterprises directly compete with the primary agricultural activity for resources such as land, labour or capital. They operate in the same market niche and may require significant investment or operational adjustments. For example, a farmer may have to choose between cultivating different crops with similar resource requirements but which serve different market demands.

  While competitive enterprises offer potential diversification and income opportunities, they require careful consideration and strategic decision making. Farmers need to assess the feasibility and profitability of each competitive enterprise, considering factors such as market demand, resource availability and their own capacity to manage multiple activities effectively.

Understanding the roles and interactions of supplementary, complementary and competitive enterprises is essential for effective farm planning and management. Farmers must carefully evaluate their farm resources, market dynamics and personal goals to determine the optimal mix of enterprises within their farm system. By diversifying income streams, capitalizing on synergies and minimizing conflicts, farmers can enhance overall profitability, resilience and sustainability.

4.5 Input–Output Markets

Traditional small family farming often uses inputs and materials originating from the farm itself, such as livestock feed, compost and farmyard manure. Market-oriented smallholder farmers tend to purchase inputs that are manufactured and specially prepared. For instance, livestock producers often require purchased feed. Although it is possible to mix some of these feeds themselves, it is often more efficient for them to be manufactured. However, the costs of purchased inputs are often higher than inputs produced on the farm. As a result, farmers need to be aware of the costs involved so that the correct decisions are taken to ensure maximum profits.

Market-oriented farming requires that farmers understand market channels, input suppliers, price variations, and buying and selling conditions. For example, when buying inputs, farmers need to be conscious of prices, delivery costs, transport, storage, expiry dates and a range of other technical factors that influence what, where and when to buy inputs.

Producing for the market depends on the nature of the market. Some markets cater for fresh, unprocessed foods, while others are more structured and formal, such as selling to agribusinesses under contract. Still others may require good packaging and an assured quality. In more formal and structured markets, farmers need to focus more on management rather than just production.

Markets can be simple, complex or sophisticated. But they all operate on the same basis; they all involve a process of sellers (farmers) offering a product to buyers in exchange for money (or some other form of currency). More complex markets involve more intermediaries between the seller and the buyer. The more complex the market, the more the farm must be run as a business.

4.5.1 Understanding input markets

Smallholder farmers rely on input markets to acquire the necessary resources for agricultural production. These markets encompass a wide range of inputs, such as seeds, fertilizers, pesticides, machinery and labour. Understanding the dynamics of input markets is crucial for smallholder farmers to make informed decisions and optimize their production processes.

In input markets, farmers encounter various stakeholders, including input suppliers, distributors and retailers. They need to assess factors such as input quality, availability, pricing and accessibility to ensure they obtain the most suitable inputs for their specific needs. For example, in regions with limited access to quality seeds, smallholder farmers may face challenges in acquiring reliable seed varieties, potentially affecting crop yields and overall profitability.

4.5.2 Exploring output markets

Once smallholder farmers have successfully produced their agricultural goods, they must navigate output markets to sell their products and generate income. Output markets encompass a diverse range of buyers, including wholesalers, retailers, processors, exporters and even direct consumers. Exploring these markets is essential for smallholder farmers to understand demand, market dynamics and potential profit opportunities.

Different agricultural products may have distinct output market channels. For instance, fresh produce like fruits and vegetables often find their way to local markets or supply chains supplying supermarkets and restaurants. Farmers may directly sell their produce at farmers' markets or establish partnerships with local retailers or food service providers. These direct-to-consumer channels offer opportunities for smallholder farmers to differentiate their products, establish relationships with customers and capture a larger portion of the retail price.

On the other hand, commodities like grains, oilseeds and cotton may follow a different path in the output markets. These products are often sold to large-scale buyers or agribusinesses, such as grain mills, oilseed crushers or textile manufacturers. These buyers aggregate and process the commodities to meet the demand of industrial or international markets. Smallholder farmers may sell their commodities through cooperatives, commodity exchanges or contracts with agribusinesses, ensuring a consistent market and fair pricing.

In addition to local and industrial markets, smallholder farmers also have opportunities in niche or speciality markets. These markets cater to specific consumer demands for organic, fair trade or sustainably produced products. Participating in niche markets requires additional certifications, adherence to specific production practices and engagement with specialized buyers or cooperatives. While niche markets may involve smaller volumes, they often offer higher prices and the potential for direct engagement with conscious consumers.

Moreover, in recent years, e-commerce platforms and online marketplaces have emerged as alternative channels for smallholder farmers. These platforms provide opportunities for direct sales, bypassing traditional intermediaries and connecting farmers with consumers in both local and international markets. Online marketplaces enable farmers to showcase their products, reach a broader customer base and leverage technology for efficient logistics and marketing.

Navigating output markets successfully requires understanding market demands, quality standards and market access requirements. Smallholder farmers need to stay informed about market trends, engage in market research, and build relationships with buyers and value chain actors. It is also important to develop strong branding and marketing strategies to differentiate products and create a competitive advantage. Markets and marketing is further explored in Chapter 6.

4.6 The Farm as Part of a Value Chain

The farm is one part of a system of organizations and enterprises involved in transforming raw materials into products that are bought by consumers. This system is called a value chain; it consists of all actors involved in linking production to final consumption, including input suppliers, processors, service providers and buyers, to name a few (Fig. 4.3).

The functions of value chain stakeholders can be direct or supporting. Direct functions include primary production, collection, processing, wholesaling and retailing. At each stage in this chain of actors, value is added until it reaches the final consumer.

For farmers to be profitable they need to be part of a competitive value chain. They can start by improving the performance of the farm business, but they need to be mindful of the entire value chain to satisfy consumer demands, ensure quality and safety and enhance their farm's profitability.

The value chain concept helps us understand how different actors in agriculture depend on each other. Understanding the value chain is essential for smallholder farmers as it provides insights into market dynamics, value addition opportunities and potential areas for improving efficiency and competitiveness. For example, in the coffee industry, the value chain includes coffee bean farmers, processors that transform the beans into roasted coffee, exporters who facilitate international trade, wholesalers who distribute to retailers and, ultimately, coffee shops or consumers that enjoy the final product. Each actor in the value chain adds value to the product at different stages, contributing to its quality, branding and market appeal.

4.7 Key Elements of the Value Chain

Figure 4.3 shows the flow of the produce from the producer to the final consumer. This is the most basic form of the value chain. In reality, it is more complex, comprised of three key elements described as follows:

- **Value chain actors:** identifying the key actors and their interactions within the value chain is crucial for smallholder farmers seeking to integrate into profitable markets. Value chain actors can include input suppliers, aggregators, processors, traders, wholesalers, retailers and even certification bodies or consumer organizations. Recognizing the roles and relationships between these actors helps farmers understand the flow of products, information and finances throughout the value chain.
- **Support services:** these play a vital role in enhancing the performance and competitiveness of smallholder farmers along the value chain. Support services encompass a range of interventions aimed at addressing the diverse needs of farmers and enabling them to overcome challenges, improve productivity and access profitable markets. Recognizing the importance of support services is key to empowering smallholder farmers and promoting sustainable agricultural development.

  For instance, agricultural advisory services provide farmers with technical advice, knowledge transfer and training on best farming practices. This support helps farmers improve their production techniques, adopt sustainable approaches and mitigate risks. Additionally, financial services such as microcredit, savings accounts and insurance products contribute to farm resilience and facilitate investments in inputs, equipment and infrastructure. However, with the expanding role of agricultural advisory services, modern advisory services increasingly find themselves providing advice to multiple actors in the value chain, not only primary producers. Since this is becoming more common, advisory services need to be prepared. This will likely require them to develop specialist capacities within the advisory service. This can be done by training existing personnel, recruiting new personnel with the relevant knowledge and skills, and/or by partnering

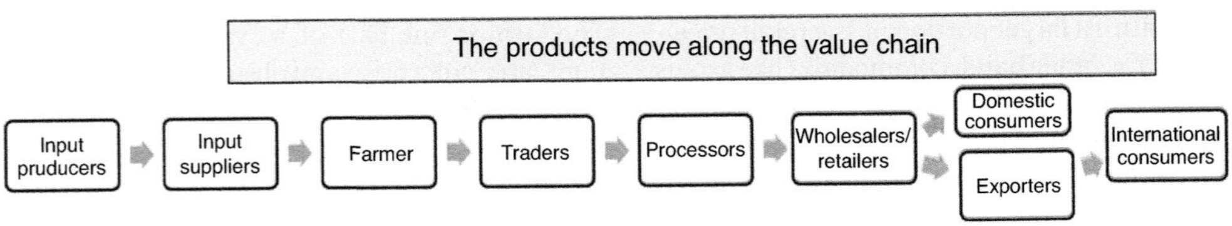

Fig. 4.3. An agricultural value chain and its actors.

with other individuals, businesses or agencies with the relevant knowledge and skills.

- **The enabling environment:** Creating a conducive enabling environment is crucial for fostering small farm commercialization and market-oriented farming. The enabling environment refers to the broader socio-economic, policy and institutional conditions that support and promote agricultural development, market integration and the growth of smallholder farmers. Understanding the key components of the enabling environment is essential for policymakers and stakeholders to design effective interventions and strategies.

  The enabling environment encompasses factors such as access to land, secure land tenure, access to finance and credit, infrastructure development, availability of market information, supportive policies and regulations, capacity building programmes and institutional support. By addressing these factors, policymakers can create an environment that encourages smallholder farmers to engage in commercial farming, access markets and enhance their livelihoods.

Figure 4.4 shows this more complex depiction of an agricultural value chain.

4.8 The Farm Decision-making Unit

The farm functions as a decision-making unit that encompasses not only production activities but also consumption patterns and dynamics. Viewing the farm as both a production and consumption unit is essential for effective management and to achieve livelihood goals. Additionally, considering non-farm sources of income and intra-household dynamics provides further insights into the complexities of farm decision making. These aspects are outlined here:

- **Unit of production:** as a unit of production, the farm involves a range of decisions related to agricultural activities, resource allocation and productivity enhancement. These decisions include choices regarding crop selection, livestock management, input utilization, technology adoption and marketing strategies. Farmers must consider factors such as resource availability, market conditions, labour availability and risk management in their decision-making processes. The goal is to optimize production, maximize profitability and ensure long-term sustainability.

  Central to smallholder farming is the recognition that consumer demand should guide farmer decision-making processes. Farmers should align their production strategies with market preferences and trends, ensuring that their products meet consumer needs. Consequently, markets become the vital transmission mechanism that bridges the gap between farmers and consumers, facilitating the exchange of goods and services.
- **Unit of consumption:** the farm itself is a unit of consumption wherein the farm household makes decisions about how to use the output and resources generated. These decisions encompass choices related to food consumption, household expenses, investment in education or healthcare, and quality of life improvements. Understanding the consumption patterns and preferences of farm households is essential for aligning production decisions with their needs and aspirations.
- **Non-farm sources of income:** in many farming systems, non-farm sources of income play a significant role in farm decision making. Farm households often engage in off-farm activities or rely on remittances, pensions or other income sources outside of agriculture. These additional income streams can impact the farm's resource allocation, investment decisions and risk management strategies. Understanding the dynamics between farm and non-farm income sources is crucial for optimizing livelihood outcomes and improving overall farm resilience.
- **Intra-household dynamics:** intra-household dynamics within farm households significantly influence decision-making processes. Gender roles, family structures and power dynamics shape how resources are allocated, how decisions are made and how benefits are distributed within the household. Recognizing and understanding these dynamics is essential for promoting equitable and inclusive farm management practices. It involves

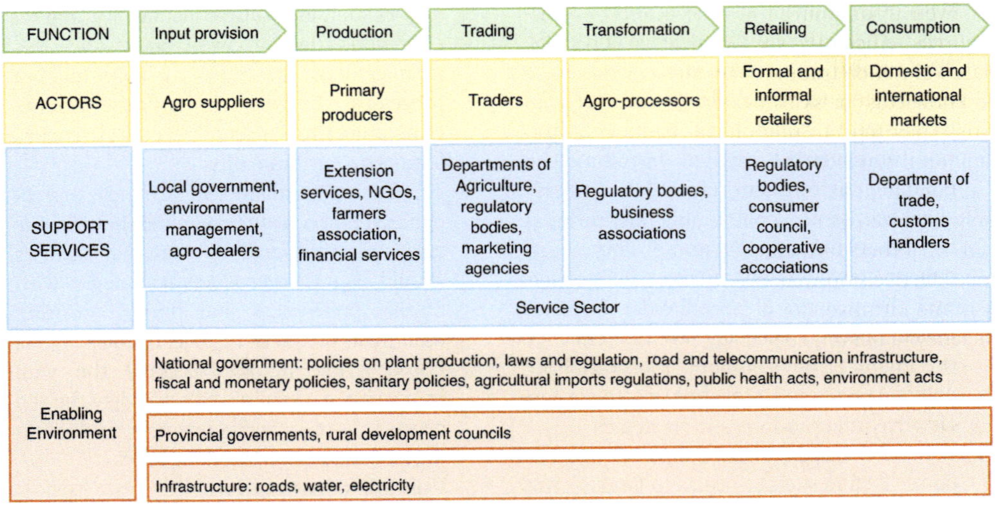

Fig. 4.4. Expanded agricultural value chain.

considering the needs, preferences and aspirations of all household members and ensuring their active participation in decision-making processes.

Examining the farm as a unit of both production and consumption, as well as considering non-farm income sources and intra-household dynamics, provides a comprehensive understanding of the complexities involved in farm decision making. Farmers need to balance production objectives with household needs, integrate diverse income sources and address intra-household dynamics to achieve sustainable livelihood outcomes. By incorporating a holistic perspective, farm management strategies can be tailored to meet the specific needs and aspirations of the farm household, contributing to improved well-being, resilience and overall farm performance.

4.9 Farm Household Decisions

Decision making starts with identifying a problem, challenge or opportunity. For instance, the introduction of a new herbicide requires that a decision must be made about whether to use it over the existing herbicide. A potential new market – an opportunity – also requires thought and analysis. Similarly, traditional problems, such as a crop not growing as expected, necessitate determining the cause of the problem and deciding on the best course of action. In all these cases, effective observation, analysis and decision making are crucial to ensure the farm follows the optimal path.

In traditional farming, decisions are often made jointly by family members and shared in various ways. In some situations, however, there may be a division of responsibility. One example is that food crops are often produced by women, while cash crops are tended by men. Similarly, in some cultures, large livestock is the responsibility of men and boys, while small stock is the responsibility of women. Frequently, individual members of the household may have independent control of some plots of land or groups of livestock.

Traditional farming recognizes the role of the farmer as producer and manager. As a producer, the farmer is a worker and deals with all aspects of farming and field operations. The traditional role of the farmer has been to take care of plants and livestock but with greater market orientation, technical know-how is not enough.

As a manager, the farmer makes complex decisions, including choosing which crops to plant in each field, what livestock to raise and deciding how to distribute tasks among the available labour. They involve choices, for example, regarding the selection and number of draught animals needed for work in the field.

Even traditional farmers aim to use their resources efficiently to achieve their family goals. Very often traditional farmers do not have the knowledge or skills to make informed decisions. They tend to rely on intuition, untested traditional practice or by copying the decisions of other farmers. Farmers, for example, often apply fertilizer at rates of application that are used by other farmers cultivating the same crop. The decision-making process is often very basic and intuitive. Box 4.3. further explores the efficiency of smallholder farms.

4.10 Farm Household Livelihood Outcomes

As noted earlier, farm households are driven by a variety of goals that extend beyond mere profit maximization. Recognizing the various goals of farm households is essential for understanding their decisions and creating suitable management strategies. These goals, as listed here, encompass food security, profit maximization, risk reduction, social aspirations and satisficing, among others.

- **Food security:** one of the primary goals for farm households is to ensure food security for themselves and their dependents. Farming provides a direct source of food, and households strive to produce enough to meet their nutritional needs. Decisions regarding crop selection, production strategies and resource allocation are often influenced by the objective of maintaining a stable and sufficient food supply.
- **Profit maximization:** another key goal for many farm households seeking to generate sufficient income from their farming activities to sustain their livelihoods, support their families and invest in future farm development. Decision making related to input use, production techniques, market access and value addition is driven by the goal of maximizing farm profitability.
- **Risk reduction:** family farm households face various risks inherent in agricultural

Box 4.3. Smallholder farmer efficiency.

In 1964, Theodore Schultz explored the efficiency of smallholder farmers by investigating their economic behaviours and productivity in developing countries. Contrary to the prevailing views at the time, which often regarded these farmers as irrational and inefficient, Schultz found that smallholder farmers are quite rational and make efficient use of their limited resources when given accurate price signals and access to markets. He argued that smallholder farmers efficiently allocate the resources available to them, such as labour and land, to maximize their utility and income generated. He observed that these farmers respond rationally to economic incentives, which include changes in the prices of agricultural inputs and outputs (Shultz, 1961, 1964). Thus, there is a strong foundation for rational decision making by smallholder farmers.

Shultz highlighted the importance of the farmer, and not just the technological advancement aimed at increasing productivity. This remains true today; and yet it is often still overlooked.

Since the 1960s, the world has become more complex. The agricultural landscape has changed vastly and is far more intricate and dynamic. While the small farmer may tend towards rationality, many are now not equipped to deal with the extent and rapidity of change. They often lack the skills to make informed decisions in light of these rapid changes. Moreover, a lack of data, along with the limited education and training of traditional smallholders and extension workers limits the expression of the rational decision making by the farmers.

To enhance decision-making capacity among smallholder farmers, two things are required: (i) training of farmers in farm management; and (ii) removal of barriers and obstacles to progress. Enhancing the human capital of farmers – through education, training and health – is crucial as it improves their ability to adapt and use technology effectively.

Removing informational and institutional constraints for small farmers, promoting market access and investing in human capital development frees the farmer to grow.

To foster more inclusive and sustainable growth of small family farmers, development strategies must expand beyond merely providing physical inputs and technology to enhancing the overall decision-making capacities and efficiency of the smallholder farmers.

production, such as weather fluctuations, pest outbreaks, market volatility and unforeseen events. Mitigating these risks is a significant goal for farm households. They employ risk management strategies, including diversification, insurance, savings and investment in resilient farming practices. Decision making is influenced by the objective of reducing risks and safeguarding livelihoods.

- **Social goals:** farm households often have social goals that extend beyond economic considerations. They strive to contribute to the well-being of their communities, preserve cultural heritage and maintain social cohesion. Decision making may involve supporting local markets, engaging in community initiatives or adopting sustainable farming practices that have positive social impacts.
- **Satisficing:** farm households often engage in satisficing, which means they seek satisfactory outcomes rather than optimal ones. They aim to strike a balance between different objectives, such as earning a decent income, maintaining a good quality of life and enjoying a fulfilling farming experience. Decision making involves finding a satisfactory level of achievement across multiple goals, considering the trade-offs and limitations inherent in their farming context. It is about creating coherence for their farm and their family.

Recognizing the multiple outcomes pursued by farm households is crucial for farm management. It requires an understanding that decision-making processes are complex and influenced by a range of factors, including economic, social and cultural considerations. Farmers as managers need to develop strategies that align with these diverse objectives, considering the farm's resources, capabilities and external constraints.

Understanding the diverse goals pursued by farm households is crucial for advisory services and extension. A comprehensive understanding of these goals means extension workers can tailor their services and recommendations to meet the specific needs and aspirations of farmers. This knowledge allows extension workers to provide targeted guidance and support that aligns with the priorities of farm households, increasing the effectiveness and relevance of their advisory services.

The goals of farmers shape their decision-making processes, influencing their choices regarding production practices, resource allocation, risk management and market engagement. Advisors who are aware of these goals can provide advice that considers the multiple objectives of farmers, helping them to make informed decisions that align with their desired outcomes. This understanding also enables extension workers to address the challenges and trade-offs that may arise when pursuing multiple goals simultaneously.

Moreover, recognizing the diverse goals of farm households fosters better communication and trust between extension workers and farmers. Agricultural advisors who demonstrate an understanding of farmers' objectives can establish meaningful and collaborative relationships, enhancing the uptake and implementation of recommended practices. By incorporating the goals and aspirations of farmers into their advisory services, extension workers can empower farmers to achieve their desired outcomes while promoting sustainable agricultural practices and improving rural livelihoods.

Ultimately, understanding and knowing the diverse goals of farm households allows advisory and extension service advisors to provide context-specific, relevant and impactful support. It enables them to tailor their services to meet the unique needs of farmers, helping them achieve their desired outcomes, whether it be food security, profit maximization, risk reduction, social aspirations or a combination of these goals. By incorporating this understanding into their advisory approaches, extension workers can contribute to the sustainable development and success of farm households and agricultural communities.

This dynamic in the small farm household is explained in Fig. 4.5, which illustrates the basic flow of inputs and outputs needed for farming operations and the maintenance of the small family farm household. This is the context within which farm and household decisions are made.

4.11 The Farm-family Cycle

The farm-family cycle is a concept that recognizes the dynamic nature of the farm household over time. Drawing on the work of scholars such as Gasson and Errington (1993) and Olsen (2004), it sheds light on the interplay between family

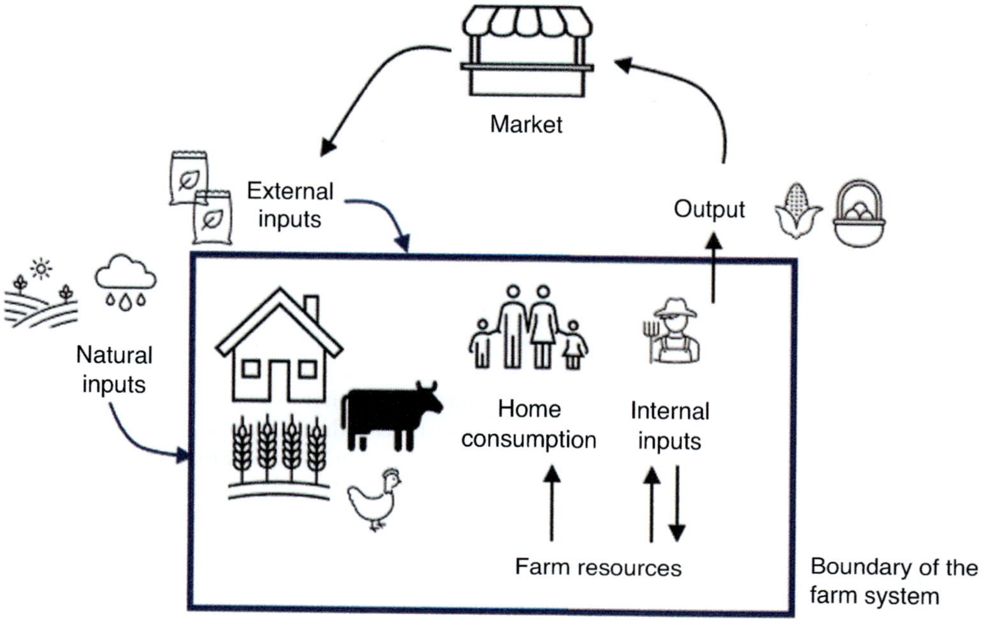

Fig. 4.5. Flow of inputs and outputs through a small family farm household. Adapted from Reijntjes *et al.*, 1992.

dynamics, life stages and farm management decisions. Understanding the farm-family cycle helps farmers and extension workers see how family dynamics affect resources, decisions and farm development. The following should be considered:

- **Family life stages:** the farm-family cycle encompasses various life stages of farm households, including starting a family, raising children, educating children and, eventually, transitioning into retirement. Each stage presents unique challenges and opportunities that shape farm management decisions. For instance, during the early stages, the focus may be on expansion and building capital, while later stages may prioritize succession planning and ensuring the well-being of older family members.

  During the child-rearing stage, the emphasis may be on securing resources for education, healthcare and housing. This stage often requires significant investment in both time and money, impacting the overall resource allocation within the farm. As the family transitions into later stages, priorities shift towards ensuring financial security for retirement and planning for the transfer of the farm to the next generation.
- **Resource allocation:** the farm-family cycle significantly impacts the allocation of resources within the farm. As families evolve and grow, the demands on resources change. For example, during the child-rearing stage, resources may be allocated towards education, healthcare and housing. In contrast, during retirement, resources may shift towards ensuring financial security and transitioning ownership to the next generation. Understanding these shifts allows farmers to allocate resources effectively and make informed decisions aligned with the needs of the family.

  Effective resource allocation also involves planning for unexpected events that might disrupt the family cycle, such as health issues or economic downturns. This requires a flexible approach to resource management, ensuring that the farm can adapt to changing family needs without compromising its operational viability.
- **Decision-making processes:** the farm–family cycle also influences decision-making

processes within the farm. Different life stages bring about varying priorities and aspirations. Farmers must consider the interests and goals of family members when making decisions. For instance, decisions regarding expansion, diversification or investment in new technologies may be influenced by the long-term vision of the family and their desired outcomes at specific stages of the cycle.

Decision making is also influenced by how different family members are involved. Younger generations might bring fresh perspectives and new technological solutions, while older generations provide wisdom and experience. Integrating these inputs is crucial for making decisions that are beneficial for the entire family and the farm's future.

- **Inter-generational transfer and succession planning:** a significant aspect of the farm-family cycle is the inter-generational transfer of the farm. Succession planning is an integral part of the farm-family cycle, involving the preparation for transferring farm ownership and management to the next generation. This process includes identifying potential successors, providing them with necessary training and experience, and setting up legal and financial mechanisms to facilitate the transition. Succession planning is not a one-time event but a continuous process that evolves with the family's needs and the farm's development. Effective succession planning ensures that the farm remains viable and can continue to thrive under new leadership. It also addresses potential conflicts by clarifying roles and expectations, thus reducing the risk of disputes that could jeopardize the farm's future.
- **Family dynamics and well-being:** the farm-family cycle recognizes that the well-being and dynamics of the family have direct implications for the management of the farm. Family relationships, communication patterns and individual aspirations can significantly impact decision making and the overall functioning of the farm. Recognizing and addressing issues related to family dynamics promotes a supportive and harmonious environment that enhances farm management outcomes.

Maintaining open lines of communication and addressing conflicts promptly can prevent misunderstandings and ensure that all family members are aligned with the farm's goals. Additionally, considering the well-being of all family members, including mental and physical health, contributes to a more productive and positive working environment.

By understanding the farm-family cycle and the interplay between family dynamics and farm management decisions, farmers and advisors can better navigate the complexities of the farm system. They can provide guidance and support that aligns with the needs and aspirations of the family at different stages of the cycle. Additionally, recognizing the importance of family well-being allows for the development of strategies that promote the holistic success of both the farm and the family.

4.12 Conclusion

The farm is a complex system that requires a holistic understanding for effective management. It encompasses various elements such as resources, technologies, markets and institutions. Supplementary enterprises provide additional income and diversify revenue streams, while complementary enterprises enhance the primary agricultural activity. Competitive enterprises directly compete for resources with the primary activity. The farm's asset base includes physical, financial, human, social and natural capital assets, each playing a vital role in decision making. Understanding the interplay of these assets and the influences on them is crucial for optimizing farm performance. The farm serves as both a production and consumption unit, with decisions influenced by off-farm income and family dynamics. In short, the farm household system will be understood as comprising the household decision-making unit, the farm crop and livestock enterprises and the off-farm income and employment sources, and finally, linkages to input sources, markets and finance.

A key aspect of market-oriented farming is the recognition that consumer demand for products should drive farmer decision making while also staying focused on the farm-family's livelihood goals. Farmers need to be responsive to market signals, adapting their production and marketing strategies to meet consumer preferences and changing trends. Markets serve as the transmission mechanism that links farmers with consumers, providing the opportunity for smallholder farmers to capture value and generate income.

Support systems play a crucial role in enabling smallholder farmers to thrive in market-oriented farming. It is important to emphasize that support services are not limited to farmers alone but extend to a range of stakeholders along the value chain. While public extension services are one advisory service provider, there is a need for diverse and tailored support services that cater to the specific needs of different actors in the value chain. Understanding the intricacies of the farm as a system is key to improving farm livelihoods and promoting sustainable agricultural development.

Understanding that the family farm is part of a larger value chain is critical to the future success of small family farms. The farmers need to learn how to engage with other actors in the value chain to benefit from the opportunities this presents. Concurrently, agricultural advisory services must also embrace the role of the small family farmer in the value chain and prepare itself to engage with other actors in the chain – providing advice, creating linkages and establishing resilient networks.

Finally, the discussion on the farm-family cycle has underscored the importance of succession planning and the inter-generational transfer of farm ownership and management. Effective succession planning is crucial for ensuring the continuity and long-term viability of farming enterprises. This involves preparing the next generation through education and experience, setting up legal and financial frameworks and fostering open communication within the family. By understanding the evolving needs and dynamics of farm families, farmers and their advisors can develop strategies that promote the holistic success of both the farm and the family, ensuring sustainable agricultural development for future generations.

References

Gasson, R. and Errington, A. (1993) *The Farm Family Business*. CABI, Wallingford, UK.

Harwood, R.R. (1979) *Small Farm Development: Understanding and Improving Farming Systems in the Humid Tropics*. Westview Press, Boulder, Colorado.

Olsen, K.D. (2004) *Farm Management: Principles and Strategies*. Iowa State Press, Ames Iowa, IA. DOI: 10.1093/erae/31.2.225. (accessed 17 March 2025).

Reijntjes, C., Haverkort, B. and Waters-Bayer, A. (1992) *Farming for the Future: An introduction to low-external-input and sustainable agriculture*. ILEIA, Leusden, Netherlands.

Schultz, T. (1964) *Transforming Traditional Agriculture*. New Haven, Yale University Press.

Additional Reading

DFID (1999) Sustainable Livelihoods Guidance Sheets. DFID: London. (http://www.livelihoods.org/info/info_guidanceSheets.html)

Kahan D. (2013) *Market Oriented Farming – an overview, Farm Management Extension Guide*. FAO: Rome. https://www.fao.org/sustainable-food-value-chains/training-and-learning-center/details-materials/en/c/274679/ (accessed 27 January 2025)

Nell W.T. and Napier R.J. (2009) Strategic Approach to Farming Success. https://www.researchgate.net/publication/23512532_Strategic_Approach_to_Farming_Success

UNDP (2017) *Guidance Note: Application of the Sustainable Livelihoods Framewsork in Development Projects*. UNDP, Panama City, Panama. Available at: https://www.undp.org/latinamerica/publications/guidance-note-application-sustainable-livelihoods-framework-development-projects (accessed 27 January 2025).

5 Modern Farm Management

Effective farm management is a cornerstone of agricultural success, enabling farmers to navigate challenges, optimize resources and achieve their desired livelihood goals. This chapter explores the concept of farm management. It emphasizes the importance of decision making in shaping the strategies farmers adopt to achieve their goals. It enquires into the historical perspective on farm management, highlighting past assumptions of farmers' inherent efficiency and the need to differentiate modern management practices from traditional beliefs and practices.

This chapter also highlights the evolving understanding that farm management is a skill that needs to be developed, rather than an innate ability. Improving farm management practices can serve as a vital entry point for enhancing farmer livelihoods and driving agricultural productivity. Additionally, the distinction between farm management and entrepreneurship is addressed, recognizing the unique aspects of both while emphasizing their interconnectedness in achieving sustainable farming operations.

The chapter introduces the reader to the key tasks of a manager, including setting objectives, acquiring external information, assessing resource utilization, engaging in forward planning and monitoring performance. It emphasizes that farm management focuses on sound decision making and effective implementation.

5.1 Importance of Effective Farm Management

Effective farm management enables small family farmers to make the most of their available resources – natural, human, financial, physical and social capital – to achieve their livelihood goals. Within a livelihoods framework, farm management serves as a means of strengthening household resilience, improving food security and increasing income-generation opportunities. At the same time, from a business perspective, farm management provides the tools and processes needed to operate the farm as a productive and profitable business enterprise.

Farm management involves a structured approach to planning, organizing, implementing and controlling farm activities. It ensures that key assets – such as land, labour, capital and technology – are allocated and used efficiently and in ways that align with both the household's livelihood strategies and the objectives of the farm business. Effective farm management supports informed decision making, allowing farmers to optimize productivity, reduce costs and manage risks. It transforms farming from a subsistence activity into a deliberate, market-oriented livelihood strategy.

Aligning farm decisions with broader household goals and with the realities of the market environment is critical. This requires farmers to assess their available resources and understand how best to deploy them – whether to increase yields, diversify income sources, reduce vulnerability or invest in long-term sustainability.

Technology plays a key role in modern farm management, particularly in enabling timely and data-driven decisions. Tools such as mobile applications, digital record-keeping, remote sensing and decision-support systems help smallholder farmers forecast yields, manage inputs more precisely and track performance. These tools enhance operational efficiency while supporting better planning and market responsiveness (see Chapter 10, for further discussion). Moreover, effective farm management builds adaptive capacity. It helps farmers anticipate

DOI: 10.1079/9781836990536.0005

and respond to external pressures – such as fluctuating prices, climate variability or policy changes – while positioning them to take advantage of emerging opportunities.

Ultimately, farm management bridges the gap between livelihood security and business growth. It not only improves short-term outcomes but also contributes to long-term sustainability, ensuring that farming remains a viable, resilient and rewarding pathway for rural households and their communities.

5.2 Understanding Farm Management and Decision Making

Farm management involves the activities farmers undertake, and decisions farmers make to efficiently operate their farms and achieve livelihood goals. It demands a systematic approach to decision making, resource allocation and goal setting, with the aim of maximizing productivity, profitability and sustainability. This section provides an in-depth understanding of farm management and highlights the significance of decision making in shaping the strategies small scale family farmers need to achieve their goals.

The earlier neglect of farm management stemmed from the widespread belief that small farmers were already operating at maximum efficiency. While, as discussed previously, farmers tend to act rationally within the constraints they face, rationality does not always equate to efficiency. As earlier chapters have shown, the skills and practices that once served farmers well are no longer adequate to cope with the increasing complexity and demands of modern farm management. Small farmers need support to make informed decisions, especially as agriculture continues to change rapidly. The rationality of small farmers' decision making needs to be reinforced by building their capacity as discussed in the previous chapters (Schultz, 1961). The future success of small family farms will not be achieved merely by the application of modern technologies – it rests in the greatest measure in building the farm management capacity of these farmers.

The key tasks of a farmer as manager revolve around setting objectives, assessing available assets and resources, acquiring relevant information about the external environment, considering potential and alternative uses of assets and resources, engaging in forward planning, systematic and methodical implementation of plans, and monitoring performance. Farm management extends beyond purely economic considerations and involves sound decision making that takes into account ecological sustainability, social impacts and community dynamics – all in the context of achieving the farm-family's livelihood goals.

Recognizing the pivotal role of decision-making in farm management, this chapter explores the fundamental decisions that farmers face. These decisions include determining what to produce based on market demand and resource suitability, choosing production methods and technologies, assessing the optimal scale of production, identifying appropriate markets and devising strategies to obtain the best prices. These decisions also include responding to unexpected, or unintended, outcomes in the process of implementing the chosen strategies and plans. This chapter also highlights the importance of gender considerations in decision making, acknowledging the diverse roles and perspectives of male and female farmers.

Farmers, in their capacity as managers, play a crucial role in shaping the future of their enterprises and livelihoods, and that of the agricultural sector as a whole. Successful farm management requires a keen understanding of the decision-making process, which involves both linear models and iterative approaches (Giles and Stansfield, 1980). The chapter further delves into the steps involved in setting goals, distinguishing between long-term strategic objectives and short-term tactical goals that guide day-to-day operations.

Recognizing the importance of management as a skill, efforts to improve farm management practices of smallholder farmers can enhance their livelihoods and contribute to the overall success and sustainability of their farming systems. By understanding the key tasks of the farmer as manager whilst developing their skills in the decision-making processes, smallholder farmers will be better equipped to make informed choices, optimize resource utilization and navigate the complexities of modern farming systems.

5.3 Historical Perspective on Farm Management

Farm management has undergone a significant evolution over the years. It has been driven by the introduction of new technologies, changing agricultural practices, national and global social and economic shifts, and a deeper understanding of the complexities involved in running successful farming businesses in an increasingly complex global society. This section explores an historical perspective on farm management, shedding light on past attitudes and assumptions, and the subsequent recognition of the need for more comprehensive management practices.

In the past, farm management practices received little attention. The assumption was that all small farmers possessed an inherent efficiency that allowed them to effectively manage their operations without the need for specialized management techniques. This oversimplification overlooked the multifaceted nature of farming (particularly in today's world), and the new challenges faced by farmers in making informed decisions regarding resource allocation, production methods and market engagement.

A major turning point in recognizing the importance of farm management came with Schultz's pioneering work (see Chapter 3), which challenged the assumption that small farmers were irrational and, therefore, inefficient. Instead, he argued that they are rational decision-makers who optimize their limited resources. However, in an evolving agricultural landscape, productivity and profitability depend on better access to informaton, improved technologies, education, training and institutional support. While acknowledging farmers' rationality, Schultz emphasized that enhanced management practices could further improve their efficiency and outcomes. His research underscored the critical role of management skills in boosting agricultural productivity, improving livelihoods and strengthening overall farm performance (Schultz, 1964).

Distinguishing modern management practices from traditional economic assumptions helps understand how farm management has evolved. While traditional views associated management solely with the technical aspects of farming, modern management encompasses a broader set of skills and considerations. It involves decision making, resource assessment, financial planning, risk assessment, marketing strategies, human resource management and the ability to adapt to changing legal, economic (including market) and socio-political conditions and environmental factors.

Recognizing that farm management is a skill that can be developed and improved marks a departure from the assumption that all farmers are already efficient managers. It underscores the importance of providing farmers with the knowledge, tools and support to enhance their management capabilities and make informed decisions that can positively impact their agricultural enterprises and adapt them to the ever-changing agricultural landscape.

5.4 Farming as a Business

The modern farm – even the small family farm – is increasingly being run as a business. As noted previously, farming as a business covers a broader range of functions than traditional farming. Farming as a business includes decisions about production, marketing, input purchasing, technology use, labour and transport. A farm business is an enterprise with linkages to input suppliers and output markets – as well as to various support services. It consists of activities which, if operating together efficiently and effectively, create value and, at the end of the process, increase profit. Profit (sometimes called the margin) is the difference between the total value and the costs of carrying out the activities: margins indicate profit.

The farm as a business can be understood in terms of the functions and activities described in Fig. 5.1. The figure shows eight areas of activity on any farm of any size. There are management activities (information management, organizational management), economic activities (finance, marketing) and, of course, production activities (land, technology, infrastructure, input supply). Each of these areas of activity has become more complex in response to the changing agricultural landscape discussed previously. They involve a plurality of expertise – chief among them is farm (business) management.

As indicated in Fig. 5.1, the farm business has three main areas of decision making: (i) production; (ii) economics; and (iii) management. Each of these areas is relevant to the decisions farmers must make for every enterprise they pursue as part of their livelihood strategy. The majority of the decisions to be made are related to production. Decisions must be made about land use, technology choices (including tools and processes), managing infrastructure, and choosing and applying inputs. Economic decision making is centred on finance and markets, and marketing. And the area of general management requires decisions about managing information and overall organizational management (which include labour management and communication). As also depicted in Fig. 5.1, all these decisions are made in the context of the social and environmental sustainability.

Collectively, the decisions made in each of these spaces determine what resources will be used, how they will be used, which products will be produced, how much will be produced, and where and how they will be financed and marketed. These decisions and their resulting activities provide the building blocks for profitable and sustainable farm businesses; they are not a collection of independent activities but are closely interdependent, all governed by the desire to achieve the livelihood goals of the farm family.

5.4.1 Management for profitability and sustainability

As we learned in the previous chapter, the farm is the primary source of livelihood for the rural farm household. In addition to providing food and in some cases energy, the farm needs to generate cash income. Therefore, it is essential that the farm business – as small as it might be – is profitable. Simply to break even is not sufficient because family labour, which could earn income off-farm, is a significant part of the operational costs of the farm business. It must be profitable – that is, it must generate more income than expenditure. The success of a farm business is undeniably linked to its profitability, as it provides the financial foundation for growth, innovation and long-term viability.

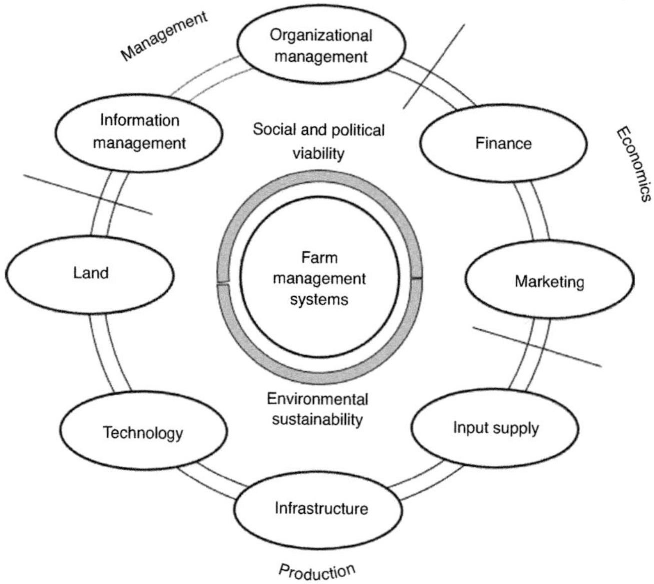

Fig. 5.1. The extension carousel showing the complexity of farm management. Adapted from Worth, 2020, unpublished material.

Relationship between management and profitability/sustainability

In today's world, the concept of profitability cannot be divorced from the notion of sustainability. Long-term success requires farmers to adopt principles of environmental care, social responsibility and economic resilience.

Environmental sustainability is vital in farm management, as farms depend on ecosystem services like clean water, fertile soil and a stable climate. Sustainable practices safeguard these resources for future generations while promoting current and long-term farm viability. Optimizing resource use is key. Precision agriculture, targeted irrigation, and responsible fertilizer and pesticide management minimize waste and pollution. Conservation farming – such as reduced tillage, cover cropping and agroforestry – enhances soil health, prevents erosion and supports biodiversity.

Climate change adaptation is crucial, especially for smallholder farmers. Climate-smart practices, including improved water management, crop diversification and renewable energy reduce emissions and build resilience. Farmers using adaptive strategies can better withstand climate risks while contributing to broader sustainability goals. By integrating sustainable techniques, farmers enhance productivity, reduce environmental impact, and strengthen the resilience of their farms and communities (FAO, 2013). Climate change and climate change adaptation are discussed in greater detail in Chapter 7.

Social responsibility is also a vital element of sustainable farm management within small family farming systems. It involves considering the well-being of the farm family, any hired casual labour, the local farming community, and the needs and expectations of the market and consumers. Small family farmers can also promote social responsibility by supporting local economies and producing high-quality, safe and nutritious food.

Economic resilience, the third pillar of sustainability, is closely intertwined with environmental and social sustainability. By adopting efficient management practices, reducing production costs and diversifying income streams, small family farmers can enhance their resilience to market fluctuations and economic uncertainties. Investing in research and innovation, accessing financial resources and collaborating with value chain partners also contribute to the economic sustainability of the farm business.

Factors affecting the profitability of the farm business

Several interconnected factors significantly influence farm profitability. Some of these factors are external to the farm, others are internal.

External factors include input-related factors such as the availability of high-quality resources, their costs and the reliability of input suppliers, which can significantly impact overall productivity and cost-efficiency. Additionally, the proximity and accessibility of markets, along with market demand and prevailing prices, dictate the potential income a farm can generate. These market dynamics are further influenced by the level of competition, and the accessibility and affordability of credit, which can enable or hinder a farmer's ability to invest in useful resources and technologies. These factors are illustrated in Fig. 5.2.

While the adoption of advanced technologies and marketing can offer farmers competitive advantage, access and affordability will limit these possibilities. Further, external factors such as weather patterns, government policies and shifts in consumer demand also play influential roles. Effective management requires understanding these factors and adapting farming strategies accordingly. By continuously assessing these factors and strategically responding to changes, farmers can optimize their farm operations to maximize profitability while mitigating the risks associated with agricultural production.

In each of the other areas outlined in Fig. 5.2, the farmer will make decisions that will impact the profitability of the farm business and, in turn, the income generated for the farm household.

Internal factors are those aspects of the farm itself that influence productivity and profitability. This includes soil condition, condition and management of equipment, labour efficiency, farm planning and organization, crop and livestock selection, and production operations such as plant density and spacing, water and irrigation management, and pest and disease management. (Fig. 5.3).

These are real-time factors since they occur on the farmer's farm within a given season. Thus, the specific factors will vary from season to season

Fig. 5.2. External factors affecting farm profitability.

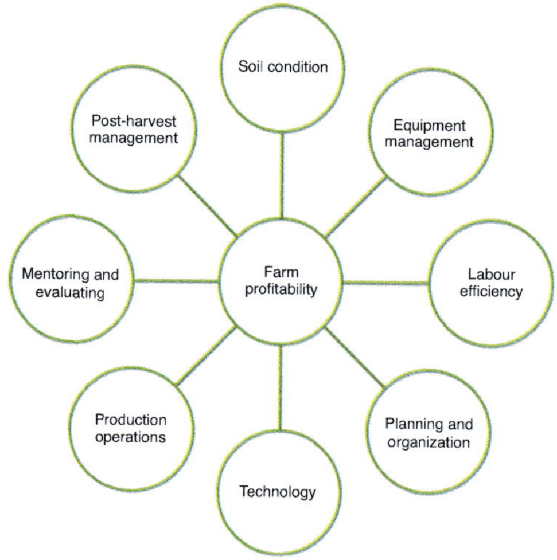

Fig. 5.3. Examples of internal factors affecting farm profitability.

and from farmer to farmer. One way to identify the relevant factors is by using a derivation network.

A derivation network is a diagram that visually represents the factors influencing profitability – the outcome of a farm or enterprise, and how they interconnect (Lloyd, 1968). It shows the relationships or dependencies that contribute to farm profit, displaying them in increasing detail, schematically. An example of a derivation network for a farm enterprise – maize – is given in Fig. 5.4. It starts with farm profit on the left. Moving to the right, we see that the profit depends on two main factors: the level of income and the level of costs. Moving again to the right, we see that the level of income depends on two main factors: the price of the maize sold and yield. Moving right again, we see that the price of the maize sold is influenced by the timing of the sale of the maize, the moisture content of the maize and the variety planted. Moving down we see that yield is also influenced by the amount of ha planted, the condition of the soil and the production system chosen by the farmer.

At the bottom half of the figure are costs. We see that costs are influenced by the amount of labour hired, the choice of inputs and the level of the application of inputs. Moving again to the right, the amount of labour hired is influenced by the availability of family labour and the production system chosen by the farmer. We also find that the choice of inputs is similarly influenced by the production system, but also by the condition of the soil. And finally, we see that the level of application of inputs is influenced by the number of weeds present in the fields and the incidence of pests and diseases. What stands out from this example, is that the variety of seed, the condition of the soil and the production system (all three under the control of the farmer) have a lot of influence, impacting both income and costs. While other factors may be equally important, the farmer will want to pay special attention to these three factors.

However, all the factors shown in Fig. 5.4 influence farm profitability. As they are internal factors, the farmer should have the ability to address them directly, unlike those of the external factors.

This example demonstrates that it is a primary function of farm management to monitor both external and internal factors and determine their potential and real impact on the profitability of the farm business. The farmer's first task is to discover whether any weaknesses exist with any of these factors or with the combination in which they occur.

Whether internal or external, the farmer needs to be aware of the factors that influence farm profitability. The farmer will then have greater command of the situation and will be able to make more informed decisions.

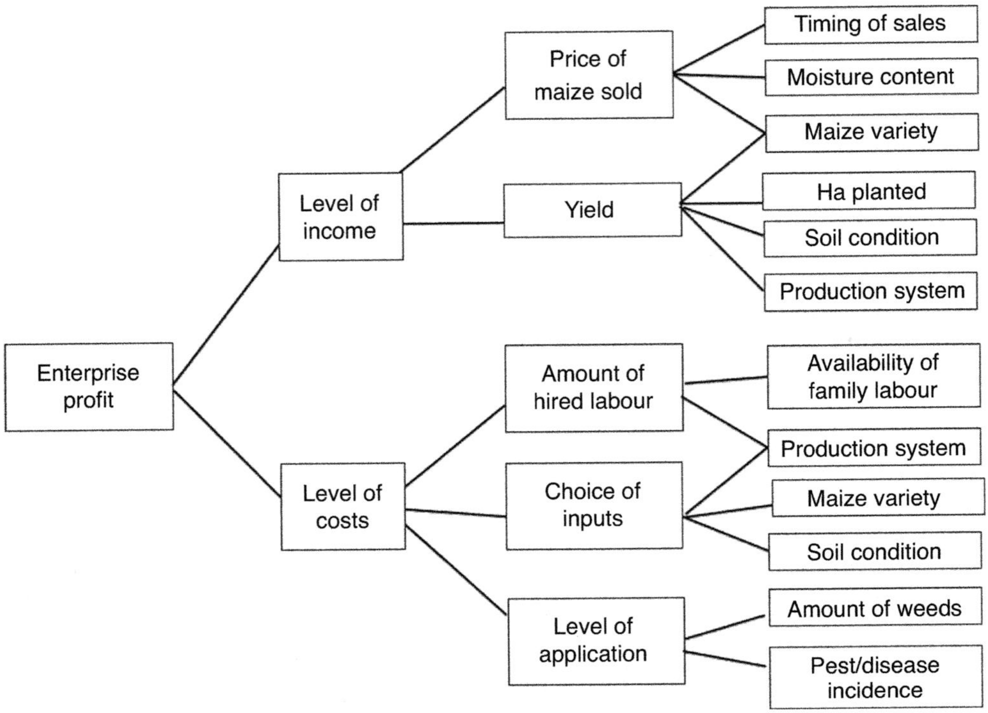

Fig. 5.4. Profit derivation network for maize production.

5.4.2 Management and entrepreneurship

An important aspect of farming as a business is the relationship between farm management and entrepreneurship. Both are important for small family farmers to achieve improved livelihoods.

Developing management competencies

Recognizing the importance of strong management competencies in farming is a key entry point for improving the livelihoods of smallholder farmers. Effective farm management requires making informed decisions, optimizing the use of limited resources and strategically planning for future success. While technical knowledge has long been the foundation of good farming, many small family farmers already possess strong technical competencies – particularly in the areas of production and input management.

Production is at the core of any farming business. Small family farmers already understand what inputs are needed for their farm enterprises and how to source and apply them effectively. However, modern farm management requires farmers to go a step further. They need to actively seek out better quality inputs, lower prices and more efficient alternatives. They must stay informed, experiment with new options and remain open to learning. Their approach to input management should be driven by a desire to improve both productivity and cost-effectiveness.

While most skilled farmers have the technical ability to produce a good crop or raise healthy livestock, market-oriented family farmers must focus on producing in the most profitable and sustainable way. They need to understand the importance of timeliness, efficiency and innovation. Rather than simply repeating what has worked in the past, they must be willing to test new production techniques, adopt improved technologies and respond quickly to changing environmental and market conditions.

No matter how well a farm produces, profitability ultimately depends on selling the output. Many small family farmers know their local markets, but to be successful managers, again, they need to go further. They must have the skills and competencies to identify the most profitable markets for each product, negotiate favourable terms, and adapt quickly to new opportunities. Marketing must be treated as an active, ongoing process, where there is a constant search for better prices, more reliable buyers and ways to add value.

Developing these management competencies requires farmers to acquire knowledge in various areas, including information management, organizational capacity, marketing management, financial management and risk management. By understanding the principles of efficient resource allocation, cost analysis and market dynamics, farmers can make informed decisions that align with their goals and aspirations. Building skills in planning, organizing and monitoring the operations of the farming system, enables smallholder farmers to adapt to changing circumstances and capitalize on emerging opportunities. Farm management tools and methods can be used to support farmers in decision making, and these are elaborated on in Part 3. By enhancing their management competencies, farmers can increase productivity, minimize risks and maximize profitability – ultimately leading to improved livelihoods and greater economic well-being.

Differentiating management from entrepreneurship

While farm management and entrepreneurship are closely related, it is important to distinguish between these two concepts. Farm management primarily focuses on optimizing resources, making informed decisions and improving operational efficiency within the existing farming system and in alignment with the livelihood goals set by the farm family. It encompasses strategic planning, financial oversight and day-to-day operations aimed at achieving desired farm outcomes.

Entrepreneurship in farming, on the other hand, is centred on innovation, risk-taking and the pursuit of new opportunities. Entrepreneurs identify gaps in the market, develop new products or services and create value through their ventures. Entrepreneurship is driven by adaptability, creativity and forward-thinking, driving agricultural growth and opening up pathways for diversification and value addition. Some have argued that all smallholder farmers are, by definition, risk-takers – constantly adapting their practices to market shifts and climate variability. While this is true to some extent, the concept of entrepreneurship is broader, and includes traits such as motivation, perseverance, confidence, goal-orientation and problem-solving. While many smallholder farmers may exhibit some of these qualities, they are by no means evenly distributed across the farming population; nor do they exist to the depth demanded by modern entrepreneurship.

Understanding the distinction between farm management and entrepreneurship helps farmers better grasp the full scope of running a successful farm business. It highlights the importance of both effective management practices and entrepreneurial thinking, for thriving in an increasingly competitive and dynamic agricultural environment. By combining sound management skills with an entrepreneurial mindset, farmers are better equipped to seize opportunities, innovate and adapt – ultimately building sustainable and resilient farm enterprises.

What distinguishes entrepreneurial farmers is their ability to apply their competencies – through proactive behaviour and a commitment to continuous improvement. In this context, what sets them apart is not necessarily superior knowledge, but a mindset focused on learning, growth and opportunity. By building on their technical strengths and further developing their management and entrepreneurial competencies, small family farmers can manage risks, improve profitability and strengthen the long-term resilience of their farming systems.

Integrating competencies

While it is useful to describe the individual competencies of a successful small family farmer, true success as a farmer-entrepreneur lies in the ability to integrate these competencies in practice. It is not simply about possessing technical or managerial skills in isolation, but about how these skills are integrated – with each other and with the farmer's available resources – to respond effectively to changing production conditions and market demands. By combining and applying these competencies in a coordinated way, the farmer-entrepreneur is better positioned to identify opportunities, adapt to change and improve the overall performance of the farm business. This integration of knowledge, skills and strategic thinking increases the likelihood of success and enhances both profitability and resilience in an increasingly dynamic agricultural environment.

5.4.3 The farm management cycle

Farm management centres on five areas: (i) diagnosis; (ii) planning; (iii) implementing; (iv) monitoring; and (v) evaluating. These are critical to managing a farm as a business. Learning how to make decisions in these five areas will put small family farmers in greater command of the resources and processes that influence food security and income generation and, ultimately, help them achieve their livelihood goals. As shown in Fig. 5.5, these five areas flow in a pattern which supports a continuous learning process about what works best for the farm as a business and for the farm family more broadly.

Diagnosis

Diagnosing means assessing the farm and household over a period and determining its current state of 'health'. For an initial diagnosis, the farmer may want to understand how the farm has produced over several seasons. Just like a doctor with a patient, a diagnosis will give the farmer important clues about the 'health' of the farm. How well is it producing? What is the condition of the farm's resources? Are resources becoming more or less productive? How profitable is the farm? Are farm goals being achieved? What problems and opportunities are there? Such a diagnosis would benefit from a careful analysis using the livelihoods framework discussed in the previous chapter. Some key questions to ask might be:

- What is the state of the farm-family's livelihood assets?
- What is influencing the assets and how have they been affected in terms of quantity and quality?
- How efficiently are the assets being used?
- What policies and institutions are influencing access and use of these assets?
- What were the goals set for the farm and the farm family? What results were achieved? Are they still relevant?
- What livelihood strategies were implemented to achieve these goals?
- What challenges were faced in implementing the livelihood strategies? What problems occurred?
- What is on the horizon? What changes are happening in the broader context, which influence policies and institutions?

In addition to assessing their own farm business, farmers need to regularly gather and analyse information about the external environment and how it affects their farming system. This includes understanding market trends, consumer preferences, technological advancements, regulatory changes and climate patterns. By being informed about external factors that influence their farming enterprise, farmers as managers can anticipate risks, identify opportunities and adapt their strategies accordingly. Accessing relevant information through market research, networking, industry publications; and engaging with extension services enhances the decision-making process. This information is needed for both diagnosis and planning. We will discuss information management later in this chapter as well as in Chapter 10, which explores data and information in detail.

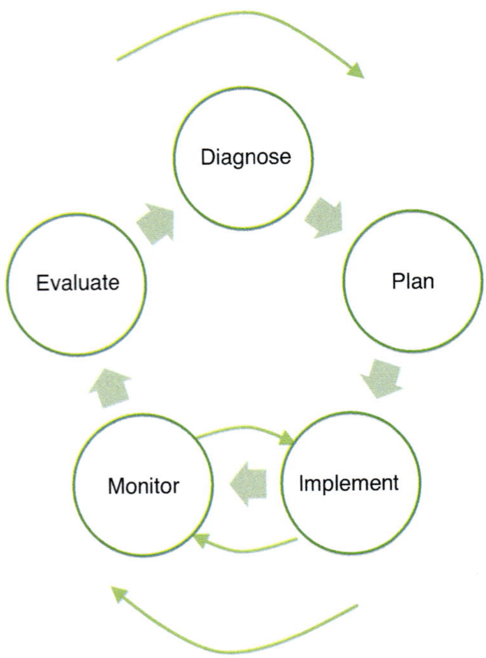

Fig. 5.5. The farm management cycle.

Planning

Once a farmer has completed the diagnosis, planning can begin. For small family farms, planning takes place at two levels. The first level involves reviewing and setting the family's long-term livelihood goals. What does the family want to achieve over time, and what are the key milestones along the way? At this stage, broad decisions are made about which livelihood strategies to pursue – both on- and off-farm. The second level of planning focuses on the on-farm strategies. Here, specific decisions are made about what to produce, how to produce it, in what quantities and how and where to market the products.

Planning should build directly on the results of the diagnosis. Some planning decisions will be based on knowledge, such as how much land, labour and other assets and resources are available. Other planning decisions are based on less certain things, such as rainfall and product prices. In addition to including decisions about resources and inputs, the plan will include estimated results in terms of yields and income, as well as the potential risks and plans to manage them.

Implementing

Implementation means putting the plan into action. Generally, this stage does not involve major decisions, as these should have been addressed during planning. However, unexpected changes – such as lower-than-expected rainfall, fluctuating crop prices, labour shortages or limited access to inputs and markets – can arise. When this happens, the operational plan may need to be adjusted to respond to these new conditions.

Monitoring

Monitoring involves tracking farm activities to see how well they align with the farm plan, linking it closely to the implementation stage. As shown in Fig. 5.5, these two stages are part of an ongoing, iterative process in the farm management cycle. For example, if the plan calls for planting 0.5 ha of cabbage, monitoring allows the farmer to check how much was actually planted. If the plan expects 100 mm of rainfall by a certain date, the farmer can record the actual amount received. Using this information, the farmer can make necessary adjustments to the plan to help achieve the desired outcomes. Monitoring also provides the farmer with the information needed to evaluate the success of the plan.

Evaluating

Evaluation means judging how well the farm performed when compared to the intended results of the plan. Did things go as planned? Were expected yields achieved? Were expected incomes reached? Evaluation also helps the farmer evaluate the plan as a whole, as well as its individual parts. This will help the farmer improve on the planning process in the next cycle or season.

Monitoring and evaluation provide essential information for diagnosing the farm at the end of the season, forming the basis for the next planning cycle. By following these steps in decision making, farmers gain a strong process for understanding and improving their farm performance. They will be able to identify what is working well and should be repeated, as well as recognize problems and take appropriate action. This approach strengthens the sustainability of the farm, improves household food security and helps make the family's income more stable and reliable.

5.4.4 Key management functions

We have explored the farm management cycle, which included the stages of diagnosing, planning, implementing, monitoring and evaluating a farm plan. This cycle offers a structured approach to managing the farm over time. Within it, the core functions of farm management – planning, organizing, directing and controlling – play a central role in ensuring the success and sustainability of small family farms.

These functions operate at two levels:

Strategic level: focused on guiding the overall management cycle and making long-term decisions.

Operational level: concerned with managing day-to-day activities and short-term decisions.

Together, these functions help farmers use resources efficiently, improve daily operations and adapt to both short-term challenges and long-term goals.

Planning: setting the foundation

Planning is fundamental to management at both strategic and operational levels. At the strategic level, planning involves setting long-term goals and developing strategies to achieve them. For small family farmers, this might include deciding which crops or livestock to raise based on local climate and market demands, determining seasonal timelines for planting and harvesting, and budgeting for inputs like seeds, fertilizer and small-scale machinery. At the operational level, planning focuses on daily tasks, such as allocating family labour, scheduling irrigation or preparing for pest control. By integrating both levels, planning ensures that resources are used effectively and that the farm can adapt to changing circumstances.

Organizing: structuring resources

Organizing follows planning and involves arranging resources and tasks to execute the farm plan effectively. At the strategic level, this might mean structuring activities to align with the availability of the farm resources, including labour, tools and equipment. At the operational level, organizing focuses on coordinating daily tasks, such as assigning roles to family members, scheduling work to avoid overlap, and ensuring tools and inputs are available when needed. This function is particularly critical in regions where farming systems are impacted by weather or market access, as it ensures that agricultural activities are completed efficiently and on time.

Directing: leading and motivating

Directing involves leading and motivating family and hired labour to implement the farm plan effectively. At the strategic level, this might include fostering a shared vision for the farm's future and engaging the family in long-term decision making. At the operational level, directing involves day-to-day leadership, including the coordination of land, labour and capital to manage production, financial and marketing activities effectively. Strong interpersonal skills and community engagement are essential for this function, as they help maintain harmony and motivation within the family and the broader community.

Controlling: monitoring and adjusting

Controlling is the final function and ensures that the farm's operations align with the farm plan. At the strategic level, this involves monitoring the long-term progress of the farm towards the goals set, such as tracking yields, assessing market performance or evaluating the impact of new practices. At the operational level, controlling focuses on daily adjustments, such as monitoring crop growth, adapting pest management practices or responding to unexpected weather changes. Effective control helps small family farmers maximize yields, sustain their livelihoods and adapt to variations in market prices, input costs and environmental conditions.

5.4.5 The farmer as manager

Effective decision making shapes farm direction, performance, profitability and sustainability. This requires that farmers see themselves as managers and dedicate themselves to developing the skills needed to fulfil this role. By adopting this managerial mindset, they will gain greater command over the factors that influence the profitability, sustainability and resilience of their farm businesses.

Importance of decision making in farm management

Decision making is a crucial aspect of farm management, as it influences various aspects of the farm operations, including production, resource allocation, marketing and risk management. Well-informed and strategic decisions enable farmers to optimize their use of resources, respond to market dynamics, capitalize on opportunities and navigate challenges effectively. In an increasingly complex and dynamic agricultural landscape, decision making plays a vital role in driving the success and profitability of farming enterprises.

Managerial ability

Managerial ability involves essential knowledge, skills and personal qualities. Knowledge about various production and marketing methods enables a farmer to identify problems, compare current methods with alternatives and gather relevant information. Being able to understand the interdependencies within the farm business is also important. For example, any decision to increase production should consider its impact on marketing. A key management skill is the ability to interpret data and draw meaningful conclusions from the information collected and analysed.

Managerial ability also includes what are often called 'soft skills', in addition to technical or 'hard skills' (see Box 5.1.). In farm management, hard skills refer to technical knowledge and abilities related to production, marketing, finance and similar areas. Soft skills, on the other hand, involve personal and interpersonal qualities, such as problem-solving, communication, confidence, adaptability, conflict resolution, a learning mindset, negotiation, critical thinking, decisiveness, follow-through and trustworthiness. These skills are essential for effective leadership and decision making in the day-to-day management of the farm.

Beyond knowledge and skills, a successful manager must possess motivation and drive to implement decisions effectively. This motivation distinguishes practical individuals from those who struggle with execution. Some people can navigate the decision-making process but fail to implement the decisions successfully. Therefore, managerial ability, encompassing both decision-making and execution, is vital in planning and achieving agricultural changes. Recognizing and developing these attributes can significantly impact the success of farm management and agricultural innovation.

Pragmatic approach to problem solving and decision making

There is a maxim that states: it is better to be approximately right than precisely wrong. Perfect precision may be desirable, but is it usually unachievable, impractical and unnecessary. It is most often the case that problems must be solved, and decisions must be made without all the information or time that one would want. This is true for many farm management decisions. The farmer will not have all the information or an endless amount of time in which to make decisions. Most decisions require an

Box 5.1. Soft management skills.

- **Problem-solving:** the ability to define a problem; determine the cause of the problem; identify, prioritize and select alternatives for a solution; and implement a solution.
- **Communication:** the ability to convey information effectively so that it is clearly understood by the recipient. This includes the ability to determine if the information is accurate, relevant, timely and suitable for the recipient.
- **Confidence:** being clear about one's capacity and limitations; having a sense of self-control and trust in your ability to handle any situation that arises – even if it means asking for help.
- **Conflict resolution:** the ability to resolve conflicts or disputes justly and fairly, and to create a greater sense of unity between conflicting parties.
- **Adaptability:** the ability to adjust to changes in the environment; being open to new ideas that may not be yours.
- **Posture of learning:** the quality of being willing to learn from others, from situations and from one's own mistakes.
- **Negotiation:** the ability to come to an agreement between two or more parties. This requires active listening, being able to see both side of an issue, understanding context and being able to persuade.
- **Critical thinking:** the ability to analyse situations, data and information objectively and arrive at an unbiased opinion or view.
- **Decisiveness:** the ability and willingness to make effective and appropriate decisions in a timely manner.
- **Follow-through:** the ability to start and complete tasks in a reliable and consistent manner.
- **Trustworthiness:** the quality of one's words matching one's actions; being reliable and worthy of trust.

element of pragmatism where the farmer deals with things realistically in a way that is based on practical considerations and experience. This means that decisions are made on qualitative rather than quantitative information or exact figures.

The risk is that the farmer may fall into a trap of not thinking outside the 'box' of their normal experience. Thus, the farmer must be committed to continually updating their problem perceptions, options and plans as new information becomes available. This ongoing reassessment is practical and necessary, given the uncertainties and complexities inherent in farming. It highlights that farmers' decision-making is not rigid but adaptable and responsive to changing conditions. At the same time, farmers must be equally committed to gathering new information and to learning – otherwise, they will stagnate.

Farmers generally prefer quick and simple decision approaches over-detailed, elaborate ones. They often test small-scale changes before full implementation, allowing for real-time adjustments based on new information. This iterative method improves efficiency and reduces the risk associated with large-scale changes. Additionally, farmers value discussions with peers and family members, using these interactions to refine their decisions. This collaborative aspect underscores the importance of community and shared knowledge in agricultural decision making, providing valuable insights that help manage uncertainties.

In practice, farmers take many shortcuts compared to simple linear or even dynamic processes. They aim for acceptable solutions rather than optimal ones, saving time and reducing mental strain. This approach involves quickly ruling out non-viable options, such as crops unsuitable for the local growing season, and selecting viable alternatives without exhaustive analysis. The pragmatic focus on efficiency acknowledges the costs and mental effort involved in decision making, ensuring that farmers can make practical and timely choices that support their operations. This method of continuous evaluation and adaptation is central to effective farm management.

In dynamic decision making, few farmers formally analyse day-to-day choices. Instead, they rely on years of hard-earned experience to develop 'rules of thumb' – mental shortcuts to making the best decision under specific conditions. These informal hints help farmers quickly determine the most effective course of action in various situations, enabling them to adapt swiftly to changing circumstances without extensive formal analysis. Each unique scenario triggers a different rule of thumb, reflecting the practical, experience-based nature of their decision-making process.

Below are some aspects to consider for enhancing management ability in the context of small family farming:

- **Skill development:** continual improvement of skills pertinent to farm management is vital. This encompasses financial management, marketing, production planning, human resource management and risk assessment. By developing these skills, small family farmers can enhance their decision-making capabilities, optimize the use of resources, and effectively respond to changes in market conditions and the agricultural environment.
- **Problem solving and adaptability:** as one of the key soft skills, the ability to solve problems and adapt to changing conditions is critical in agriculture, where farmers frequently encounter issues like market variability, climatic unpredictability and pest infestations. Developing problem-solving skills and adaptability allows farmers to effectively tackle these challenges, implement innovative solutions and capitalize on growth opportunities.
- **Knowledge acquisition:** staying informed about the latest agricultural advancements, best practices, emerging trends and changes in legislation and policy is fundamental for effective farm management. Small family farmers should be knowledgeable about new technologies, understand market dynamics, be aware of the impacts of climate change and familiarize themselves with relevant regulatory frameworks. Gaining this knowledge enables farmers to make well-informed decisions, adopt sustainable agricultural practices and explore new opportunities to increase their farms' productivity and profitability.

For many small family farmers, limitations in resources, access to information, and technical expertise can pose significant hurdles. Providing access to education and training programmes focused on essential aspects of farm management – such as effective production techniques, financial literacy, marketing strategies and understanding risk – can empower farmers to better manage their operations. Enhanced management skills lead to improved decision making, increased productivity and, ultimately, more sustainable and profitable farming practices. These capabilities are crucial for small family farmers to thrive and improve their livelihoods.

5.5 The Decision-making Process and Decision-making Frameworks

As we have noted, farm management is all about making decisions – both strategic and operational decisions. Understanding the decision-making process and the steps involved helps small family farmers become more methodical and systematic in making decision. And it is often very helpful to have a framework for making decisions. This will help create coherence and consistency among the many decisions to be made.

5.5.1 Understanding the decision-making process

Modern farm management requires that farmers face a multitude of decisions in their day-to-day operations. As a reminder, some of the key management decisions include determining what to produce, selecting production methods, deciding on the scale of production, identifying markets, setting pricing strategies and managing financial resources. These decisions require an understanding of market trends, consumer preferences, resource availability and risk management. Decision making involves gathering information, evaluating options, assessing risks and choosing actions that align with family objectives.

There are different theories of decision making. The standard rational approach is to carry out several steps in a simple linear order. A common list of tasks is as follows:

- Clearly articulate the issue (problem, challenge, opportunity) requiring a decision.
- Gather relevant reliable data and information.
- Analyse the data and information.
- Generate alternatives including their specifications.
- Evaluate alternatives, in terms of viability and relevance to the goals of the business, and weigh their relative value.
- Choose one or more alternatives.
- Create a plan to implement the alternative(s).
- Implement the plan.

These steps are interdependent and not strictly linear. For example, while generating alternatives usually follows the gathering and analysis of data, evaluating those alternatives may reveal the need to revisit or collect additional information. The process often involves a combination of sequential and iterative actions until a decision is made and put into practice. In some cases, evaluating alternatives may uncover new problems, challenges or opportunities, leading to revised or entirely new options. What matters most is that the process remains methodical and is based on sound evidence.

Caution must be taken not to over formalize the process as this will stifle creativity and the ability to be objective and see new possibilities. Similarly, the process should not be so cumbersome and detailed that it inhibits or even paralyses the decision-making process. No decision will be perfect, and each will have risks and come at a price. The process is intended to help the decision maker arrive at plausible solutions.

Figure 5.6 shows a 'flowchart' of the decision-making process. It traces the main decision areas of identifying a decision opportunity, generating alternatives, making the decision and implementing the decision.

5.5.2 Farm management decision frameworks

It can be useful to have or develop a decision-making framework to help visualize and compare choices to be made. Whether simple

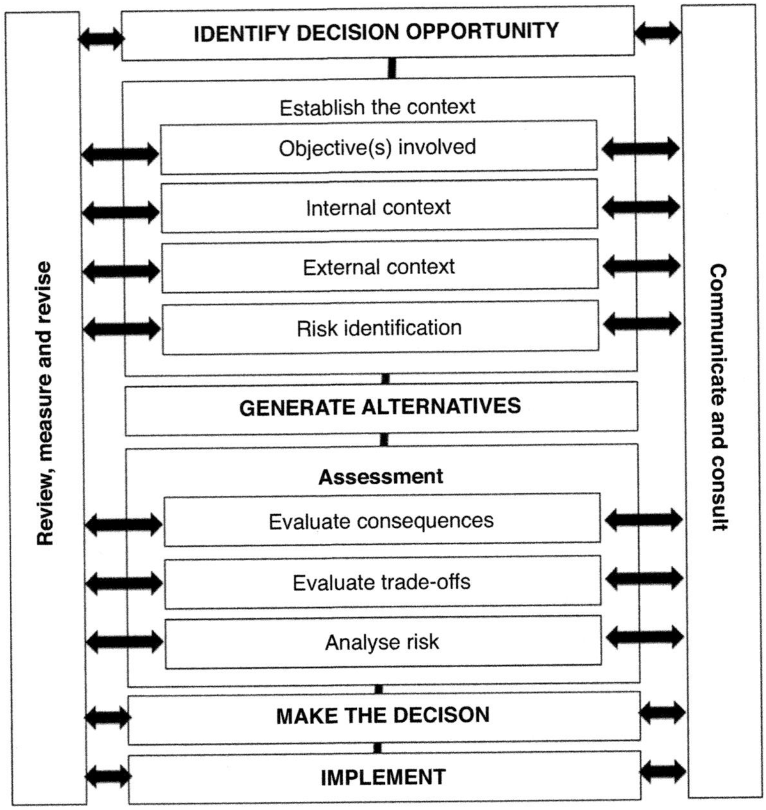

Fig. 5.6. The decision-making process. Based on Castle and Nelson, 1987.

or complex, the decision-making framework should support systematic thinking to make informed decisions. It involves identifying and evaluating different options based on criteria such as profitability, risk, sustainability and personal values. A framework helps farmers as managers consider the trade-offs and priorities associated with each decision, facilitating a structured approach to decision making. By utilizing decision-making tools and frameworks, farmers can enhance their managerial ability to assess alternatives, weigh their implications, and make choices that align with their farm's objectives and long-term sustainability.

Based on the livelihoods framework discussed in Chapter 4, and the farm management cycle discussed in this chapter (Section 5.4.3), we have created a farm management decision framework that marries the requirements of good management with the development and implementation of livelihood strategies.

Figure 5.7 provides a holistic view of the elements and interactions necessary for effective decision making. Small family farm decision-makers can use this framework as a guide to the range and content of the decisions they should consider in the process of managing their farms.

The framework starts with diagnosis. As indicated earlier, here the farm family reviews what assets they have and/or have access to. Then they can consider what is influencing their farming operations as well as their assets. This would include things like weather, markets, government policies and regulations, etc. After that they can think about their specific challenges, including past performance of the farm, current and potential markets, availability of family labour, input supply issues and demands on their finances.

Next the farmer (with the support of the family) can plan; starting with their goals, then considering what livelihood strategies they could adopt, based on the diagnosis. To do this,

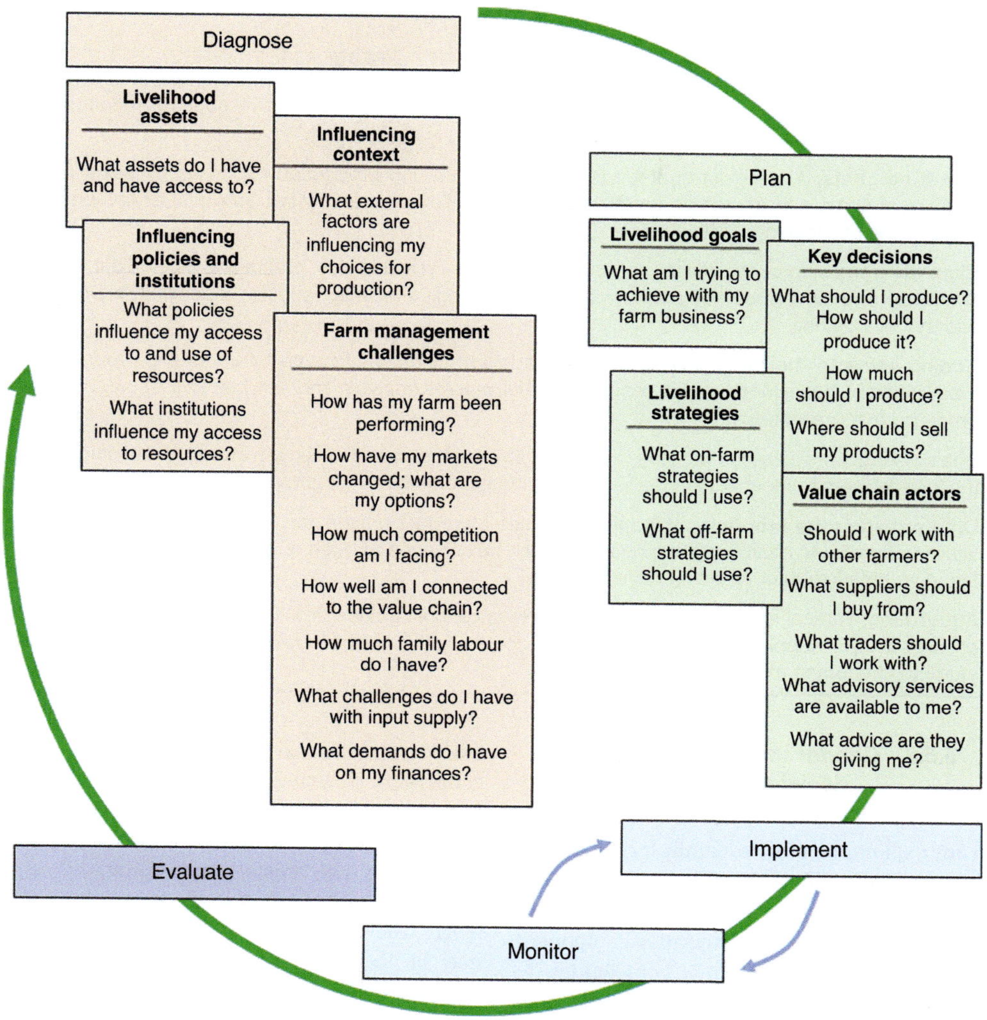

Fig. 5.7. A livelihoods farm management decision framework.

they can follow the systematic decision-making process discussed in Section 5.5.1 of this chapter. In the process of doing this, they consider the key decisions regarding:

- what to produce;
- how to produce it;
- how much to produce; and
- where to sell it.

An example of strategic goals is given in Box 5.2.

They must also consider the best use of family labour. Answers to these questions will guide the family in choosing their livelihood strategies both on and off the farm, such as:

- Large stock production.
- Small stock production.
- Rain-fed system.
- Irrigated system.
- Off-farm income options.

Of course, once the plan has been developed, the famer will implement, monitor and evaluate the plan as a part of good farm management practice. See Box 5.2. tells a story of better decision making.

Box 5.2. Examples of strategic goals.

Diversifying: Diversification involves increasing the number of products produced and sold by employing under-used resources and meeting local demand. It spreads income sources, reduces risk and brings new knowledge and skills to the farm.

Lowering costs: A low-cost strategy focuses on minimizing production expenses by using more efficient systems and affordable inputs. Lower costs allow farmers to sell products at competitive prices while maintaining profitability.

Expanding the size of the business: farm expansion increases sales, profits and assets, but is only effective if the farm is already profitable. Growth strategies include capacity expansion, replication and modernization.

Adding value to the farm enterprise: value addition increases income by improving products to meet different buyer needs. This requires careful market analysis and adjustments to production, processing or presentation.

Specializing: specialization focuses on a few enterprises, ensuring maximum attention and efficiency. It allows farmers to meet market demands with larger volumes and improved product quality.

Differentiating the product: differentiation makes a product unique and more appealing to buyers, allowing farmers to charge higher prices. This can be achieved through quality improvements, branding, direct marketing or targeting niche markets.

Integrating: vertical integration involves controlling multiple stages of the value chain, such as processing or input supply. Horizontal integration occurs when farms collaborate at the same production stage to strengthen market position.

5.5.3 Gender in farm management decision making

In any exploration of small family farm management in low-income countries, it is critical to address the issue of gender. As noted in Chapter 2, women comprise a substantial component of small family farming. Their contribution to family livelihood goals, agricultural production and food security cannot be overestimated. Yet their needs differ significantly from those of men, and their role in farm management decision making varies widely across low-income countries. In many cases, women have less access to productive assets and less power in decision making overall (see Box 5.3.).

While women and men equally possess the capability to apply modern farm management principles, gender plays an important role in practice – especially in terms of access to resources. Farm management, as a theoretical framework, is not affected by gender. However, farm decision making is shaped by gender roles and social norms, which influence decisions related to crop selection, resource use, marketing and investment. To be effective, management advice must be carefully tailored to address gender-specific issues.

Research consistently highlights the importance of considering women's roles in agriculture. This means recognizing not only their equal rights but also the particular challenges they face. For example, while women may be included in farm management training, such programmes often fail to account for the time constraints imposed on women by domestic responsibilities such as childcare – responsibilities rarely borne by men. These roles are complex and shaped by deep-seated gender norms. Even when decisions appear to be made jointly, imbalances can persist. The involvement of women in decisions related to production, marketing and the use of income varies widely, and these differences significantly affect farm planning and management. When men control cash crops, for instance, their preferences may dominate resource allocation, sidelining crops or activities important to women. Addressing such disparities is essential for promoting equity, improving productivity and achieving sustainable agricultural development.

Empowering women – through education, training and better access to credit and

Box 5.3. The Pérez family's path to better farm decisions through diagnosing and planning.

In the Andean highlands of Peru, the Pérez family – Carlos, his wife Rosa and their two children – faced a difficult decision. On their small farm, they had always relied on growing potatoes, and keeping a few guinea pigs for extra income, but erratic weather, rising costs and fluctuating market prices had made it harder to sustain their livelihood.

Determined to improve the future of their farm, Carlos and Rosa sat down one evening with a notebook and a simple diagram from an agricultural extension officer. It was a farm management decision framework – a tool to help them assess their situation, explore options and plan wisely.

Step 1: Diagnosis – understanding their resources and challenges
First, they reviewed their assets:

- Land: 2 ha of sloping land, with some areas suitable for irrigation.
- Labour: the family worked the farm together but the eldest son was considering taking a job in town.
- Market access: a small local market nearby but better prices were available in the city.
- Livestock: a few pigs and chickens for household use and occasional sales.

Next they considered external factors:

- The region had been experiencing unpredictable rains.
- Transport to larger markets was expensive.
- Input prices (fertilizers and seeds) had increased.
- Government extension workers had introduced new climate-resilient crop varieties.

Finally, they identified key challenges:

- Their potatoes sometimes yielded poorly due to erratic rainfall.
- The local market offered low prices, reducing profits.
- With their son possibly leaving, they would have less family labour.

Step 2: Planning – setting goals and choosing strategies
Carlos and Rosa discussed their long-term goals:

- Increase income stability.
- Reduce dependency on unreliable weather.
- Find ways to use their labour efficiently.

Using the framework, they explored possible strategies:

- What to produce? They considered switching to a mix of potatoes and quinoa, the latter being a crop that fetched good prices and was more drought resistant.
- How to produce it? They learned about conservation farming techniques to improve soil moisture retention.
- How much to produce? They decided to allocate 1 ha to potatoes and 1 ha to quinoa, balancing risk and income potential.
- Where to sell? Instead of relying only on the local market, they would cooperate with other farmers to share transport costs and sell in the larger city market.
- Labour decisions – with their son possibly leaving, they considered raising more guinea pigs, a high-value product requiring less labour than crop farming.

Moving Forward
With their plan in place, the Pérez family felt more confident about their farm's future. They had used a structured approach to make informed choices, balancing risk, sustainability and financial needs. With support from agricultural extension officers and their community, they were ready to take the next steps towards a more resilient and profitable farm business.

Box 5.4. Considering women. From Ngoma-Kasanda and Sichilima, 2016.

Socio-economic characteristics such as household assets, distance to the field, household headship, as well as cultural variations, have a significant influence on the decision making of women. Therefore, understanding the socio-economic characteristics of targeted beneficiaries is crucial for predicting how a market intervention will impact men and women.

Factoring in gender

Gender sensitive programming poses special challenges in agricultural interventions, and it is, therefore, imperative that empirical evidence from gender aware research is used in developing interventions. Understanding the factors that cause women's decision-making power to change in different crop sectors, at different levels of decision making, becomes critical in paving the way for gender sensitive programming and implementation.

resources – can significantly improve the performance of small family farms. This does not imply treating women's needs as separate from those of men but, rather, recognizing that the path to empowerment may differ. Effective programmes must reflect women's specific realities, aspirations and constraints. Encouraging collaboration between women and men can lead to better outcomes, including greater adoption of innovations, improved practices and stronger market participation. There is no one-size-fits-all solution to gender dynamics in farm decision making. The goal should be to foster a deeper understanding of women's roles, opportunities and constraints within each local context.

5.6 Strategic and Tactical Goals

Setting goals is an integral part of farm management. Goals provide direction and purpose, guiding the decision-making process and resource allocation. However, some goals are strategic while others are tactical. It is important that farmers understand the difference between strategic goals and tactical goals.

Strategic goals focus on long-term aspirations and the farm's vision. These goals typically span several years and focus on broad objectives such as market expansion, diversification, succession planning or sustainability initiatives. Strategic goals help farmers define the overall direction of their operations and shape their decision making. The farm-family's livelihood goals are usually strategic goals (see Box 5.4.).

On the other hand, *tactical goals* are short-term objectives supporting strategic goals. Tactical goals may include targets for crop yields, production efficiency, cost reduction or improvement to marketing strategies. These goals are specific, measurable, achievable, relevant and time-bound (SMART), allowing farmers to track progress and make necessary adjustments along the way. These goals are embedded in the farm plan as a part of the strategies chosen for the farm business.

By differentiating between strategic and tactical goals, farmers can align their decision-making with the broader vision of their farm while also focusing on immediate priorities and performance indicators. This strategic–tactical goal-setting approach ensures that farmers have a well-defined roadmap for their operations, enabling them to make informed decisions that support long-term viability and success.

5.7 Importance of Trade-offs and Balances

One important concept in farm management is understanding trade-offs and balances. Farmers as managers often face competing priorities and limited resources, requiring careful decisions for optimal outcomes. For example, balancing the need for short-term profitability with long-term sustainability, or considering the trade-off between maximizing production and minimizing environmental impact.

Trade-offs may also arise when allocating resources between different enterprises or when making decisions that impact the farm's overall strategic goals. Good farm management practices involve carefully weighing these trade-offs, considering the potential risks and benefits and making informed decisions that strike the right balance for the specific farm and its unique circumstances.

Furthermore, achieving trade-offs and balances requires a holistic approach, considering not only economic factors but also social and environmental dimensions. Sustainable farm management involves integrating all these aspects – economic viability, social well-being and environmental stewardship – whilst recognizing their interconnectedness. By understanding the importance of trade-offs and balances, farmers can make informed decisions that align with their farm's objectives, enhance productivity and profitability, and contribute to the long-term sustainability and resilience of their agricultural operations.

5.8 Ethical Considerations

Ethical considerations in small family farm management are essential for ensuring sustainable livelihoods, environmental stewardship and community well-being. Ethical farm management creates coherence among economic, social and environmental responsibilities. One of the key ethical concerns for smallholder farmers is the fair and responsible use of labour. Many small farms rely on family labour, seasonal workers or community-based labour-sharing systems, making it vital to ensure safe working conditions and equitable labour arrangements. While setting formal wages may not always be feasible, ensuring reasonable workloads, avoiding child labour in hazardous activities, and fostering fair treatment within families and communities are crucial ethical considerations.

Environmental sustainability is another major ethical responsibility. Small family farmers often depend directly on local natural resources for their livelihoods, making the sustainable use of soil, water and biodiversity essential for long-term productivity. Ethical farming in this context involves preventing soil degradation, practicing crop rotation, reducing reliance on harmful chemicals and adopting climate-smart techniques to safeguard the environment for future generations (Berger, 2019). These relationships are illustrated in Fig. 5.8.

While small family farmers may have limited influence over large-scale animal welfare policies, those involved in livestock farming should strive to provide adequate nutrition, shelter and humane treatment for their animals. Proper disease management and ethical breeding practices also contribute to the well-being of farm animals. Access to markets and fair-trade principles are also ethical considerations. Small family farmers often face unfair pricing, market exploitation and lack of bargaining power, making cooperative farming, transparent pricing and ethical trading relationships critical for improving their livelihoods. Strengthening collective action and farmer organizations can help ensure that smallholders receive fair compensation for their produce. These actions are elaborated on in Chapter 8. By prioritizing fair labour practices, environmental conservation, responsible animal care and equitable market participation, small family farmers can contribute to a more just, sustainable and resilient agricultural system.

5.9 'Good' Farm Management Practices

Good farm management practices are essential for achieving optimal results and long-term success. While specific practices may vary depending on the farm's context and goals, the following are some illustrative examples of key practices:

- **Responding to the market:** responding to market demand enables small family farmers to align their production with profitable opportunities, ensuring better prices and reducing the risk of surplus losses. By integrating market trends into decision making, family farmers can diversify crops and improve value addition, ultimately enhancing farm household income. This market-oriented approach not only strengthens financial stability but also promotes long-term sustainability by encouraging adaptability, innovation and strategic investments in farm operations.
- **Efficient resource management:** careful management of land, water, labour and

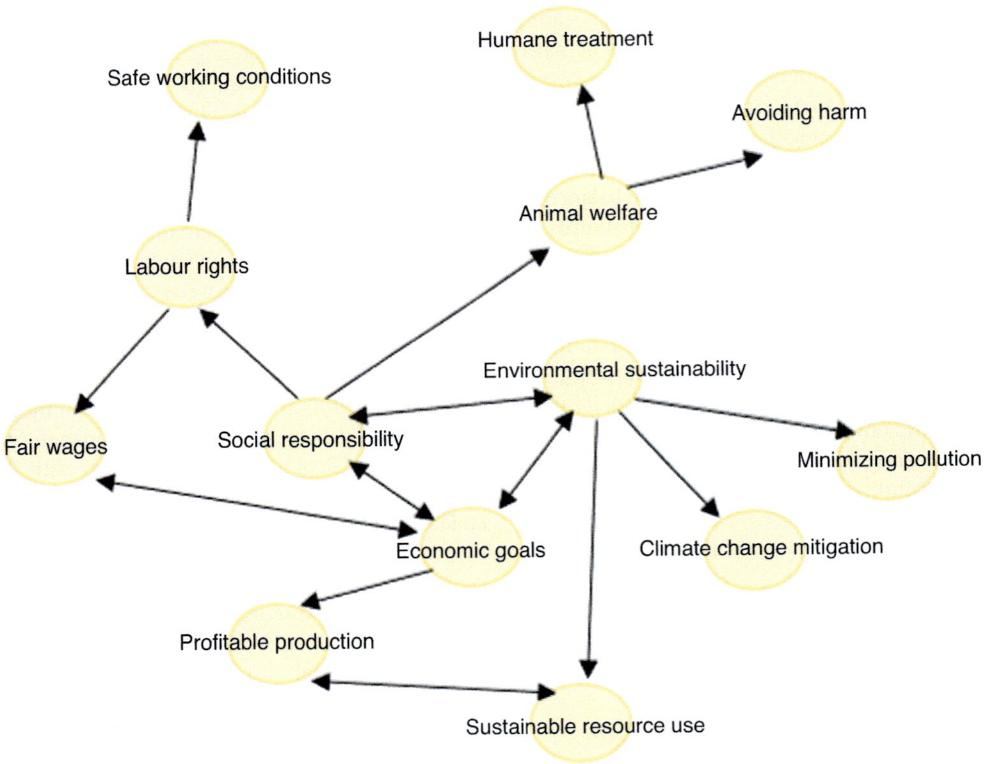

Fig. 5.8. Ethical considerations in farm management.

other resources to maximize productivity and minimize waste. This includes practices such as soil conservation, water management, integrated pest management and efficient use of machinery and labour.

- **Crop rotation and diversification:** rotating crops and diversifying production to improve soil health, reduce pest and disease pressures and enhance resilience to market fluctuations. Crop rotation can also optimize nutrient utilization and reduce the reliance on chemical inputs.
- **Record-keeping and data analysis:** maintaining accurate records of inputs, outputs, expenses and production activities to track performance, identify trends and make informed decisions. Analysing data can provide valuable insights into cost efficiency, yield variability market trends and other factors influencing farm profitability.
- **Risk management:** implementing strategies to identify, assess and manage risks associated with weather events, market fluctuations, pests, diseases and other uncertainties. This may include insurance coverage, diversification of crops or income streams, adoption of climate-smart practices and contingency planning.
- **Continuous learning and adaptation:** being informed about new technologies, research findings and best practices through training programmes, workshops and networking with other farmers facilitated by agricultural advisors. Being open to learning and adapting strategies based on changing circumstances can enhance farm management effectiveness.
- **Sustainable environmental management:** incorporating sustainable practices that promote biodiversity, soil conservation, water conservation and other measures

to protect the environment and natural resources for future generations.

In summary, implementing good farm management practices is crucial for optimal farm performance and ensuring long-term success. Through practices such as responding to market demands and changes, efficient resource management, crop rotation and diversification, record-keeping and data analysis, risk management, continuous learning, and environmental stewardship; farm managers can enhance productivity, profitability, income and sustainability. Additionally, understanding trade-offs and balances enables managers to make informed decisions that align with their farm's goals and values, considering economic, social and environmental factors all within an ethical framework.

5.10 Strengthening Support Services for Modern Farm Management

For small family farmers to succeed in managing their farms as modern, competitive enterprises, they must have access to a comprehensive set of support services. These include tangible services – such as input, finance, veterinary care and mechanization – as well as advisory and information services that guide farm-level decision-making. Equally important is the development of farmers' own management skills, enabling them to operate their farms more strategically and efficiently.

As we have seen, modern farm management requires a diverse range of competencies: decision making, planning, implementation, financial and risk management, monitoring and evaluation, marketing and human resource management. Building these skills allows farmers to optimize the use of limited resources, increase profitability, and achieve both short- and long-term objectives. As farming becomes increasingly integrated into value chains and exposed to market and climate risks, these capabilities are critical for long-term sustainability.

Access to timely, relevant and practical farm management advice is central to this process. Advisory services – whether public or private – provide smallholders with the tools, technical guidance and insights needed to respond to the challenges and opportunities facing smallholder farmers. These services are most effective when they are tailored to local contexts and delivered in ways that empower farmers to take control of their business decisions.

Strong collaboration between public and private sectors is essential to ensure that support services are available and aligned across the entire value chain – from input supply and production to processing and market access. Governments, agribusinesses, financial institutions, NGOs and farmer organizations all have a role to play. When they work together, they can create an enabling environment where smallholders receive coordinated support that builds their farm management capacities and integrates them into more profitable and sustainable agricultural systems. These relations will be further elaborated on in later chapters.

Inclusivity must be at the heart of support services for smallholder farmers. As noted previously, women in particular face significant barriers to accessing resources and training. For advisory services to be effective, they must recognize and respond to the roles women play in farm decision making within their specific local and cultural contexts. Despite their vital contributions, women are often underserved by advisory systems that are typically designed with male farmers in mind. This not only limits the effectiveness of these services but also risks reinforcing existing inequalities. One practical approach to closing this gap is the recruitment and training of more female extension workers. Female advisors are often better positioned to understand and address the diverse needs of women farmers – whether they are young women, mothers, heads of households or rural entrepreneurs (FAO, 2021). Ensuring that advisory services reflect and respond to these realities is essential for promoting inclusive, equitable and effective support for farm management.

Ultimately, for small family farms to succeed in today's evolving agricultural environment, investments must extend beyond services and technologies to include people. Strengthening advisory support and developing the management capacities of farmers is essential in building resilient, market-oriented and sustainable farming businesses.

5.11 Conclusion

Farm management is essential for smallholder farmers, as it provides the foundation for rational decision making and sustainable growth. Historically, assumptions that small farms were inherently efficient have limited the attention devoted to management practices. Today, however, it is crucial to move beyond these traditional assumptions towards modern approaches that explicitly strengthen the management capacities of farmers.

Effective farm management is not an inherent skill possessed by all farmers but, rather, a distinct set of competencies that must be actively developed and refined. Improving these management skills serves as a key entry point for enhancing farmer livelihoods. By equipping farmers with necessary tools and knowledge, they can become empowered to make informed decisions, optimize resource use and successfully navigate agricultural markets.

Furthermore, ethical considerations constitute a crucial aspect of farm management, ensuring that farming practices do not compromise the well-being of workers, animals or the environment. Farmers must balance their economic goals with social responsibility – addressing labour rights, fair wages and sustainable resource use – to promote long-term sustainability and foster a just food system.

Central to effective farm management is the application of the farm management cycle, which guides farmers in developing farm plans aligned with their strategic livelihood goals, as outlined in the livelihoods framework. This systematic approach supports investigative processes and informed decision making, enhancing the abilities of farmers to realize desired outcomes.

Entrepreneurship also plays a significant role in modern farm management, distinct from traditional management practices. While management focuses primarily on decision making and implementation, entrepreneurship involves innovation, risk-taking and the identification of new opportunities. Recognizing this distinction enables farmers to adopt a holistic approach that combines effective management practices with entrepreneurial thinking, thus driving sustainable growth.

Key managerial tasks – such as setting clear objectives, acquiring relevant external information, considering alternative resource uses, forward planning and monitoring performance – remain essential for sound decision making and effective implementation in farm management. Decision making itself is pivotal, with farmers continually addressing critical choices, such as what, how, and how much, to produce, where to market products, and how to achieve optimal pricing. Importantly, incorporating a gender-sensitive approach to decision making is vital, recognizing the significant role that gender plays within agricultural contexts, as well as their particular requirements and circumstances.

References

Berger, J. (2019) Farm management. In: Kaplan, D.M. (ed.) *Encyclopedia of Food and Agricultural Ethics*. Springer, Dordrecht, Netherlands. Available at: https://dokumen.pub/qdownload/encyclopedia-of-food-and-agricultural-ethics-2nd-ed-978-94-024-1178-2978-94-024-1179-9.html (accessed 7 February 2025).

Castle, E. N., Becker, M. H. and Nelson, G. L. (1987) *Farm business management: The decision-making process*, 3rd edn. Macmillan Publishing, New York City, New York. https://books.google.com/books/about/Farm_Business_Management.html?id=QwAbAQAAMAAJ

FAO (2013) *Climate-Smart Agriculture: Sourcebook*. FAO, Rome. Available at: https://www.fao.org/climate-smart-agriculture-sourcebook/en/ (accessed 3 February 2025).

Giles, T. and Stansfield, M. (1980) *The Farmer as Manager*. Oxford, Blackwell, UK.

Lloyd, D. (1968) A systematic approach to farm business analysis without accounts data. Study No.4 University of reading, Department of Agriculture.

Ngoma-Kasanda, E. and Sichilima, T. (2016) *Gender and Decision Making in Agriculture: A Case Study of the Smallholder Groundnuts Sector in Zambia*, Lusaka, Zambia. Musika Development Initiatives: https://beamexchange.org/resources/1165/January30

Schultz, T. (1964) *Transforming Traditional Agriculture*. Yale University Press. Available at: https://academic.oup.com/ej/article-abstract/74/296/996/5249777

Schultz, T.W. (1961) Investment in human capital. *The American Economic Review* 51(1), 1–17. Available at: https://www.jstor.org/stable/1818907 (accessed 7 February 2025).

Additional reading

Castle, E.N., Becker, M.H. and Nelson, G.L. (1987) Farm business management: The decision-making process, 3rd edn. Macmillan Publishing, New York City, New York.

FAO (2021) Making extension and advisory services work for women. FAO, Rome, Italy. Available at: https://openknowledge.fao.org/items/e12bbf92-fc19-4b85-80ac-fc25792d465a (accessed 30 September, 2025).

Johnson, D. T. (1982) The Business of Farming: A Guide to Farm Business Management in the Tropics. London, Macmillan.

Kay, R., Edwards, W. and Duffy, P. A. (1994) *Farm Management*. 3rd edn. McMillan Publishing, New York.

Nuthall, P. L. (2010) *Farm Business Management: The Human Factor*. Blackwell Publishing, Oxford.

Olsen, K. D. (2004) *Farm Management: Principles and strategies*. 2nd edn. McGraw-Hill, New York.

6 Essentials of Management

While technical expertise and improved production methods are crucial, it is the ability to make informed decisions and manage resources effectively that ultimately drive profitability and sustainability. This chapter advances the discussions from earlier sections of the book, taking them to a different level where management skills become central to decision making. It highlights how effective and efficient management is at the core of transforming small family farming into a viable business, ensuring long-term success in an increasingly complex agricultural environment.

The chapter begins by examining the essential functions of management that drive success in farm businesses. These include production management, ensuring the efficient use of inputs, technology and sustainable farming practices; marketing management, which involves identifying profitable markets, pricing strategies and value chain integration; and financial management, covering investment planning and cost control. Additionally, the chapter explores labour management, ensuring workforce efficiency and motivation; risk management, which helps mitigate uncertainties such as climate variability and price fluctuations; and organizational and information management, which support structured farm operations and data-driven decision making.

Beyond outlining these management tasks, the chapter emphasizes the need to strengthen management capacity among small family farmers. It discusses strategies for improving managerial skills, including training programmes, peer learning and the adoption of decision-support tools. Developing strong management skills enables farmers to optimize resource use, streamline production, and create effective marketing and financial strategies.

In today's rapidly evolving agricultural landscape, farmers face growing complexities such as rising input costs, climate challenges and changing market demands. Effective farm management has the potential to equip small family farmers to navigate these uncertainties, seize new opportunities, and build resilient, profitable businesses. By mastering these essential management principles, small family farmers can enhance their competitiveness, improve their livelihoods and contribute to a more sustainable agricultural sector.

6.1 Farm Management Tasks

The factors that affect profitability discussed in Chapter 5, relate directly to the business functions of the small family farm. It is in these spaces that the farmer must be aware of the costs and potential incomes. Decisions made in these management areas directly affect profitability.

All farms, regardless of size, share the same basic business management tasks. These include production management, marketing management, financial management, organizational management, labour management, information management and risk management (Fig. 6.1). These tasks are not carried out in isolation but in an integrated manner. Being part of the overall farm system, each of these business tasks is interconnected – what happens in one area, affects, and is affected by, what happens in other areas. For example, production is influenced by marketing, and marketing is influenced by production.

Further, each of the key management functions discussed in the previous chapter – planning, organizing, directing and controlling – are applied in each of the business tasks. Crop production, for

DOI: 10.1079/9781836990536.0006

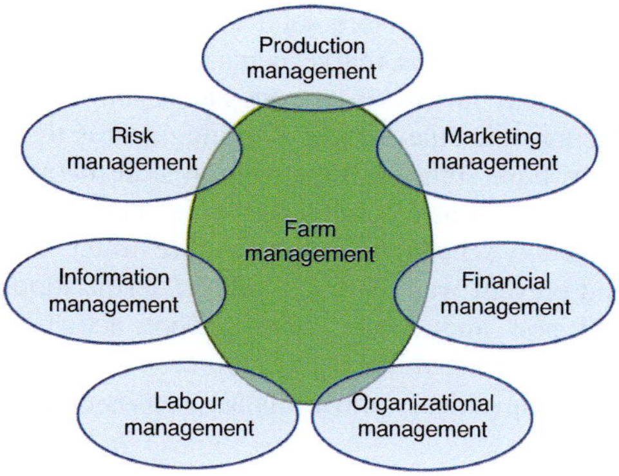

Fig. 6.1. Farm management tasks.

example, needs planning to decide which crops to grow, organizing to gather the inputs, directing to produce and market the crop and control to decide whether an insecticide application is needed. Likewise, financial management needs planning to decide how much credit is needed, organizing to obtain credit, directing to use the credit when needed and control to check that the correct credit terms have been used.

6.2 Production Management

Production management is the cornerstone of successful farm operations, encompassing the planning, coordination and execution of activities that transform inputs into agricultural outputs. For small family farmers, effective production management is essential to optimize resource use, maximize yields and ensure the sustainability of their farming systems. This section explores key decisions and practices in crop management, including cropping systems, production scheduling, input procurement and enterprise policies. By making informed choices in these areas, farmers can enhance productivity and potential profitability, reduce risks and align their operations with both market demands and environmental sustainability.

6.2.1 Crop management decisions

Effective crop management is essential for maximizing productivity and sustainability and running the farm as a business. To achieve optimal yields and maintain the health of their land, farmers must make well-informed decisions across various aspects of crop production. Each production decision – from crop selection to pest control – has financial implications that directly affect farm profitability, risk management and long-term sustainability. Farmers must choose crops that align with market demand, adapt well to local climate and soil conditions, and fit within their resource constraints. Strategic choices about planting times and crop rotations not only boost productivity but also preserve soil fertility, reducing the need for costly inputs over time.

Water management and integrated pest control are equally important business decisions. Efficient irrigation and targeted pest management reduce input costs, protect yields and help avoid losses due to drought or infestation. These approaches enhance productivity while lowering production risks, leading to more stable cash flows. Adopting innovative and sustainable farming practices is both an environmental and economic decision. Conservation agriculture, for instance, improves soil health and water retention, which can reduce input use and increase yields over the long term. Similarly, investing in climate-smart technologies helps farmers adapt to changing weather patterns, protecting their assets and securing future income. Similarly, integrating organic farming principles supports biodiversity, reduces environmental toxins and enhances soil fertility, making it a crucial component of sustainable agricultural systems. Together, these practices not only support the ecological balance but also enhance the long-term viability and profitability of farming operations.

Cropping systems

The cropping system is a sub-set of the overall farming system and refers to how crops are arranged. Farmers need to make decisions regarding the choice of cropping systems and the intensity of crop production based on various factors such as soil type, climate, market demand and resource availability. Different cropping systems, such as monoculture, crop rotation, intercropping and agroforestry, offer unique benefits and challenges. Farmers must assess the suitability of each cropping system to their specific circumstances and objectives.

Factors such as pest and disease management, nutrient cycling, water use efficiency and labour requirements should be considered when selecting the most appropriate cropping system. Additionally, crop rotation can help break pest and disease cycles, improve soil fertility and reduce weed pressure.

Cropping intensity

Cropping intensity refers to how frequently crops are grown on a particular piece of land within a given time period. Farmers need to determine the optimal cropping intensity based on factors such as soil fertility, market demand and the availability of resources. This decision affects the yield potential, risk of pest and disease outbreaks and soil nutrient depletion. By carefully managing cropping intensity, farmers can maintain soil health, minimize crop failure risks, and optimize the use of available resources.

Production scheduling

Production scheduling involves planning the timing and sequence of various production activities on the farm (Fig. 6.2.). Farmers need to make decisions regarding the scheduling of planting, irrigation, fertilization, pest and disease management and harvesting, to ensure optimal crop growth and minimize production risks. By considering factors such as crop maturity, climate conditions, market demand and resource availability, farmers can develop an effective production schedule. Timely planting ensures that crops have sufficient time to grow and mature under favourable conditions, while appropriate irrigation scheduling helps optimize water use and prevent water stress. Similarly, fertilization and pest and disease management activities need to be scheduled at the right time to ensure maximum efficacy and minimize crop losses.

Input procurement and selection

Input procurement and selection refer to the process of acquiring and choosing the necessary resources for crop production, including seeds, fertilizers, pesticides and machinery. Farmers need to make decisions regarding the procurement of inputs based on their quality, availability, cost and compatibility with their production systems, as well as considering social and environmental factors. Selecting high-quality inputs is crucial for ensuring optimal crop performance and minimizing potential negative impacts on the environment. Farmers should carefully evaluate the attributes of different input options, such as seed varieties with desired traits, fertilizers with balanced nutrient content and pesticides with low environmental

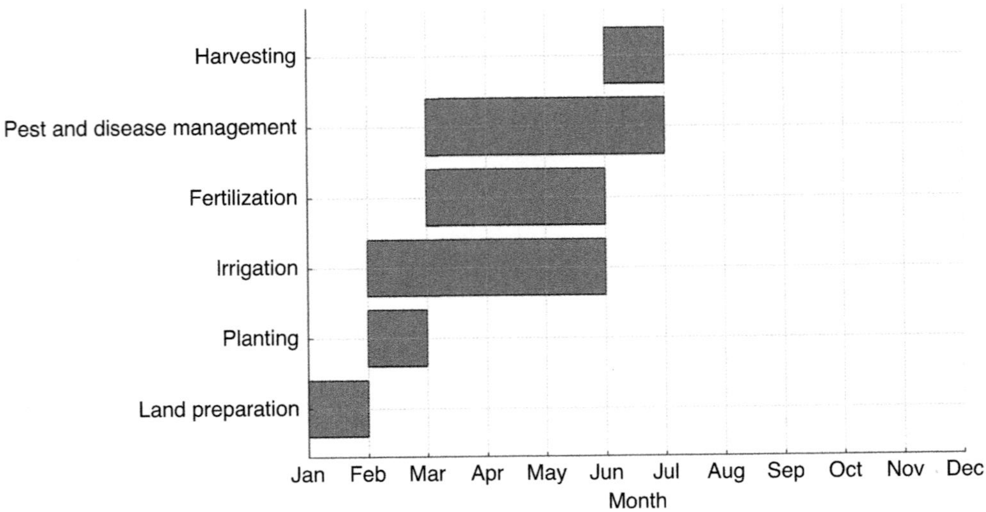

Fig. 6.2. Seasonal chart of key production activities.

toxicity. Additionally, considering factors such as input cost-effectiveness, reliability of suppliers and compatibility with organic farming practices is essential when making procurement decisions.

Crop enterprise policies

Crop enterprise policies encompass the strategic decisions and guidelines that govern crop production activities on the farm. Farmers need to establish policies regarding aspects such as crop selection, hectarage allocation, yield targets, quality standards and marketing strategies. Crop selection should be based on market demand, profitability and the suitability of different crops to the local growing conditions. Land area allocation decisions involve determining the appropriate distribution of land among different crops to optimize resource use and minimize risks. Yield targets and quality standards help set performance standards and ensure consistency in crop production. Finally, marketing strategies guide farmers in identifying target markets, building relationships with buyers and promoting their crops effectively.

6.2.2 Livestock management decisions

Effective livestock management is a critical component of operating a farm as a business. Each decision made in managing livestock – whether related to breed selection, nutrition, health or reproduction – directly affects productivity, profitability and long-term sustainability. Farmers must evaluate breed or species choices based on market demand, adaptability to local conditions, and their overall production goals. Selecting the right livestock ensures efficient resource use and better alignment with income-generating opportunities. Nutrition management is equally strategic. Developing cost-effective feeding plans, optimizing grazing systems and ensuring consistent feed availability are essential for maintaining animal productivity and minimizing input costs. Investing in disease prevention and control – through vaccination, regular monitoring and biosecurity protocols – helps avoid costly outbreaks and losses, ensuring the continuity of farm operations. These practices are not just about animal health but also risk management and cost containment. Reproduction management decisions, such as breeding schedules, artificial insemination and calving or farrowing planning, are essential for maintaining consistent production cycles and meeting market demand. They also improve herd genetics and productivity over time. Animal welfare practices, including adequate housing, hygiene and humane handling, support long-term productivity and market reputation. Healthy, well-cared-for animals perform better and attract premium prices, especially where consumer expectations around welfare are rising. By approaching livestock management through a business lens, farmers can optimize outputs, control costs, reduce risks and enhance the overall profitability and competitiveness of their farms.

Livestock systems

Livestock systems refer to the different ways in which livestock are managed and integrated into the farm business. Farmers need to consider various factors when deciding on the most suitable livestock system, such as available land, capital investment, labour resources, market demand and environmental considerations. Different livestock systems include extensive grazing, intensive confinement, rotational grazing and mixed farming systems. Each system has its own advantages and challenges in terms of resource use efficiency, environmental impact, labour requirements and profitability. Farmers must assess the compatibility of each system with their available resources, production objectives and the welfare of the animals involved.

Livestock enterprise policies

Livestock enterprise policies encompass the strategic decisions and guidelines that govern the management of livestock enterprises on the farm. Farmers need to establish policies regarding aspects such as herd or flock size, breed selection, breeding goals, feeding management, health management and marketing. Determining the optimal herd or flock size involves considering factors such as available land and resources, market demand and the carrying capacity of the farm. Breed selection decisions are based on the desired traits, adaptability to local conditions and market preferences. Breeding goals guide farmers in improving the genetic quality of their herds, focusing on traits such as growth rate, milk production or disease resistance. Feeding protocols ensure that animals receive balanced and appropriate nutrition for their respective stages of growth and production. Health management policies encompass vaccination programmes, regular check-ups and disease prevention measures. Finally, marketing strategies help farmers identify target markets, establish supply chains and promote their livestock products effectively.

6.3 Marketing Management

As noted in Chapter 4, market-oriented smallholder family farmers use input markets when they need to purchase inputs for production and output markets to sell their farm products. In this section, we will look more closely at marketing and output markets.

Marketing is a fundamental component of farm business management, ensuring that agricultural products reach the right consumers at the right time and price. Effective marketing helps farmers convert production into profit, emphasizing the need for a market-oriented approach. Understanding market demand, product positioning, pricing and distribution strategies enables farmers to increase their competitiveness and secure sustainable incomes.

A key aspect of marketing is the marketing chain, which describes the journey of a product from the farm to the final consumer. This chain includes activities such as assembling, sorting, packaging, storing, transporting, processing and selling. Each step adds value but also incurs costs, making efficiency in the marketing chain critical to farm profitability. Farmers who understand these processes can make informed decisions about where and how to sell their products for maximum returns.

Despite its importance, many smallholder farmers face common marketing problems that hinder their ability to access profitable markets. These challenges include inadequate market information, difficulty in reaching minimum quantity standards, seasonal supply fluctuations, and limited access to transport and storage facilities. Many farmers also lack the technical knowledge to meet market quality requirements, leading to price penalties or rejection of their produce. Addressing these challenges requires better access to information, improved infrastructure and cooperative approaches that enable farmers to collectively market their products.

To navigate these challenges, farmers must consider different marketing types and channels. Marketing options range from selling directly to consumers at farm gates and local markets to supplying wholesalers, retailers, processors and export markets. Each channel has distinct advantages and disadvantages. Direct sales offer higher prices but require strong customer relationships and logistical effort, while contract farming and cooperative marketing provide stability but may limit pricing flexibility. Choosing the right marketing channel depends on factors such as product type, volume, quality requirements and transportation costs.

Key marketing decisions involve pricing, market segmentation, branding and promotions. Smallholder farmers must determine whether to sell at prevailing market prices or adopt differentiated pricing based on quality, organic certification or niche markets. Additionally, marketing decisions include considerations such as storage timing to take advantage of seasonal price fluctuations and leveraging digital tools for price forecasting and customer engagement.

By developing a structured approach to marketing management, smallholder farmers can overcome barriers, increase their bargaining power and secure higher incomes. This section will further explore marketing in depth, providing strategies to help farmers make informed decisions that improve their market participation and long-term profitability.

Marketing management is discussed in greater detail in Section 6.11, this chapter, while market planning is explored further in Chapter 13.

6.4 Financial Management

Financial management is a critical aspect of farm management that involves making informed decisions about acquiring, allocating and utilizing financial resources. For small family farmers, financial management is often based on practical experience rather than formal strategies. However, even basic financial practices – such as estimating income, tracking expenses and managing cash flow – are essential for maintaining farm stability and ensuring long-term sustainability.

One of the most challenging management tasks is managing cash flow. Cash flow is the movement of money (cash) in and out of the farm, reflecting the cash income from sales and other sources, minus cash expenditures like input costs, labour and loan payments, over a specific period. Cash refers to money immediately available to pay expenses, while income is money earned from sales, regardless of when received. A farm can be profitable on paper but still struggle financially if it lacks cash to cover day-to-day costs. Also, a loan injects cash into the farm business, but the loan is not income for the farm.

Another major challenge smallholder farmers face is access to finance. Limited financial resources restrict their ability to invest in better seeds, fertilizers, irrigation systems and other productivity-enhancing technologies. Farmers also need funds for working capital, particularly to cover seasonal expenses before harvest, as well as for capital investments in equipment, storage facilities or livestock. Without access to credit, many farmers rely on informal lenders, often at high interest rates, making it harder to achieve financial security.

A fundamental part of financial management is capital structure, which determines the balance between self-financing and external credit. While using personal savings minimizes risk, many farmers require loans to expand operations or improve efficiency. Managing debt effectively is crucial, as high repayment burdens can threaten farm viability. Smallholder farmers must carefully evaluate loan terms, repayment schedules and potential returns before taking on debt.

Sources of finance and credit vary widely. Informal sources, such as family loans and community lending groups, provide flexibility but may offer limited amounts. Formal sources, including banks, cooperatives, microfinance institutions and government credit programmes, can provide structured financial support but often have strict eligibility criteria. Many small farmers struggle with collateral requirements, making it difficult to access traditional loans. Cooperative lending and group financing models can help overcome these barriers by pooling resources and spreading risk.

Effective financial decision making involves planning for both short- and long-term needs. Working capital management ensures farmers have enough liquidity to cover daily expenses, while investment decisions – such as purchasing new equipment or expanding production – must be based on profitability and long-term sustainability. Additionally, financial analysis helps farmers assess their operations by tracking income, expenses and debt levels. Even simple record-keeping, such as maintaining a ledger of income and costs, can significantly improve financial awareness and decision making.

Debt repayment capacity is another key consideration. Farmers must assess their ability to meet loan obligations without jeopardizing their farm's sustainability. Poor debt management can lead to financial distress, forcing farmers to sell assets or cut back on essential inputs, reducing productivity in the long run.

By improving financial literacy, accessing appropriate credit sources, and adopting sound financial practices, smallholder farmers can strengthen their resilience and enhance their farm's profitability. We will discuss financial management in greater detail in Section 6.12, this chapter.

6.5 Organizational Management

Organizational management refers to the overall management of the various farm production systems, financial and marketing systems. Organizational management involves decisions ensuring farm operations run smoothly and efficiently. It includes record-keeping, inventory management, time-management and other related operational activities, as described here:

- **Record-keeping**: record-keeping is essential to good organizational management. Farmers need records of their production activities, including input purchase and application, agricultural operations, yields, labour and all post-harvest activities. In addition, they need records related to marketing and sales, and a range of financial records including income and expenses. This topic will be further elaborated on in Chapter 10.

- **Inventory management**: farmers generally have three kinds of inventories: production inputs; equipment, machinery and infrastructure; and harvested products. It is vital for the farmer to keep track of what they have, where it is and the condition it is in. When the farm is small and its agricultural activities simple, inventory management may also be simple. But as the complexity of the farm and its activities grow, inventory management will also become more complex.
- **Time management**: time is one of the most limited of all the farmer's resources. Farmers need to be very skilled at making the best use of their own time and that of others. It means being able to determine the logical sequence of activities and to prioritize activities to ensure the most important and most critical activities are given the time they require, and that less important activities do not take up unnecessary time.

Building organizational capacity is key to ensuring the farm runs smoothly, coherently and profitably.

6.6 Labour Management

Labour management is a crucial aspect of farm management, particularly for small family farmers who rely primarily on family labour, supplemented by casual workers during peak periods such as planting and harvesting. Efficient use of family and hired labour is essential for productivity and sustainability in small farms.

6.6.1 Family labour dynamics

In small family farming, family labour is often the backbone of the operation. Small family farmers must assess the availability and skills of family members who can participate in various farm activities. This assessment includes understanding each family member's capabilities, willingness and the amount of time they can dedicate to farm tasks. Efficiently managing this labour pool requires clear communication about farm goals, individual responsibilities and the scheduling of tasks to align with the farm's needs. This ensures that all members are aware of their roles and contribute effectively, minimizing conflicts and the duplication of efforts.

6.6.2 Labour planning

Labour planning is an important component of managing a small family farm, especially when considering the fluctuations in labour demand throughout the year and the varied sources of labour available. Small family farmers need to carefully assess their annual labour needs, considering peak periods such as planting, weeding and harvesting, which require increased manpower. During these times, the limited labour available from the farm family often needs to be supplemented by hiring casual labour. This approach ensures that there are enough hands to handle the increased workload without overwhelming the regular workforce. The topic of labour planning is detailed in Chapter 13.

Another aspect of labour planning is giving specific attention to children and youth. Many countries are adopting stricter laws regulating child and youth labour. Further, small family farmers will need to consider both the education of their children and the possibility of them leaving the farm to pursue life and work away from the farm. Both dynamics affect labour planning, both short- and long-term.

6.6.3 Casual labour integration

During peak labour demands, small family farmers often need to supplement family labour with casual workers. This necessitates an understanding of the local labour market, including the availability, cost and reliability of hiring additional hands. Farmers need strategies to integrate casual labour with family efforts smoothly, ensuring that all workers are well-coordinated and productive. Proper training and clear instructions are vital to maximize the efficiency of temporary workers and ensure they meet the farm's standards. As with the case with children and youth, many governments are adopting stricter farm labour laws which will directly affect hiring and paying casual labour.

6.6.4 Managing workloads and conflict

Managing a mixed workforce of family and casual labour involves not only aligning workloads with agricultural timelines but also navigating interpersonal dynamics. Clear delineation of responsibilities helps in managing expectations and reducing potential conflicts. Regular family meetings to discuss farm operations can foster a collaborative atmosphere and help address any grievances that may arise, promoting a harmonious work environment. Conflict resolution mechanisms should be in place to handle disputes constructively without disrupting farm operations.

6.6.5 Fostering a collaborative work environment

Creating a collaborative and supportive work environment is essential for effective family labour management. Encouraging teamwork, mutual respect and open dialogue among family members helps in building strong relationships and promoting a positive work atmosphere. Regularly seeking input from family members and involving them in decision-making processes fosters a sense of ownership and shared responsibility, increasing their motivation and engagement in farm activities.

6.6.6 Conflict resolution

Managing family labour also involves recognizing and addressing potential conflicts that may arise in a family-run business. Conflicts can emerge due to differences in opinions, work styles or personal dynamics. It is important to establish mechanisms for resolving conflicts in a fair and constructive manner. Implementing strategies such as active listening, compromise, and seeking professional mediation, if needed, can help in maintaining harmonious relationships and ensuring smooth farm operations.

6.6.7 Work–life balance

It is crucial for small family farmers to maintain a healthy work-life balance. This challenge can be particularly acute in family-run farms where the boundaries between home and work are often blurred. Farmers should encourage time off and ensure that family members can engage in non-farm activities – including education and training – that are essential for mental health and overall well-being.

By focusing on these aspects, small family farmers can effectively manage their labour resources, leading to improved farm efficiency, increased productivity, and enhanced family and worker satisfaction. This holistic approach to labour management is essential for sustaining the viability of small family farms.

6.7 Information Management

All farm management decisions require information. Some is gathered on the farm such as yields, rainfall, soil conditions, etc. Other information is gathered off-farm including market information, weather forecasts, financial data, information about technological developments, information about potential pests and disease outbreaks, and information about activities and opportunities along the value chain. Reliable and relevant information is essential for making farm management decisions. As the farm becomes more market oriented, the farmer will require more and more varied information. Farmers need skills to find, gather, analyse and interpret information effectively. Analytical skills enable farmers to analyse complex information, evaluate various options and make informed decisions based on sound judgment (Kay *et al.*, 2020).

In addition, the integration of digital tools and technologies is transforming how farmers manage information, which will be discussed in detail in Chapter 10. By leveraging these technologies, farmers can enhance their decision-making processes and increase their resilience against unforeseen challenges such as climate variability or fluctuating market prices.

Furthermore, fostering a culture of collaboration and knowledge sharing among farmers, agricultural cooperatives and local organizations can significantly improve access to valuable information. By sharing insights and experiences, farmers can collectively address common issues, explore innovative practices and adapt to ever-evolving agricultural landscapes.

Educational programmes that focus on developing information management skills can empower small family farmers to utilize available resources effectively and sustainably, ultimately contributing to the long-term viability of their farms.

6.8 Risk Management

Previously discussed are some of the risks involved in farming. Managing risk is a critical function of farm management and requires taking management decisions to minimize their impact on farm operations and the desired outcomes. It involves analysing potential risks, developing strategies to manage and mitigate those risks, implementing risk management plans, and continuously monitoring and adjusting these plans as needed. The scope of risk management for small family farmers extends to all aspects of agricultural activities, including production, marketing, finance, human resources and the overall business environment. It encompasses activities such as identifying and evaluating climate-related risks, developing strategies to manage market price volatility, implementing financial planning and budgeting practices and ensuring the health and safety of the family farm workforce.

With the inherent risks in agricultural operations, farmers employ various strategies and risk management measures to protect their farm business stability and minimize potential losses. Risk can be categorized as production, marketing, financial, human and institutional. Production risk stems from uncertainty of factors that affect the quantity and quality of farm produce. Marketing risk exists because of the variability of product prices and the uncertainty of future market prices. Financial risk occurs when money is borrowed to finance the farm business. Institutional risk occurs because of unpredictable changes in the provision of services. And human risk refers to the risks to the farm business caused by human illness and the personal situation of the farm family.

Strategies have been developed to cope with all types of risks. Often farmers experience the threat of different risks occurring at the same time. Risk reducing strategies are often used in combination with one another, because no single strategy can cover all of the risk likely to be encountered. Farmers need to consider the risks simultaneously and to develop an integrated approach for better management. They need to recognize the advantages and disadvantages of each risk management option, both individually and in combination. Individual farmers should select an appropriate strategy based on their goals, attitudes towards risk and their personal and financial situations. Table 6.1 identifies some common sources of risk and possible risk management strategies.

All the strategies discussed are aimed at generating greater security for the farmer. However, the farmer must determine if the benefits gained outweigh the direct or implied cost of the strategy. Making this determination involves a number of steps:

1. Identify the most appropriate strategy.
2. Determine the degree to which risk may be reduced (i.e. the benefit).
3. Identify the cost of implementing the strategy.
4. Weigh up the costs and benefits.

Then the farmer is in a position to decide whether to implement the strategy.

Farmers should develop a broad range of strategies that consider the advantages and disadvantages (benefits and costs) of each risk management option individually and in combination. Key questions that can help a farmer make these decisions:

- What risks does the farmer face? What is the likelihood that these unfavourable events will occur? What are the consequences of these risks?
- What risk management strategies are available to the farmer?
- What effect do the risk management strategies have on the events or consequences faced?
- What are the economic benefits and costs of the options?
- How do these benefits and costs vary for each of the farm enterprises?
- How do the best risk-reducing options fit together?
- How do they affect one another in terms of costs and benefits, and in terms of generating new risks?

Table 6.1. Sources of risk and possible management strategies.

Sources of risk	Possible management strategies
Production and technical risk	Risk-reducing inputs (e.g. drought resistant varieties)
	Risk-reducing technologies (e.g. irrigation)
	Selecting low-risk activities (e.g. low risk crops and livestock)
	Farming system flexibility
	Crop and livestock diversification
	Climate-smart farming practices
	Growing the same crop in different spatial locations
	Staggered planting
	Generating income from off-farm activities
	Reserves of inputs and produce
	Share leases
	Custom farming
Marketing or price risk	Spreading sales
	Direct sales
	Contractual agreements to sell produce and buy inputs
	Forward pricing
	Market price information
	Value-adding and market differentiation
	Selling and leasing assets
	Credit
Financial risk	Holding liquid assets
	Managing the phasing of investments
	Contingencies
	Insurance
Institutional risk	Traditional institutions and social arrangements
	Producer groups
	Cooperatives
	Government policies and support programmes
Human and personal risk	Labour planning
	Labour productivity
	Training and skill development
	Succession planning and farm labour management
	Health and nutrition

When farmers explore these and similar questions, they will be in a better position to decide on the range of risk management strategies that are most effective for their farm. These strategies should consider their household and farm goals, their attitude towards risk, and their unique family, household and farm situations (see Box 6.1.).

Box 6.1. An example of combining risk management strategies.

For much of his life, farmer Miguel has been growing food for his own family consumption. At times he has managed to produce surpluses of grain which he sold in the local market. The income from these sales has slowly built up, allowing him to diversify into a higher value farm enterprise: egg production.

Owing to poor rains last year, the maize crop failed and some of his chickens died after contracting Newcastle disease. Farmer Miguel was worried about what to do over the next season. He needed to take decisions that would not be risky and threaten the family's food security. Farmer Miguel decided to use drought resistant seed for his maize crop. He also decided to vaccinate his remaining layers (risk reducing inputs: technologies). The farmer was also concerned about the price he was likely to receive from the sale of maize at harvest time. He also enquired about the kind of prices he could get for his eggs in different markets (market price information). He decided to sell some of his maize in the local market (direct sales) but negotiated an agreement with a local trader who he knew well (building trust) for a fixed price at the next harvest season (forward pricing). He also learned that some farmers had formed a producer group for egg marketing and were supplying local schools in the area. The schools were also paying a fixed price with a written agreement drawn up (contract pricing) for the eggs bought. Farmer Miguel also discovered that the group provided its members with veterinary services and group members assisted each other at peak times in collecting eggs. In this way he spread his marketing risk.

6.9 More About Marketing Management

As suggested in our initial discussion on marketing management, a key premise of farm business management is that to increase the chance of making a profit on the farm, farmers need to orient their thinking towards the market. Market orientation implies understanding the demand for goods and products and making farm business decisions according to the conditions of this demand. Marketing is an important part of any successful farm business. It connects what the buyer wants with what producers (farmers) can produce and supply at a profit. Marketing can identify new demands and guide farmers towards satisfying this demand.

6.9.1 The basics of marketing

Marketing tells the farmer what to produce and how to make products and services available to buyers in the most desirable and efficient way. Through linking production with marketing, small-scale farmers can learn what adjustments they must make in their production system to better meet buyer demand. Buyers can be final consumers buying for their own personal or family consumption or, alternatively, for further processing or reselling (processors, wholesalers and retailers).

Marketing also includes the different services involved in moving produce from the farm to the point of consumption. The services involved consist of:

- assembling products in small quantities from a large number of scattered producers;
- sorting products into various sizes, degree of ripeness and other similar characteristics;
- packaging to make handling easy and prevent loss due to bruising;
- transporting products to consumers;
- storing products after harvest for future consumption;
- selling and buying when products are bought and sold and change ownership several times;
- processing produce to form new products such as butter, bread, jam and crisps; and
- retailing, or distributing produce to several retailers who in turn sell to the consumers.

With an effective marketing strategy, farmers can ensure continuing profits. For marketing to be successful, all participants in the process must make a profit. Farmers will only grow crops and raise livestock beyond their immediate needs and provide goods for sale when they think they can earn a profit in doing so. The same is true of the traders, processors and distributors.

Three important elements of marketing include:

1. **The priority of the buyer**: marketing begins not with the product, but with what the buyer wants. This requires some market investigation. There is no point in producing something that buyers do not want. Farmers need to understand what buyers really want.
2. **A process of selection**: the farmer should decide to whom to sell. Then the product must be geared to that buyer(s). The buyer could be a wholesaler, a retailer or a final consumer in the market. This element of marketing determines how and where the product is sold. One consideration is, of course, the cost involved with each option and the need to make a profit.
3. **Relationship building**: good marketing includes working with the buyer, whether they are a trader or final consumer. Buyers should not be cheated. Strong relationships need to be built up with the buyer. This means that agreements will be honoured and commitments to supply products will be kept. It means delivering consistent qualities. Trust between the farmer and buyer must be built and nurtured.

6.9.2 Reaching more lucrative markets

Most farmers are price-takers. They produce and sell in a competitive market with many buyers and sellers and have very little influence on the price. They can sell as much as they want at the market price. Their marketing challenge is to cope with fluctuations in supply and demand during the year. In such an environment, successful marketing implies selling as much as possible at times when higher prices are available. Strategies for reaching this goal may include investing in crop storage or greenhouses or establishing pre-purchase agreements with certain buyers. In some, very exceptional cases, a farmer may be able to set prices freely, when offering products with very little competition. In economics, these are called price-makers. However, there are clearly limits to the prices that can be established, as consumers will look for substitutes for a given product if the prices are outside of their reach or willingness to pay.

An option for improving income through increased prices is for farmers to produce differentiated products which can be sold in markets that pay a price premium for characteristics such as quality, appearance, packaging or specific favourable production techniques. In these cases, the marketing challenge is to get a higher price by making products different from other similar products. Competing in differentiated or niche markets requires significantly more management capacity, and often requires joining with other farm households to supply enough products to attract buyers.

Many small-scale farmers in low-income countries experience limitations when attempting to reach higher-value markets. Some of the common problems found in these countries are:

- difficulty reaching minimum quantity standards due to small-scale production;
- difficulty providing a consistent supply due to inherent production risk, seasonality of production and limited farm size;
- lack of marketing knowledge and access to marketing information;
- poor, or non-existent, infrastructure facilities; and
- limited investment capacity.

Difficulty reaching minimum quantity standards due to small-scale production

Small family farms often have limited production volumes. Limited volume makes it difficult to cover marketing costs for reaching more distant markets. In addition, many more lucrative markets require minimum quantities that an individual farmer or farm household cannot meet. As a result, many farmers end up selling their produce to traders who come to the farm gate or in local markets.

Difficulty providing a consistent supply due to inherent production risk, seasonality and limited farm size

Many market opportunities require that farmers can provide a steady supply of a product throughout the year. However, this is difficult for small family farmers who are often dependent on rain-fed agriculture, and thus on weather cycles. Limited farm size also restricts farmers' capacity to provide a continual supply, as do production losses.

Lack of marketing knowledge and access to marketing information

Many farmers have very little exposure to competitive markets and marketing. Therefore, they are unaware of the needs and requirements of these markets and the role they play in determining price. This can be seen by the lack of appreciation of the quality, size, appearance and packaging of products which play a major role in the market, in addition to a lack of understanding of the vulnerability of supply and demand situations, and of the costs and risks involved in marketing the produce.

The availability of reliable information about market demand, desired product characteristics and alternative distribution channels is very important. Lack of information restricts the small farmer's ability to maximize income. Smallholder farmers commonly do as their neighbours do because they have no other source of information. This behaviour frequently results in oversupply of a product and low returns to producers. Price and market information are sometimes provided by traders who come to the village to buy produce, or by government market information services.

Farmers need information from the market that will allow them to make the appropriate changes in farming activities. Without proper information, farmers probably cannot adjust production or the product to make use of marketing opportunities.

Poor storage and transportation infrastructure

Traditionally, small farmers have produced basic foods in cycles dictated by weather conditions. This has led to production cycles in which large amounts of production at certain times of year lead to lower prices. Lack of storage facilities leads to farmers selling their produce at harvest time when prices are lowest. In addition, poor road systems make transportation to markets difficult. Farmers in marginal and remote areas of a country may be restricted by transport options and costs because of a lack of economies of scale. Transportation costs are often high, but they can be reduced with improved roads and by efficient collection and distribution systems. Lack of markets and/or processing facilities deprive the farmers of a good return for their produce.

Limited investment capacity

Many farms are limited in their capacity to generate a cash surplus that can be ploughed back into farming as an investment. When there is little cash income, the farmer's investment capacity is limited. Farmers often need credit to diversify production towards higher value crops and livestock products.

6.9.3 The marketing chain

The different steps involved in moving produce from the farm to the consumer is generally called the *marketing chain*. It comprises the post-harvest elements of the value chain. All transfers of produce involve marketing activities in some form or other. All activities involve costs. At the simplest level, the cost involved may just be the time taken by the farmer to walk to a nearby market and stay there until all their vegetables are sold. At the most complex level, a product may be stored for long periods, transported long distances and processed several times before reaching the form in which it is finally sold.

Why is the price of a product in a shop or retail market often much higher than the price paid to the farmer? The costs involved with marketing are not always fully understood. We can understand that traders or processors spend money on transport or packaging, but there are many other less obvious costs. Because these costs are not always visible, those people doing the marketing are often accused of making too much profit. People look at prices paid to the farmers by traders and compare them with the prices consumers pay for the same product and assume that the farmers and consumers are being exploited. Sometimes, of course, traders do make very high profits but many times they make small profits or even losses. Clearly, unless they make an overall profit, traders will not want to continue in business. This is bad for both consumers and farmers.

Generally, the more complex and lengthier the marketing chain, the higher the marketing costs. If a farmer lives 5 km from a market they will normally receive a higher share of the final price than one who lives 20 km away, because of lower transport costs. A producer of a perishable crop, such as tomatoes, is likely to receive a lower share of the final price than the producer of a non-perishable crop, such as pumpkins, because some of their crop may be rotten by the time it reaches market. In comparing farmer and consumer prices, we need to be fully aware of all the costs involved.

Different types of costs that can be involved in the marketing channel are:

- **Product preparation and packaging costs:** consist of two components – (i) the movement of produce to the farm gate or house; and (ii) all the costs associated with packaging.
  While packaging is a major cost, the costs of trying to save money on packaging can be much greater. Poor quality packaging may increase losses due to product damage. It may also make the product less attractive to the buyer, reducing the price that he or she is prepared to pay.
- **Handling costs:** at all stages in the marketing chain, produce will have to be packed and unpacked, loaded and unloaded, put into store and taken out again. Each individual handling cost will not amount to much, but the sum total of all such handling costs can be substantial.
- **Transport costs:** once packed, the produce is then transported. This transport cost could include produce transported on boats, trucks, personal mobility vehicles, etc. Sometimes transport costs are a simple matter to calculate because the farmer or trader pays a set price per kilo to the transporters. Other times produce is carried on a 'per container' basis or farmers/traders may hire a complete truck and transport a variety of crops. It does become difficult to calculate a trader's actual transport costs if they own their own vehicles. It is best to calculate the cost based on how much it would be to rent a similar vehicle.

6.9.4 Common marketing problems

Market information is ranked as the most important problem in marketing. The lack of current price information causes problems for farmers who want to market some of their production. It makes it difficult to develop sensible farm plans. As a result, there are frequent surpluses on the market; these surpluses decrease prices.

When farmers are questioned, they usually list the following concerns:

- they lack market information;
- there is a lack of expertise and information and a shortage of extension officers to convey information;
- farmers produce low volumes and often have quality problems with their produce as well, leading to poor returns;
- there is minimal government support;
- there is market flooding (over-supply);
- there is a lack of local marketing outlet infrastructure, e.g. a lack of roadside stalls;
- they lack the technical know-how on packaging and grading;
- there is an inconsistent supply of farm products;
- there is little contact between producers and buyers;
- there is a lack of transport, or the farmers are unable to afford the fees charged and/or cannot obtain financing to pay the fees;
- they lack access to storage facilities and pack-houses;
- prices of equivalent imported items are low; and
- they experience ongoing discrimination in some marketing channels.

6.9.5 Marketing types and channels

Market types

Farmers are situated at different distances from the market. The nearest market is the local market, and this is usually in the vicinity of the farm. If the farmer sells to the trader, the market can even come to the farmer. The farmer may sell produce at the side of the road to travelling customers. This is another local market. Sometimes farmers choose to travel further away and sell their produce in the nearest town. This is also a local market. Some farmers may transport their produce to the capital or other larger cities depending on domestic demand and transport costs. The higher value and often more specialized produce can be sold to supermarkets, hotels and catering companies. Some produce might be sold to processing plants and transformed into canned produce and juices. Alternatively produce can be exported to international markets. These are some of the market options that farmers have.

But few farmers can sell their own products in large city markets, unless they live close to them. Even fewer sell their produce on international markets. These markets are too far away. An individual farmer has no way of contacting buyers in those markets. They may not have the means of transporting produce there. They often do not have the know-how or the facilities for all the operations that are needed. The volume of production is often not large enough to justify performing post-harvest operations; it would be inefficient to do so.

Most small family farmers must sell their products at the farm or in the local market. Their incentive to produce commodities for sale rather than only for their own subsistence therefore depends upon the prices they can get locally. These prices depend partly on the efficiency of the marketing system linking local markets to those in the city.

Marketing channels

A number of marketing channels exist for farmers. They can:

- market directly from their farms to the surrounding communities;
- supply processing units;
- directly supply various retail outlets, such as supermarkets and hotels;
- market through farm or market stalls;
- sell through contracts to commercial farmers or processors; and
- add value to their own produce and market products through the various marketing channels mentioned above.

Several markets and marketing channels are described in Table 6.2.

6.9.6 Marketing decisions

Marketing decisions in farm management are essential for positioning agricultural products effectively and achieving success in the marketplace. Farmers must carefully consider various factors to make informed choices that align with their business goals and customer preferences. Following are some key aspects of marketing decisions in farm management (Kotler and Armstrong, 2010):

- **Marketing channels**: farmers need to select the most appropriate marketing channels to reach their target customers, as previously discussed. Understanding the advantages and disadvantages of different channels, as well as the preferences of the target market, helps farmers determine the most effective way to distribute products.
- **Pricing strategies**: farmers need to determine appropriate pricing strategies that align with market conditions, production costs and customer perception of value. This may involve competitive pricing, value-based pricing or other pricing approaches based on market research and cost analysis.
- **Product promotions**: effective product promotions help create awareness, generate interest and stimulate demand for agricultural products. Farmers should develop compelling promotional strategies that highlight the unique features, quality and benefits of their products. This may include advertising, public relations activities, social media campaigns, product demonstrations or participation in trade shows and events. By effectively communicating the value proposition of their products, farmers can attract and retain customers.
- **Product differentiation and branding**: farmers can differentiate their products from competitors by emphasizing unique attributes, such as organic certification, sustainable production practices, superior quality or local origin. Creating a strong brand image and identity helps build customer loyalty and recognition. Farmers should carefully develop branding strategies that align with their target market and effectively communicate the value and values associated with their products.

Table 6.2. Types of markets and marketing channels.

Farm-gate marketing

Farm-gate marketing is where consumers come to the farm to buy produce. This may be retail, where neighbours or others come to the farm to purchase vegetables, eggs or poultry, for example. It could also be wholesale where products are sold in bulk to an intermediary who then sells to final consumers, processors or other actors in the chain. Wholesale prices are usually lower than retail prices

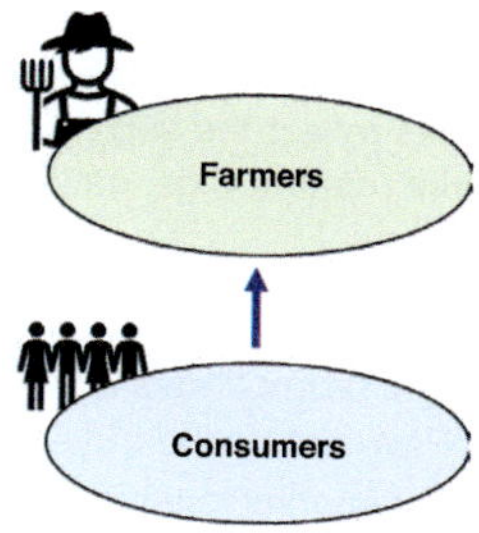

Advantages

- There are no transport costs
- Produce can be sold by the farmer him or herself, thus costs are reduced even if prices may be lower
- This is better suited to the small-scale farmer

Disadvantages

- The farmer must accept the local price for their produce
- The farmer may not be well located to sell the product

Village marketing with farm stalls

Village marketing entails the farmers in a village community organizing to sell their produce at a farm stall located in the village or along a main road. Farmers may sell as individuals, or they may group together to sell collectively. Unlike farm-gate marketing, the farmers take their products closer to the customer

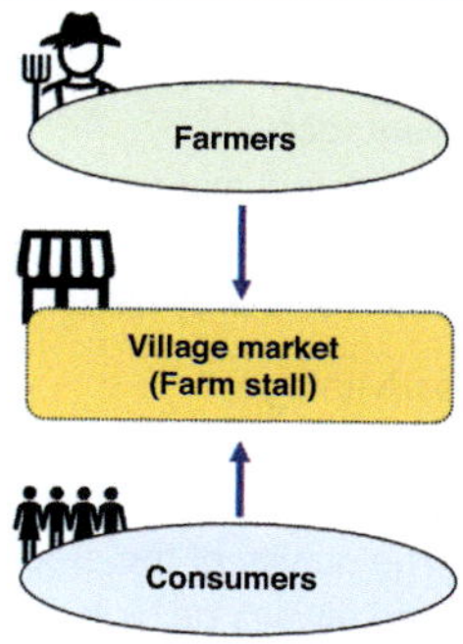

Advantages

- Larger markets can be exploited
- Farmers can take advantage of more favourable prices
- Price fluctuations are generally small

Disadvantages

- Transport of the produce may be difficult
- The quality of the produce may have to be higher as the consumer in the market may be fussier
- A constant supply of produce must be available to satisfy the needs of the market

Produce markets

Produce markets are set up in larger centres mainly for the sale of vegetables and fruit. These markets usually seek larger quantities of specific grades of produce. Farmers can use agents at the market to sell their produce

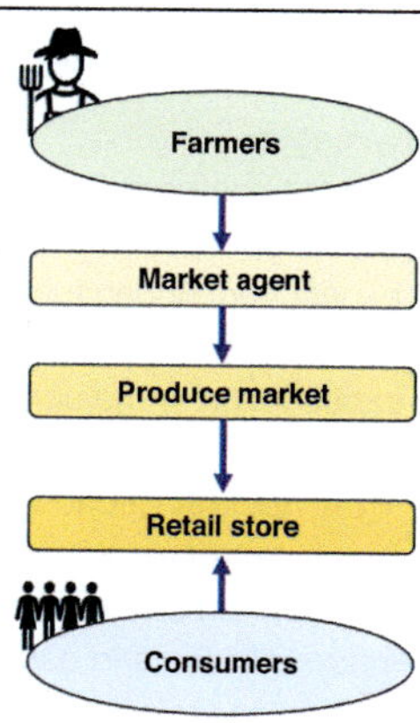

Advantages

- Farmers and farmer groups can take advantage of higher prices in times of short supply, if they have produce available
- The market is able to sell large quantities of farmers' produce
- The farmers can employ the services of an agent to perform the task of marketing

Disadvantages

- Quality, packaging and presentation are very important, and produce must conform to accepted grade and packaging standards
- The farmers need to be confident that they can cover the higher marketing costs, including the agent's commission

Continued

Table 6.2. Continued

Stock sales

Stock sales are an auction system where the sellers offer animals for sale and buyers offer a price for the animals. The seller may decide whether to accept the price offered by the buyer. The prices are not fixed. Generally, prices reflect the supply and demand position both locally and within the entire market. Pigs, cattle, goats and other animals are commonly marketed this way

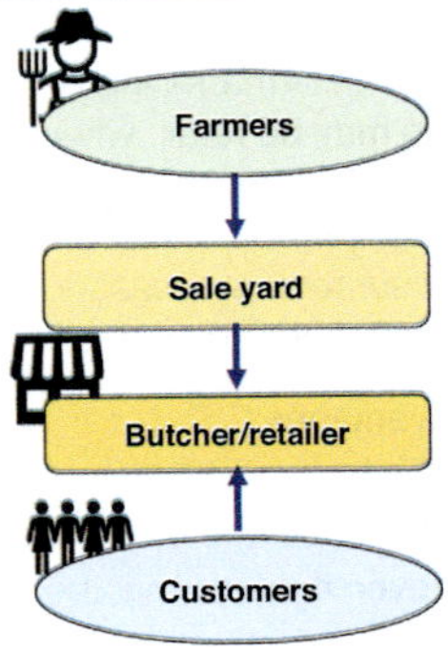

Advantages

- The promotion is done on behalf of the farmer
- Payment by the buyer is guaranteed
- The market is larger than the local market
- Small farmers have access to these sales

Disadvantages

- The seller may not get the price they want for the animal
- Prices may be lower than the 'market price'

Contract marketing

With contact marketing, the farmer sells directly to the retailer at prices, quantities and qualities agreed to in advance.

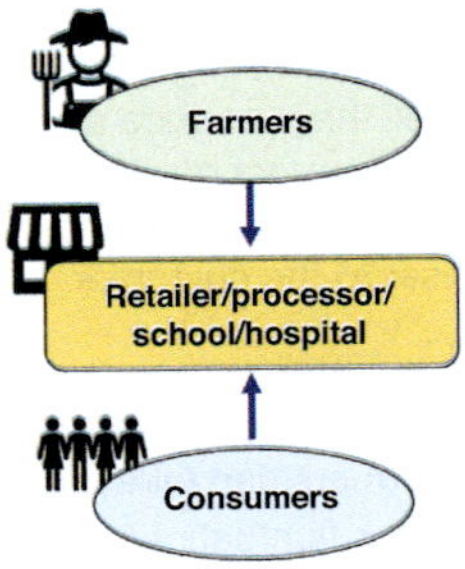

Advantages

- Marketing margins could be reduced and thus the producer could obtain a higher price for the product
- The volume of sales is guaranteed to the farmers

Disadvantages

- The farmers must ensure that they have sufficient produce of acceptable quality to supply the customer/retailer at all times
- The quality of the produce must meet the specific terms of the contract and always be high
- If the farmers cannot meet the needs of the retailer, the farmers will have to buy-in produce to make up the order of quantities required

Community-supported marketing (CSM)

In this form of marketing, a farmer, or group of farmers supported by a neighbourhood or community, agrees to provide produce at a set price or a proportion of yield based on the degree to which the community have supported their costs of production

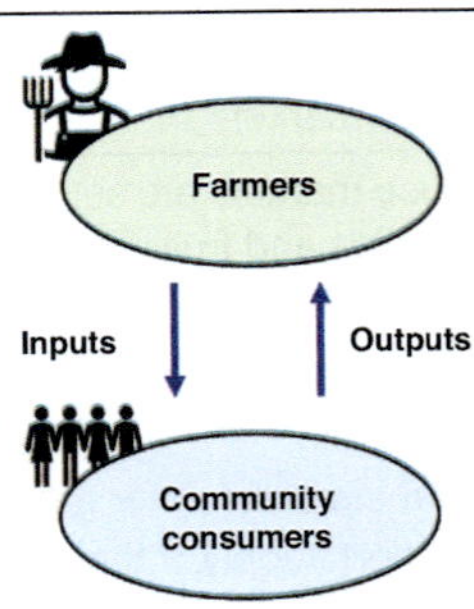

Advantages

- CSM strengthens the relationship between producers and consumers. Consumers often value knowing the source of their products, and producers benefit from understanding the needs and preferences of their customers
- Since consumers typically commit to purchasing in advance (often through subscription or membership fees), producers gain financial stability
- Consumers who invest in community-supported systems are often more loyal, not only because of their financial commitment but also because of the trust and relationship built with the producer
- CSM systems typically offer fair and transparent pricing. Producers avoid intermediaries and earn a higher share of the final price
- CSM tends to support local, organic or sustainable practices, as consumers are often motivated by environmental and social values

Continued

Table 6.2. Continued

Community-supported marketing (CSM)

- By directly engaging with consumers through a subscription or community model, producers can lower their marketing expenses
- Producers can get direct feedback from their consumers, allowing them to improve product quality and even introduce new offerings that cater specifically to the community's preferences and needs

Disadvantages

- Once a producer commits to supplying a certain amount of goods to the community, they are under pressure to meet those obligations
- In agricultural contexts, CSM often passes on some of the risks of production failure (e.g. crop loss) to consumers, who may not receive the expected quantity or quality of goods in poor seasons
- Consumers may have limited control over what they receive in CSM programmes. This can lead to dissatisfaction if expectations are not managed
- Coordinating the logistics of regularly distributing products to consumers can be a challenge, especially for small producers
- CSM often relies on local consumers, which can limit the size of the market. Producers who want to expand beyond local communities may face challenges scaling their operations while maintaining the community-oriented nature of the system
- Establishing a CSM system requires an initial investment in terms of time and resources. Producers may need to develop new ways of managing subscriptions, payments and logistics
- CSM relies on good communication between producers and consumers. Producers must consistently provide updates and manage expectations, especially when there are changes in production or delivery schedules
- CSM models work best in local or regional contexts, and it can be challenging to scale them to larger or more geographically dispersed markets

Collective marketing

Farmers may choose to market collectively. Farmers' associations may get together and jointly market their crop. This form of marketing is one of the basic functions of a cooperative. Products sold through collective marketing can be sold at a village farm stall market or a produce market, or even through contract marketing if the farmer group is formally organized

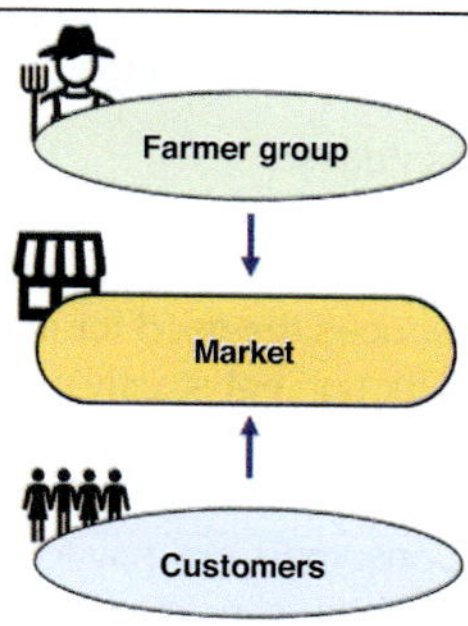

Advantages

- Producers can negotiate better prices and terms with buyers, as they can offer larger quantities of goods, making them more attractive to bulk buyers or retailers
- Pooling resources allows for shared costs in marketing, transportation, packaging and even production inputs. This can lower individual costs and improve profitability
- Collectively, producers can reach larger markets that may be inaccessible to smaller-scale individual farmers. This can include international markets or larger retail chains
- Collectives can standardize the quality of products through shared quality controls, ensuring uniformity, which is often required by buyers. This can enhance the reputation of the collective's products

Continued

Table 6.2. Continued

Collective marketing

- By working together, farmers can spread the risks associated with production, marketing and pricing, reducing the vulnerability of individual members to market volatility
- Collective marketing initiatives often attract government or donor support for training, capacity building and infrastructure, which can enhance productivity and market linkages

Disadvantages

- Managing a collective requires effective coordination and communication among members. Disputes over pricing, profit sharing or quality standards can arise, and not all members may equally participate
- While some costs may decrease through economies of scale, organizing and maintaining a collective can incur significant transaction costs, such as administrative expenses and time invested in meetings and decision making
- Poor leadership or lack of transparency can lead to mistrust within the collective. Weak governance can result in mismanagement of funds or unequal benefit distribution among members
- Individual producers may have to conform to collective decisions on pricing, marketing strategies, or product standards, limiting their ability to respond independently to market opportunities or adjust their production methods
- Collectives may become too dependent on certain buyers or markets, which can be risky if these buyers decide to source elsewhere or if demand fluctuates significantly
- Establishing a collective often requires an initial investment in infrastructure, such as storage facilities, transport or packaging systems. This can be a barrier for smaller groups without external support or funding
- There may be resistance to collective marketing due to traditional practices or social divisions among community members

Online marketing/e-marketing

Online marketing or e-marketing is where farmers use the internet as a platform to sell their products. There is a range of options for e-marketing. There are several free and more pay-for applications available to farmers. Examples of such apps can be found in Box 6.2.

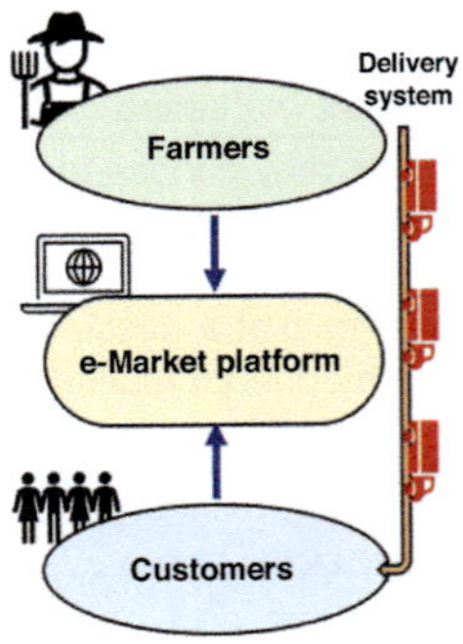

Advantages

- Farmers advertise their products on platforms where consumers can buy them directly. This eliminates the need for physical marketplaces and reduces the need for middlemen
- Although not strictly e-marketing, these tools help farmers gather and analyse data analytics for better crop management. This improves yield predictions which affect marketing decisions
- Consumers pre-pay for a portion of the harvest, receiving fresh agricultural products periodically. This model offers farmers upfront capital and ensures a dedicated market
- Farmers can locate and buy their production inputs directly. This can make the procurement process more efficient
- E-marketing can give farmers easier access international markets

Disadvantages

- Many rural areas do not have reliable internet connectivity or digital infrastructure. This limits both adoption and use of e-marketing options

Continued

Table 6.2. Continued

Online marketing/e-marketing
• Many farmers and smallholders do not have sufficient digital skills to use e-marketing effectively • Lack of access to financial services and inefficient digital payment systems limit the ability of farmers to conduct transactions smoothly on e-market platforms • Many low-income countries lack consistent and clear e-marketing regulations and policies or policies that inhibit internet access. This can discourage investment and innovation in e-marketing

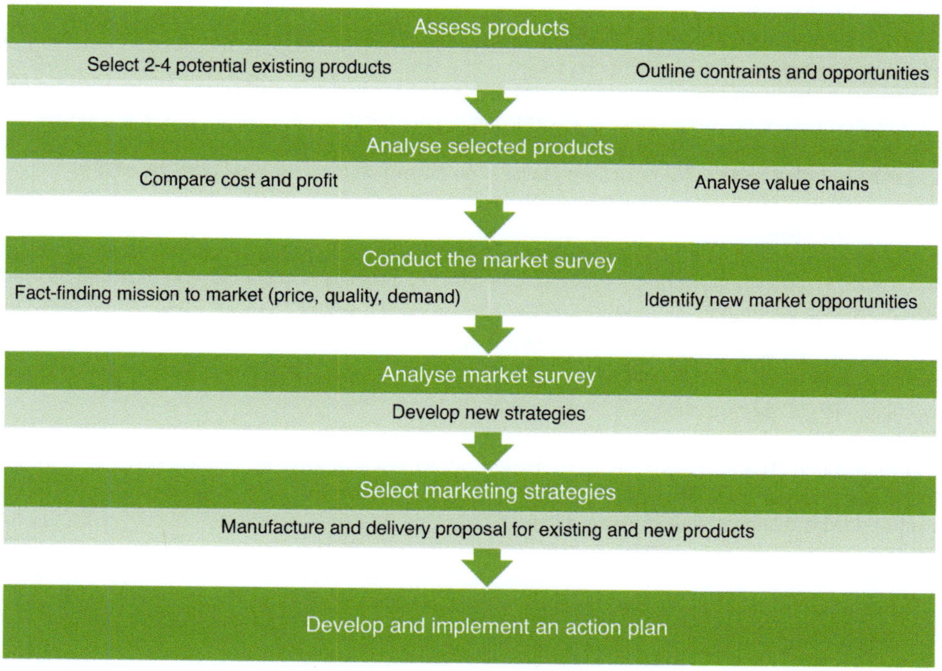

Fig. 6.3. Steps in conducting market surveys. Adapted from Poitevin and Hossain, 2006.

- **Market surveys**: conducting market surveys and gathering customer insights is crucial for making informed marketing decisions. Farmers should understand their target market's preferences, needs and purchasing behaviours. This information helps in developing tailored marketing strategies, identifying market trends and adapting to changing customer demands. Regular market surveys enable farmers to stay competitive, anticipate customer preferences and identify opportunities for growth.

Market analysis, strategy and planning

An important step in Fig. 6.3 involves market analysis. Effective marketing management is a critical component of farm business success; and market analysis provides data and information needed to develop marketing strategies and create comprehensive marketing plans. Market analysis forms the foundation of effective marketing management. It involves understanding customer preferences, identifying target markets, analysing market trends and evaluating the competitive landscape.

Box 6.2. Agricultural e-marketing apps in India for agricultural produce. From Medium, 2018.

SmartCrop
While trading farm produce, SmartCrop is an online marketplace that allows users to buy and sell crops throughout India. Farmers can post their products and attract more buyers, saving them precious time and money. They can get better profits than is typical. According to the market prices, the product entries can be removed or re-activated by the users. As these portals reduce the middleman, buyers and sellers can privately chat to negotiate prices safely.

Dhaan Mandi
Exchanges for various commodities have been offered through digital platforms for meeting marketing needs. Dhaan Mandi (crop market) is a free mobile app which acts as an online market for crops. It is a one-stop marketplace that is open to all to buy or sell crops across India. This app allows the end user to advertise their products both in English and Hindi.

Agribuzz – AgriApp
Agribuzz – AgriApp is a free mobile application targeted at agriculture practitioners, which provides a platform in which to connect and bring together the farming community, and helps them in selling, buying and exchanging agriculture commodities and services locally without middlemen, via an ad/listing posted from their mobile device. People can connect with each other to buy or sell in over 12 categories and 110 sub-categories. Agribuzz has a chat feature that allows the seller or buyer to chat using ready-made text templates, share details by email and discuss the deal over the phone.

Digital Mandi India
Digital Mandi India is a mobile application that helps an end user to check the latest Indian wholesale market prices of agricultural commodities across different states and cities. An end user can browse through various commodity and states categories. It is simple to reach the selected mandi price for commodities, making it user friendly. The data is synced with the Indian government portal https://enam.gov.in/web/–Powered by NIC.

Gramseva: Kisan (Mandi Prices)
Gramseva: Kisan (Mandi Prices) application is geared towards farmers and provides market data for farm produce. The app gives real-time market price from the government's website, data.gov.in. Graphs are used to showcase pricing trends which can enable informed decision making. Prices saved by the user can be shared via email and SMS.

Mandi Trades
Agricultural commodity trading platform connecting farmers with traders and enables direct transactions between traders and farmers. Mandi trades is an app where you can discover mandi prices, price alerts, which food produces are in demand. Commodity prices and farm produce data is sourced from the Government of India. It includes geo-tagging of farm produce that enables map-based identification of procurement source. Direct transactions take place between farmers and traders, thus avoiding middlemen.

Agri Market
Agri Market is a user-friendly agriculture application with an easier way to discover market prices for crops with the help of GIS. The application lets a user identify the market price of the crops within 50 km of the device's location.

By conducting market analyses, farmers can gain insights into the needs and preferences of their target customers. They can identify emerging trends, understand the dynamics of the market and anticipate changes in demand. This information enables farmers to make informed decisions about their marketing efforts and develop strategies that resonate with their target audience.

Based on the findings of market analysis, farmers can develop marketing strategies that effectively position their products in the market. These strategies may involve targeting specific customer segments, emphasizing unique product attributes or focusing on specific geographical markets. The goal is to differentiate the farm's products from competitors and create a competitive advantage. By understanding customer needs and market trends, farmers can tailor their marketing strategies to meet the demands of the target market effectively.

By incorporating market analysis, marketing strategies and marketing plans, farmers can effectively promote their products and connect with their target customers. This comprehensive approach to marketing management enables farmers to position their products effectively, reach their target markets and differentiate themselves from competitors. It maximizes the impact of their marketing efforts and increases the chances of success in a competitive market.

By making well-informed marketing decisions, farmers can effectively position their products, attract the right customers and achieve sustainable profitability. This requires a deep understanding of the target market, continuous market monitoring and the ability to adapt marketing strategies to changing market dynamics. Successful marketing decisions in farm management contribute to building a strong customer base, increasing brand value and ensuring long-term success in the agricultural marketplace.

The four Ps: product, price, place and promotion

When conducting market analyses and developing marketing strategies and plans it is helpful to use the four Ps of marketing: product, price, place and promotion (Kotler and Keller, 2016).

1. **Product**: farmers need to carefully consider the attributes and features of their agricultural products to meet the needs and preferences of their target customers. This involves decisions on product quality, packaging, labelling and any unique selling points that differentiate their products from competitors.
2. **Price**: setting the right price for agricultural products is crucial to achieve profitability while remaining competitive in the market. Farmers must consider factors such as production and marketing costs, market demand, competitor pricing and perceived value by customers when determining the optimal price.
3. **Place**: decisions regarding the distribution channels and market reach are essential for farmers to ensure their products are available to consumers when and where they want them. Farmers need to assess the most appropriate channels, such as direct sales at farmers' markets, selling through local retailers or exploring online platforms, to reach their target customers effectively.
4. **Promotion**: promotional activities are aimed at creating awareness, generating interest and persuading potential customers to purchase agricultural products. Farmers must develop marketing campaigns that highlight the unique features and benefits of their products. This may include advertising, social media marketing, participation in trade shows or engaging with local communities.

Price setting

Price setting for small family farmers often involves navigating the constraints of being 'price takers' rather than 'price makers' as discussed previously. As also noted previously, there is potential for small family farmers to transition towards becoming 'price makers' through the sale of niche or specialty products that cater to specific market segments. By focusing on products such as organic produce, heritage crop varieties, or other niche products, small family farmers can create a unique market position. To effectively set prices for these products, farmers need to thoroughly understand their production costs, which include inputs like seeds and fertilizers, and more complex costs like labour and possibly higher post-harvest handling expenses.

Market demand is a pivotal element in setting prices for niche products. In line with the previous discussion about market analysis, small family farmers should engage in at least basic market research to gauge consumer interest and willingness to pay for products that are marketed as organic, eco-friendly or socially responsible. This can involve direct interactions with consumers at local markets or collaboration with cooperatives that conduct such research collectively. Understanding market demand helps in setting a price that reflects the true value of the product to consumers, often allowing for a premium over conventional products.

Competitor analysis also becomes crucial in niche markets. Farmers need to identify not just local but sometimes international competitors, assessing the prices of similar niche products to ensure they remain competitive. They should consider factors that add value to their products, such as reduced chemical use or enhanced biodiversity, and communicate these effectively to consumers to justify potentially higher prices.

Lastly, the perceived value of these niche products by customers plays a significant role in price setting. Small family farmers must effectively communicate the unique attributes and benefits of their products, whether through labelling, storytelling or direct consumer engagement. This helps in building a brand perception that can justify premium pricing, aligning with the values and expectations of consumers. Effective use of simple marketing tools and platforms, as discussed previously, can aid small family farmers in highlighting the distinctiveness of their products, increasing their leverage as price makers in specialized markets.

The following are examples of actions farmers can take to improve the price of their products.

- **Produce larger eggs and creamier milk**: large eggs and creamier milk may bring higher prices compared with small eggs and milk with a lower cream content. But it is important that farmers consider the extra costs involved in the production process and compare them with the extra money that the produce will bring in. If the profit is higher than that previously obtained, the extra effort will be worth it.
- **Organize a marketing group with other farmers**: this can increase bargaining power and help farmers deal with buyers and negotiate contracts. Marketing groups may also be able to offer their members other benefits such as better market information.

Farmers acting in groups may also be able to improve packaging. In some countries, farmers working together have successfully increased prices by using a brand name or by stressing the geographical origin of the product.

Farmers acting individually to improve their prices may not gain unless they work in a group. A single farmer improving quality will have to sell to the same trader as all other farmers. The trader will probably not pay a premium as it will not be possible to keep good quality and poor quality produce separate.

Price forecasting

Price forecasting is a critical component of farm management, particularly for small family farmers who face the challenge of constant market volatility. Understanding future market trends enables these farmers to make informed decisions about when to sell their crops to maximize profits and minimize losses. However, forecasting requires access to market data and analytical skills that many small family farmers do not possess. Therefore, support from extension services or business advisors is crucial in providing the necessary data and helping interpret market signals.

Small family farmers can benefit significantly from simplified forecasting models provided by agricultural extension services. These models can analyse historical market data, supply and demand dynamics, and other relevant market indicators to predict future price trends. By incorporating local market conditions and specific crop data, these services can offer predictions tailored to the farmers' specific contexts. For instance, knowing when market prices are likely to peak can help farmers plan their harvest times and sales strategies more effectively, thus ensuring they receive the best possible prices for their produce.

Extension workers can assist farmers by organizing workshops to train them in basic forecasting techniques and the use of simple mobile apps that provide regular market updates and price trends. Additionally, these support entities can facilitate the development of community-based programmes where farmers collectively share and analyse market information, thereby enhancing their collective understanding and capability to make strategic marketing decisions.

By leveraging such collaborative approaches and technology, small family farmers can navigate market complexities more effectively, leading to improved profitability and sustainability of their farming enterprises.

Product selling strategies

Product selling strategies encompass the tactics and techniques used by farmers to sell their agricultural products effectively. These strategies involve identifying the most appropriate sales channels, developing strong relationships with buyers, and employing effective sales and negotiation techniques.

Farmers may choose to sell directly to consumers through farmers' markets, community-supported agriculture programmes or online platforms. Alternatively, they may engage with wholesalers, retailers or food service businesses to reach a wider market. Each selling strategy requires careful consideration of customer preferences, market dynamics and the desired level of control over pricing and distribution.

Linking physical data with price and cost information

Effective marketing management requires farmers to link physical data, such as production quantities, quality standards and input usage, with price and cost information. By integrating these data sets, farmers can assess the profitability of their agricultural products more accurately.

Farmers need to collect and analyse physical data related to their production processes, yields and product quality. This information is then combined with price and cost data to evaluate the overall profitability of their farm business. Linking physical data with price and cost information enables farmers to make data-driven decisions regarding production levels, pricing strategies and cost optimization.

Input procurement

Marketing management extends beyond the sale of agricultural products to include the procurement of inputs required for production. Farmers must make informed decisions when procuring inputs such as seeds, fertilizers, pesticides and equipment. This involves evaluating supplier options, comparing prices, considering quality and reliability, and negotiating favourable terms.

Post-harvest management

Post-harvest management is an essential component of marketing management that directly impacts the value and marketability of agricultural products. It encompasses a range of activities and practices aimed at preserving the quality, freshness and nutritional value of harvested crops or livestock.

Effective post-harvest management begins with proper handling practices. Farmers need to ensure gentle handling of harvested produce to minimize physical damage and bruising. Sorting and grading are important steps that involve classifying products based on their size, shape, colour and quality. This enables farmers to segregate products with varying characteristics, allowing for better market segmentation and price differentiation.

Packaging protects products from damage and enhances their presentation and appeal to buyers. Farmers should choose appropriate packaging materials that provide adequate ventilation, protection from pests and diseases, and are suitable for the specific product. Properly labelled packaging with relevant product information can also contribute to consumer trust and facilitate marketing efforts.

Maintaining the quality and freshness of agricultural products is essential for maximizing their value. Cooling and temperature control are commonly used techniques to slow down the natural deterioration process and extend the shelf life of perishable products. Where possible, farmers should invest in appropriate post-harvest infrastructure such as cold storage facilities or refrigerated transport to ensure optimal temperature conditions during storage and transportation.

Timely delivery to buyers is crucial to preserve the quality and market value of agricultural products. Farmers need to establish efficient logistics and transportation systems to minimize delays and ensure prompt delivery. This includes proper coordination with transportation providers, monitoring shipping schedules, and implementing effective inventory management practices to avoid overstocking or shortages.

Market demand and quality standards are important considerations in post-harvest management. Farmers should align their handling and storage practices with the specific requirements of their target markets. This may include adhering to specific certifications, meeting food safety regulations, or conforming to industry standards for grading and quality assurance. Understanding market preferences and quality expectations help farmers maintain product integrity and meet customer requirements.

Value addition

Value addition is another crucial aspect of agricultural management that enables farmers to enhance the value and profitability of their products. It is closely related to post-harvest handling. By transforming raw agricultural commodities into higher-value products or incorporating additional features and services, farmers can capitalize on market opportunities and gain a competitive edge.

From a management perspective, value addition requires careful analysis and decision making. Farmers need to assess market demand and consumer preferences to identify value-added opportunities that align with their production capabilities and resources. Understanding the needs and preferences of the target market is essential for developing value addition strategies that align with customer expectations and support premium pricing.

Cost considerations are another crucial aspect of value addition from a management perspective. Farmers need to analyse the financial implications of implementing value addition strategies, including the costs associated with processing, packaging, branding and marketing. It is important to conduct cost-benefit analyses to evaluate the potential returns on investment and assess the profitability of value-added products.

Furthermore, effective value addition management involves developing unique product attributes and branding strategies to differentiate farmers from their competitors. This may involve creating a distinct brand identity, emphasizing product quality, promoting sustainability or ethical production practices, or offering customized or niche products that cater to specific consumer segments. Farmers should invest in branding and marketing efforts to effectively communicate the value proposition of their value-added products to the target market.

Implementing value addition strategies also requires effective supply chain management. Farmers need to establish relationships with suppliers, processors, distributors and retailers to ensure smooth coordination and timely delivery of value-added products to the market. Efficient supply chain management helps in reducing costs, minimizing wastage and maintaining product quality throughout the value addition process.

6.10 Financial Management

As we noted in our earlier discussion, financial management is a critical aspect of farm management that involves making informed decisions regarding the acquisition, allocation and utilization of financial resources. In the context of small family farmers, who often do not maintain detailed financial records, making sound financial management decisions is crucial for maintaining stability and ensuring the sustainability of their farms. For small family farmers, financial management often relies more on practical knowledge and an intuitive understanding of their farm's needs and market conditions rather than formal financial strategies. It is important for these farmers to focus at least on basic financial practices such as estimating income from crops and expenditures on inputs like seeds and fertilizers. This can help them manage cash flow more effectively and make informed decisions about when to sell their produce or invest in farm improvements.

6.10.1 The need for finance

Small family farmers frequently face significant barriers in accessing the financial resources necessary to enhance their productivity and sustainability. The lack of access to finance restricts their ability to invest in improved agricultural technologies, such as high-quality seeds, fertilizers and efficient irrigation systems, which are essential for increasing crop yields and improving food security. Moreover, without financial resources, small family farmers struggle to expand their agricultural activities or diversify their crops, limiting their potential for growth and increasing their vulnerability to economic and environmental shocks.

Funds are needed as working capital for seasonal needs. The demand for liquidity is especially pronounced before harvest or to cover operating costs. Credit may also be required for consumption needs including wedding and funeral requirements, etc. Farmers often demand consumption credit as they must incur both production and consumption over the crop season before cash is generated by the sale of products after harvest. The denial of consumption credit by lending institutions could force farmers to borrow from private money lenders-cum-traders with the attendant problems of divided loyalty, high rates of interest and other forms of exploitation. Finance is also needed for longer-term capital investments.

Capital structure

Capital structure is an important consideration for small family farmers, who must carefully decide how much to rely on debt vs how much they can finance through their own resources, or equity. This balance affects not only the cost of capital but also the level of risk the farm is exposed to and its long-term financial health. Small family farmers need to evaluate their ability to take on debt, considering factors such as the interest rates available to them and the terms of repayment, which must align with their cash flow patterns and agricultural cycles.

Mobilization of credit

Farmers, especially small farm families, often face challenges in accessing and managing funds, which are critical for purchasing inputs, acquiring equipment and sustaining daily operations. Small family farmers primarily rely on a combination of formal and informal financial sources. Formal sources might include loans and credit lines from banks or agricultural credit agencies, often offering terms tailored to agricultural needs such as seasonal repayment schedules. However, due to barriers such as stringent collateral requirements or complex application processes, these sources may not always be accessible to small farm families. Other sources of credit might be microfinance institutions, cooperatives, associations, agricultural input dealers and non-government organizations.

Investment opportunities

Evaluating investment opportunities is essential for expanding operations, adopting new technologies or diversifying the farm business. Investment management, however, is particularly challenging for small farm families, who must carefully assess the viability and benefits of potential investments against their limited financial resources. Choices might include purchasing new seeds, small-scale equipment or making minor infrastructural improvements. Additionally, understanding the proper management of working capital – like negotiating for cash discounts on bulk purchases of supplies – can lead to significant savings and improve cash flow. Farmers need to assess the potential returns, risks and payback periods of different investment options. This includes analysing the expected cash flows, considering the time value of money and comparing alternative investments to make informed decisions that align with their business goals.

It should be noted that credit requirements are part of a dynamic process. Requirements change in the course of market and enterprise development. As activities become directed away from subsistence towards a market economy, products change – often becoming more capital intensive. This is sometimes accompanied by an increased need for long-term loans involving larger sums of money. For example, as farm size expands, the viability of tractors improves, and such an investment becomes more realistic. However, the investment is likely to require a long-term loan.

The consequences of limited financial access extend beyond the immediate economic constraints faced by small family farmers. They also hinder broader developmental goals. When small family farmers lack the means to invest in their operations, there is a direct impact on agricultural productivity and, consequently, on national food supplies. This situation can exacerbate food insecurity and poverty in rural areas, where agriculture is often a primary source of income. Additionally, without the capital to implement sustainable farming practices, there is a greater risk of environmental degradation, such as soil depletion and deforestation, which further jeopardizes the long-term viability of agricultural livelihoods in these regions.

To address these challenges, it is imperative that financial services tailored to the unique needs of small family farmers are developed and expanded. This includes providing micro-loans, credit facilities, and insurance products that are accessible and appropriately designed for small-scale agriculture. Additionally, small family farmers must have the capacity to take appropriate management decisions regarding the use of finance. Advisory service providers are necessary to support these farmers to access the capital needed to invest in their farms.

6.10.2 Sources of finance and credit

At times, farmers struggle to finance their enterprises. This is true for both large and small-scale farmers. When this situation arises, the farmer will need to consider carefully their options before making a decision.

- **Self-financing**: some farmers have savings that they use in times of need. The savings can be money from farm activities or off-farm sources. Savings are a preferred type of financial capital because they belong wholly to the household (no one outside the household has a claim on them). In addition, by using savings the farmer does not have to rely on others and is less complicated than borrowing money.
- **Credit financing**: when using savings is not feasible, farmers may have to resort to credit or borrowed money to finance farm activities. Farmers can get credit from formal or informal sources.

The sources of credit financing can be informal or formal. Informal sources, such as community lending groups, or loans from family and friends, play a significant role in the financial lives of small family farmers. These sources usually offer more flexibility and quicker access to funds without the bureaucratic hurdles of formal banking systems (see Box 6.3.). Efficient allocation and management of these funds are crucial. Farmers need to prioritize expenditures and investments that offer the most significant returns or essential inputs to avoid high-interest debt traps and maintain financial health.

Farmers can also obtain credit from formal sources such as banks, cooperative societies some non-governmental organizations and traders or money lenders. Credit from any of these sources mentioned can be taken for different lengths of time varying from short- to long-term (see Boxes 6.4. and 6.5.).

Box 6.3. Advantages and disadvantages of informal sources of credit.

Advantages:

- They are convenient, available locally, require no documentation and can provide credit quickly.
- The informal moneylender has local knowledge to help in appraising household credit needs and creditworthiness.
- There is little risk of default because the lender is generally well placed to apply pressure on the borrower to ensure payment.
- Administrative costs are low.
- With some types of informal credit such as the caretaking of livestock and the promise of tree crops, the risks of borrowing are shared between the two parties involved.
- Interest rates may be low.

Disadvantages:

- The borrower feels he has an obligation to the lender and loses his/her independence.
- There are few alternatives to choose from.
- Only short-term and relatively small loans are available.
- Interest rates may be high.

Box 6.4. Sources of formal financing.

- **Banks**: provide agricultural credit to farmers. In some countries, banks provide credit through their country branches to the farmers' credit and saving associations, which then extend the credit to their members.
- **Some non-governmental organizations**: provide loans and inputs on credit to farmers.
- **Cooperative societies**: give loans to their members or organize loans for them when they need to buy supplies.
- **Government organizations**: Some governments set up a special agricultural credit organization to lend money to farmers. Borrowing terms at such institutions are normally fair and at low interest rates.
- **Money lenders**: Private lenders normally ask very high interest rates, which becomes prohibitive to farmers.

Box 6.5. Loan duration.

- Short-term loans cover up to 1 year, generally for inputs.
- Medium-term loans are for 2–5 years, for equipment, stores etc.
- Long-term loans are 5 years and above for irrigation, land improvement, perennial crops and buildings.

6.10.3 Accessing (getting) financial capital/credit

For many small family farmers, access to formal credit can be limited, making it essential to consider alternative financing options such as micro-loans, government subsidies or cooperative lending programmes that are often more tailored to the unique needs of smaller agricultural operations. These sources can offer more favourable terms, such as lower interest rates or grace periods that align with harvest seasons. By understanding and carefully selecting the right mix of financing, small family farmers can ensure they maintain a sustainable operation that manages risk effectively while still having the potential to grow and invest in improvements.

When accessing financial credit there are conditions to be met by the farmers. Factors that are normally considered are:

- **Viability of the proposed plan or previous success**: viability refers to the likelihood that the plan will succeed. When a plan is viable, it means the farmer should be able to pay back the loan. Viability is shown in the plan, budget and financial records. If a plan is not shown to be viable, then moneylenders are not likely to offer credit.
- **The character and ability of the farmer**: lenders need to know whether the farmer is reliable, hardworking and experienced. They investigate past records to see if they have borrowed funds before.
- **Security**: in case the farmer is not able, or fails, to pay back money borrowed, the lenders will require some form of security or collateral for the loan. Security is anything promised to be given to the lender to sell if the borrowed money cannot be repaid. Security can be in the form of crops, cattle or buildings. The farmer needs to negotiate this carefully with the lender.

  In many countries in West Africa, security is seldom accepted because it is too difficult to obtain. Instead, banks generally accept fixed registered property for security, which most family farmers do not possess. For this reason, in most cases in West Africa, credit is provided only when there is shared responsibility (like cooperatives).
- **The cost of lending**: borrowing money costs money. Lending money is a business; the lender has costs and needs to make a profit, so will charge interest on the loan. Interest is the price paid for the use of someone else's money, just as rent is the price for use of land.

Box 6.6. Advantages and disadvantages of credit and loans.

Advantages:

- Credit makes money available to farmers to buy inputs at times when they do not have enough cash to buy them.
- Credit makes it possible for a farmer to start different enterprises or to expand existing enterprises. This gives them the chance to increase their profitability.

Disadvantages:

- A farmer may not have anything to offer as collateral security. They cannot offer their land as security unless they have legal title to the land.
- Small family farmers usually need small loans. Many lenders are not interested in making small loans because they are not as profitable as large loans. Lenders prefer to lend large amounts to commercial farmers.
- Lenders prefer to lend money to farmers who are near the lender's offices. Farmers are often located in remote and difficult to reach areas. The costs of travel and administration are high. When rural branches are opened, they may do much less business than town branches.
- Many farmers lack knowledge and experience of formal application procedures. They are discouraged by the need to complete application forms and other documents, which naturally make the process long.
- Many lenders believe that agricultural production is too risky and that the chances are high that the farmer will fail to pay back the loan.

Before agreeing to take out a loan, and especially before signing any documents, the farmer needs to be very clear about the rate of interest being charged on the loan. The farmer should ask the lender to make this clear and to show how much money they will have to pay to borrow the money. If they are not happy with the interest rate, they should talk to the lender about getting a lower rate. In the end, it is the farmer who carries the full responsibility of the loan. The farmer must pay it back plus the interest (see Box 6.6.).

As a result of these limitations, formal credit agencies rarely operate in rural areas without government support and promotion. Farmers are often forced to rely on informal sources.

6.10.4 Capital in different enterprises

Tree crops

The capital cost involved in tree production is equal to the cost of purchasing or establishing the tree plus the cost of growing the tree up to the stage of full production. When the tree dies, it should be replaced by a new tree. Taking good care of trees and controlling diseases and pests is very important to protect the money that has been spent growing the tree. If the tree dies before it has produced anything, all the money is lost.

Livestock

The capital cost involved in livestock production is equal to the cost of purchasing (if it is purchased) plus the cost of keeping that animal up to the stage it reaches full production. Whether the animal is being kept for meat, milk, eggs or some other product, the animal should be kept fit and healthy so that it will produce efficiently. Only in this way can the farmer get high gross margins through their livestock. Again, the investment must be protected. If the animal dies before producing, the money is lost.

Buildings

The cost of buildings can be kept very low if the farmer and their family do most of the building. Careful repair and maintenance of buildings will make the buildings last longer and this will reduce depreciation (or decrease in value). Where possible and practical, a farmer can hire a building. If a building is hired, it should be put to maximum productive use. An empty building is a waste of money. However, there is no point in filling a building with produce where it will be losing value. If the produce can be sold at a good price, it should be sold.

Machinery and equipment

Investment in machinery and equipment can be very profitable if the farmer can manage the finance and if it increases profitability. Mechanization can be costly both in terms of buying it and in terms of repairs and maintenance.

Mechanization through animal traction

In many countries animal traction with draft animals is a simple way of mechanizing. The investment includes the animals and the equipment they will pull such as a plough, a planter, a cart and other farm implements.

Motorized machinery and equipment

- **Intermediate power-driven equipment**: this includes two-wheel tractors and light four-wheel tractors. Farmers can also invest in small-scale post-harvest equipment like threshers, mills, oil-presses, etc. to add value to their products before taking them to market.
- **Heavier power-driven equipment**: this includes tractors, combine harvesters, trailers, etc. Hiring is an option that can be used by farmers, which saves on cost of buying the machine. For example, hiring a tractor to plough the land or a threshing machine for maize. Hiring is often cheaper than buying.

Cooperative use of machinery or sharing is another option. Shared use of equipment by two or more farmers has often been more successful than cooperatives. In Kenya, for example, sharing often means one farmer brings a cart, another one the oxen, and another the yoke. In other cases, one farmer brings the tractor, and another farm brings the threshing machine.

Digital farming technology

This includes tools such as GIS, drones and sensors as well as digital instruments such as computers and mobile phones. These are increasingly becoming a part of the regular equipment needed by modern market-oriented farmers. It is possible to both buy or hire this equipment. In each case the farmer will need to assess the cost against the improvements in profits.

6.10.5 Financial management decisions

Some key aspects of financial management decisions for small family farmers include the following.

Important factors affecting long-term capital decisions

Capital is always limited and should be allocated where it generates the highest returns. Once invested in long-term assets such as livestock, perennial crops, orchard trees or buildings, capital becomes fixed and cannot be easily converted into other forms that might offer better profitability. These durable investments reduce financial flexibility, as the capital is committed and no longer available for alternative uses that could potentially yield a higher rate of return.

Current assets and liabilities for financial management

Effective financial management on the farm begins with a clear understanding of current assets and current liabilities, as these directly influence short-term liquidity and decision making. Current assets include cash, accounts receivable, inventory (such as harvested crops or livestock ready for sale) and inputs like seed or fertilizer that will be used within the current season (see Box 6.7.). These assets are essential for meeting day-to-day operational needs and responding to market opportunities. On the other hand, current liabilities represent the farm's short-term financial obligations – such as unpaid bills, wages, loan repayments or input credit – that must be settled within the same period.

Monitoring the balance between current assets and current liabilities is crucial for managing cash flow, maintaining creditworthiness and avoiding financial stress. For example, if liabilities exceed assets, the farm may face difficulty

Box 6.7. Current assets and liabilities.

Current assets include:

- cash on hand or in the bank;
- accounts receivable (money owed by buyers for produce sold);
- inventory of harvested crops, seeds or fertilizers;
- livestock for sale (if they are to be sold within the year); and
- supplies that will be used within the year (like feed or fuel).

Current liabilities include:

- short-term loans or credit due within the year;
- accounts payable (money owed to suppliers or service providers);
- wages payable to farm workers;
- taxes payable; and
- input purchases on credit, like fertilizer or seeds bought but not yet paid for.

covering expenses, risking default or disruptions in operations. Conversely, maintaining a healthy surplus of current assets provides flexibility to invest in new opportunities, negotiate better terms with suppliers, or withstand unexpected setbacks such as crop failure or price drops.

Understanding this relationship allows farmers to make informed financial management decisions – such as when to borrow, how much working capital is needed or whether it is viable to expand production. It also supports timely decisions on selling produce, paying debts and purchasing inputs. In essence, managing current assets and liabilities is not just about bookkeeping; it is a critical part of running a financially sound and resilient farming business.

Working capital management

Working capital, in the context of farm business management, refers to the funds available to cover the farm's day-to-day operational expenses. It is calculated by subtracting current liabilities from current assets. These assets and liabilities typically include cash, input inventories, accounts receivable, unpaid bills, short-term loans and other obligations due within the farming year.

Working Capital = Current Assets – Current Liabilities

Sufficient working capital is vital for the smooth day-to-day operation of a farm and serves as a buffer against unexpected financial shocks. It allows farmers to manage routine expenses without relying heavily on external borrowing, which is especially important for small family farms operating on tight margins and with limited financial reserves. Effective working capital management involves the careful planning, monitoring and control of cash flow, inventories and short-term obligations. By maintaining adequate liquidity, farmers can cover essential expenses such as seed, fertilizer and labour, avoid production delays, and take advantage of timely opportunities – such as purchasing inputs when prices are favourable. In addition, managing working capital effectively helps build financial resilience, improves creditworthiness and strengthens the long-term sustainability of the farm business. A critical component of this process is maintaining tight control over accounts receivable and payable – ensuring timely collection of payments from buyers and negotiating practical payment terms with suppliers. By closely managing these elements, small family farmers can avoid cash flow problems, prevent financial overextension and ensure continuous operation throughout the agricultural cycle.

Debt repayment capacity

Debt repayment capacity is crucial for small family farmers, who need to ensure they can meet their financial obligations without compromising their farm's sustainability. This involves having the skills to carefully analyse the farm's cash flow, the profitability of their agricultural activities, and their ability to generate sufficient income to cover debt payments. Understanding their repayment capacity helps small family farmers make prudent borrowing decisions, secure loans with terms that are realistic and manageable and avoid taking on debt that could become unsustainable. This strategic financial planning is vital for maintaining the viability of their farm operations and ensuring long-term economic resilience.

Financial analysis and record-keeping

In the small family farming context, where formal financial records are often scarce, it is crucial for farmers to adopt basic financial management practices. Small family farmers can benefit from simplified financial analysis to understand their farm's financial health and make informed decisions. This could involve maintaining basic records of income and expenditures, even in a simple ledger or notebook, which can help them track their cash flow and understand their financial position. There are now also digital apps and programmes designed specifically for farm financial management that the farmer can use. Whichever method is used, regularly maintaining and reviewing basic records can help farmers identify areas where costs can be reduced, income can be increased and overall farm efficiency can be improved. This rudimentary form of financial tracking is crucial for small family farmers to manage their resources effectively and plan for future investments or expenses. The topic of record-keeping is discussed in detail in Chapter 14.

Financial monitoring and control

Regular monitoring of their financial situation, even through simple methods like keeping a daily record of sales and expenses in a notebook, can significantly help farmers adapt to market changes and plan for future needs. Developing these habits allows farmers to gauge the financial health of their operations without needing complex financial reports. They can then use this information to avoid overextending themselves with debt and to seize opportunities that could lead to better profitability and sustainability of their farming activities.

To support these financial decisions, small family farmers benefit from basic financial management training and tools. Simple cash flow tracking methods, understanding basic financial statements, and learning to forecast and plan for financial needs can dramatically improve a farmer's ability to manage debt and invest wisely. Development programmes and non-governmental organizations often play a critical role in providing such education and resources, helping bridge the gap between small family capabilities and the demands of effective farm financial management.

6.11 Strengthening Management Capacity

Strengthening farm management capacity is essential for enabling small family farmers to run their farms as successful businesses. As we have seen in this chapter, effective farm management involves more than just technical know-how; it requires the ability to make strategic decisions across a range of interlinked areas – production, finance, marketing, labour, risk and information management. Building this capacity among farmers is a prerequisite for enhancing resilience, increasing profitability and achieving long-term sustainability.

To meet these demands, training and extension services must move beyond technical recommendations and focus explicitly on developing farmers' managerial skills. This includes building competencies in planning, organizing, directing and controlling farm operations – core functions of management. Training should also cover enterprise budgeting, resource allocation, labour supervision, marketing strategies and record-keeping. When farmers are equipped with these skills, they can make informed choices, adjust to changing conditions and better engage with value chains and markets.

Equally important is the capacity of agricultural advisers, both public and private, who support farmers with information and guidance. Advisers must themselves be trained in farm business management to effectively coach and mentor farmers in developing business plans, evaluating investment options, managing cash flow and making strategic production choices. Without this capacity, advisory services risk focusing solely on production, without addressing the broader business environment in which farms operate.

Farmers' organizations, such as cooperatives and producer groups, also require management capacity-building. These organizations play a vital role in input procurement, collective marketing, information dissemination and negotiating with buyers. Strengthening the governance, financial management and leadership skills of farmer organization managers and their members enhances collective bargaining power, improves transparency and helps smallholders access services and markets more efficiently.

Public–private collaboration is key to ensuring that the full range of support services – inputs, finance, veterinary care, information and advisory services – are available and well-coordinated across the entire value chain. Effective partnerships can help overcome service delivery gaps and promote integrated solutions that are responsive to the needs of farmers.

Ultimately, building strong management capabilities among small family farmers – supported by well-trained advisers and effective farmer organizations – is fundamental to creating a more resilient, competitive and business-oriented agricultural sector. Capacity-building must be an ongoing process, inclusive of all farmers, and adapted to the diverse contexts, challenges and opportunities that smallholders face.

Equally important is the need to cultivate a culture of continuous learning and adaptation among both farmers and advisory services. In an agricultural landscape shaped by shifting markets, climate change and technological advancement, the ability to learn, adapt and innovate is essential for long-term competitiveness and resilience. These issues will be further developed in Part 4.

6.12 Conclusion

Throughout this chapter, we have explored the broad based nature of farm management for smallholder farmers, emphasizing the critical components of production, marketing, financial, organizational and labour management, as well as information and risk management. Effective production management is foundational for any farm business, as it involves the application of best practices and modern technologies to enhance crop and livestock yields, and improve product quality. By adopting advanced agricultural techniques and sustainable practices, small farmers can achieve higher productivity, reduce environmental impact and meet market standards, which are crucial for maintaining competitiveness in the agricultural sector.

Marketing management has been highlighted as equally crucial, focusing on the necessity of understanding market demands and effectively promoting and selling farm products. Farmers must stay attuned to consumer preferences, market trends and pricing strategies to maximize their income. Effective marketing strategies involve not only direct sales and traditional markets but also exploring digital platforms and value-added products. By diversifying their marketing approaches and building strong relationships with buyers, processors and retailers, small family farmers can enhance their market presence, reduce dependency on single market channels and ensure better returns for their produce.

Financial, labour, organizational and information management are also pivotal aspects covered in this chapter. Financial management involves meticulous planning and control of financial resources, enabling farmers to make informed investment decisions, manage costs effectively and ensure the financial health of their farm operations. This includes understanding credit options, managing loans and utilizing financial tools for budgeting and forecasting.

Labour management, on the other hand, focuses on the optimal utilization of both family and hired labour to boost productivity and sustainability. Efficient labour management practices ensure that tasks are performed timely and efficiently, which is essential for maintaining the smooth operation of the farm. By investing in training and development, and fostering a positive work environment, smallholder farmers can improve labour productivity and retention, contributing to the overall success and sustainability of the farm enterprise.

Organizational management refers to the wide range of decisions that farmers make regarding the many systems on the farm.

Effective organizational management is critical to the successful running of a farm. Information management involves the ability to locate, gather, analyse and interpret information needed to make sound farm management decisions.

Risk management is a critical aspect of farm business management, helping small family farmers anticipate and mitigate uncertainties related to production, markets, finance, human resources and external factors such as climate change. By adopting a proactive approach to risk, farmers can implement strategies such as enterprise diversification, climate-smart farming, contract agreements and financial planning, to reduce vulnerabilities. Understanding and assessing different types of risk – whether production-related, financial, institutional or human – allows farmers to develop integrated strategies that enhance resilience and sustainability. Effective risk management not only protects farm operations from unexpected disruptions but also contributes to long-term profitability and stability in an increasingly unpredictable agricultural environment.

The success of any farm depends on sound management. This applies equally to small family farms.

References

Kay, R.D., Kay, R.D., Edwards, W.M. and Duffy, P.A. (2020) *Farm Management*, 9th edn. McGraw-Hill Education, New York. Available at: https://www.mheducation.com/unitas/highered/sample-chapters/9781260002195.pdf

Kotler, P. and Armstrong, G. (2010) *Principles of Marketing*. Pearson Education, Indianapolis, Indiana.

Kotler, P. and Keller, K.L. (2018) *Marketing Management*. Pearson Education, Indianapolis, Indiana. Available at: https://www.pearson.com/se/Nordics-Higher-Education/subject-catalogue/marketing/Kotler-Keller-Marketing-Management-Global-Edition-16e.html

Medium (2018) *7 Bests Apps for 3-Market Platforms for Farmers. Medium Website*. Available at: https://medium.com/@sumana.T/7-best-apps-for-e-market-platforms-for-farmers-324dd668270b (accessed 3 February 2025).

Poitevin, B. and Hossain, S. (2006) *Marketing Extension: A Powerful Process in 6 Steps*. Livelihoods, Empowerment and Agroforestry Project (LEAF), Bangladesh.

Additional Reading

Barry, P.J., Ellinger, P.N., Hopkin, J.A., and Baker, C.B. (2011) *Financial Management in Agriculture:* Pearson: London

Boehlje M and V. Eidman (1984). *Farm Management,* John Wiley and Sons, Hoboken, New Jersey

Drucker P. (1982) *The Practice of Management*. Harper Perennial, New York, New York

Giles T. and M. Stansfield, (1980) The Farmer as Manager. George Allen and Unwin

Johnson D. (1982) *The Business of Farming: A Guide to Farm Business Management in the Tropics*. Macmillan International. New York.

Olson, K.D. (2004) *Farm Management: Principles and Strategies*. Iowa State Press. Available at: https://academic.oup.com/erae/article-abstract/31/2/225/503838?redirectedFrom=fulltext (accessed 17 March 2025).

Warren M. (1998) *Financial Management for Farmers and Rural Managers*, Blackwell Science

7 Adapting to Climate Change

Climate change represents one of the most pressing challenges confronting small family farms, significantly affecting their agricultural productivity, economic stability and long-term sustainability. These farmers are disproportionately vulnerable to the impact of climate change due to their reliance on rain-fed agriculture, limited access to resources and constrained adaptive capacity. Rising temperatures, increased frequency of extreme weather events, unpredictable rainfall patterns and expanding pest and disease pressures have direct consequences on yields, incomes and livelihoods. Moreover, climate change exacerbates existing vulnerabilities, intensifying the difficulties faced by farmers who already contend with limited resources and market uncertainties.

As previously highlighted, climate change is a risk and a key concern for small family farmers. To effectively respond to the challenge of climate change, small family farmers must make informed and strategic management decisions. The ability to evaluate options and proactively adopt sustainable practices is essential for maintaining productivity, profitability and resilience under climate variability. Central to this approach is the need to incorporate climate-smart agriculture practices, sustainable production systems, efficient irrigation methods, appropriate mechanization and innovative digital technologies into farm operations. These practices are further described in the following sections. Yet, these adaptation measures are complex, often requiring careful assessment and management skills that many small family farmers have not yet fully developed.

This chapter looks at the decisions that small family farmers must make to adapt successfully to these climate-induced changes. It explores how integrating sustainable agricultural practices should be informed by economic analysis and supported by customized advisory services. Only in this way can the capacity of farmers to redress negative climate-induced impacts be improved. The chapter further highlights the importance of developing farm economic skills and competencies among smallholder farmers through the support of agricultural advisers who can provide guidance in decision making, resource allocation and risk management. By doing so, it positions effective management decisions as central to ensuring farm viability, productivity and sustainability in the face of growing climatic uncertainty.

7.1 The Impact of Climate Change on the Farming System

The immediate effects of climate change manifest primarily in reduced crop productivity and farm income. Rising temperatures cause heat stress in plants, shortening growth cycles and reducing yields, particularly for staple crops such as maize, rice and wheat. Rainfall variability introduces further unpredictability into crop planning, resulting in increased vulnerability to droughts and floods, both of which severely compromise agricultural production and productivity. Additionally, these altered weather patterns create more favourable conditions for pests and diseases to occur, significantly increasing their prevalence and intensity. Farmers, consequently, face escalating pressures from pest infestations and plant diseases and this often requires the application of additional inputs and invested management efforts to protect crop health and maintain yields.

Increased frequency of extreme weather events and prolonged dry spells, in particular,

DOI: 10.1079/9781836990536.0007

drive up input costs, particularly for irrigation, pest management and soil fertility improvement. Simultaneously, yield reductions and crop failures significantly lower farm income, diminishing farm profitability and threatening long term sustainability. Smallholders also face heightened market risks as fluctuations in agricultural production influence market supply and prices, creating further uncertainty and financial instability. In these ways, climate change intensifies economic vulnerability among smallholders, potentially pushing them deeper into food insecurity and poverty unless strategic adaptations are undertaken in time.

Resource availability is another critical area impacted by climate change. Water scarcity has become a pressing issue, as rainfall variability reduces the reliability of traditional water sources, making irrigation and crop planning increasingly challenging. Degradation of water quality, driven by factors such as salinization, soil erosion and runoff contamination, further restricts access to clean, usable water for agriculture. These water-related constraints directly undermine farmers' ability to sustain agricultural productivity and secure sustainable household livelihoods.

Understanding these immediate and long-term impacts underscores the urgent need for strategic and tactical farm management responses to mitigate climate risks, safeguard resources and maintain farm productivity and profitability over time.

7.2 Sustainable Production Systems for Long-term Viability

A sustainable farm system is essential for ensuring the long-term viability, profitability and resilience of small family farms in the face of climate change. These systems consider ecological, economic and social dimensions, aiming to minimize environmental harm while maximizing productivity and economic returns. By adopting sustainable and climate-smart practices, smallholder farmers can reduce greenhouse gas emissions, conserve natural resources, protect biodiversity and enhance soil health. These practices also help build resilience to climate variability and extreme weather events, ensuring the sustained provision of the benefits that people obtain from natural ecosystems (ecosystem services) vital to agricultural productivity and natural resource sustainability.

7.2.1 Agroecological approaches and regenerative agriculture

Agroecology offers a holistic framework that integrates ecological principles into many agricultural practices. It promotes biodiversity, strengthens natural processes and improves resource efficiency through methods such as agroforestry, intercropping, cover cropping and integrated pest management (IPM). These practices enhance soil structure, increase water retention and improve nutrient cycling while supporting beneficial organisms and ecosystem services. These practices are elaborated on in the sections that follow.

Regenerative agriculture builds on these agroecological principles by actively restoring degraded ecosystems. It combines soil health enhancement, biodiversity conservation and climate resilience into a unified strategy aimed at revitalizing farmland and increasing productivity. The long-term economic gains of regenerative practices are substantial. Reduced dependence on chemical inputs lowers production costs, while improved soil health contributes to more consistent yields over time, thus enhancing income stability.

7.2.2 Soil health and nutrient management

Healthy soils form the backbone of sustainable agriculture. Practices such as conservation tillage, organic matter incorporation, crop rotation and nutrient balancing improve soil fertility, structure and water-holding capacity. These practices not only reduce erosion and nutrient leaching but also limit greenhouse gas emissions. Smallholder farmers who shift to organic nutrient sources and adopt precision fertilizer application techniques often see a noticeable reduction in input costs and improved crop performance, particularly during periods of irregular rainfall. Well-managed soils also promote stronger root development, which reduces the risk of crop failure and supports higher yields under both favourable and challenging conditions.

Soil testing and precision nutrient application technologies are becoming more accessible through digital extension services and mobile-based platforms, elaborated upon in later chapters. These tools support informed decision making, allowing farmers to tailor nutrient applications to crop needs – improving fertilizer-use efficiency and protecting the environment while optimizing input expenditures.

7.2.3 Sustainable pest and disease management

Integrated pest management (IPM) relies on combining biological, cultural and chemical control methods to manage pests sustainably. Practices include using pest-resistant varieties, rotating crops to break pest cycles, encouraging natural predators and using pesticides only as a last resort. This reduces chemical inputs, protects beneficial organisms and maintains ecological balance, contributing to sustainable and resilient production systems. The economic benefits of IPM are widely recognized. Farmers who adopt biological pest control methods often significantly reduce their reliance on chemical pesticides, which helps lower input costs and supports healthier farm ecosystems. Reducing pesticide use also decreases health risks for farmers and minimizes environmental contamination.

7.2.4 Conservation agriculture

Conservation agriculture (CA) is a well-established climate-smart approach grounded in three core principles: (i) reduced tillage; (ii) permanent soil cover; and (iii) crop rotation. Reduced tillage preserves soil structure, improves water retention and lowers erosion. Permanent soil cover using mulch or cover crops protects against evaporation and adds organic matter. Crop rotation disrupts pest and disease cycles, improves nutrient availability and reduces dependence on chemical inputs. These principles work synergistically to preserve soil structure, improve water retention and suppress erosion.

Medium and large-scale farmers who adopt CA practices often achieve higher yields compared to those using conventional tillage, especially during periods of low rainfall. Although the impact on smallholder farmers is less proven, there is evidence in some contexts of improved productivity and household food security and the generation of surplus produce for sale in local markets. The economic advantages of CA generally include reduced fuel and labour costs due to fewer field operations, decreased fertilizer needs as soil organic matter improves, and more effective pest and disease management through crop diversification. These combined benefits support climate adaptation while boosting long-term farm profitability and resilience (see Box 7.1.).

7.2.5 Sustainable mechanization

To complement sustainable farming practices, small family farmers can adopt sustainable mechanization – machinery and tools tailored to small-scale operations that minimize environmental impact. These can include two-wheel tractors and direct-seeding equipment, which can significantly reduce labour requirements and increase the area of land that smallholder farmers are able to cultivate. Improved planting precision from such tools often leads to more uniform crop stands and better yields. Sustainable mechanization is especially critical for small farms dealing with erratic weather patterns and labour shortages. When properly managed, it contributes to productivity gains while preserving resources for future generations. The topic of mechanization management is explored in more detail in Section 7.7.

7.2.6 Water-efficient irrigation and climate-smart water use

Water-efficient irrigation methods such as drip, micro-sprinkler and precision irrigation technologies are essential to manage scarce water resources in a sustainable way. These systems can increase water-use efficiency by up to 90% compared to traditional flood irrigation (Wu *et al.*, 2023). For small family farms, improved

Box 7.1. Conservation agriculture and regenerative agriculture.

Key differences between regenerative and conservation agriculture:

1. Scope
   - Regenerative agriculture takes a more holistic, ecosystem focused approach, encompassing soil health, biodiversity, carbon sequestration and livestock management.
   - Conservation agriculture focuses more specifically on soil conservation through reduced tillage, crop cover and rotations, with a narrower scope on addressing soil erosion and degradation.
2. Intensity of practices
   - Regenerative agriculture seeks to restore ecosystems and reverse environmental damage, aiming to improve the entire health of the land over time.
   - Conservation agriculture aims to sustain current levels of productivity while preventing further degradation, focusing on practices that reduce soil erosion and maintain fertility.
3. Animal integration
   - Regenerative agriculture often integrates holistic livestock management as a key component, using animals to enhance soil health and biodiversity.
   - Conservation agriculture typically focuses more on plant-based systems and does not always include livestock management as a core practice.
4. Carbon sequestration
   - Regenerative agriculture is explicitly focused on carbon sequestration in soil and plant biomass to mitigate climate change.
   - Conservation agriculture indirectly supports carbon sequestration by reducing soil disturbance and enhancing organic matter but this is not its primary goal.

How they relate:

- Both approaches aim to create more sustainable farming systems by improving soil health, reducing the need for chemical inputs and increasing biodiversity.
- Both practices emphasize the importance of minimal soil disturbance (no-till or low-till practices) and crop diversity, which are essential for soil fertility and reducing environmental degradation.
- Cover crops and crop rotations are central to both systems, helping to maintain soil health, suppress weeds and enhance nutrient cycling.

water management means greater control over planting schedules, reduced risk of crop failure and lower energy costs from pumping. Farmers who adopt solar-powered drip irrigation systems can reduce water usage while also lowering energy requirements, particularly by decreasing reliance on fuel-powered pumps. These savings contribute to improved profit margins over time. Additionally, combining irrigation scheduling with tools like weather forecasts and soil moisture sensors enables farmers to apply water only when needed, helping to avoid both under- and over-irrigation – practices that can negatively impact yields and increase production costs. On-farm irrigation practices are also elaborated on later in Section 7.5.

7.2.7 Biodiversity conservation and ecosystem services

Maintaining biodiversity within agricultural landscapes is also crucial for sustainability. Biodiversity supports pollination, pest control and nutrient cycling – key ecosystem services for productive farming. Strategies such as planting native species, establishing buffer zones and creating wildlife habitats improve environmental health, reduce dependency on external inputs and contribute to long-term farm viability. Farms that incorporate greater crop and habitat diversity often experience improved pollination, reduced dependence on synthetic fertilizers and pesticides, and lower overall input costs.

Promoting biodiversity within farming systems also strengthens resilience, as diversified farms are typically more capable of withstanding environmental and economic shocks.

7.2.8 Agroforestry

Agroforestry – the integration of trees with crops and/or livestock – offers multiple economic and ecological benefits. Trees improve soil structure, reduce erosion, support biodiversity and sequester carbon. They also buffer temperature extremes, conserve water and provide additional income through products like timber, fruit and fodder, helping to stabilize yields and incomes amid climate variability. Farmers who integrate trees into their croplands through natural regeneration or agroforestry practices often see improvements in soil quality and increased crop yields. These systems also help buffer farms against the impacts of drought and market fluctuations, promoting greater economic stability and long-term sustainability.

Collectively, these practices form the basis of regenerative agriculture – a strategy that not only adapts to climate change but actively restores degraded agricultural ecosystems. As we have discussed previously, the economic benefits for farmers that adopt climate smart agricultural practices are especially impactful in the increasingly uncertain climate where resilience and adaptability are essential for survival.

Sustainable production systems, therefore, represent both an environmental necessity and an economic opportunity for small family farmers. Their successful implementation depends not only on access to appropriate technologies and practices but also on farmers' capacity to make informed management decisions, supported by advisory services, market incentives and enabling policies.

7.3 Management Decisions to Respond to Climate Change

Informed farm management decisions are central to helping small family farmers effectively respond to the escalating challenges of climate change, particularly in low-income countries. Smallholders must make critical choices regarding resource allocation, production practices, crop selection and investment in appropriate technologies and infrastructure. The ability to evaluate options carefully and respond proactively to changing conditions can significantly enhance farm resilience, maintain or improve productivity and protect economic viability in the face of uncertainty.

Effective farm-level decision making requires a clear understanding of available adaptation strategies. As previously noted, small family farms often face financial constraints and limited access to advanced technologies, making them more vulnerable to climate shocks. In such contexts, efficient resource use becomes paramount. Farmers must adopt a strategic approach that balances short-term productivity with long-term sustainability while optimizing resource allocation and mitigating climate risks.

A key priority for farmers is the adoption of climate-smart agricultural practices, such as crop rotation, conservation tillage and organic soil amendments, as described previously. These practices improve soil fertility, enhance water retention and reduce reliance on chemical inputs – boosting productivity while cutting input costs and limiting environmental degradation. Efficient water management through techniques like rainwater harvesting, drip irrigation and mulching is also vital to reducing drought stress and stabilizing yields. These technologies are discussed in more detail in Section 7.4, Chapter 7.

The digital agricultural technologies, referred to previously, further support climate adaptation by enabling more informed decision-making. These innovations are transforming agriculture, making it more adaptive and accessible for farms of all sizes. Management decisions need to be taken on the appropriateness of these technologies, as well as the costs and benefits of their application. Chapter 10, provides a deeper exploration of these technologies.

Another key management strategy is diversification. By varying crops, livestock and income sources, farmers can reduce their exposure to risks from climate-related disruptions. For instance, cultivating crops with different climatic tolerances increases the likelihood of achieving at least partial yields during adverse conditions. Similarly, integrating livestock into cropping systems supports nutrient cycling, provides alternative income streams and enhances overall farm stability. Beyond core production, engaging in agro-processing, agro-tourism or off-farm employment helps further stabilize household income and builds financial resilience against extreme weather or market fluctuations.

Climate risk management must also be a cornerstone of farm planning. Strategies such as crop insurance, diversified income portfolios, climate forecasting and adaptive financial planning enable farmers to anticipate climate variability and reduce economic losses. These tools help farmers manage uncertainty and maintain stability in the face of production variability caused by changing weather patterns.

Equally important is community collaboration and knowledge sharing. By engaging with peers, local organizations and extension services, farmers can access shared resources, adopt new technologies and learn about locally effective adaptation practices. Strong networks enhance the capacity of smallholders to implement climate-resilient strategies and to advocate for supportive policies. Participation in community-based resource management initiatives also provides access to information on climate trends and strengthens farmers' collective bargaining power in markets.

These interconnected strategies form a comprehensive framework for climate adaptation. They enable small family farmers to strengthen resilience, increase productivity and pursue sustainability amid growing environmental challenges. Importantly, farmers must also weigh the trade-offs involved in adopting these strategies, ensuring that choices align with their long-term goals and resource capacity.

Ultimately, farm management under the conditions of climate change must be dynamic, adaptive and forward-looking. By combining sustainable production practices, financial risk mitigation and technological innovation, small family farmers can enhance their ability to adapt to climate change while securing the future viability of their farms and livelihoods.

7.4 Digital Agricultural Technologies and Climate Change

The world has moved into the digital age, and digital technologies affect almost every aspect of social and economic life. This is also true for agriculture. As introduced in Chapter 1, digital farming technologies involve the use of data-driven tools and technologies to enhance farming efficiency, productivity and sustainability. As will be discussed in Chapter 10, digital technologies play a pivotal role in farm management decision making. In general, the overall aim of digital agricultural technologies is to enhance and increase the efficiency and accuracy of decision making.

However, some digital agricultural technologies are particularly relevant to climate change adaptation. Thus, in this section, we briefly review a few of the technologies that can assist farmers address climate change challenges. These technologies leverage advancements such as sensors, Geospatial Information Systems (GIS), drones, data analytics and machine learning, to monitor and manage crop health, soil conditions, irrigation and pest control in real time – all of which help farmers make decisions related to the impact of the changing climate and weather patterns. These technologies also include a growing range of ultra-modern technologies such as the Internet of Things (IoT); precision agriculture; smart irrigation; aerial imaging; SMS alerts; artificial intelligence (AI) and on-farm sensors.

By providing precise, actionable insights, digital farming technologies help farmers make more informed decisions, optimize the use of resources such as water, fertilizers and pesticides, reduce environmental impacts and improve yields. These innovations, which are transforming agriculture and making it more adaptive to climate change, are now scalable for farms of all sizes.

A brief overview of the various technologies is set out in Box 7.2.

Table 7.1 offers another perspective of these technologies, indicating their primary function and the key climate adaptation benefits they offer.

Despite their promise, there are many challenges involved when applying digital technologies to smallholder farming. There is a prevailing digital divide, high investment costs, technical complexity (including literacy issues), the availability and cost of internet access, privacy and security concerns, and the negative environmental impact of some of these technologies.

One way to address some of these challenges is through (government) investment in infrastructure and training, creating a supportive policy environment specifically supporting digital solutions that are tailored to local conditions and small farmer needs (Parra-Lopez, 2024). However, while advanced technologies (e.g. AI) show promise, simpler solutions like mobile alerts often have faster adoption and immediate benefits, particularly with respect to climate change adaptation.

7.5 On-farm Irrigation Management

On-farm irrigation is becoming increasingly critical for adapting to climate change and ensuring sustainable agricultural production. As rainfall patterns grow more erratic and water resources more limited, efficient and effective irrigation management is essential to maintain productivity, reduce environmental degradation and improve farm resilience. Smallholder farmers must adopt strategies that optimize

Box 7.2. Overview of digital technologies for climate change.

Internet of Things (IoT): the IoT is a system that connects sensors and devices that monitor soil, weather and crop conditions in real-time. This allows farmers to automate and optimize processes like irrigation and fertilization.

Precision agriculture: this technology uses GIS, sensors and data analytics to apply inputs like water, fertilizers and pesticides more efficiently. This helps reduce waste and increase productivity, even in challenging climates.

Smart irrigation systems: this a specific IoT application of connected devices that automate water delivery based on soil moisture data. This reduces water waste and ensures crops receive adequate water during droughts or periods of stress.

On-farm sensors: these are affordable, solar-powered devices placed in fields to monitor soil moisture, temperature or humidity. They send alerts when conditions deviate from optimal ranges, enabling timely irrigation or pest control to combat climate stresses.

Remote sensing: this refers to satellite- or drone-based technology that tracks large-scale environmental changes such as droughts, deforestation and crop development. It helps farmers adjust their agricultural practices.

Geospatial information systems (GIS): GIS combines mapping technologies with data analysis to help farmers monitor soil health, land use and crop performance. This supports more efficient and sustainable land management.

Drones and aerial imaging: these technologies provide aerial views of farmland for monitoring crop health, detecting pests or diseases, and assessing soil conditions. They enable early intervention and targeted treatment.

SMS alerts: these are text messages delivering real-time weather forecasts, pest warnings or market prices to basic mobile phones. They help farmers prepare for extreme weather (e.g. storms, droughts) and reduce crop losses, even without internet access.

Mobile weather apps: these applications provide farmers with real-time, location-specific weather forecasts, including temperature, rainfall and extreme weather alerts. They help farmers plan planting, irrigation and harvesting schedules more effectively, reducing risks from unpredictable climate conditions.

Artificial Intelligence (AI) and machine learning: these are very modern technologies. They help analyse vast amounts of data to predict crop yields, pest outbreaks and climate impacts, and offer tailored advice to optimize farming decisions. While they are applicable to general farm management, they also have an important role to play in adapting to climate change.

Big data analytics: in this context, big data analytics refers to processing large datasets on climate, soil and crop performance to offer actionable insights for improving farm management and long-term climate adaptation strategies.

water-use efficiency, preserve water quality and mitigate the impacts of salinity. These practices are not only vital for enhancing yields but also for safeguarding the long-term viability of small family farms (see Box 7.3.).

A variety of irrigation systems are available to meet these needs, each with distinct agronomic and economic considerations. Common systems include flood, sprinkler, drip and centre-pivot irrigation. For smallholders, the selection of a system depends on factors such as the suitability of the technology for their plot size and location, as well as technical aspects like crop water requirements, soil type, topography and water availability. Economic factors are equally important – particularly initial investment costs, ongoing operational expenses and expected returns. For example, flood irrigation may be inexpensive to install on flat land but is typically inefficient, leading to water waste and increased long-term costs. In contrast, drip irrigation, while more expensive upfront,

Table 7.1. Digital technologies for climate change adaptation.

Technology	Primary function	Key climate adaptation benefits
Precision Agriculture	Optimizing water/ input use in erratic climates	• Reduces fertilizer waste by 20–40% through GIS-guided application, cutting N_2O emissions. • Optimizes water use in drought-prone areas via soil moisture sensors. • Lowers fuel costs by minimizing tractor passes (fewer CO_2 emissions).
Smart Irrigation Systems	Designed specifically for water scarcity	• Prevents crop failure during droughts with automated, soil-responsive watering. • Slashes water use by 30–60% compared to flood irrigation. • Mitigates groundwater depletion in water-stressed regions.
Remote Sensing	Tracks climate impacts (droughts, floods) in real time	• Provides early warnings for floods/droughts via real-time field monitoring. • Identifies pest outbreaks linked to warming trends (e.g. locust swarms). • Tracks soil degradation from extreme weather, guiding restoration efforts.
Drones	Early detection of climate-related stressors (droughts, floods, pests)	• Prevents yield loss by spotting drought stress more quickly than ground surveys. • Guides water use via thermal imaging of soil moisture, cutting irrigation waste. • Assesses damage post-extreme weather (e.g. hailstorms), accelerating coping strategies.
Mobile Weather Apps	Deliver climate-specific alerts (storms, heatwaves)	• Avoids planting/harvest losses with hyperlocal storm/ heatwave alerts. • Adjusts crop choices using seasonal climate forecasts (e.g. drought-resistant varieties). • Reduces pesticide waste by timing applications to post-rainfall pest surges.
AI/Machine Learning	Predicts climate-related pest outbreaks/yield losses	• Predicts yield losses from climate shifts (e.g. rising temperatures). • Recommends resilient crops based on future climate models. • Optimizes microclimate management in greenhouses/ polyhouses.

Box 7.3. Irrigation management decisions.

Deciding whether to introduce irrigation into a small family farming system requires careful consideration of multiple management factors. The technical feasibility of irrigating specific plots depends on soil type, water availability and the practical challenges of delivering water to crops efficiently. Farmers must also determine the optimal timing of irrigation, both within the growing season and daily, to maximize water-use efficiency and crop productivity. Another critical aspect is the financial investment required for irrigation infrastructure. The upfront costs can be substantial, often posing a significant challenge for small-scale farmers. However, these costs should be evaluated over multiple years, considering the potential for long-term benefits, including increased yields, improved crop reliability and reduced water costs. While the initial years of irrigation may result in financial losses, sustained productivity gains can lead to greater profitability in the long run. Given these financial and technical complexities, a thorough economic analysis is essential to assess the feasibility of the investment. Farmers need strong business management skills to determine whether irrigation is a worthwhile long-term strategy for their farms. In Part 3, we delve deeper into farm management tools that can assist farmers in making informed decisions about irrigation investments and overall resource allocation.

offers higher water-use efficiency and yields, potentially justifying the investment – especially in water-scarce areas.

For many smallholders, however, the cost and complexity of installing irrigation systems can appear prohibitive. Nonetheless, the long-term benefits of improved water management – such as increased productivity, greater crop reliability and more stable incomes – can make irrigation a practical investment. These benefits become more accessible when supported by affordable, scalable solutions or cooperative arrangements that allow farmers to share infrastructure and costs. Access to concessional finance and technical assistance is also a significant factor in the decision-making process.

To maximize the benefits of irrigation, smallholder farmers must also focus on optimizing water application. Effective irrigation scheduling and monitoring – aligned with crop growth stages, soil conditions and weather forecasts – can significantly improve efficiency. For smallholders, this requires training and ongoing extension support to build capacity in irrigation management. Precision practices such as deficit irrigation, mulching and timely application can conserve water while maintaining yields. For smallholder farmers this requires training and extension support in irrigation management.

Technologies like soil-moisture sensors, mobile-based weather forecasts and decision-support apps offer opportunities for more targeted irrigation. However, these tools often require strong farm management skills, reliable infrastructure and technical support – conditions that are not always available in smallholder settings. Where feasible, such technologies can be transformative, but their adoption must be context-appropriate and accompanied by capacity-building.

Water-saving practices – such as mulching, efficient on-farm distribution systems and rainwater harvesting – offer practical, low-cost alternatives or complements to high-tech solutions. These can reduce water loss, lower energy use and improve soil moisture retention, contributing to both environmental and economic sustainability.

Economic viability remains central to irrigation decision making. The costs associated with water, energy and infrastructure must be carefully weighed against potential gains in crop quality, marketability and income stability. Efficient irrigation reduces input costs and enables reinvestment in other areas of farm development, including diversification and risk mitigation.

However, irrigation itself is increasingly vulnerable to climate pressures. Rising temperatures, shifting rainfall patterns and prolonged droughts threaten water availability and reliability. In this context, the sustainability of irrigation systems becomes a pressing concern for smallholders.

To remain viable, small family farms must transition to irrigation strategies that are not only efficient but also climate resilient. This involves a combination of modern technologies and traditional practices adapted to current environmental realities – such as adjusting planting times to rainfall patterns, using drought-tolerant crops and adopting low-cost water harvesting methods. On-farm water storage solutions, such as lined ponds or tanks, can buffer against dry spells and stabilize water supply.

Climate change presents both challenges and opportunities for smallholder irrigation. While it increases water scarcity, it also highlights the strategic importance of investing in irrigation as a key pillar of resilience. Water-efficient technologies, combined with improved scheduling tools and collaborative water management, are essential. Equally vital are financing mechanisms, policy support and farmer-led initiatives that empower smallholders to manage water as a shared and strategic resource.

To sum up, the future of irrigation for small family farms will be shaped by the need to conserve water, enhance efficiency and adapt to growing climate unpredictability. By integrating affordable, water-efficient technologies, improving irrigation management skills and promoting locally appropriate solutions, smallholders can sustain productivity and strengthen their resilience. In a resource-constrained agricultural environment, informed irrigation management is not only a strategy for climate adaptation, but a fundamental driver of farm profitability and long-term sustainability.

The critical considerations for effective irrigation management are summarized in see Box 7.4.

7.6 Climate Change Challenges for Irrigation

As rainfall patterns become more erratic and water resources more limited, the need for efficient and effective water management has never been greater. Water is a precious resource, especially in regions affected by climate change, where rainfall patterns are becoming increasingly uncertain. Maximizing water use efficiency has significant economic implications.

Figure 7.1 provides a framework for understanding how water use on small family farms must adapt in response to climate change. It highlights key strategic areas – from adopting efficient irrigation systems and digital monitoring tools to implementing water harvesting, alternative sources and climate-smart financing – that collectively address growing water scarcity and variability. This integrated approach can guide the transition towards more resilient, efficient and sustainable water management practices at the farm level.

7.6.1 Shift towards water-efficient irrigation systems

Small family farms will increasingly adopt water-saving technologies as water scarcity becomes more pressing. Traditional flood irrigation methods will be replaced by more efficient irrigation systems such as drip irrigation, sprinkler and micro-irrigation. These technologies will

Box 7.4. Critical irrigation management decisions.

Farmers must consider several key decisions to maximize irrigation efficiency and climate resilience:

- **Choosing the right irrigation system:** evaluate initial cost, water availability, energy requirements and long-term economic returns.
- **Timing of irrigation:** align watering schedules with crop water needs, soil moisture status and weather forecasts.
- **Monitoring and evaluation:** regularly monitor soil moisture and crop condition to fine-tune irrigation schedules and water application rates.
- **System maintenance:** regularly inspect and maintain irrigation equipment to avoid leaks, ensure uniform distribution and minimize resource waste.
- **Financial assessment:** consider total cost-benefit analysis, including initial investment, operating costs, potential yield increases and long-term sustainability impacts.

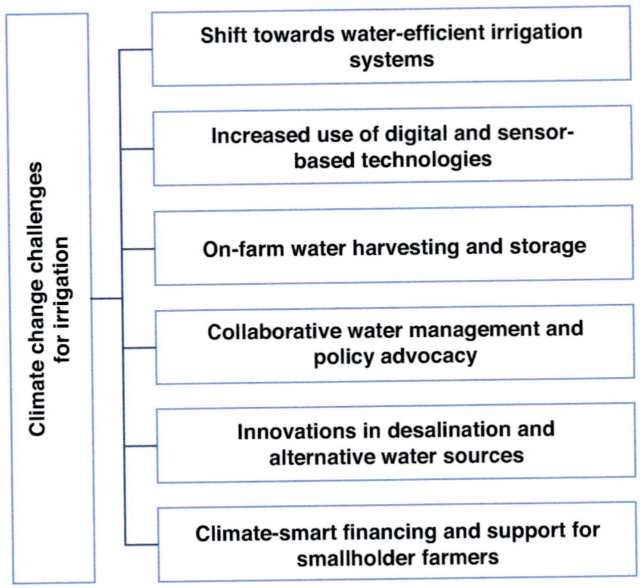

Fig. 7.1. Responses to climate change challenges for irrigation.

help small farmers maximize the productivity of limited water resources.

7.6.2 Increased use of digital and sensor-based technologies

Smart irrigation systems will become more common as farmers look for ways to optimize water use and respond to climate variability. These include:

- **Soil moisture sensors:** provide real-time data on soil water content, helping farmers apply water only when and where needed.
- **Remote monitoring and automated systems:** allow farmers to control irrigation remotely and respond quickly to weather and soil conditions using mobile apps and computer systems.
- **Data-driven decision making:** through satellite imagery, weather forecasts and climate models, farmers can plan irrigation schedules more effectively.

By using data to inform their irrigation decisions, small family farms can reduce water waste and respond more flexibly to climate conditions.

7.6.3 On-farm water harvesting and storage

As climate change disrupts traditional rainfall patterns, more small farms will implement rainwater harvesting and on-farm water storage systems to capture and store water during wet periods. Stored water can be used during droughts or times of water scarcity. Key methods include:

- **Building ponds and reservoirs**: farmers can capture excess rainwater or runoff for later use.
- **Water recycling:** wastewater from farming activities, such as cleaning or processing, can be treated and reused for irrigation.

These strategies offer small farmers greater water security and reduce dependence on unreliable external sources.

7.6.4 Collaborative water management and policy advocacy

Given the increasing competition for water resources, small family farmers may need to participate in collective water management initiatives and advocate for policies that ensure equitable water access. This could involve:

- **Water user associations:** small farmers can collaborate with other stakeholders to manage shared water resources more sustainably.
- **Advocating for sustainable water policies:** pushing for government policies that support access to water for small-scale farmers; and invest in water-saving technologies and protect water rights.

In many areas, water scarcity will demand collaborative community approaches to manage water fairly and efficiently.

7.6.5 Innovations in desalination and alternative water sources

As traditional freshwater sources become more stressed, small family farms near coastal areas may turn to desalination technologies to treat seawater for irrigation. Similarly, there may be increased investment in:

- **Greywater use**: treating and reusing household wastewater for agricultural purposes.
- **Saline water use:** utilizing brackish or saline water for cultivating salt-tolerant crops, either independently or in combination with freshwater leaching to manage soil salinity.
- **Treated wastewater for irrigation:** in urban or peri-urban areas, using treated wastewater from cities for agricultural irrigation is likely to become more common.

While these technologies are more expensive, advances in affordability and accessibility may make them viable options for small farmers individually and as farmer groups in water-scarce areas.

7.6.6 Climate-smart financing and support for small farmers

Adapting to climate change can require significant investment, which may be a barrier for small family farms. However, the future of irrigation will likely be supported by:

- **Subsidies and grants:** governments and international organizations may increase financial support for small farmers to adopt water-saving technologies and practices.
- **Climate insurance:** farmers could access climate insurance products that protect them from crop failures due to water shortages or drought, providing a safety net while they transition to more resilient practices.

To successfully navigate these water-use challenges posed by climate change, small family farmers must combine innovative water-efficient technologies with robust community-based management approaches. By embracing such comprehensive strategies, they can significantly enhance resilience, maintain agricultural productivity and secure their livelihoods in the face of increasing climate uncertainty.

Farmers can implement various water conservation practices to minimize water losses and ensure that water is used where it is most needed. Soil moisture monitoring tools, efficient water distribution systems and the strategic use of mulching can significantly reduce water consumption, thereby lowering operational costs. Additionally, techniques like deficit irrigation, where water is applied in quantities less than the crop's full requirements, can help manage water resources more sustainably, particularly in regions where water scarcity is a pressing concern. This approach not only conserves water but also reduces the costs associated with water extraction, energy use and potential penalties for overuse of water resources.

To meet these challenges, farmers must optimize water efficiency, maintain quality and manage salinity impacts. These efforts are essential not only for enhancing crop yields but also for sustaining the long-term viability of agricultural practices in a changing climate. The economic aspects of farm management decision making clearly play a crucial role, as the costs associated with irrigation systems, water use and energy must be carefully balanced against the potential gains in productivity and sustainability.

By adopting these forward-looking irrigation management practices, small family farms can significantly enhance their adaptive capacity, secure long-term access to water and ensure continued agricultural viability. In today's competitive agricultural landscape, water management is not only a matter of environmental responsibility but also a fundamental pillar of economic resilience and sustainable development.

7.7 Sustainable Mechanization

The importance of farm mechanization in modern farm management and the commercialization of small family farming is a crucial element in transforming small family farms into commercially viable businesses capable of competing in the global market. Mechanization helps small family farmers tackle climate challenges, access new markets and improve their economic resilience. Moreover, mechanization, done in the correct way, is integral to the shift towards more sustainable farming practices. When, as mentioned earlier, mechanization supports the implementation of conservation farming techniques such as zero-tillage, it can improve soil health and water retention, reduce erosion and decrease the carbon footprint of farming operations – all of which foster sustainability.

7.7.1 Why mechanize?

The primary reason for mechanization among small family farmers is the potential increase in productivity and efficiency it offers. Mechanization allows for the timely completion of farming tasks, which is crucial during planting and harvesting periods when timing affects crop yields. For example, the use of tractors for preparing land or seed drills for planting can drastically reduce the time required for these activities compared to manual methods. This allows farmers to cover more ground quickly and effectively. Mechanization also reduces the physical strain on farmers, who traditionally spend long hours performing strenuous tasks. Another reason to mechanize is that mechanical equipment can help mitigate some of the risks posed by the increasingly unpredictable weather patterns due to climate change.

7.7.2 Environmental impact of mechanization

The impact of mechanization on the environment can be profound, influencing everything from soil health to energy consumption. The use of heavy machinery can lead to soil compaction, which diminishes soil aeration and water infiltration, negatively affecting crop growth. However, when implemented thoughtfully, mechanization can also contribute to better soil management and reduced greenhouse gas emissions. For example, minimum tillage machines can reduce the number of passes needed over a field, lessening soil disturbance, fuel consumption and emissions. Moreover, precision agriculture technologies like GIS and sensor-based systems, enable farmers to apply water, fertilizers and pesticides more efficiently, minimizing runoff and pollution.

Sustainable management recognizes the fragility of the natural environment within which the farm business operates. Given that small-scale farmers typically focus on profit and rarely consider the impact on the environment, sustainable management is a challenge. Nevertheless, farmers are increasingly becoming aware of the need for sustainability of their farm system to ensure long-term agricultural productivity, cost savings and greater profitability. It encompasses adopting farm machinery and practices that not only enhance farm output but also conserve natural resources and reduce the ecological footprint of farming activities. Sustainability and profitability are not conflicting business goals.

7.7.3 Farm management decisions for mechanization

To achieve effective, sustainable mechanization, farmers should follow clear guidelines for informed decision making. Firstly, it is essential to assess mechanization needs accurately, considering farm scale, cropping systems, existing infrastructure and labour availability. Mechanization must align closely with farm capacity to maximize operational efficiency and minimize unnecessary investment.

For small family farmers, making informed farm management decisions is essential. Economic considerations for mechanization investments include conducting thorough cost-benefit analyses to evaluate initial purchase costs, maintenance and operational expenses, expected productivity improvements and overall financial returns. Farmers should carefully assess affordability, financing options and potential payback periods, prioritizing investments that offer the highest long-term value and resilience.

Farmers face complex decisions regarding the selection and use of machinery, such as the number and size of tractors, draft livestock and equipment needed; whether to purchase or hire mechanization services; when to replace machinery and livestock; and whether to opt for new or used machinery. Each decision involves balancing various forms of capital – human, natural, social and financial – and determining the optimal combination with land and labour to maximize profitability.

Purchasing farm machinery is a significant investment. Farmers must evaluate their specific needs, consider the expected return on investment and assess their budget. It is crucial to research and compare different machine models for reliability, fuel efficiency, maintenance needs and after-sales support. Financing options such as leasing or loans should also be considered.

Choosing the right machinery involves assessing the farm's terrain and crops, and the machinery's capacity, ease of use and maintenance needs. Four-wheel tractors may offer greater power for large or challenging terrains, while two-wheel tractors could be better for smaller fields. Implement selection must also match the machinery's capabilities with the tasks at hand, emphasizing the need for equipment that can operate under varied weather conditions and meet diverse cropping needs.

Farmers must also decide whether to purchase, lease or hire machinery – each option has implications for financial capacity and long-term planning. Sole ownership may be ideal for those who frequently use their machinery, whereas hiring or contracting can be beneficial for seasonal needs without the costs of full ownership. Forming cooperatives or machinery rings can also reduce costs and expand access to various equipment.

There are business opportunities in hiring out machinery during non-peak periods, which can help offset ownership costs. Key considerations include setting competitive prices, determining service terms and managing equipment logistics.

Efficient use of mechanization is essential for financial success, demanding not only technical but also management skills. Efficient machinery use throughout the year is crucial; this includes scheduling to maximize use, maintaining equipment to prevent breakdowns and training operators for efficiency. Financial considerations must also be addressed, including calculating costs for purchase, operation, maintenance and replacement, and weighing these against expected productivity gains and labour cost reductions.

By prioritizing sustainable mechanization, small family farmers can improve their operational efficiency and environmental impact, paving the way for a more resilient and productive agricultural sector. This strategic approach not only helps in tackling immediate agricultural challenges but also contributes to broader environmental and socio-economic objectives, reinforcing the role of small family farms in sustainable development. Some of these decisions will be elaborated on further.

7.7.4 Assessing mechanization needs

For small family farmers, assessing the need for mechanization is a crucial step in aligning farm operations with both economic and sustainability goals. This section delves into how farmers can evaluate their mechanization requirements, the financial implications and adapt strategies based on the size and nature of their farms.

Determining the need for mechanization

The decision to mechanize should start with a clear understanding of the specific challenges and opportunities within a farmer's operations. Small family farmers need to consider factors such as the time taken to perform tasks manually vs mechanically, the availability of labour and the frequency of certain tasks. For instance, if labour shortages are a recurring issue during peak agricultural seasons, mechanization can offer a reliable solution. Additionally, tasks that are highly labour-intensive and repetitive, such as ploughing or harvesting, are prime candidates for mechanization. Farmers should evaluate their typical farm activities, identifying areas where machines could enhance efficiency, reduce drudgery and potentially increase crop production.

Evaluating the economic viability of mechanization

Assessing the economic impact of investing in farm machinery is crucial for small family farmers. They should calculate the initial costs of acquiring equipment, ongoing maintenance expenses, potential increases in productivity and reductions in labour costs. This analysis should also consider the lifespan of the machinery and the possibility of selling it in the future. Additionally, the impact of mechanization on crop yield and quality plays a crucial role in this evaluation. By comparing the costs of mechanization with the projected benefits, farmers with the requisite farm management skills can make informed decisions that balance their financial capabilities with their operational needs (see Chapter 9, Section 9.2.4).

Farm size and mechanization strategies

The size of a farm significantly influences the type and scale of mechanization that is feasible and beneficial. Small plots may not benefit from large, heavy machinery that is designed for vast acreages due to issues of manoeuvrability and cost-effectiveness. Instead, smaller, more versatile equipment might be more suitable. For very small operations, sharing machinery services or renting equipment can be cost-effective strategies. This allows farmers to access modern technology without the high costs of ownership. Additionally, the layout and topography of the land should guide the choice of machinery to ensure that the equipment can operate efficiently without harming the soil structure or productivity.

Each of these considerations requires farmers to make thoughtful decisions about when, how and to what extent, to mechanize. By carefully assessing their needs, analysing the costs vs benefits and considering the specific characteristics of their farm, small family farmers can effectively integrate mechanization into their farm management practices, thereby enhancing their productivity and long-term sustainability.

As is evident from the previous discussion, mechanization for small family farms faces several challenges, including cost, availability and appropriateness of technology. However, there are realistic and adaptable mechanization options that can improve productivity, reduce labour and increase farm resilience. The key is to focus on appropriate technology – solutions that are affordable, scalable and tailored to the needs of smallholder farmers.

Some of the economic considerations and management decisions for selecting mechanization technologies for small family farms are as follows:

- **Two-wheel (walking) tractors:** two-wheel tractors are versatile, affordable compared to larger four-wheel models, and particularly suited to small, irregular plots. Economically, farmers must consider the initial purchase cost, maintenance expenses, fuel efficiency and potential labour savings. Though more affordable than larger machines, individual ownership can still represent a substantial investment. Therefore, cooperative ownership or rental schemes are practical strategies to mitigate initial costs and enhance affordability.
- **Small-scale harvesting tools:** semi-mechanized or mechanized harvesting tools such as small motor-driven threshers or crop-specific harvesters significantly reduce labour costs and post-harvest losses. Farmers should assess economic gains from reduced losses, labour savings and improved timing efficiency against training and maintenance requirements. Investment decisions should include the calculation of return on investment based on expected yield improvements and labour reduction.
- **Pedal or hand-powered tools:** tools such as pedal-powered threshers, hand-held seeders or small tillers can help improve the efficiency of planting, weeding and post-harvest handling. These low-cost, manually operated tools are highly economical and accessible for small-scale and low-income farmers. Their affordability, ease of maintenance and independence from fuel or electricity costs are strong economic incentives. However, the trade-off involves increased labour intensity and slower operational efficiency compared to motorized equipment. Management decisions must balance initial affordability with long-term productivity considerations.

- **Solar-powered technologies:** solar-powered technologies such as water pumps, grain dryers and lighting offer substantial long-term economic advantages due to significantly reduced operating costs and reliance on renewable energy sources. Although the initial investment can be high, running costs are reduced, and payback periods are usually relatively short. Farmers will need to evaluate payback periods, anticipated savings on fuel or electricity and potential productivity increases. Availability of financing or subsidies for renewable energy should also factor into decision making.
- **Small-scale processing equipment:** investing in small processing units (e.g. mills; grinders; mobile units for fruits, vegetables, dairy) enables value addition directly on-farm, enhancing income streams and reducing transportation costs. Farmers need to consider initial costs, ongoing operational expenses, market opportunities for processed products and the profitability of value-added activities. Access to these technologies can be limited by cost, but cooperatives or sharing schemes can mitigate these costs. Cooperative purchasing and shared ownership models can significantly lower individual financial burdens and facilitate broader access.
- **Cooperative machinery sharing and rental systems:** given the financial constraints of small family farms, cooperative ownership or machinery rental systems allow farmers to access machinery without the burden of full ownership costs. It reduces individual financial strain and enables access to more advanced technologies. Cooperative models or machinery rental systems are economically beneficial as they reduce individual capital expenditure, distribute maintenance and operational costs, and increase access to advanced equipment. Management considerations include establishing clear terms for equipment sharing, scheduling, responsibility for repairs and equitable cost allocation among members.

In summary, mechanization on small family farms is possible, but it must be affordable, appropriate and scalable to the realities of small farm sizes and limited financial capacity. Successful mechanization for small family farms requires economic feasibility analyses and strategic management decisions addressing affordability, appropriateness to farm size, operational and maintenance costs, and potential productivity gains. Effective strategies typically combine low-cost manual tools, affordable motorized equipment, renewable energy-based technologies and cooperative machinery-sharing models, all aimed at enhancing farm productivity and resilience to climate change. Mechanization, when combined with water-saving and sustainable farming practices, can significantly increase farm productivity and resilience to climate change.

7.8 Climate Change Adaptation: Challenges and Opportunities

Farmers often face difficult choices when balancing climate adaptation with market-oriented production. These trade-offs depend significantly on factors such as their production system, market access and the availability of support services. Farmers seeking higher market returns may prioritize high-yield, high-input crops tailored to consumer demand, yet these are highly vulnerable to climate shocks. In contrast, adaptation strategies aimed at enhancing resilience may not yield immediate financial returns.

Market integration frequently encourages specialization in high-value cash crops. Although this specialization can increase incomes, it exposes farmers to greater climate-related risks. Intensified input use – such as synthetic fertilizers, pesticides and irrigation – can improve yields in the short term but may degrade soil and water resources over time. Conversely, sustainable practices such as organic inputs, conservation agriculture and water-efficient techniques provide long-term benefits but often require upfront investments and management adjustments.

For many small farmers, immediate income often outweighs long-term sustainability considerations, especially when financial support for adaptation is lacking. Without incentives like higher prices for sustainably grown products or easy access to credit, farmers may hesitate to shift away from conventional practices. Addressing these trade-offs thus requires policies and market mechanisms that integrate climate resilience into smallholder farm business decisions, enabling farmers to remain competitive while safeguarding their livelihoods.

Market-oriented farming and climate adaptation can mutually support each other if planned strategically. Smallholder farmers integrating climate-smart practices into their production systems can build long-term resilience while maintaining or even enhancing profitability. Sustainable soil and water management practices mitigate climate risks and tend to improve yields and reduce input costs over time. Diversification – in terms of crop selection and market engagement – can spread risks and tap into emerging consumer preferences for sustainably produced and climate-resilient agricultural products.

Value chain stakeholders upstream from farm level are increasingly recognizing the need for climate-smart supply chains and are providing incentives for farmers adopting adaptive practices. Certification schemes, carbon credit markets and impact-driven financing are creating new income streams for farmers implementing sustainable land-use approaches. Governments and development organizations play a key role by aligning agricultural policies with climate-resilient production through targeted subsidies, insurance mechanisms and research investments into climate-adapted crops and livestock systems. Digital innovations further strengthen these synergies, enabling farmers to access real-time climate data, financial services and market intelligence, leading to better decision making and enhanced farm viability.

Farm management must extend beyond merely optimizing production, particularly given climate-related risks. It requires a strategic vision integrating technical and business aspects while carefully considering trade-offs between short- and long-term goals. For example, a smallholder farmer growing maize might diversify with drought-tolerant crops such as sorghum or integrate livestock systems, thereby reducing vulnerability to climate variability. Such strategies may initially affect short-term profitability but ultimately enhance income stability and market opportunities.

In this way, the relationship between profitability and sustainability is deeply interconnected. Short-term financial gains provide farmers the resources to invest in practices improving soil health, conserving water and diversifying production systems – crucial steps in building resilience against climate change. While such investments might strain short-term profitability, they yield significant long-term benefits. Conversely, sustainable management protects natural resources, reduces environmental risks, lowers input costs and enhances overall farm stability, ensuring dependable long-term returns.

Adapting effectively to climate change thus requires farmers to think holistically and adopt an integrated approach to farm management. In this way, farmers must be mindful that small changes in one part of the farming system often have repercussions elsewhere. Consequently, selecting climate-resilient practices necessitates balancing short-term profitability with long-term sustainability. Maintaining the farm's natural resource base and adopting a business mindset ensures economic sustainability and profitability over time.

By adopting sustainable practices and managing farms as dynamic enterprises, smallholder farmers can create resilient systems capable of withstanding climate shocks while promoting economic and ecological benefits. Integrating short- and long-term perspectives into farm management enables farmers to adapt successfully to changing conditions, secure their livelihoods and contribute meaningfully to broader sustainability goals.

7.9 Support of Agricultural Advisory Services to Adapt to Climate Change

Agricultural advisory services, especially through the efforts of frontline agricultural extension workers, play a vital role in helping small family farms make economically sound decisions for climate change adaptation. Acting as critical intermediaries between research and farmers, these services translate scientific research, new technologies and practical know-how into actionable guidance tailored to local conditions and the needs of farmers.

Extension workers have a role to play in communicating climate risks and appropriate adaptation strategies, empowering farmers to proactively manage their farms, adopt new practices and enhance resilience to climate variability. They support farmers in assessing diverse climate-smart adaptation options, guiding them through complex decisions regarding suitable technologies, sustainable production methods, irrigation, mechanization and crop diversification. With input from crop and livestock specialists, extension workers recommend climate-smart practices and technologies, evaluating potential risks and opportunities associated with adaptation investments. To effectively advise on the selection of adaptation practices, extension workers must be trained in farm management economics, enabling them to assess economic viability to guide them in investment decisions.

With these skills, extension workers can guide farmers in selecting crop varieties that are both resilient to climate conditions and economically viable, considering market demands and input costs. By advising on optimal planting times and water management strategies, these extension workers can assist farmers to minimize risks, avoid costly mistakes, optimize resources and enhance long-term profitability.

Agricultural extension workers can also provide invaluable assistance by supporting the implementation of tailored irrigation systems and, in particular, water-efficient drip irrigation technologies. Despite initial costs, these technologies offer substantial long-term savings through reduced water and energy usage. These extension workers can educate farmers about the economic benefits of timely and precise irrigation scheduling and moisture monitoring, helping to avoid over- or under-irrigation, and preventing crop losses and reduced incomes. By optimizing water use, farmers can reduce their input costs, improve their crop productivity and increase their overall farm profitability.

Agricultural extension workers – trained in mechanization – assist farmers in making informed decisions relating to mechanization management. Support of farm management advisers is often needed to evaluate the cost-effectiveness of different options. By working together, they can provide advice on selecting machinery that fits the size of the farm plots available to the smallholder farmer to ensure that investments in equipment yield the best possible returns. Additionally, the technical specialists can help farmers understand the economics of machinery maintenance and operations, guiding them to avoid unnecessary expenses and extend the lifespan of their equipment. Extension workers can also facilitate the organization of cooperative models, where small farmers are encouraged to pool resources to share machinery, reducing individual costs and making mechanization more accessible to a broader number of farmers and more economically viable.

The need for extension workers to possess these farm management skills is elaborated on further in Chapter 16.

7.10 Conclusion

This chapter highlighted the significant impacts of climate change on small family farms, underscoring the essential role that informed management decisions play in successful adaptation. Climate-related risks such as unpredictable rainfall, rising temperatures and increased pest pressures threaten farm productivity, profitability and overall sustainability. To effectively address these challenges, farmers require strong farm management skills and competencies to evaluate available options and implement strategies tailored to their specific conditions.

Central to this approach is the integration of climate-smart agricultural practices, including conservation agriculture, efficient irrigation systems, appropriate mechanization and digital agricultural technologies. However, the successful adoption and effective use of these practices depend heavily on the ability of farmers to critically assess costs, benefits and potential risks. Such informed decision-making processes ensure that resources are allocated efficiently, production is optimized and environmental sustainability is maintained.

Agricultural advisers equipped with the requisite management skills play a pivotal role in building the capacities of farmers. They assist in diagnosing farm-level issues, evaluating technological options and facilitating access to critical information and market opportunities. Advisers with strong managerial expertise not only deliver technical guidance but also empower farmers to make strategic business decisions that balance economic viability with environmental stewardship. Ultimately, informed farm management decisions, enhanced farm management competencies among farmers, and strong, skilled advisory services collectively form the foundation for successful climate adaptation and the sustained viability of small family farms. Investments in these areas not only help smallholders manage immediate climate challenges but also position them to thrive amid long-term climatic and market uncertainties.

References

Parra-Lopez (2024) Integrating digital technologies in agriculture for climate change adaptation and mitigation: State of the art and future perspectives. *Computers and Electronics in Agriculture* 226, November 2024. Available at: https://www.sciencedirect.com/science/article/abs/pii/S0168169924008032 (accessed 4 February 2025).

Wu, P., Wang, Y., Li, Y., Yu, H. and Shao, J. (2023) Optimizing irrigation strategies for sustainable crop productivity and reduced groundwater consumption in a winter wheat-maize rotation system. *Journal of Environmental Management* 348. DOI: 10.1016/j.jenvman.2023.119469.

Additional Reading

Barrow C. (1987) Water Resources and Agricultural Development in the Tropics. Longman Development Studies. Routledge: London, UK

Baudron F, D. Kahan *et al.* “Re-examining appropriate mechanisation in Eastern and Southern Africa: two-wheel tractors, conservation agriculture, and private sector involvement” Food Security, May 2015.

Binswanger P. (1978). The economics of tractors in south Asia: an analytical review. Agricultural Development Council and ICRISAT.

FAO (2021) Scaling Up Climate-Smart Crop and Mechanisation Systems to Promote Sustainable Crop Production in Sri Lanka and Zambia. Rome: Food and Agriculture Organisation of the United Nations. Available at: https://openknowledge.fao.org/handle/20.500.14283/cb5116en/. (accessed 09/02/2025)

FAO (2021) Small-scale Agricultural Mechanisation Hire Services as a Business Enterprise. Rome: Food and Agriculture Organisation of the United Nations. https://elearning.fao.org/course/view.php?id=711. (accessed 09/02/2025)

FAO (2013) Climate-Smart Agriculture: Sourcebook. Rome: Food and Agriculture Organisation of the United Nations. Available at: https://www.fao.org/climate-smart-agriculture-sourcebook/en/. (accessed 09/02/2025)

FAO Development Series 2 (1997) Small-Scale Irrigation for Arid Zones: principles and options, FAO: Rome. https://www.fao.org/4/W3094E/w3094e00.htm (accessed 09 February 2025)

Kahan D. *et. al*. Thinking outside the plot: Insights on small-scale mechanisation from case studies. *Journal of Development Studies*, 54(11): 1–16. DOI: 10.1080/00220388.2017.1329525 https://www.researchgate.net/publication/317987851_Thinking_Outside_the_Plot_Insights_on_Small-Scale_Mechanisation_from_Case_Studies_in_East_Africa (accessed 09 February 2025)

Mrema, G.C., Baker, D., Kahan, D. (2008). Agricultural Mechanisation in Sub-Saharan Africa: Time for a new look. Agricultural Management, Marketing and Finance Occasional Paper 22. FAO, Rome, Italy. https://www.fao.org/4/i0219e/i0219e00.htm (accessed 17 March 2025)

Pingali P, Bigot, Y., and Binswanger, H.P. (1987) Agricultural Mechanisation and the Evolution of Farming Systems in Sub-Saharan Africa. World Bank, Washington, D.C., US. http://documents.worldbank.org/curated/en/347731468914757627 (accessed 17 March 2025)

Reijntjes C. and B. Haverkort and A. Waters-Bayer (1992) Farming for the Future. Macmillan. https://www.researchgate.net/publication/240489297_Farming_for_the_future_An_introduction_to_low_-_external_input_and_sustainable_agriculture (accessed 17 March 2025).

Sims, B. and Kienzle, J. (2006). Farm power and mechanisation for small farms in sub-Saharan Africa. FAO, Rome, Italy. https://openknowledge.fao.org/items/3ccc1135-9251-4d3a-86b7-1f8ef5c6012a (accessed 17 March 2025).

8 Partnerships and Linkages for Agricultural Commercialization

This chapter explores the challenges small family farmers face due to the rise of agricultural commercialization and the growing influence of large agribusinesses. It highlights the need for small-scale farmers to build partnerships and linkages with agribusinesses to improve their efficiency and profitability. By looking at different types of partnerships and linkages, the chapter outlines practical strategies that small farmers can use to access markets and join modern value chains. To succeed in this changing environment, small farmers must take part in partnerships and share resources. These joint efforts can boost their competitiveness, expand market access and improve their bargaining position. Through such partnerships, smallholders can face the challenges of agribusiness control and work towards long-term growth. Central to this is the need for farmers to develop strong business management skills and join farmer organizations that connect them with modern agribusiness networks.

8.1 Expansion of the Post-production Sector and Agribusiness Dominance

This section looks at the challenges and opportunities of agricultural commercialization. It highlights the rising role of agribusiness and the need for small farmers to form partnerships, share resources and adopt market-focused strategies to grow sustainably.

8.1.1 The inevitable dimension of agricultural commercialization

As agricultural markets have become increasingly globalized, the post-production sector has expanded significantly to meet the growing demand for processed, packaged and market-ready agricultural products. This growth has led to more agribusinesses focusing on post-production, with the resources, infrastructure and market links needed to process and distribute farm products efficiently and at scale. The post-production sector plays a critical role in the agricultural value chain, covering various activities that occur after the primary production of agricultural goods. These include processing, packaging, distribution and marketing, all of which contribute to adding value and transforming raw agricultural products into market-ready goods.

This growth in the post-production sector is closely linked to the commercialization of agriculture. Historically, agriculture was driven by farmers producing primarily for their own consumption, with surpluses sold in local markets. However, the rise of globalization, along with advances in transportation, communication and trade, has facilitated the integration of agricultural markets on a global scale. As a result, farmers are increasingly focused on producing agricultural goods for broader and more distant markets. This transformation has been driven by several factors, including a growing demand for food and agricultural products, changing consumer preferences, and the pursuit of economic growth and development, as described in Chapter 1.

DOI: 10.1079/9781836990536.0008

To thrive in this evolving landscape, small farmers must recognize the value of partnerships and linkages with agribusinesses. These partnerships allow smallholder farmers to access shared infrastructure, negotiate better prices, collectively market their products and strengthen their position within the value chain. Furthermore, such collaborations can enable small farmers to expand their businesses by incorporating post-harvest value-adding activities.

8.1.2 Growing dominance of agribusiness in driving the agricultural sector

Agribusiness enterprises cover activities such as input supply, processing, exporting, trading and retailing. Typically, these entities operate on a large scale and engage in vertical integration, streamlining the entire agricultural value chain from production to consumer delivery. They involve a range of players – large farms, processors, distributors, retailers and input suppliers – working together to make agriculture more efficient and productive. This integrated approach not only optimizes resource use but also supports sustained agricultural development by connecting different segments of the market (see Box 8.1.).

One defining feature of agribusiness enterprises is their market power. The significant market presence of agribusinesses allows them to influence market trends and conditions. Due to their scale and resources, they can often dictate terms and conditions along the value chain. Agribusiness enterprises can influence pricing mechanisms, establish quality and safety standards, and even shape consumer preferences through marketing and advertising strategies.

The financial resources of medium- and large-scale agribusiness enterprises give them a competitive advantage. They have access to capital, enabling them to invest in modern technologies, infrastructure, and research and development. This financial aptitude allows them to adopt innovative farming practices, implement efficient processing and packaging methods, and establish strong distribution networks. As a result, they can optimize productivity, reduce costs and maintain a competitive edge in the market.

Supermarkets, in particular, play a crucial role in modern agribusiness, particularly in enhancing market access for small family farmers. These retail giants are part of what Reardon *et al.* (2018) refers to as the 'supermarket revolution', which has been transforming the food market landscapes in low-income countries by integrating small family farmers into more modern, efficient value chains. Supermarkets can provide small family farmers with reliable and consistent market outlets, helping them shift towards more commercial operations.

8.1.3 Implications for small family farmers

The increasing dominance of agribusiness enterprises presents both opportunities and challenges for small family farmers. On one hand, these enterprises provide access to larger markets, modern technologies and improved infrastructure, enabling farmers to enhance productivity and profitability. As agribusinesses gain more market power, smallholders often struggle with less bargaining power, fewer market options and increased pressure to meet strict standards. To navigate these complexities, small farmers must develop strong farm business management skills, form strategic partnerships and adopt market-oriented approaches that enhance their resilience and competitiveness.

While linkages with agribusinesses improve market access and productivity, farmers must invest resources and adapt their production methods to meet buyer demands. Formal contracts often require farmers to commit land, labour and inputs exclusively to one buyer, limiting diversification and increasing financial risks. For example, a farmer in an exclusive deal with an agribusiness might have to give up other income sources, like growing different crops or selling in local markets, which limits their ability to manage risks.

Expanding production can initially boost profitability but beyond a certain threshold, increasing investments in land, labour and inputs yields diminishing returns. Over-investment may lead to inefficiencies, financial strain and unsustainable operations. To avoid these pitfalls, small farmers must develop business management competencies to assess

Box 8.1. Defining the private sector stakeholders in the African agri-food system. From AGRA, 2024.

The African agri-food system is a dynamic and diverse landscape that includes a wide array of private sector stakeholders, each playing a crucial role in shaping the continent's food security and economic growth. These stakeholders can be broadly categorized into two main groups: (i) micro, small and medium enterprises (MSMEs); and (ii) medium to large enterprises (MLEs).

MSMEs: form the backbone of Africa's agri-food private sector, accounting for approximately 85% of the volume of activity within agri-food value chains (AGRA, 2024). These enterprises include small-scale input suppliers, agrodealers, wholesalers, processors and retailers, often operating informally within rural and urban markets. Despite their importance, MSMEs often face significant challenges, including limited access to finance, inadequate infrastructure and regulatory barriers.

MLEs: though fewer in number compared to MSMEs, they are pivotal in scaling agricultural production and reaching broader markets. These enterprises typically engage in both domestic and regional markets, with some extending their operations to extra-African markets. MLEs are often involved in more capital-intensive activities such as large-scale processing, logistics and exporting. They play a significant role in enhancing Africa's competitiveness in the global market by managing larger volumes of food production, imports and exports, particularly in processed foods and high-value commodities.

Both MSMEs and MLEs are essential to the functioning of the agri-food system, with MSMEs primarily supporting domestic food consumption and MLEs driving export growth and handling large-scale food processing. The interplay between these stakeholders is vital for the sustainable development of Africa's food systems, particularly as the continent continues to urbanize and demand for diverse and processed foods increases.

profitability, manage costs effectively and optimize resource use.

As we have previously noted, supermarkets are increasingly influential in agricultural markets, offering stable demand, broader consumer reach and potentially higher prices for small farmers. Joining supermarket supply chains means farmers must meet high standards for quality, consistency and food safety. This often requires investing in better infrastructure, technology and post-harvest handling. Although supermarkets offer stable markets and lower transaction costs, many small farmers find it hard to join due to limited support, lack of funds and difficulty meeting required volumes and quality. Without appropriate support, they risk exclusion from formal markets and dependence on lower-margin, informal trading networks.

One of the most effective ways for small farmers to strengthen their market position and overcome agribusiness dominance is through collective action. Additionally, small farmers are increasingly seeking strategic partnerships across the value chain to overcome market challenges. Collaborating with input suppliers, service providers, financial institutions and agribusiness firms can facilitate access to capital, training and advanced technology, helping small family farmers improve productivity and efficiency.

In summary, although agribusiness partnerships can improve farm efficiency through better technology, streamlined supply chains and organized logistics, these advantages must be carefully balanced against the potential risks involved.

The following sections will explore various forms of partnerships and strategies that small farmers can adopt to successfully navigate commercialization and ensure their continued integration into the agribusiness sector.

8.2 Farmer Organizations in Support of Smallholder Farmers

Farmer organizations (including cooperatives) are essential to empowering smallholder farmers and enabling them to overcome the challenges relating to market access, input supply, financial constraints and resource management. Farmer organizations allow members to pool resources, share knowledge and collaborate, strengthening their market position.

One of the primary benefits of farmer organizations is the ability to reduce costs and increase efficiency by working together (known as economies of scale). By pooling resources such as machinery, storage facilities and transportation, smallholder farmers can reduce individual costs and improve overall efficiency. For instance, farmer organizations and cooperatives often negotiate bulk purchases of inputs like seeds, fertilizers and pesticides, securing better prices and ensuring consistent quality for their members. Similarly, shared access to machinery and infrastructure enables farmers to adopt modern farming practices and reduce operational costs.

Collective action also enhances farmers' bargaining power in markets. By aggregating their produce, farmer organizations can negotiate better prices and terms with buyers, ensuring fair compensation and access to larger markets that would be challenging for individual farmers to reach. Collaborative marketing strategies, such as joint branding and value addition through processing, further increase the competitiveness of smallholder farmers in both domestic and export markets. Successful examples include farmer organizations investing in cold storage facilities or transportation systems to maintain product quality and expand market reach.

Farmer organizations also significantly improve farmers' access to finance. By pooling resources and collectively negotiating with financial institutions, they can secure better terms for loans and credit. Farmer organizations often act as intermediaries, helping members get financial services that suit their needs. They also give guarantees to lenders, making it less risky to lend to smallholder farmers. Access to finance enables these farmers to invest in high-quality inputs, modern equipment and advanced technologies to increase productivity and, potentially, profitability (see Box 8.2.).

Training and capacity building are integral to the success of these farmer organizations. By being organized into groups they provide structured platforms for knowledge sharing and technical support. Farmer organizations often collaborate with extension services, NGOs and private sector actors to deliver training programmes.

Farmer organizations also contribute to social and environmental sustainability. By promoting practices that conserve natural resources, reduce environmental degradation and enhance resilience to climate change, they support the broader goals of sustainable development. Collaborative initiatives, such as developing farmer clusters that specialize in specific crops or value-added activities, allow small farmers to achieve higher quality standards, attract larger buyers and access niche markets. These clusters facilitate the exchange of best practices, innovations and knowledge, driving overall improvements in production and profitability. This model is elaborated on later in the chapter.

The case studies described in Boxes 8.3. and 8.4. demonstrate the potential of farmer organizations (and cooperatives) to empower small farmers and effect positive change. However, notwithstanding these benefits, farmer organizations face challenges in ensuring effective governance and decision making within these collective entities. Balancing the diverse interests and perspectives of individual farmers can be complex and requires transparent and inclusive processes. In addition, securing sustainable funding, managing administrative tasks and building capacity among members are ongoing challenges. Limitations may also exist in terms of market access, especially when competing with larger agribusinesses. Limited bargaining power and access to distribution networks can hinder small farmers from penetrating lucrative markets. Overcoming these limitations requires strategic partnerships, strong market linkages and continuous capacity building.

8.3 Strengthening Market Integration through Horizontal and Vertical Linkages

To compete in today's commercialized agriculture, small family farmers need to join broader market systems through teamwork and value chain partnerships. This section looks at two main ways: (i) horizontal linkages, where farmers join forces to boost their strength and bargaining power; and (ii) vertical linkages, where they connect with agribusinesses and other value chain players to access markets, services and resources.

Box 8.2. Benefits of collective action – farmer organizations.

- Improve the bargaining power of resource-poor farmers by pooling their products, allowing them to negotiate better with other businesses.
- Reduce costs through bulk purchase of supplies and through lowering net costs for individual members.
- Expand existing and develop new market opportunities by offering larger quantities of assured types and quality to attract buyers.
- Improve product or service quality by adding value to products through improved facilities and equipment.
- Increase income by expanding the members' bargaining power when negotiating prices directly with buyers.
- Create employment for skilled and unskilled labour to work for the group business.
- Help weaker members that cannot do business on their own to improve their business by helping them gain access to better markets.
- Eliminate middlemen from transactions through increased bargaining power in negotiating better prices with buyers directly.

Box 8.3. The Amul Cooperative, India. From Pallathadka *et al.*, 2022.

The Amul Cooperative in India stands as one of the most iconic and successful examples of collective action in the dairy sector. Established in 1946 in the small town of Anand in Gujarat, Amul started with just a few milk producers that came together to form a cooperative society. Their goal was to eliminate exploitation by middlemen and to ensure dairy farmers received a fair price for their milk. Over the years, the cooperative grew exponentially, transforming into a major player in the Indian dairy industry and pioneering what became known as the 'White Revolution', or Operation Flood, which made India one of the largest producers of milk in the world.

Amul's success is built on its unique model of cooperation and empowerment. By pooling resources and sharing infrastructure, the cooperative has been able to process large quantities of milk and produce a wide range of dairy products. This collective approach allowed small-scale dairy farmers to leverage modern processing technologies and infrastructure, which would have been unaffordable individually. The cooperative also collectively negotiates with buyers, ensuring that farmers receive better prices for their milk.

Amul's innovative practices have provided its members with access to modern dairy technology, including cold chain systems, automated processing plants and quality control measures, which have significantly improved the efficiency and quality of dairy production. The cooperative has also invested in marketing and branding, creating a strong identity for Amul products, both domestically and internationally. These efforts have translated into improved incomes for millions of small dairy farmers, who now benefit from regular income, access to veterinary care and training in best practices for dairy farming.

The impact of Amul extends beyond just the economic benefits. The cooperative has played a crucial role in empowering rural communities, especially women, who make up a significant portion of the dairy farming workforce. Through its vast network, Amul has enabled these women to achieve financial independence and contribute to the economic growth of their communities.

8.3.1 Horizontal linkages: building networks and farm clusters

Horizontal linkages involve creating farmer networks and clusters for collaboration, resource sharing and collective action within specific sectors or geographical areas (see Box 8.5.). These linkages also enable farmers to pool essential resources such as machinery, irrigation systems and storage facilities, reducing costs and improving operational efficiency. Similar to farmer organizations, these networks enhance coordination in production, knowledge exchange and collective negotiation for better market

Box 8.4. The cooperative coffee movement, Latin America. From Giovannucci and Koekoek, 2003; Raynolds *et al.*, 2004; Bacon, 2005.

The cooperative coffee movement in Latin America has significantly transformed the livelihoods of small coffee farmers by fostering collective action and direct market access. These cooperatives, which spread across countries like Colombia, Peru and Nicaragua, have empowered farmers to bypass traditional intermediaries that often captured a large share of the profits; instead establishing direct trade relationships with international buyers, particularly in the specialty coffee market. By doing so, the cooperatives have secured fair and stable prices for their members' coffee, providing them with more predictable incomes that are often higher.

Beyond securing better prices, these cooperatives have also made substantial investments in improving the quality of their coffee. Many cooperatives work to get organic certification and support sustainable farming, making their coffee more appealing to eco-conscious buyers who are willing to pay more. This emphasis on quality and sustainability has allowed small coffee farmers to tap into the lucrative specialty coffee market, where demand for high-quality, ethically produced beans is strong.

Through these initiatives, small family farmers have not only increased their income but improved their overall livelihoods. The cooperative model has provided them with access to better resources, including training in sustainable agricultural practices and investments in community infrastructure, such as coffee processing facilities. As a result, these farmers are better positioned to compete in the global coffee market and to strengthen the long-term viability of their operations.

terms, allowing smallholders to strengthen their bargaining power and compete more effectively in commercialized agricultural markets.

One of the greatest benefits of horizontal integration among small family farms is how it facilitates bridging and bonding. Horizontal integration among small family farms strengthens bonding among the farming families by deepening trust and cooperation within farmer groups, enabling collective action. At the same time, it fosters bridging among farmers by connecting them to external networks, including markets, financial institutions, research organizations and policymakers. By combining these two forms of social capital, farmers strengthen their resilience and improve their chances to tap into economic opportunities, leading to better productivity, market access and long-term success.

As discussed in Section 4.3 (on livelihoods) in Chapter 4, social capital is a vital livelihood asset for small family farmers, providing access to resources, knowledge and opportunities that might otherwise be out of reach. Strong networks of trust and cooperation enable farmers to share risks, mobilize support during crises and collectively negotiate better terms in markets and finance (Fig. 8.1.). Social capital, built through collaboration and mutual support, helps farmers remain resilient against challenges like climate change and price swings. It also promotes long-term sustainability through sharing knowledge, encouraging innovation and influencing policies.

Farm clusters, however, refer specifically to geographic concentrations of interconnected agricultural businesses, suppliers and support institutions centred around a particular commodity or related set of activities (see Box 8.6.). Inspired by industrial development models, clustering facilitates technology diffusion, improves access to extension services and lowers transaction costs (Schmitz and Nadvi, 1999; Gálvez-Nogales, 2010). Clusters can form naturally in areas with favourable conditions like water, fertile land or nearby markets; or can be planned by the government or private sector. They boost productivity, innovation and market access, creating a supportive space for small businesses, agribusinesses and service providers to grow.

Horizontal collaboration among farmer organizations is widely recognized as a strategy for aggregating production and achieving economies of scale. Clusters improve market access, drive innovation and create job opportunities, contributing to economic growth, poverty reduction and food security in rural areas. Beyond cost-sharing, clustering involves selling larger volumes of produce to buyers, which supports collective marketing efforts. This enables farmers to negotiate better prices, access larger markets and reduce transaction costs. These networks also facilitate the exchange of best practices and

Box 8.5. Horizontal linkages among small family farmers.

- **Collective production:** small family farmers collaborate to produce similar crops or livestock, allowing them to standardize quality, optimize land use and improve efficiency.
- **Bulk input purchases:** farmers pool resources to buy seeds, fertilizers and other inputs, in larger quantities, reducing costs and improving access to quality supplies.
- **Group savings and credit:** groups of farmers establish savings and credit schemes, enabling easier access to loans, reducing financial risk and fostering investment in farm improvements.
- **Joint borrowing:** farmers collectively apply for loans, using group guarantees to improve credit access, reduce individual risk and secure better lending terms for agricultural investments.
- **Collective marketing:** farmers coordinate sales efforts to negotiate better prices, access larger markets and reduce transaction costs by aggregating their produce.
- **Resource sharing:** farmers share essential resources such as land, water, equipment and storage facilities to reduce costs and enhance productivity.
- **Joint investments:** farmer groups invest collectively in infrastructure, machinery or processing facilities, improving efficiency and value addition opportunities.
- **Knowledge sharing:** farmers exchange best practices, market information and farming techniques to enhance productivity and sustainability.
- **Group technology innovation:** farmer networks collaborate to develop, test and refine new agricultural technologies suited to their local conditions.
- **Group learning:** farmers engage in structured peer learning through training sessions, field demonstrations and experience-sharing initiatives.
- **Policy advocacy:** farmer groups collectively voice their needs and concerns to policymakers, advocating for better policies, access to resources and market support.

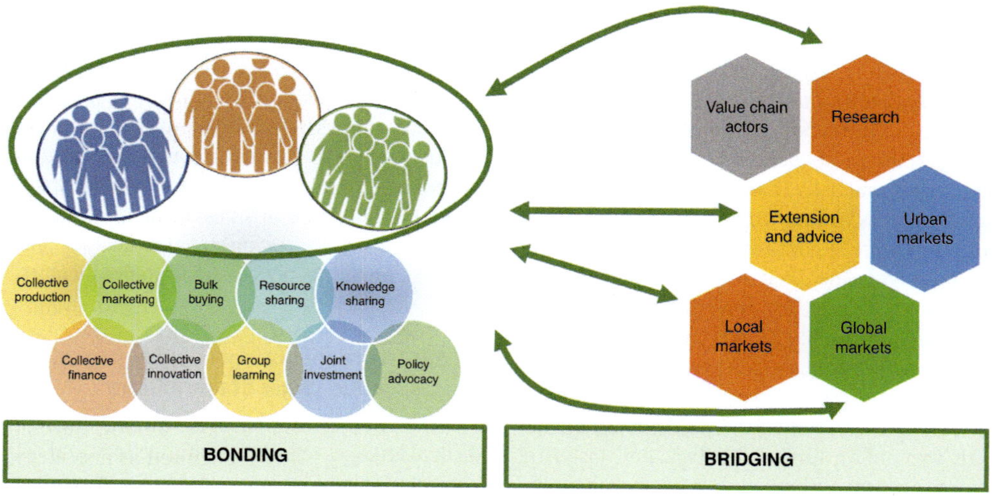

Fig. 8.1. Horizontal integration of small family farmers fosters bridging and bonding.

innovative farming techniques, promoting crop diversification and reducing reliance on monocropping. The cluster approach strengthens interactions between farm enterprises, service providers and support institutions, driving productivity and diversification into higher-value production. Recognizing these advantages, governments have incorporated clustering into agricultural policies, supporting it as an effective tool for integrating small family farmers into value chains in a sustainable and competitive manner (Requier-Desjardins *et al.*, 2003).

Box 8.6. Clusters within Africa's agri-food sector. From AGRA, 2024.

- **Spontaneous clusters:** These clusters develop organically without direct intervention from the government or external entities. They are typically driven by local entrepreneurs and farmers who take advantage of favourable conditions such as natural resources, infrastructure and market access. An example is the fish farming cluster in Kebbi State, Nigeria, which emerged around the confluence of the Niger River and the A1 highway. This cluster has seen rapid growth in fish production and related activities, driven by the availability of water resources and access to markets.
- **Managed clusters:** These are clusters that are deliberately developed and managed by governments, private sector entities or public-private partnerships. Managed clusters often involve the creation of agro-industrial parks or special economic zones designed to concentrate agricultural processing and related services in a specific area. The aim is to attract investment, improve infrastructure and enhance the competitiveness of the agri-food sector.

Agricultural clusters promote both horizontal and vertical linkages, enabling smallholders to benefit from bundled services and improved access to markets and information. By fostering relationships with supporting institutions, private sector actors and other stakeholders, clustering encourages investment in infrastructure, improved service provision and greater productivity. See Boxes 8.7. and 8.8. for examples of clustering. Additionally, clusters enhance the diversification of farming systems into high-value-added production, improving income stability and market positioning for smallholder farmers.

Despite these benefits, clusters also face challenges, such as inadequate infrastructure, limited access to finance and weak stakeholder coordination. Spontaneous clusters often struggle with insufficient support and formal structures, restricting their ability to scale. Managed clusters, while more structured, require significant investment and coordination. To ensure success, policymakers must create an enabling environment by investing in infrastructure, expanding financial access and fostering stronger partnerships between key stakeholders.

Agricultural clusters have been widely adopted in Latin America but have faced greater challenges in Asia, where informal structures and weaker linkages among stakeholders have made clustering for small-scale production more difficult to establish. In countries such as the Philippines, Thailand and India, clusters have emerged, but many lack critical mass and external support, requiring greater facilitation to reach full potential. These experiences highlight the importance of strong horizontal linkages in expanding market access, reducing costs and fostering collective resilience. The success of well-managed clusters serves as a model for other farmers and stakeholders, demonstrating how cooperation and collective action can unlock new market opportunities and enhance competitiveness.

Participation in horizontal linkages requires small family farmers to develop key business and financial management skills. Effective decision making in these groups depends on members' ability to analyse financial data, assess risks and manage collective investments responsibly. The case studies illustrate how well-managed producer organizations have successfully entered high-value markets by implementing sound financial and business strategies, further strengthening the economic resilience and market competitiveness of smallholder farmers.

8.3.2 Vertical linkages: engaging with value chain stakeholders

Vertical linkages refer to the connections and collaborations small farmers establish with various stakeholders within the agricultural value chain. Developing linkages requires coordination between the different stakeholders and organizing the various stages of production, processing and distribution in a harmonized manner. Partnership relations are essential for integrating small farmers into larger markets, allowing them to participate in more complex supply chains and gain access to inputs, technical expertise and market opportunities. It often includes strategic

Box 8.7. NorminVeggies, the Philippines. From Uy, 2011.

This case in the Philippines involves the organization of small vegetable producers into producer groups that are ultimately aggregated into a producer association, the Northern Mindanao Vegetable Producers Association (NorminVeggies). The vegetable producer groups, after being organized, were encouraged to form horizontal clusters for collective marketing. The clusters provided a supply base to consolidate product volumes and variety and result in greater access for members to diversified and predictable markets. NGOs, in alliance with local government units, provided post-harvest and marketing services to small farmers through partnership arrangements with the producers' association. Technical support was provided by GEM/USAID* in partnership with the department of agriculture. The clustering strategy developed by the project followed eight sequential steps aimed at linking farmers to the market. The steps included: (i) site selection, partnership building and formation of a working group; (ii) product supply assessment and product selection; (iii) market chain study; (iv) cluster formation; (v) cluster plan formulation; (vi) test marketing; (vii) scaling up; and (viii) cluster strengthening.

The scheme required strong coordination by NorminVeggies of interrelated activities – production scheduling, post-harvest requirements, logistics (land transport, seaport and airport operations), grower and buyer communication, invoicing, payment collection and sale remittances to the producers. A management team was formed to handle these various responsibilities, funded through market facilitation fees charged to producers for these forward integrated services. NorminVeggies, in effect, acts as a 'market facilitator' linking clusters of producers directly to buyers. The company is accountable for product quality and reliability and assuring producers of the prices they receive for produce sold. Via clustering, produce could be more easily traced to the farm and to specific growers. The system has proved to be attractive to agri-food companies in the Philippines, introducing mechanisms for quality control, quick responses to buyer feedback and implementation of market innovations. The system, in short, aims at mitigating the risks faced by dispersed small producers by providing continuous supplies of raw material.

*USAID was an international development agency of the government of the USA. GEM is the Growth with Equity in Mindanao Programme in the Philippines.

Box 8.8. Kaffrine horticultural cluster, Senegal. From Tarchiani *et al.*, 2020.

The Kaffrine horticultural cluster in Senegal is a powerful example of how small-scale farmers can overcome the challenges of land degradation and climate change through collective action and sustainable practices. In the face of worsening soil conditions and increasing climate variability, farmers in the Kaffrine region came together to form a horticultural cluster. This cluster enabled them to share knowledge, pool resources and collaborate on implementing innovative farming techniques that were more resilient to the changing environmental conditions.

Within this framework, farmers introduced a range of soil conservation techniques, such as contour farming and the use of organic mulches, which helped to prevent erosion and restore soil fertility. They also embraced agroforestry systems, integrating trees into their farming practices to enhance biodiversity, improve soil health and provide additional sources of income. In addition, the cluster promoted water management strategies, including the construction of small reservoirs and the adoption of drip irrigation systems, which helped to optimize water use in an area where water scarcity is a growing concern.

The collective efforts of the farmers in the Kaffrine horticultural cluster not only led to improved land productivity but also significantly increased their resilience against the impacts of climate variability. By working together, the farmers were able to produce higher quality horticultural products that met market demands, thereby enhancing their access to local and regional markets. This increased market access has translated into better incomes and greater economic stability for the farmers involved, demonstrating the power of community-driven approaches to sustainable agriculture.

alliances – agreements between independent entities to achieve a shared objective.

One prominent form of vertical linkage is *contract farming*, where small family farmers enter agreements with larger commercial farmers or agribusinesses, such as input suppliers, traders, agro-processors and exporters. These arrangements allow small farmers to gain access to resources and markets that would otherwise be out of reach.

Contract farming significantly improves small farmers' production capacity and market access. Under such agreements, farmers commit to producing specified crops or livestock based on predefined conditions – such as agreed prices, quantities and quality standards. The contracting agribusiness or larger farm often provides necessary inputs like seeds, fertilizers and technical guidance, to ensure farmers meet market demands. This model offers numerous advantages to small family farmers, including secure market access, which reduces the risks associated with price volatility. Furthermore, contract farming provides farmers with technical and financial support, helping them enhance productivity and meet stringent market standards. For instance, Kenyan small family farmers growing high-value crops like French beans for international markets benefit from this model, as do dairy farmers in India who partner with cooperatives to improve milk yields and stabilize income through consistent pricing. See Boxes 8.9. and 8.10. for more examples.

Contracts range from informal verbal agreements to formal written arrangements. These can include market-specific agreements, where buyers commit to purchasing farmers' produce, to production management contracts,

Box 8.9. Inclusive business models. From SNV Nepal, personal communication, 2013.

Inclusive business models are approaches that integrate small family farmers into agricultural value chains, ensuring they gain fair market access, resources and stable incomes. By fostering partnerships through contract farming, cooperatives and fair trade, these models provide training, financing and support, helping farmers improve productivity, reduce risks and build resilience while businesses benefit from a reliable and sustainable supply of produce.

The inclusive business approach aims to:

- Reduce poverty by selecting cash crops and forest products with high return potential and promising employment opportunities for smallholder farmers and forest users.
- Increase social and environmental responsibility by mainstreaming corporate social responsibility and implementing codes of conduct in production, processing and marketing.
- Ensure social inclusion by addressing the needs of marginalized and socially excluded people involved in selected value chains.

The process includes low-income households in value chains as suppliers, distributors and consumers. It starts by identifying key businesses along the value chain and then builds connections with small farmers and vulnerable households. Producer–buyer contracts are advocated, linking production to guaranteed markets while ensuring a high level of transparency. The contracts usually specify cultivation practices, the quantity and quality to be supplied, the price offered and delivery times. Attention is given to developing niche and differentiated products. The approach is original, combining poverty alleviation with profits and business development. Effort is made to create competition and avoid market distortion and monopolistic behaviour of private companies.

Value chain development involves the following steps:

1. Selection and assessment of prospective participating agribusinesses and their business proposals based on standard criteria.
2. Formulation of value chain business plans, supported by signed implementation agreements between the involved parties.
3. Implementation of value chain business plans with each party taking up its role and responsibilities, including contract formulation, production and processing and contract fulfilment.
4. Participatory monitoring of progress where adjustments are made, and impact is evaluated.

Box 8.10. PepsiCo in India. From PepsiCo, 2020.

In India, one notable example of contract farming is the collaboration between PepsiCo and smallholder farmers in the production of potatoes used for the company's snack products, such as Lay's chips. Under this agreement, PepsiCo provides farmers with high-quality potato seeds, technical guidance and support in managing crop production. Farmers are guaranteed a stable price for their produce, reducing the risks associated with market price fluctuations. In return, the farmers agree to produce potatoes that meet the company's specific quality and quantity standards. This arrangement benefits smallholder farmers by providing them with a reliable income, access to advanced agricultural techniques, and the assurance that their produce will be purchased.

PepsiCo's contract farming initiative has not only improved the livelihoods of thousands of smallholder farmers but also strengthened the company's supply chain by ensuring a consistent supply of high-quality potatoes. In addition to offering technical assistance, PepsiCo helps farmers adopt sustainable farming practices, such as efficient water management and crop rotation, which have led to increased yields and better resource utilization. This contract farming model exemplifies how large agribusinesses and smallholder farmers can work together to create mutually beneficial outcomes, providing smallholders with access to resources and markets they might otherwise struggle to reach.

where buyers actively participate in production decisions. Resource-providing contracts, where buyers supply key inputs and supervize production, are another model. In contrast to contract farming, *vertical integration*, where a single entity owns the production, processing and distribution stages, allows for complete control over the supply chain but is less common in small family contexts.

There are, however, various models of contract farming based on collaboration between farmers and buyers with advisory services and training often embedded within commercial sales. An innovative contract farming scheme is the K-Farm carambola programme in Malaysia, which seeks a more integrated relationship between farmers and buyers. K-Farm is a fruit export company that in the past was a leading commercial farming operation specializing in tropical fruit production. As it developed as an export producer of carambola (starfruit), a specialized exotic product, it recognized the need for volume to enhance the competitive position of the firm in the European export market. A concerted strategy was followed to develop the supply chain network for carambola, ensuring volume and quality while adhering to global good agricultural practice (GAP) standards (see Box 8.11.). The approach aims at integrating small individual farmers into the carambola supply chain, thereby ensuring control of the quality of production, the safe management of pesticides and the provision of consistent and regular supplies of raw material. The scheme requires that farmers enter formal contractual arrangements with K-Farm. K-Farm represents the producers externally to upstream buyers. The company provides the growers with regular extension and training support and is responsible for the continuous monitoring of the farming activities. While contracts are used, the close relationship with the producers and the ongoing assistance strengthens the bond between both parties. There is mutual interest to maintain the relationship – the producers receive advisory support and secure export outlets, while K-Farm can be assured of volume, quality and consistency of produce. Efforts have been made by K-Farm to increase the profit margins of farmers by reducing the costs of purchased inputs and securing premium prices. The attractive economic benefits attained were sufficient to integrate small producers into the supply chain. Advisory support provided to the producers was also an incentive for them to continue with the scheme. This included market price information, farm planning and management assistance, advice on quality and safety and the testing of new production technologies. While contract farming relations often suggest two separate value chain stakeholders engaged in a collaborative relationship, the more successful cases of interlocking linkages, as depicted by K-Farm, demonstrate the benefits of closer integration between the two.

Box 8.11. Good agricultural practice (GAP).

GAP is a certification system for agriculture, specifying procedures (and attendant documentation) that must be implemented to create food for consumers or further processing that is safe and wholesome, using sustainable methods. While there are numerous competing definitions of what methods constitute good agricultural practice, there are several broadly accepted schemes that producers can adhere to.

Another example of private sector-led advisory support is HJS Condiments Limited in Sri Lanka. The initiative began in 1988, with the creation of Sunfrost Ltd., a company producing and marketing gherkins and semi-processed pickles for export. Owing to the high labour costs of production, the company decided to set up a contract farming scheme with small-scale farmers. This operated for over 5 years after which the company decided to form HJS Condiments to increase value addition by processing pickles and diversifying into other fruits and vegetables. The scheme works with over 12,000 outgrowers and the company provides extension advice, credit (in kind) and an assured market outlet. The ratio of extension workers per farmer is one to a hundred, a manageable ratio, and each farmer is visited on average twice a week during the production season. Extension advice is provided free of charge with the costs recovered at the time of settlement when the products are delivered (Swanson and Rajalahti, 2010).

There are many other examples of these emerging private sector extension systems for labour-intensive, high-value crops being exported from countries such as India, Thailand and Vietnam. In nearly all these cases, the firm absorbs the cost of advisory services, because such services are essential for maintaining product quality and meeting international quality standards.

These case studies show that advisory services play an important role in facilitating vertical coordination, particularly in cases requiring public–private collaboration, where costs are often borne by the public sector. The role of the farm management specialist or agribusiness advisor is critical here to support not only farmers but agribusiness operators in developing their business skills and facilitating access to markets and finance. This role is elaborated on in Part 4 of this book.

8.3.3 Benefits and challenges of contract services for small farmers

Contract farming and agribusiness linkages present both opportunities and challenges for small farmers. On the positive side, these partnerships provide secure market outlets, reduce the risks of price fluctuations and offer access to essential resources such as inputs and technical knowledge. However, there are also potential downsides. Power imbalances between small farmers and larger agribusinesses may lead to unfair contract terms, while over-dependence on a single buyer could limit farmers' autonomy and bargaining power. Dependency on a single buyer or long-term contracts can reduce flexibility and expose farmers to price volatility, contract restrictions and shifting market conditions. To reduce risks, small farmers should diversify their income across different markets and value chains, negotiate fair and clear contracts, and improve their financial skills to understand long-term business sustainability. Adapting to agribusiness requirements often requires investments in improved farming techniques, post-harvest handling and modern inputs, but these investments must be carefully planned to ensure real economic benefits rather than simply increasing costs. Addressing these challenges requires careful attention to contract fairness, transparency and the equitable distribution of benefits along the value chain.

The dairy value chain in Kenya is a good example of contract farming. The value chain has undergone significant transformation through the formation of strategic partnerships between small dairy farmers, processors and retailers. These collaborations have been instrumental in creating reliable and efficient supply chains, which have directly benefited small-scale dairy farmers by providing them with better access to markets and improving the quality of their milk.

One of the key outcomes of these partnerships has been the establishment of higher milk quality standards. Through training and technical support provided by processors and other stakeholders, smallholder farmers have been equipped with the knowledge and tools to produce higher-quality milk that meets market demands. This focus on quality has not only helped farmers to secure better prices for their products but has also opened opportunities for them to access more lucrative markets, including urban centres and export markets.

Additionally, by forming cooperatives and farmer groups, these small dairy farmers have been able to enhance their bargaining power when negotiating with buyers and processors. This collective approach has allowed them to negotiate more favourable terms, ensuring that they receive a fair share of the profits generated along the value chain. As a result, many small dairy farmers in Kenya have experienced improved incomes, greater economic stability and enhanced livelihoods.

These developments in the dairy value chain have also contributed to the overall growth of the dairy sector in Kenya, making it one of the most dynamic agricultural sub-sectors in the country. The success of these partnerships underscores the importance of collaborative efforts in overcoming challenges faced by small family farmers and in driving sustainable agricultural development (Bebe *et al.*, 2003; FAO, 2019; Technoserve, n.d.).

8.3.4 Contract negotiation: power relationships and impact on smallholder farmers

Contract negotiations between smallholder farmers and agribusinesses are often influenced by power imbalances. Agribusinesses often have more financial resources, market knowledge and bargaining power, which can put smallholder farmers at a disadvantage. Understanding these power dynamics is essential for smallholder farmers to navigate contract negotiations effectively and protect their interests.

Power imbalances in negotiations can significantly disadvantage small farmers. They may face challenges such as unfair contract terms, unequal distribution of risks and rewards and limited decision-making power. These imbalances can hinder the ability of small farmers to negotiate favourable terms, achieve sustainable profitability, and maintain their autonomy within the value chain. It is important to address these impacts and strive for more equitable power relationships to ensure the long-term viability of small farmers.

The additional case studies in Boxes 8.12. and 8.13., further demonstrate the effectiveness of partnerships and resource sharing in contract negotiation. They highlight the role of cooperatives, farmer associations and farmer groups, in empowering small farmers and enabling them to secure fair contracts (see Box 8.12.). By learning from these examples and implementing similar strategies, small farmers can strengthen their position in contract negotiations, improve their income and ensure their long-term sustainability (see Box 8.13.).

In conclusion, while there are advantages and disadvantages to consider, successful case studies exemplify how well-designed and equitable contract services can contribute to the growth and sustainability of small-scale farming enterprises.

8.4 Small Family Farm Empowerment Opportunities

Empowering small family farmers is essential for improving livelihoods, enhancing food security and supporting sustainable agriculture. There are strategic initiatives beyond those discussed in the preceding sections that can help overcome the many challenges they face and create new opportunities for growth.

One key strategy is linking small family farmers to institutional markets, such as national school feeding programmes, which provide stable demand and reliable income. Additionally, inclusive business shareholder models, commodity associations and joint ventures enable farmers to participate more equitably in agribusiness, strengthening their role in value chains. These approaches not only enhance market access but also build resilience by promoting financial sustainability, innovation and collaborative decision making.

This section explores various pathways to empowering small family farmers, highlighting successful case studies and practical strategies that can be adapted to different contexts. Through these mechanisms, farmers can improve their economic stability, strengthen

Box 8.12. Fostering success in contract negotiation for small farmers.

Strengthening farmer organizations and cooperatives: collective action through farmer organizations and cooperatives can strengthen the bargaining power of smallholder farmers, allowing them to negotiate from a position of greater influence. By uniting and presenting a cohesive front, smallholder farmers can exert more leverage and secure fairer contracts. To further support this, capacity-building efforts are essential to help small-scale farmers develop skills in contract negotiation, legal literacy and understanding market dynamics, all of which can improve their ability to advocate for their interests effectively.

Access to market information and legal support: smallholder farmers need access to reliable market information and legal support to understand their rights, obligations and the available alternatives during contract negotiations. This knowledge equips them with the tools to make informed decisions and advocate for fair terms.

Building trust and long-term relationships: developing trust and establishing long-term relationships with agribusiness partners can contribute to more collaborative and equitable contract negotiations. Trust facilitates open communication, shared goals and mutual understanding, leading to agreements that benefit both parties.

Box 8.13. Lessons around contract negotiations.

Lessons from the poultry industry in Brazil demonstrate how small poultry farmers have successfully organized themselves into associations to negotiate more effectively with large poultry processing companies. These associations have allowed small farmers to leverage their collective strength, enabling them to secure more favourable contract terms that include better prices for their poultry, reliable input supply arrangements and access to technical support and expertise. This collective organization has empowered the farmers, providing them with greater bargaining power, reducing their dependency on volatile market conditions and ensuring more stable and predictable income streams. The success of these farmer associations in Brazil underscores the importance of collective action in enhancing the negotiating power of smallholders, leading to improved livelihoods and fostering a more equitable and sustainable poultry industry (Wilkinson, 2004).

In Kenya, small vegetable farmers have been effective in forming farmer groups and associations to enhance their negotiating power with agribusinesses. These groups serve as a platform for collective marketing, joint purchasing of inputs and improved access to extension services, enabling farmers to secure better prices and reliable markets. By organizing collectively, these farmers have significantly improved their productivity and profitability, benefiting from economies of scale and stronger bargaining positions. Moreover, these farmer groups have facilitated access to credit and financial services, empowering small farmers to invest in necessary infrastructure and technology to meet the quality and timing standards demanded by agribusinesses. This collective approach has not only strengthened the farmers' economic position but also promoted long-term sustainability and growth within the agricultural sector (Markelova *et al.*, 2009).

their bargaining power and contribute more effectively to national and global food systems.

8.4.1 Family farm linkages to institutional markets

Linking small farmers to institutional markets, like school feeding programmes, is an effective strategy for improving rural livelihoods. One prominent example of this approach is Brazil's Zero Hunger Program (Fome Zero), launched in 2003. This comprehensive initiative aimed at combatting hunger and poverty while supporting family farmers by integrating their produce into public procurement programs. Key components of the Zero Hunger Program, such as the Food Purchase Program (PAA) and the National School Feeding Program (PNAE), created guaranteed markets for family farmers by mandating the purchase of locally produced food for school meals and other social programmes. This approach not only improved the nutritional quality of meals but also provided small farmers with a reliable source of income, reducing their exposure to the volatile private market. According to research, 13 million Brazilians were lifted out of poverty through Fome Zero (Silva *et al.*, 2010).

The PAA Africa programme, modelled after Brazil's PAA, is an extension of this successful strategy into five African countries: Mozambique, Malawi, Senegal, Niger and Ethiopia. Launched in 2012 through a partnership between the Brazilian government, FAO and the World Food Programme, PAA Africa sought to replicate the success of Brazil's Zero Hunger Programme by connecting small family farmers to institutional markets, such as school feeding initiatives. The programme aimed to support sustainable agricultural production, improve food security and empower local economies in these countries by facilitating public procurement from local farmers. By strengthening the links between small farmers and institutional buyers like schools, PAA Africa helped farmers secure steady incomes and encouraged them to improve the quality and quantity of their agricultural products.

The Brazilian experience, and its replication in Africa, illustrates the importance of aligning policies and institutions to support family farmers. Central to these efforts is ensuring that small family farmers have access to the resources needed to meet institutional demand. This involves improving farmers' access to credit, technical assistance and extension services. Brazil's public policies also incorporated credit lines specifically for women and young farmers, along with rural insurance schemes to protect farmers from risks associated with agricultural production. The comprehensive support package helped small family farmers in Brazil boost their productivity and enabled them to meet the quality standards required by institutional markets, ensuring the sustainability of the school feeding programmes.

However, challenges remain in scaling up such initiatives to reach more farmers and expand institutional markets. In Brazil, despite the success of the Zero Hunger Program, not all family farmers benefit equally, and coverage gaps persist. Similarly, in Africa, while PAA Africa has made progress in linking farmers to school feeding programs, many small farm families still lack access to the technical and financial support needed to fully participate in institutional markets. Addressing these challenges requires continued collaboration between governments, international organizations and civil society to create policies that support small family farmers in both supply and demand, ensuring that they can contribute to national food security while benefiting from stable, institutional markets.

8.4.2 Shareholder models

As we have seen thus far, empowering small farmers to become active participants in agribusiness linkages is crucial for their sustained growth and development. As previously noted, empowerment can be significantly enhanced through the formation and strengthening of farmer organizations in their various forms. One effective way for farmers to benefit equally is through shared ownership (shareholder models) of their business ventures. Farmer organizations that operate on a shareholder model allow members to have a stake in the success of the enterprise. This not only aligns the interests of the farmers with the organization's goals but also ensures that profits are equitably distributed among those who contribute to the business. When farmers own a part of the enterprise, they are more likely to be invested in its success and more willing to adopt innovative practices that can enhance productivity and profitability (see Box 8.14.).

The shareholder model offers several key advantages. By holding shares, small family farmers have a direct financial incentive to see the business thrive, as their returns are tied to the enterprise's overall performance. This model fosters a sense of ownership and responsibility among members, encouraging them to actively participate in decision-making processes and governance. Additionally, the reinvestment of profits back into the organization can lead to further development, such as improved infrastructure, access to better technologies and enhanced training programmes. These reinvestments create a virtuous cycle of growth and sustainability, benefiting all farm organization members.

Moreover, the shareholder model can help mitigate risks and provide a safety net for small family farmers. In times of market volatility

Box 8.14. The shareholder model.

The shareholder model refers to a business structure where individuals or entities (such as farmers, investors or agribusinesses) own shares in an agricultural enterprise. These shareholders collectively own the business, and their financial contributions, risks and returns are proportional to the number of shares they hold. The model allows for the pooling of resources, shared decision making and shared profits or losses, based on the performance of the business.

Key features of the shareholder model:

- **Ownership of shares:** shareholders buy or are allocated shares in the agricultural business. These shares represent their ownership stake and entitle them to a portion of the profits.
- **Shared risk and reward:** profits and losses are distributed among shareholders based on the number of shares they own. This distributes the financial risks of farming (e.g. crop failure, market fluctuations) across multiple stakeholders.
- **Collective decision making:** shareholders typically have voting rights proportional to their shareholding, allowing them to participate in key business decisions, such as the selection of crops, investment in infrastructure or expansion of the business.
- **Access to capital:** the shareholder model enables agricultural businesses to access capital from a variety of sources, including individual investors, cooperatives or larger companies. This is particularly useful for funding large-scale projects like buying machinery, irrigation systems or processing facilities.
- **Professional management:** in larger shareholder agricultural businesses, the day-to-day management is often handled by professional managers rather than the shareholders themselves, allowing for more efficient and specialized management.

or poor harvests, the collective strength of a farmer organization can offer stability through shared resources and mutual support. Diversified income streams generated through collaborative ventures, such as processing facilities or value-added products, can also provide additional financial resilience. This model ensures that farmers are not merely passive participants but are integral to the business's success, sharing both the risks and rewards of the enterprise.

A prime example of a farmer-owned venture that has provided substantial benefits to its members is BRAC in Bangladesh. BRAC, originally established as a development organization, has evolved into a large-scale, farmer-owned enterprise that includes multiple agricultural ventures. BRAC's approach integrates small farm families into various agribusiness activities, providing them with the necessary support to improve their productivity and market access. Farmers involved with BRAC benefit from training, access to quality inputs and a guaranteed market for their produce. The organization's structure ensures that profits are reinvested into community development projects, further enhancing the livelihoods of its members.

BRAC's model demonstrates how farmer-owned enterprises can drive agricultural innovation and economic development. By focusing on inclusivity and shared ownership, BRAC has empowered small family farmers to become active participants in the market, improving their income stability and resilience. This approach can be replicated in other contexts to promote sustainable agricultural practices and economic empowerment. It highlights the importance of creating business models that not only involve farmers as producers but also as stakeholders who benefit directly from the value added through agribusiness activities.

8.4.3 Commodity associations

Commodity associations unite producers, processors, traders and others involved in marketing specific agricultural products or 'commodities', such as coffee, cotton, dairy or grains. These associations aim to support their members by addressing common challenges, improving product quality, enhancing market access and advocating for favourable policies.

Boxes 8.15. and 8.16. give examples of key functions of commodity associations which

Box 8.15. Examples of commodity associations.

Ghana Cocoa Board (COCOBOD): provides marketing, price stabilization and extension services to smallholder cocoa farmers. They work to improve cocoa quality, promote fair trade and ensure farmers get better prices.

East Africa Grain Council (EAGC): helps grain producers and traders improve grain quality, access regional markets and navigate policy issues. The EAGC promotes standardized grading and storage techniques to reduce post-harvest losses.

Colombian Coffee Growers Federation (FNC): supports over 500,000 small coffee producers in Colombia. It focuses on improving coffee quality, promoting Colombian coffee worldwide and ensuring that farmers get fair prices for their coffee. They also invest in rural development projects for coffee-growing communities.

Box 8.16. Key functions of commodity associations.

Advocacy and representation: commodity associations represent the interests of their members in discussions with government agencies, industry bodies and international organizations. They advocate for policies and regulations that benefit their commodity sector, such as subsidies, export incentives or improved infrastructure.

Market access and promotion: associations often help members access larger markets by promoting the commodity both domestically and internationally. They may organize trade fairs, marketing campaigns or partnerships with buyers to ensure better visibility and market penetration for their commodity.

Standardization and quality control: to maintain competitiveness, commodity associations set quality standards and ensure that members adhere to them. This helps protect the reputation of the product in both local and global markets.

Research and development: many commodity associations fund research to improve crop yields, pest control, disease management and processing methods. They work to introduce innovations that improve productivity and profitability for their members.

Training and capacity building: associations offer training on best practices in production, harvesting, processing and storage. They help farmers adopt modern techniques and sustainable methods to enhance efficiency and product quality.

Advisory services and information and knowledge sharing: commodity associations provide their members with up-to-date market information, such as price trends, trade policies and buyer demands. This helps farmers and traders make informed decisions on production and sales.

Access to finance and inputs: some associations help members access financial services such as loans, credit or subsidies. They may also negotiate bulk purchases of inputs like seeds, fertilizers and equipment, reducing costs for members.

Risk management and climate resilience: commodity organizations often support climate adaptation strategies, such as promoting climate-resilient crops, insurance schemes and disaster preparedness plans to help farmers cope with extreme weather events.

Collective bargaining power: By organizing farmers into a unified group, associations strengthen negotiation power with buyers, reducing price exploitation by middlemen and ensuring fairer pricing structures.

Value addition and processing support: some associations help develop local processing facilities to add value to raw commodities (e.g. milling grain, roasting coffee or processing dairy), enabling farmers to earn higher profits from finished goods rather than selling raw produce.

Logistics and supply chain coordination: commodity organizations may facilitate storage, transportation and distribution networks, ensuring that products reach markets efficiently and with minimal post-harvest losses.

are pivotal in advancing the interests of small family farmers within specific agricultural markets. By uniting a diverse group of stakeholders – including farmers, traders, processors and exporters – these associations facilitate a collaborative approach to overcoming common challenges and seizing market opportunities. They often take on multiple roles beyond advocacy, including involvement in policy discussions, research and development, trade negotiations and promoting quality standards.

The function of commodity associations is crucial in ensuring that small family farmers are not left at a disadvantage in the agricultural value chain. By representing a collective voice, these associations have the leverage to influence policy decisions that can lead to improved market conditions for the commodity in question. This may include negotiating better trade terms, enhancing access to necessary agricultural inputs, or advocating for regulatory changes that benefit small-scale producers.

Moreover, as indicated in Box 8.17., commodity associations often take an active role in addressing logistical challenges that affect the commodity's value chain. This might involve improving storage facilities, enhancing transportation networks or streamlining export processes, all of which can significantly reduce post-harvest losses and increase profitability for small family farmers. Although less common, some commodity associations also provide direct advisory services and market information, empowering farmers with the knowledge needed to make informed decisions about crop management, market timing and pricing.

Through these activities, commodity associations help to stabilize market access and improve the economic outcomes for their members. They facilitate the pooling of resources, sharing of knowledge and collective action, which can be particularly beneficial in negotiating with large buyers or entering new markets. By doing so, they enhance the competitive position of small family farmers, promoting more equitable growth and sustainable agricultural practices within the communities they serve.

Finally, as is also indicated in Box 8.17., commodity associations are powerful instruments for addressing risk, especially risk associated with the manifold impacts and challenges of climate changes discussed in detail in Chapter 7. A key role of commodity organizations might be to enhance climate resilience by promoting climate-resilient farming practices. Access to climate information and early warning systems is essential for farmers to prepare for weather-related disruptions. If equipped with the necessary resources, they could provide timely weather forecasts, seasonal climate predictions and early warning alerts through platforms like radio programmes, mobile services and farmer cooperatives. This could enable farmers to make proactive decisions, such as adjusting planting schedules or taking preventive measures against anticipated droughts or pest outbreaks.

Beyond on-farm adaptations, commodity organizations have the potential to advocate for policies that support climate resilience. They, in some cases, influence governments and financial institutions to incentivize climate-smart agriculture, such as subsidies for organic fertilizers, tax reductions for water-efficient technologies and financial support for sustainable farming practices. Additionally, they may be able to push for stronger disaster relief programmes and ensure the inclusion of smallholders in climate finance mechanisms like carbon credit markets.

Commodity organizations are well placed to facilitate access to financial tools like climate risk insurance, low-interest loans and credit schemes that can help farmers recover from losses and invest in climate adaptation technologies. These developments are increasingly occurring in many low-income countries. Moreover, with improved infrastructure, commodity organizations have the potential to help farmers store surplus produce, overcome logistical challenges and protect against the economic impacts of climate change. In this way they are well placed to contribute to long-term sustainability.

8.4.4 Joint ventures

Joint ventures between small farmers and larger agribusinesses help overcome challenges and capitalize on agricultural opportunities. These partnerships typically involve farmers organized into farmer associations or cooperatives, which

Box 8.17. The role of commodity associations in supporting smallholder farmers.

African Fine Coffees Association (AFCA) – Africa: the AFCA is a prominent commodity association representing coffee sectors across 11 African countries, including Uganda, Kenya, Ethiopia and Rwanda. Established in 2000, AFCA plays a pivotal role in supporting smallholder coffee farmers by promoting African specialty coffees in global markets and advocating for policies that benefit coffee producers.

AFCA facilitates training programmes aimed at improving coffee quality through workshops on cupping, grading and marketing. It also organizes national coffee cupping competitions and the annual African Taste of Harvest competition, which highlights the finest coffees from member countries. These efforts ensure that smallholder farmers can meet international quality standards, thereby accessing premium markets and securing better prices for their produce.

Moreover, AFCA actively engages in policy advocacy, working with regional and national governments to create a favourable business environment for coffee farmers. This includes promoting fair trade practices, supporting certification processes and coordinating policy dialogues to address challenges faced by coffee producers. By providing these services and advocating for the interests of its members, AFCA enhances the livelihoods of smallholder coffee farmers and strengthens the overall coffee value chain in Africa (African Fine Coffees Association (AFCA), 2025).

Horticulture commodity associations – Asia: in Asia, horticultural associations have similarly played a crucial role in supporting small-scale farmers, particularly in countries like India and Thailand. These associations bring together farmers, traders, processors and exporters to address common challenges and improve market access for horticultural products. They offer essential services such as training in sustainable farming practices, quality improvement and access to market information, which help smallholders meet the demands of both domestic and international markets.

Horticultural associations in Asia also engage in advocacy to influence policy decisions that benefit their members. They work closely with government bodies to secure subsidies for inputs, streamline export processes, and improve infrastructure for post-harvest handling and transportation. By representing the collective interests of smallholder farmers, these associations help ensure that they receive fair prices for their products and are not disadvantaged in the value chain (Interréseaux Développement Rural, 2024).

collaborate with agribusiness firms to enhance the value chain from production to marketing, especially for products that require a long time to mature and significant investment.

Such joint ventures are beneficial as they allow small family farmers to access more advanced processing and marketing facilities, which might otherwise be beyond their financial reach. They also provide agribusiness companies with a stable supply of agricultural products and the potential to influence farming practices directly, ensuring quality and sustainability.

At the local level, these partnerships often extend beyond just the farmers and companies to include various stakeholders such as local government bodies, NGOs and international donors. This broad collaboration aims to create platforms that facilitate open dialogue and inclusive decision-making processes. These platforms help ensure that the concerns of small family farmers, particularly those most vulnerable, are addressed and their voices heard.

Such joint ventures not only improve the economic standing of small family farmers by opening up new markets and providing them with better bargaining power but also contribute to community development. By bringing diverse resources and expertise together, these partnerships foster innovation and resilience, making agricultural practices more responsive to both market demands and environmental challenges. Moreover, they facilitate the spread of best practices and technologies among farmers, boosting overall productivity and sustainability in the agricultural sector.

Engaging in joint ventures requires careful planning and commitment from all parties involved. It involves navigating complex agreements on profit sharing, investment responsibilities and operational roles. Successful joint ventures are based on transparency, mutual respect and a shared vision for sustainable development, which are essential for maintaining long-term and fruitful collaborations (see Box 8.18.).

Box 8.18. Examples of joint ventures in low-income countries.

Rwanda – Gikonko rice joint venture: the Gikonko joint venture in Rwanda is a partnership between the Union of Rice Cooperatives (UCORIBU) and ICM Rwanda Agribusiness, an Australian firm, aimed at improving rice production and market access for local farmers. UCORIBU holds a 40% stake, while ICM owns 60%, combining local knowledge with technical expertise and investment. Between 2011 and 2015, the partnership led to a 40% increase in paddy rice production, allowing farmers to earn stable incomes and receive bonuses for meeting delivery targets. The venture provided farmers with key benefits such as extension services, interest-free credit, agricultural inputs and guaranteed market access. However, challenges included poor seed quality, inadequate technical assistance and water shortages, with recommendations for a valley dam to address irrigation issues. Despite early financial difficulties, the Gikonko rice mill, a product of the venture, recovered from losses and issued RWF42 million in dividends within 3 years. This joint venture illustrates how strategic partnerships between farmer cooperatives and private investors can enhance agricultural productivity, increase rural incomes and create sustainable value chains. While difficulties remain, the model demonstrates the potential for inclusive agribusiness collaborations to drive long-term economic benefits for smallholder farmers (Nshimiyimana, 2016).

Vietnam – Olam coffee joint venture: this venture in Vietnam is a prime example of a successful partnership between a global agribusiness and smallholder farmers. Olam works with thousands of smallholder coffee farmers to improve productivity, sustainability and market access. Through its farmer information system, Olam provides farmers with training on sustainable farming practices, access to high-quality inputs and financing options. The company also supports farmers in obtaining certifications such as Rainforest Alliance and UTZ, enabling them to access premium international markets.

This collaboration has led to significant improvements in coffee quality and yields, directly boosting farmer incomes. Olam's focus on sustainability ensures that farming practices are environmentally friendly, reducing deforestation and water usage. Additionally, the joint venture strengthens the supply chain for Olam, ensuring a consistent and high-quality coffee supply.

The initiative highlights the mutual benefits of such partnerships: smallholder farmers gain knowledge, resources and market access, while Olam secures a sustainable and reliable supply of coffee. This model has the potential to be replicated in other regions, promoting economic development and environmental stewardship in low-income countries (Olam, 2016).

8.5 Prerequisites for Successful Market Integration

Successful market integration requires trust, effective decision making and strong farm business skills. Without these, well-intentioned efforts and initiatives to integrate small family farms into markets will not be self-sustaining.

8.5.1 The primacy of mutual trust

It is important to note that the success of the linkages we have discussed rests in one of the key principles of good management: trust. Mutual trust is essential for successful farmer-agribusiness linkages. In many cases, this trust allows farmers and agribusinesses to engage in business with deferred payments, trusting each other to fulfil their commitments without written contracts. In livelihood terms (see Chapter 2), trust is a critical social asset.

Trust is more easily established in linkages involving smaller private-sector operators, often including cross-border trade, where verbal agreements are common. However, newer business relationships, which typically involve parties without prior interactions, often lack this foundational trust, making it challenging to maintain reliable business arrangements. This is exacerbated in settings where farmers have historically received government support with no repayment obligation, setting a precedent that undermines the importance of honouring commercial commitments.

To bridge the trust gap in newer market linkages, several strategies can be effective. For example, in the Malawian paprika industry, companies have fostered trust by involving local farmers in monitoring contracts through elected Field Assistants. Similar initiatives in Thailand have shown success as well. Establishing independent arbitration and forming industry associations that include farmer and company representatives can also help build trust. Prompt and reliable payments by companies significantly reduce mistrust and demonstrate commitment, which is further reinforced when companies make long-term investments in local infrastructure. From the farmers' perspective trust can be developed by ensuring that side-selling is constrained, and agreements are maintained.

Moreover, engaging farmers directly through meetings with buyers and visits to processing plants and markets can foster a deeper understanding and trust. However, challenges like side-selling, where farmers bypass contracts for potentially higher market prices, can erode trust. Addressing these issues often requires substantial effort from extension workers and other service providers to nurture sustainable and trusting relationships between farmers and companies, emphasizing long-term benefits over short-term gains. Flexibility in contracts and the possibility for renegotiation in response to significant price or cost changes can also help maintain fair and trustworthy business relations.

8.5.2 Decision making for market integration

For small family farmers to successfully integrate into modern markets, they must make a wide range of strategic decisions at both the individual and collective levels. These decisions shape their ability to meet market demands, access resources and enhance competitiveness, whether operating independently or as part of a farmer organization, cooperative or cluster.

At the individual level, farmers must determine which market-oriented crops or livestock enterprises to focus on based on demand, profitability and risk assessment. This involves analysing market trends, price fluctuations, quality requirements and input costs, to select viable enterprises. Additionally, they must adopt post-harvest practices such as sorting, grading and packaging to meet agribusiness specifications, ensuring their produce is competitive. Strategic resource allocation is another critical consideration, including investing in improved seeds, irrigation systems, quality testing equipment or digital tools for precision farming – all of which, as we have seen before, enhance product quality and operational efficiency.

For those engaged in collective action, decision making extends beyond farm-level operations to strategic coordination within farmer organizations. These groups play a vital role in pooling resources, mobilizing finance, aggregating produce and negotiating better market terms. Joint marketing strategies require decisions on produce aggregation, contract negotiations and delivery coordination, ensuring quality consistency and reliability for buyers. Additionally, input mobilization involves selecting suppliers based on cost, reliability and quality, with farmer groups purchasing inputs in bulk to secure better prices and ensure equitable distribution among members.

Financial decision making is equally important in collective models. As we saw before, farmer organizations act as intermediaries between their members and financial institutions, negotiating loans, setting repayment terms and securing guarantees that make credit more accessible. Effective financial planning also involves determining how to invest in shared assets such as tractors, irrigation systems, cold storage and processing units. These investments require careful consideration to ensure fair allocation, proper maintenance and cost-effectiveness, allowing farmers to access mechanization and value-added processing without the high costs of individual ownership.

Beyond internal decision making, smallholder groups must also engage with external stakeholders, including government agencies, NGOs, agribusiness firms and research institutions. These relationships help secure technical training, policy support, access to subsidies and exposure to new technologies. Through these external engagements, farmer organizations enhance their bargaining power, expand market opportunities and drive innovation within their value chains.

The ability to make well-informed, strategic decisions at both individual and group levels strengthens smallholders' market positions and long-term viability. However, effective decision making requires strong farm business management skills to assess risks, allocate resources efficiently and maximize profitability. The following section explores the core farm business management competencies that enable small family farmers to navigate modern markets successfully.

8.5.3 Essential business and management skills for market integration

For small family farmers to integrate effectively into agribusiness linkages – both horizontally and vertically – they must develop strong business, negotiation and farm management skills. These competencies are critical for securing profitable contracts, improving market access, managing production efficiently and ensuring financial stability. Without them, farmers risk entering unfavourable agreements, facing financial instability and struggling to meet buyer requirements, ultimately limiting their long-term sustainability.

A solid foundation in farm business management is essential, encompassing financial literacy, cost management, risk assessment, strategic planning and contract negotiation. Farmers who understand pricing structures, track production costs and calculate break-even points are better positioned to negotiate fair contract terms, reduce financial vulnerability and maintain stable income streams. Additionally, contract formulation skills allow farmers to assess agreements, identify risks and advocate for fair pricing, payment schedules and delivery terms, ensuring their businesses remain competitive and sustainable.

Beyond financial skills, quality control, supply chain management and production planning play a crucial role in successful market integration. Many agribusiness buyers enforce strict quantity, quality and delivery standards, requiring farmers to optimize resource use, align production with market demands and maintain consistent product quality. Farmers who develop expertise in business planning can efficiently manage production cycles, meet buyer expectations and improve profitability. Advisory services and capacity-building programmes that integrate farm management and contract negotiation training have proven effective in helping small farmers navigate complex agreements and build sustainable agribusiness partnerships.

The impact of these skills is evident in Kenya's small-scale dairy sector, where many farmers initially sold raw milk at low prices to informal traders. However, through business training initiatives led by dairy cooperatives, farmers learned how to analyse costs, improve milk quality and negotiate better prices with processors. As a result, they achieved higher incomes, greater market stability and long-term sustainability, demonstrating the transformative power of business education in farming.

Beyond individual farm operations, as we have seen, farmer organizations play a vital role in enhancing financial stability, strengthening bargaining power and improving access to resources. By pooling resources for bulk input purchases, collective marketing and financing arrangements, these organizations help farmers reduce costs, increase profitability and stabilize income. However, their success is largely dependent on strong governance, financial transparency and effective leadership. Without these elements, farmer organizations risk inefficiencies, mismanagement and internal conflicts, which can undermine their ability to compete in modern agribusiness markets.

Leadership and management skills are essential for the longevity and success of these organizations. Unfortunately, such skills are often lacking within the membership, making it necessary to hire professional managers. However, this becomes financially feasible only when the organization is large enough to cover the associated costs.

In addition to technical, managerial and financial management skills, entrepreneurial capabilities are key to navigating agribusiness linkages successfully. While not every farmer naturally possesses entrepreneurial instincts, capacity-building interventions – such as training in business management, financial planning and value-adding activities – can improve their ability to make informed decisions and enhance their agricultural ventures. Collaborations with logistics providers and value chain actors can further streamline product distribution and ensure compliance with agribusiness supply chain requirements.

Balancing the benefits of contract agreements with operational flexibility is another critical aspect of decision making. Farmers must carefully weigh the security of long-term contracts against the opportunity costs of forgoing other potentially profitable ventures. Strategic diversification and risk assessment are essential in maintaining adaptability and resilience in an evolving market environment.

Continuous learning and skill development – through training programmes, farmer field schools and exposure to market-oriented production practices – allow smallholders to improve product quality, meet the stringent standards set by agribusiness buyers and enhance their competitiveness (see Box 8.19.). By focusing on business management, quality production and strategic market engagement, small family farmers can strengthen their position within agribusiness networks, increase profitability and secure a sustainable future in commercial agriculture.

In conclusion, mastering farm management and business skills is critical for small family farmers to access modern agribusiness systems. By investing in capacity-building programmes, strengthening farm organizational governance and ensuring financial literacy, small-scale family farmers can build resilience, enhance profitability and secure a sustainable future in commercial agriculture.

8.6 The Role of Advisory Services

As small family farmers transition to higher-value enterprises and seek to add value to their products, agricultural advisory services serve as key intermediaries, facilitating critical linkages between farmers and agribusiness stakeholders. These services provide smallholders with the information, guidance and technical expertise needed to navigate the complexities of the agricultural value chain. By bridging the gap between farmers and value chain actors, advisory services enhance the capacity of farmers to access markets, acquire resources and pursue value addition opportunities.

Advisory services also play an enabling role by connecting farmers with relevant partners along the value chain and fostering relationships that support collaboration, market entry and innovation. In doing so, they help small family farmers engage more effectively with opportunities that might otherwise be beyond their reach.

To further strengthen these connections, local platforms and stakeholder forums have emerged as standard mechanisms for promoting engagement and dialogue. Often facilitated by decentralized government offices or NGOs, these forums create structured spaces where stakeholders can build trust, negotiate roles and align goals. They serve as a foundation for balancing competition with collaboration and lay the groundwork for collective action and shared growth within agricultural communities.

The role of NGOs and government extension services is particularly important in helping to establish and sustain these platforms – especially in areas where small family farmers may lack the capacity to initiate or maintain such partnerships on their own. These actors often provide the external support needed to strengthen local capabilities and to develop functioning service markets that allow farmers to fully participate in evolving networks.

Crucially, effective service delivery within these networks should be driven by service providers, adopting a demand-led approach that is responsive to the real needs and priorities of smallholder farmers. This ensures that services are relevant, practical and capable of fostering long-term engagement, impact and sustainability.

Box 8.19. Uganda farmer field schools. From Davis *et al.*, 2012.

Farmer field schools in Uganda have played a pivotal role in equipping small farmers with essential knowledge and skills in post-harvest handling, packaging and quality control. Through practical, hands-on training and field demonstrations, these programmes have enabled farmers to significantly enhance their ability to meet stringent quality standards. As a result, farmers have accessed new market opportunities and achieved higher profits, directly improving their livelihoods and economic stability. The success of the farmer field schools underscores the importance of targeted educational initiatives in empowering smallholders to compete more effectively in the market and to increase their income.

While networking and stakeholder partnerships are increasingly essential, market opportunities alone are not sufficient to drive meaningful collaboration – especially in underdeveloped markets where incentives for change are weak or absent. In such cases, deliberate efforts are needed to create effective coordination mechanisms that compensate for market shortcomings and catalyse the development of inclusive, farmer-centred partnerships.

These coordination mechanisms play a pivotal role in facilitating engagement, knowledge exchange and cooperation among a broad set of stakeholders. Partnerships must extend beyond farmer-to-farmer collaboration to include service providers, training institutions, research organizations, NGOs, the private sector and government bodies. Such broad-based networks are essential for enhancing knowledge sharing, promoting innovation and expanding smallholder access to technologies, services and information.

These considerations are further elaborated on in Part 4.

8.7 Conclusion

Agricultural commercialization presents a critical pathway for small family farmers to enhance their livelihoods but brings both significant opportunities and notable challenges. The increasing dominance of agribusiness enterprises in agricultural markets means that smallholder farmers must proactively engage in strategic partnerships and adopt market-oriented practices to remain competitive. Effective market integration depends significantly on strengthening horizontal and vertical linkages. Horizontal collaboration, through farmer organizations and clusters, facilitates collective action, improves bargaining power, reduces costs through economies of scale and allows smallholders to access larger, more profitable markets. Vertical linkages, particularly through contract farming and strategic alliances with agribusinesses, offer farmers essential resources such as technical assistance, stable market access and improved technology.

However, these partnerships require careful management and robust negotiating skills to ensure that small family farmers are not disadvantaged by power imbalances with larger agribusiness entities. Transparent agreements, fair pricing and equitable risk-sharing mechanisms are essential for safeguarding farmers' interests and enhancing profitability. Additionally, farmers must develop core farm business management competencies, including financial literacy, risk assessment and strategic planning, to effectively navigate and benefit from modern market systems.

The role of farmer organizations is paramount. Cooperatives and commodity associations have demonstrated their capacity to empower smallholders economically and socially, providing necessary infrastructure, securing finance, ensuring quality standards and enhancing market access. Case studies from diverse regions highlight the substantial benefits realized when small farmers participate actively in governance and decision-making processes, reinforcing the importance of building strong, sustainable organizational structures.

Advisory services play a crucial intermediary role, facilitating interactions among farmers, agribusinesses and supporting institutions. These services not only provide vital technical support but also enable farmers to access and utilize information effectively, ensuring they meet stringent market demands for quality and consistency. To optimize these benefits, farmers must have access to continuous capacity-building opportunities and a supportive policy environment that encourages equitable market participation.

In conclusion, successful agricultural commercialization demands an integrated approach, combining strategic partnerships, effective organizational structures, strong business management skills and supportive advisory services. By leveraging these strategies, small family farmers can navigate the complexities of modern agribusiness, transitioning from subsistence operations to sustainable, profitable and market-oriented enterprises. Building resilient farming communities relies on empowering smallholders through working together, better market access and fair partnerships, helping them to stay active and succeed in the growing commercial agriculture sector.

References

African Fine Coffees Association (AFCA) (2025) About us. Available at: https://afca.coffee/about-us/ (accessed 7 November 2024).

AGRA (2024) Accelerating the Private Sector for Food Systems transformation in Africa. Africa Agriculture Status Report 2024, AGRA, Nairobi, Kenya. Available at: https://agra.org/publications/ (accessed 20 February 2025).

Bacon, C.M. (2005) Confronting the coffee crisis: Can fair trade, organic, and specialty coffees reduce small-scale farmer vulnerability in Northern Nicaragua? *World Development* 33, 497–511. Available at: https://www.researchgate.net/publication/222646249_Confronting_the_Coffee_Crisis_Can_Fair_Trade_Organic_and_Specialty_Coffees_Reduce_Small-Scale_Farmer_Vulnerability_in_Northern_Nicaragua (accessed 20 February 2025).

Bebe, B.O., Udo, H.M.J., Rowlands, G.J. and Thorpe, W. (2003) Smallholder dairy systems in the Kenya highlands: Cattle population dynamics under increasing intensification. *Livestock Production Science* 82(2–3), 211–221. Available at: https://www.sciencedirect.com/science/article/abs/pii/S0301622603000137 (accessed 20 February 2025).

Davis, K., Nkonya, E., Kato, E., Mekonnen, D.A., Odendo, M. *et al.* (2012) Impact of farmer field schools on agricultural productivity and poverty in east Africa. *World Development* 40–2, 402–413. DOI: 10.1016/j.worlddev.2011.05.019.

FAO (2019) Developing sustainable value chains for small-scale livestock producers. FAO Animal Production and Health Guidelines No.21, Rome. Available at: http://www.fao.org/3/ca5717en/ca5717en.pdf (accessed 20 February 2025).

Gálvez-Nogales, E. (2010) *Agro-Based Clusters in Developing Countries: Staying Competitive in a Globalized Economy*. FAO, Rome, Italy. Available at: https://www.fao.org/4/i1560e/i1560e00.htm (accessed 17 March 2025).

Giovannucci, D. and Koekoek, F.J. (2003) *The State of Sustainable Coffee: A Study of Twelve Major Markets*. International Coffee Organisation, London, UK. Available at: https://mpra.ub.uni-muenchen.de/17172/ (accessed 20 February 2025).

Interréseaux Développement rural (2024) *Rapport d'activité* 2024 [PDF]. Interréseaux Développement rural. Available at: https://www.inter-reseaux.org/ressource/rapport-dactivite-2024-inter-reseaux-developpement-rural/ (accessed 8 August 2025).

Markelova, H., Meinzen-Dick, R., Heillin, J. and Dohrn, S. (2009) Collective action for smallholder market access. *Food Policy* 34–1, 1–7. Available at: https://www.sciencedirect.com/science/article/abs/pii/S0306919208000730?via%3Dihub (accessed 21 February 2025).

Nshimiyimana, C. (2016) Assessing the impact of Gikonko Joint Venture Farming on rice famer's income and access to market. case of UCORIBU farmers, 2011–2015. Master's Thesis, University of Rwanda. Available at: http://hdl.handle.net/123456789/189 (accessed 17 March 2025).

Olam (2016) Olam instant coffee centres on growth in Asia, Europe. Olam Insights 2, 2016. Available at: https://www.google.com/url?sa=t&source=web&rct=j&opi=89978449&url=https://www.olamgroup.com/content/dam/olamgroup/files/uploads/IRfiles/Olam_Insight_June_v6_HR_NOCROPS-1.pdf&ved=2ahUKEwjH_pvmztKLAxXhUEEAHY4ZOu4QFnoECAwQAQ&usg=AOvVaw3yldt3zPBqQdvGKoeRp_pp (accessed 20 February 2025).

Pallathadka, H., Palla Thadka, L.K., Devi, S.K.K. and Singh, S.K. (2022) Reasons for the success story of amul: An empirical study of customers' opinions. *Integrated Journal for Research in Arts and Humanities* 2(6), 252–259. Available at: https://ijrah.com/index.php/ijrah/article/view/163 (accessed 17 February 2025).

PepsiCo (2020) Sustainability Report. PepsiCo online document. Available at: https://www.pepsico.com/our-impact/sustainability/report-downloads (accessed 17 February 2025).

Raynolds, L.T., Murray, D. and Taylor, P.L. (2004) Fair trade coffee: Building producer capacity via global networks. *Journal of International Development* 16–8, 1109–1121. DOI: 10.1002/jid.1136. (accessed 20 February 2025).

Reardon, T., Echeverria, R., Berdegué, J., Minten, B., Liverpool-Tasie, L.S.O. *et al.* (2018) Rapid transformation of food systems in developing regions: highlighting the role of agribusiness and private sector. *Agricultural Systems* 172(1). DOI: 10.1016/j.agsy.2018.01.022. (accessed 20 February 2025).

Requier-Desjardins, D., Boucher, F. and Cerdan, C. (2003) Globalization, competitive advantages and the evolution of productive systems: Rural food processing and localised food systems in Latin American countries. *Entrepreneurship and Regional Development* 15(1), 49–67. DOI: 10.1080/08985620210144983. (accessed 20 February 2025).

Schmitz, H. and Nadvi, K. (1999) Clustering and industrialization: Introduction. *World Development* 27–4, 1502–1514. DOI: 10.1016/S0305-750X(99)00072-8. (accessed 17 March 2025).

Silva, J.G. da, Del Grossi, M.E. and de França, C.G. (eds) (2010) *The Fome Zero (Zero Hunger) Program: The Brazilian Experience* (NEAD Special Series No. 13). Brazilian Ministry of Agrarian Development; Food and Agriculture Organization (FAO), Brasília. Available at: https://www.fao.org/docrep/016/i3023e/i3023e.pdf (accessed 8 August 2025).

Swanson, B.E. and Rajalahti, R. (2010) *Strengthening Agricultural Extension and Advisory Systems: Procedures for Assessing, Transforming, and Evaluating Extension Systems* (Agriculture and Rural Development Discussion Paper No. 45). The World Bank, Washington, DC. Available at: https://documents.worldbank.org/en/publication/documents-reports/documentdetail/873411468159312382/strengthening-agricultural-extension-and-advisory-systems-procedures-for-assessing-transforming-and-evaluating-extension-systems (accessed 8 August 2025).

Tarchiani, V., Vignaroli, P. and Zucchini, E. (2020) Assessing the sustainability of horticultural farms in Central Senegal: An adaptation of the IDEA method. *Journal of Agriculture and Sustainability* 13(6). Available at: https://www.infinitypress.info/index.php/jas/article/view/1923 (accessed 17 March 2025).

Technoserve (n.d.) East Africa Dairy Development. Technoserve. Available at: https://www.technoserve.org/fight-poverty/projects/east-africa-dairy-development/ (accessed 20 February 2025).

Uy, J.C. (2011) Increasing market chain participation of small producers: The norminveggies case. In: *Food for All: Investing in Food Security in Asia and the Pacific – Issues, Innovations and Practices*. Asian Development Bank, Mandaluyong City, Philippines. Available at: https://www.adb.org/publications/food-all-investing-food-security-asia-and-pacific-issues-innovations-and-practices (accessed February 2025).20

Wilkinson, J. (2004) The food processing industry, globalisation and developing countries. *Journal of Development and Agricultural Economics* 1–2, 184–201. Available at: https://www.researchgate.net/publication/5021663_The_Food_Processing_Industry_Globalization_and_Developing_Countries (accessed 20 February 2025).

Additional Reading

Arias, P., Hallam, D., Krivonos, E. and Morrison, J. 2013. *Smallholder integration in changing food. markets*. Rome, FAO.

CIAT. 2012. *Linking Smallholders: a guide on inclusive business models*. Available at: http://dapa.ciat.cgiar.org/linking-smallholders-aguide-on-inclusive-business-models/.

Experiences, Occasional Paper 18, FAO, Rome, Italy. DOI: 10.13140/RG.2.1.1662.0642 (accessed 02/03/25)

Graziano da Silva, J., Del Grossi, M.E. and de França, C.G. eds. (2010). *The Fome Zero (ZeroHunger Program): The Brazilian experience*. Brasilia, FAO and the Ministry of Agrarian Development.

Kahan D. (2007b) Business services in support of farm enterprise development: A review of relevant

Kahan D. (2011) *Market-oriented advisory services in Asia. A review and lessons learned*. FAO, Regional Office for Asia and the Pacific, Bangkok,

Nederlof, S., Wongtschowski, M. and van der Lee, F. (2011) *Putting Heads Together: Agricultural Innovation Platforms in Practice*. Amsterdam, KIT.

Shepherd, A., Cadilhon, J. and Gàlvez, E. (2009) *Commodity Associations: A Tool for Supply Chain Development?* Agricultural Management, Marketing and Finance Occasional Paper. Rome, FAO.

World Bank. (2009) *Agribusiness and innovation systems in Africa*. Washington, DC.

Part 3: Decision Support Tools

This part of the book lays out the theoretical framework underpinning the use of economic concepts and principles crucial to farm management decision making. It then introduces practical methodologies and tools for applying these concepts to the small farmer context. Part 3 also explores various management techniques that are applicable across the farm management cycle of diagnosis, planning, monitoring and evaluation. Emphasis is placed on how these methods can be tailored to small family farms, especially in low-income countries, to enhance the skills, competencies and overall farm management capabilities of farmers, leading to improved productivity and sustainability.

9 Essential Economic Concepts and Principles in Farm Management

Economics plays a pivotal role in enabling smallholder farmers to manage their farms as businesses. Without an economic perspective, farmers may make poor decisions – such as assuming that producing more always means better results or using faulty accounting that gives a false picture of their costs and benefits. Economics provides the tools and logic to understand how resources are allocated within the complex and dynamic systems of smallholder farming. While traditional farming often relies on intuition, economic analysis introduces rigour and clarity and leads to more effective decision making.

This chapter highlights the importance of economic concepts and principles as the foundation for sound farm management. For small family farms, especially in an increasingly competitive and resource-constrained agricultural environment, the ability to make informed decisions can determine the difference between success and failure. Market fluctuations, limited inputs and evolving challenges require farmers to approach their operations strategically. By applying economic principles, farmers can better allocate resources, optimize production and enhance the overall sustainability of their farms.

The chapter explores how economic principles intersect with both the technical and management aspects of farming. These principles help smallholder farmers navigate the complexities of decision making in market-oriented operations, where balancing profitability with sustainability is essential.

Economics focuses on creating wealth through the efficient use of limited resources, including physical, financial, natural, human and social assets. For smallholders operating under resource constraints, strategic and tactical decision making becomes vital for achieving their livelihood goals.

The text emphasizes the importance of informed decision making in key farm business areas such as production, marketing, finance and resource allocation. Understanding these principles is critical not only for farmers but also for extension workers who support smallholder farming. Extension workers can complement their technical expertise with economic and management skills, enabling them to provide holistic support that promotes efficiency and profitability.

In Chapter 8, we identified critical decision areas for farm management, including:

- **Production decisions:** land use, technology choices, infrastructure development and input supply.
- **Finance decisions:** access to credit, budgeting and investment planning.
- **Market decisions:** pricing strategies, market selection and sales planning.

Each of these decision areas involves specific economic concepts and tools, which will be explored in this chapter. The following sections provide an overview of the most important economic principles and their application to farm management. We will begin by introducing general economic concepts that underpin effective decision making and then delve into the tools and strategies applicable to each of these critical decision-making areas. Through this exploration, the chapter aims to equip farmers and agricultural advisors with the knowledge needed to enhance competitiveness, profitability and sustainability in smallholder farming systems.

DOI: 10.1079/9781836990536.0009

9.1 Economic Concepts for Decision Making

9.1.1 Gross income

The value of a crop after harvest extends beyond the income generated from its sale; it also includes the value of produce consumed by the farm household or stored for future use. This total value, often referred to as the 'value of output', is a critical measure for farmers as it reflects the full economic benefit of their production efforts. Understanding the total gross income is essential not only for calculating profitability but also for planning future seasons and making informed market decisions (see Box 9.1.).

Gross income = (Quantity sold + Quantity consumed + Quantity stored) x Price

The gross income is not limited to cash transactions; it also includes the value of unsold produce that is either consumed by the farm family or stored for later use. To accurately capture this value, farmers can use the market price as a convenient method of valuation, by asking themselves, 'What would this produce have sold for if it had been taken to market?' Alternatively, a more precise approach would be to consider what the family would have had to pay if they purchased the food instead of producing it themselves. However, in many rural areas where the difference between selling and buying prices is minimal, using the market price is often a practical and accurate approximation.

By accounting for these factors – produce sold, produce consumed by the family and produce stored – farmers can arrive at a true estimate of the total value of their production. This holistic view is crucial for understanding the economic impact of their farming activities, helping them to assess profitability accurately and to plan effectively for the future.

9.1.2 Costs of production

Understanding the costs associated with production is essential for making informed decisions about which enterprises to undertake and how to allocate resources. Costs of production are typically divided into *variable costs* and *fixed costs*.

Variable costs are those that fluctuate with the level of production (see Box 9.2.). These include inputs like seeds, fertilizers, feed and hired labour. For example, if a farmer decides to plant more acreage of a particular crop, the costs for seeds and labour will increase accordingly. Managing these costs effectively is crucial for maintaining profitability.

Fixed costs, on the other hand, do not change with the level of production (see Box 9.3.). These include expenses such as the depreciation of machinery, property, taxes, insurance and the cost of permanent labour. Even if production decreases, these costs remain constant, which means they must be covered regardless of output levels. Fixed costs are particularly important to consider when making long-term investment decisions or when scaling operations. They can represent a significant portion of total costs, especially in more mechanized or capital-intensive farming systems.

The *total costs* are the *aggregate expenses* incurred in operating a farm, encompassing both variable costs (which fluctuate with production levels) and fixed costs (which remain constant regardless of production volume).

By thoroughly understanding both variable, fixed costs and total costs, farmers can make more strategic decisions about which enterprises are most financially viable. For instance, if the fixed costs of maintaining livestock are high, a farmer

Box 9.1. An example of gross income.

After the maize harvest, a farmer had 50 bags of maize. She sold ten bags of maize at $5.50 each and earned $55. She put 20 in her storage shed; four bags for cow feed and chicken feed; 14 bags for her family and two to give to a farmer friend. If the farmer wanted to know the value of her maize production, she would have to add together the bags of maize that she sold in the market, the bags she put in storage, the bags she used for feed and the bags she gave away and multiply the total by the market sales price. This would be 50 bags x $5.50 = $275

Box 9.2. Examples of variable costs.

A farmer must hire labour for weeding and harvesting. If the farmer increases the area that needs to be weeded or increases the number of times the land is to be weeded, the cost of hired labour will also increase.

Similarly, the amount of labour needed for the harvest is linked to the yield. If a low yield is attained the amount of hired labour at harvest time will also be low. If a high yield is attained the labour costs will be higher.

The same is true of other inputs. If the farmer decides to increase the amount of land planted to maize, the amount of seed and fertilizer applied will increase, thus increasing the farmer's costs.

Box 9.3. Examples of fixed costs.

A farmer has a small storeroom for fertilizer, seed, animal feed and farm tools. Any costs associated with the storeroom (e.g. maintaining or cleaning it) are shared by all the farmer's enterprises. These costs are not affected by production or yield. Whether production is increased or decreased, or the yield is high or low, the costs are fixed. It would be difficult to divide such costs and allocate them to the farmer's individual enterprises.

Concerning draught power and equipment, most of the costs of keeping a tractor, draught cattle and farm equipment remain the same whether the item is or is not fully used. A tractor can be used for a mix of farm operations, cultivating a crop, transporting feed for livestock and even transporting people to town (although this is a very expensive form of transport). The cost for different activities cannot be easily allocated to any one enterprise. Portions of fixed costs, such as fuel or hours of draught animal use, can be allocated between enterprises but this requires good information, which is often unavailable to smallholder farmers.

might choose to allocate more resources to crop production if it offers a better return on investment.

9.1.3 Gross margin

Gross margin is an important measure in farm management, calculated by subtracting variable costs from the total gross income. This figure provides a clear indicator of the profitability of different crops or livestock enterprises, enabling farmers to make strategic decisions about resource allocation. By analysing gross margins, farmers can directly compare the profitability of various farming activities, helping them determine where to focus their efforts for maximum return.

When deciding between different enterprises, calculating gross margins becomes essential (see Box 9.4.). For instance, if a farmer is considering whether to continue with a certain crop or switch to another, comparing the gross margins of the two crops offers valuable insight. Since fixed costs – like equipment depreciation – are likely to remain constant regardless of the enterprise on a small family farm, the focus should be on the variable costs and the gross income. Gross margin analysis allows the farmer to see how changes in these factors impact profitability.

Gross margin = Gross income – Variable costs

9.1.4 Depreciation

Depreciation is a critical but often overlooked aspect of farm management (see Box 9.5.). When farmers invest in durable assets like machinery, buildings or even livestock, these assets lose value over time due to wear and tear or obsolescence. Depreciation spreads the cost of durable assets, like machinery, over their useful lifespan. For example, if a tractor is expected to last 10 years, its cost should be spread over that period rather than being fully accounted for in the year it was purchased. This spreading of costs ensures that each year's financial performance reflects a more accurate picture of profitability.

9.1.5 Enterprise profit

Enterprise profit is the net income generated by a specific farm activity or enterprise after

Box 9.4. An example of gross margin.

A farmer who produces a crop worth $600 at a variable cost of $100 generates a gross margin of $500 ($600 – $100).

If one crop's gross margin significantly outperforms another, it may be wise to expand the more profitable crop while reducing or even discontinuing the less profitable one. This approach not only guides resource allocation but also helps in assessing the impact of market prices on overall profitability, leading to better planning and risk management. Ultimately, using gross margin as a tool enables farmers to make informed decisions about enterprise selection and potential changes, ensuring that any shift in production strategy contributes positively to the farm's financial health.

Box 9.5. Depreciation based on useful life.

A method is used to spread the costs of durable capital items over their useful life. It gives us a fairly accurate idea of what it costs to use the item for a year. Consider the following example: the cost of a tractor is $35,000. It has a useful life of 7 years. Therefore, each year, one seventh of the cost of the tractor is taken off its value and added to the enterprise costs.

$$Depreciation = \frac{Purchase\ price}{Useful\ life} = \frac{\$35{,}000}{7\ years} = \$5,000\ per\ year$$

Each year for 7 years, $5000 will be a fixed cost to the farm.

This fixed cost remains on a yearly basis until the tractor comes to the end of its useful life.

Incorporating depreciation into cost calculations is essential for assessing the true cost of production and, by extension, the profitability of different enterprises. It also plays a key role in financial planning and investment decisions. For instance, understanding how depreciation will affect future costs can help farmers decide when it might be time to replace old equipment or invest in new technology. Ignoring depreciation can lead to an overestimation of profitability and, consequently, poor financial decisions.

By carefully selecting enterprises based on comparative advantage, managing production costs, and accounting for depreciation, farmers can allocate resources more effectively. This strategic approach not only enhances profitability but also ensures the long-term sustainability of their farming operations, enabling them to remain competitive in an ever-changing agricultural market.

accounting for all associated costs. It is calculated by subtracting both variable costs and allocatable fixed costs from the total income generated by that enterprise.

This measure provides insight into the financial performance of individual farm activities, enabling farmers to assess which enterprises are most profitable and make informed decisions about resource allocation and operational adjustments.

For example, if a vegetable farming enterprise generates $10,000 in gross income, incurs $4000 in variable costs, and $2000 in fixed costs, the enterprise profit would be:

Enterprise profit = gross income – variable costs – fixed costs

Enterprise profit = $10,000 – $4,000 – $2,000 = $4,000

Note that this is the same as subtracting the fixed costs from the gross margin.

This calculation helps determine the profitability of the vegetable farming enterprise, guiding decisions on whether to expand, modify or discontinue the activity. Understanding enterprise profit is crucial for effective farm management, as it allows for the identification of profitable activities and the optimization of the farm's overall financial performance.

9.1.6 Farm profit

Farm profit (sometimes called net farm income) refers to the money remaining after the variable costs and the fixed costs are paid. Each enterprise has a gross margin, which, as noted before, is determined by subtracting the variable costs of the enterprise from the gross income. The total gross margin on a farm is the sum of the gross margins of all enterprises. But remember, this does not include fixed costs, which must still be paid. The money to pay for the fixed costs comes from the total gross margin.

Farm Profit = total gross margin of all farm enterprises – total fixed costs

If the amount obtained by subtracting fixed costs from the total gross margin is positive, there is a profit. If the amount obtained is negative, there is a loss. Because fixed costs do not vary much with changes in production, it is almost always the case that if farmers can increase the gross margin on their farms, they will also increase profits. Further, because the smallholder farmer usually has few fixed costs, the total gross margin is almost the same as total profit.

9.1.7 Cash flow

As we saw in Chapter 6, cash flow can be simply understood as the movement of money (cash) in and out of the farm. This reflects the cash income from sales and other sources, minus cash expenditures like input costs, labour and loan payments, over a specific period.

9.1.8 Economies of scale

Economies of scale are cost advantages gained when farm operations increase in size. Economies of scale are achieved when the cost per unit of production or output marketed is reduced as the scale of the activity increases. Savings (economies) can be achieved by spreading costs over a larger scale of operation. Economies of scale can also be achieved when farmers organize themselves into groups to buy inputs, obtain capital or market produce (see Box 9.6.). These group activities can bring about a cost reduction per unit of inputs purchased or produce marketed. Economies of scale can also come about through specialization. As output increases there is a possibility of using more specialized management and introducing mechanization to take the place of labour in field operations. Increases in scale of output allow savings to occur in the time spent on these tasks. Understanding and leveraging economies of scale is particularly important when evaluating market opportunities and considering investments in new technologies or expanded production capacity. The following are some examples of economies of scale:

- Groups of farmers involved in marketing could share transport arrangements to sell their produce. This very often results in lower unit costs. If each individual farmer were to transport produce separately the amount of produce transported would be smaller and per unit transport costs would be higher than those of the collective effort.
- Farmers can look for ways to produce collectively. They could share a tractor, infrastructure for livestock production or mechanical harvesting.
- Collective buying of inputs would also have similar results, resulting in economies of scale. By ordering in bulk, the farmers would be able to buy their materials at reduced prices and transport them more cheaply.

9.1.9 Risk

Farming is inherently risky because many future events cannot be predicted with complete certainty. This uncertainty affects the decisions farmers make, especially regarding how much and what type of inputs to use. For example, in rain-fed agriculture, where farmers might face poor rainfall once every 5 years, they may be reluctant to apply large amounts of fertilizer due to the risk that the rains won't come. Risk arises when the outcomes of decisions – such as yields or prices – are unpredictable, and it directly influences both the cost and quantity of inputs used. Since farm profits are always uncertain, farmers must consider the various factors that contribute to this uncertainty. Although not all factors carry the same level of risk, each plays a role in shaping outcomes. As a result, farmers often find themselves balancing the goal of maximizing profits with the need to minimize risk (see Box 9.7.).

Farmers need to make decisions that take these risks into account. As discussed in Chapter 6, risk can be managed by the farmer adopting the correct strategies and actions.

Box 9.6. Awa's maize farm: growing stronger together.

In a small village in northern Ghana, Awa, a hardworking mother of three, struggled to make a living from her 1 ha maize farm. Every planting season, she bought seeds and fertilizer from a local trader in small amounts, paying high prices because she couldn't afford bulk purchases. To prepare her land, she rented her neighbour's ox-plough, but the fees were rising each year. When harvest time came, she had no choice but to sell her maize at the local market, where prices were low and unpredictable. No matter how hard she worked, her profits were small.

One day, a fellow farmer told Awa about the local farmers' cooperative. By joining, she could buy inputs in bulk at lower prices and share farming equipment with others. Encouraged by the idea, she became a member. That season, the cooperative purchased fertilizer and seeds in large quantities, reducing the costs for everyone. Instead of hiring an ox-plough, Awa and her fellow farmers pooled their resources to rent a tractor, which prepared their land faster and more efficiently.

At harvest time, instead of the farmers selling maize individually, the cooperative collected produce from all members and negotiated a deal with a large buyer in the city. This time, Awa earned a much higher price per kg, increasing her income significantly. With her extra earnings, she invested in better seeds for the next season and even saved for her children's school fees.

Awa's story shows how economies of scale can help small farmers reduce costs, improve efficiency, and access better markets – turning a small farm into a sustainable business through collective action.

Box 9.7. Risk vs uncertainty.

Economists often differentiate between risk and uncertainty. Risk is associated with events for which a probability of occurrence can be attached. For example, the risk of no rain in a given period can be calculated based on past rainfall patterns. Uncertainty, on the other hand, refers to events for which no objective way to assign a probability exists. An example is the prediction of prices. For most practical purposes, however, the two concepts can be used interchangeably.

9.1.10 Sustainability

Sustainability in farming refers to the ability of agricultural systems to meet current food and income needs without compromising the ability of future generations to meet their own needs. It involves balancing environmental health, economic viability and social equity.

Sustainable farming has three pillars: (i) environmental sustainability; (ii) economic sustainability (including profitability); and (iii) social sustainability (Fig. 9.1).

Profitability is central to economic sustainability. A sustainable farm must be profitable, but not at the expense of degrading the environment or exploiting labour. Sustainability must balance profit with social and environmental responsibility.

9.2 Economic Concepts for Production Decisions

Smallholder farmers regularly make production decisions that shape both their livelihoods and the long-term sustainability of their farms. These decisions – what to produce, how much and how to produce it – are influenced by market opportunities as well as the practical constraints of limited resources.

This section introduces key economic principles that support informed and efficient decision making. Modern farming is about generating income to sustain the household. While prices may signal what is profitable, actual profit depends on managing production costs relative to returns. Understanding how inputs (such as land, labour and capital) translate into outputs (crops and livestock products) is essential to making sound choices.

For smallholders, whose resources are often scarce and whose risk tolerance is low, the efficiency of production is critical. Three core economic relationships help guide production decisions:

- **Factor–product relationship:** examines how varying levels of a single input (e.g. fertilizer) influence output. It helps identify

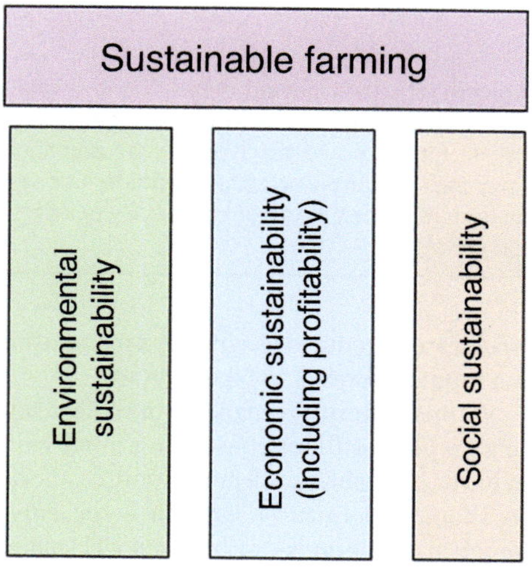

Fig. 9.1. The three pillars of sustainable farming.

the input level that maximizes profit – not necessarily yield.

- **Factor–factor relationship:** focuses on selecting the most cost-effective mix of inputs. Farmers must weigh options such as family labour versus mechanization or organic compost versus chemical fertilizer based on costs, availability and expected outcomes.
- **Product–product relationship:** involves allocating resources among different enterprises. A farmer might compare returns from maize vs goats – or from combining both – to decide the most profitable mix under existing resource constraints.

By applying these concepts, farmers can make more strategic decisions about:

- the types and combinations of farm enterprises to produce;
- the most profitable levels of production;
- the best use of land, labour and capital; and
- the choice of production methods and technologies.

Effective resource allocation underpins all production decisions. With limited land, labour and capital, smallholders must prioritize where inputs will deliver the greatest return. This means assessing the profitability of different options and allocating resources accordingly to avoid waste and maximize productivity. Economic efficiency is central to the decision-making process, but decisions often include taking into account farm household needs, cultural values, leisure and ecological concerns. These affect choices such as:

- what varieties or crops to grow;
- which markets or channels to use;
- when to sell produce;
- how to reduce risk; and
- which practices to adopt for sustainability.

These decisions cannot be easily separated from each other because they are all affected by the limited resources available to the farmer. The decision to be made between different products and their level of production is affected by how they will be produced, which, in turn, is influenced by how much is to be produced. This also affects how much of any resource should be used.

An economic perspective helps smallholders navigate these complexities. It encourages decisions that align household goals with market conditions, ultimately leading to improved efficiency, reduced costs and better livelihoods.

9.2.1 Opportunity cost concept

Every decision a farmer makes about resource allocation involves an *opportunity cost* – the value of the next best alternative that is forgone when a particular choice is made. Opportunity cost is a fundamental concept in economics that reflects the trade-offs inherent in any decision-making process. For farmers, understanding opportunity cost is crucial for making economically sound decisions (see Box 9.8.). For example, if a farmer decides to allocate more land to growing wheat instead of maize, the opportunity cost is the profit that could have been earned from growing maize on that land. Similarly, if a farmer invests capital in purchasing new equipment, the opportunity cost might be the other potential uses of that capital, such as expanding livestock operations or improving soil fertility.

Weighing opportunity costs involves considering both the tangible and intangible benefits and drawbacks of different choices. Farmers must assess not only the immediate financial returns but also the long-term implications of their decisions. For instance, choosing

Box 9.8. An example of an opportunity cost.

A farmer grows maize and earns $55. If the farmer had grown tomatoes instead of maize, the farmer could have earned $95. In this example, the opportunity cost of growing maize is $40. This is the difference between the two incomes ($95/$55), which the farmer could have gained by planting tomatoes instead of maize. Whichever crop the farmer planted, the farmer would have made money, but the point is that more money would have been made from tomatoes than from maize. By planting maize, the farmer gave up the opportunity to earn an additional $40.

to invest in a crop that offers higher short-term profits might come at the opportunity cost of not investing in a more sustainable practice that could yield benefits over the long term.

Opportunity cost also plays a role in time management on the farm. If a farmer spends a significant amount of time on a particular task, the opportunity cost is the other productive activities that could have been undertaken during that time. For example, spending time repairing equipment might come at the cost of not planting crops at the optimal time, potentially leading to lower yields.

To make informed decisions, farmers must consider all possible options and evaluate the potential returns of each, taking into account the opportunity costs. This approach ensures that resources are allocated in a way that maximizes overall farm profitability and sustainability.

9.2.2 Maximizing profit vs maximizing yield

Farmers often face an important distinction between maximizing *yield* and maximizing *profit*. While high yields may appear desirable, they do not always result in the greatest profit. Achieving the highest possible yield per hectare or per animal often requires substantial additional inputs, which may not be economically justified. The *economically optimal yield* is usually lower than the maximum yield, as the extra cost of increasing production tends to outweigh the value of the additional output.

Extension workers frequently promote practices aimed at increasing yields. However, higher yields can come with higher input costs that reduce profitability. For example, applying more fertilizer to boost maize yields may not be profitable if the additional income from the extra maize does not exceed the added cost of fertilizer. In such cases, producing slightly less maize using fewer inputs can result in higher profits.

Farmers must recognize that focusing solely on yield without considering input costs can harm profitability. Effective resource allocation should prioritize *profitability* – measured through gross margins – rather than yield alone, to ensure sustainable economic returns.

By clearly distinguishing between yield and profit maximization and making informed resource allocation decisions, farmers can improve economic efficiency, increase profitability and support the long-term sustainability of their farms.

9.2.3 Productivity

Productivity measures the quantity of output produced per unit of input. It is a measure of how much is produced on the farm, such as tons of maize per hectare or litres of milk per cow (see Box 9.9.).

It is a straightforward ratio of output divided by input. Output measurements are expressed in physical terms (e.g. tons of grain, litres of milk) or value terms (e.g. revenue from crop sales). Input measurements include land (hectares), labour (person-days), capital (machinery, buildings) and intermediate inputs (fertilizers, seeds). Productivity is a key indicator of the performance and competitiveness of the farming system.

9.2.4 Efficiency in resource use

Efficient resource use is essential for enhancing farm profitability and sustainability. In smallholder farming, where resources are often limited and expensive, efficiency means producing more with less: maximizing output while minimizing cost and waste. It involves getting the most value from available resources,

Box 9.9. Input productivity indicators.

- Land productivity = output per hectare (e.g. tons of maize per hectare (t/ha))
- Labour productivity = output per labour unit (e.g. kilograms of rice per agricultural worker)
- Capital productivity = output per unit of capital invested (e.g. net returns per dollar invested in irrigation)

Box 9.10. Technical and economic efficiency.

Technical efficiency: means producing the maximum possible output from given inputs. For instance, a farmer who can produce higher crop yields per hectare without increasing input costs is technically efficient. This form of efficiency often involves adopting best practices, using advanced technologies and ensuring that all resources are utilized to their full potential.

Economic efficiency: on the other hand, focuses on maximizing the financial returns from the resources used. It involves not just producing the most output but doing so at the lowest possible cost relative to the value of the output (income). For example, a farmer may achieve economic efficiency by choosing a crop variety that, while yielding slightly less, requires significantly fewer inputs, thereby reducing overall costs and increasing profit margins.

Box 9.11. Measures of efficiency.

Efficiency and land: land is a natural resource. If it is the most limiting factor, the farmer has no more land to use. If this is the case, the farmer will want to measure efficiency in terms of land by determining the farm's gross margin per unit of land (e.g. hectare or acre) or the profit per unit of land. To improve the efficiency of the farm, the farmer should consider ways to increase yields to try to obtain higher returns from the produce.

Efficiency and water: the most limiting factor may be water. A farmer may consider deep ploughing, which is good to maintain water in the soil. To save water, the farmer could also construct canals and transplant vegetables on raised beds. In this way, the farmer uses less water per hectare and makes more efficient use of this scarce resource.

Efficiency and labour: assume that labour is the limiting resource. This means that labour is in short supply. When this is the case, the farmer will want to measure efficiency in terms of labour by determining the farm's gross margin or profit per labour input or person-day. To improve the efficiency of the farm, the farmer should look for ways to increase production using the same amount of labour or produce the same amount using less labour. This can be done by introducing a labour-saving technology that reduces the amount of time spent on farm operations.

Efficiency and capital: capital is nearly always in short supply and needs to be used efficiently. If capital is the most limiting factor, the farmer will want to measure efficiency in terms of capital by determining the farm's gross margin or profit per unit of capital, such as per $100 or whatever value is most appropriate. To improve the efficiency of the farm, the farmer would look for ways to increase production, perhaps by investing money in high yielding seeds (instead of ordinary seeds) or by investing money in a more efficient technology.

whether through *technical efficiency* (using inputs effectively to increase output) or *economic efficiency* (maximizing profit relative to input costs) (see Boxes 9.10. and 9.11.).

Efficiency in resource use is crucial for minimizing waste and avoiding unnecessary costs. Farmers should strive to continuously improve their efficiency by monitoring their resource use, seeking out innovations that can lower costs and adapting their practices to changing circumstances. Efficiency not only enhances profitability but also contributes to the sustainability of the farm, ensuring that resources are used responsibly and conserved for future use.

Effective resource allocation, a focus on efficiency and strategic substitution are all essential components of successful farm management. By using limited resources wisely, working efficiently and choosing the best alternatives, farmers can boost productivity and profits, ensuring that their farms remain sustainable over the long term. By carefully analysing substitution possibilities using these economic principles, farmers can make strategic and informed decisions on input combinations. Such careful consideration ensures resource-use efficiency, cost minimization and enhanced farm profitability.

Improving efficiency in resource use helps farmers make better decisions, reduce production costs and achieve higher profitability – even when working with limited means. It also promotes more sustainable farming practices by reducing waste and conserving vital resources for future use.

While productivity and efficiency are closely related concepts in agriculture, they are not the same. The following is a comparison illustrating the difference between the concepts (see Box 9.12.). This is key to improving farm performance.

For example, a farmer might increase productivity by using more fertilizer to boost yields. But if the cost of the extra fertilizer is greater than the value of the additional harvest, economic efficiency declines, even though productivity has increased.

Thus, farmers need to find a balance: push for high productivity while carefully managing input costs to ensure strong profits (economic efficiency). This balance helps them make the best use of their resources while achieving the highest possible return on investment.

9.2.5 Factor–product relationships: intensity of production

In the pursuit of maximizing profits, farmers face the critical task of optimizing the allocation of their scarce resources, such as land, labour and capital. Effective resource allocation involves making informed decisions about how to distribute these limited resources across various farming activities to achieve the highest possible returns.

The factor–product relationship describes how varying the level of a single input (factor), such as fertilizer, labour or water, influences the output or product (crop yield, livestock productivity). Farmers regularly face decisions about the optimal intensity of input use, aiming to achieve the highest profit rather than merely maximizing production.

Box 9.12. Comparative definitions.

	Productivity	Efficiency
Definition	A measure of how much output is produced from a given input or set of inputs.	A measure of how well inputs are used – either technically (maximizing output) or economically (maximizing profit).
Focus	Quantity of output (e.g. tons of maize per hectare, litres of milk per cow).	Quality of input use (e.g. minimizing waste, balancing costs and returns).
Key question	How much are we producing?	Are we using our resources in the best possible way?
Measurement	Often measured as output per unit of input (e.g. yield per hectare, output per worker).	Measured as technical efficiency (max output from inputs) and economic efficiency (max profit from input–output balance).
Implication	High productivity does not always mean high profit or sustainability.	High efficiency aims to achieve optimal output at minimal cost, improving profitability and sustainability.

Law of diminishing returns

The relationship between inputs and outputs is governed by the law of diminishing returns. This principle helps farmers determine the optimal level of input use to maximize profits and is useful in assessing output potential from a single plot or the entire farm. When variable resources – such as labour, water, seed and fertilizer – are applied to a fixed area of land, the law of diminishing returns shows that, beyond a certain point, each additional input results in smaller increases in output. Eventually, the cost of the extra input exceeds the value of the additional yield it generates. The marginal concept refers to the additional output gained from using one more unit of input, helping to clarify how incremental changes affect production. This principle also applies in more complex production scenarios involving multiple inputs. Given the range of available production techniques, the farmer must select the one that delivers the highest profit.

The law of diminishing returns states that, after a certain threshold, each additional unit of input produces progressively smaller gains in output. While initial increases in inputs may significantly raise yields, continued increases become less effective and eventually uneconomical. Understanding this law enables farmers to optimize input use, reduce waste and improve profitability.

Table 9.1 shows an example of the marginal concept used to produce maize when adding different levels of fertilizer. This is the first step in deciding how much fertilizer to use. In this example, when no fertilizer is used, there is little yield (total product = 0.5). When 1 bag of fertilizer is applied, the total product of maize is 1.0 bag, and the marginal product is 0.5 bags. When the quantity of fertilizer is increased from 1 bag to 2 bags, the yield increases to 2.5 bags. This gives a marginal product of 1.5 bags of maize (2.5–1.0 = 1.5). The marginal product (or added yield) is calculated as the difference between different levels of total production. This appears in the third column.

When the amount of fertilizer is increased from 2 to 3 bags, the total product increases to 3.4 bags, but the marginal product has decreased to 0.9 bags. The returns to the amount of fertilizer being applied are diminishing (decreasing). As fertilizer is increased further, we can see that while the total product continues to increase, the marginal product continues to diminish, clearly demonstrating the law of diminishing returns.

The next step is to look at this information in terms of income and costs (see Table 9.2). To determine optimal fertilizer use, farmers must translate yield increments into economic terms. The following economic analysis is presented using market prices and fertilizer costs.

Assuming the price of a bag of maize is \$20. The cost of fertilizer is \$8 per bag.

In Table 9.2 we can see that when the production of maize increases from 1 bag to 2.5 bags, the total gross income or value product rises from \$20 to \$50. The costs of production (fertilizer) also rise from \$8 to \$16. The marginal value product of the maize also increases, from \$20 to \$30 (\$50 – \$20) and the marginal costs rise from \$0 to \$8 (\$16 – \$8).

Determining optimal fertilizer application rate for maximum profitability

How much of each resource the farmer uses depends on the cost compared with the return. The most economically rational choice is the

Table 9.1. The marginal product for maize when adding different levels of fertilizer.

Quantity of fertilizer (bags)	Total product of maize (yield) (bags)	Marginal product of maize (bags)
0	0.5	0
1	1.0	0.5 (1.0–0.5)
2	2.5	1.5 (2.5–1.0)
3	3.4	0.9 (3.4–2.5)
4	4.0	0.6 (4.0–3.4)
5	4.5	0.5 (4.5–4.0)

Table 9.2. Costs and returns for maize.

Total maize production (bags)	Total gross income (value product) ($) (a)	Marginal value product ($)	Quantity of fertilizer used (bags)	Total cost of fertilizer ($) (b)	Marginal cost of fertilizer ($)	Gross margin ($) (a–b)
0.5	10	0	0	0	0	10
1.0	20	20	1	8	8	12
2.5	50	30	2	16	8	34
3.4	68	18	3	24	8	44
4.0	80	12	4	32	8	48
4.5	90	10	5	40	8	50
4.9	98	8	6	48	8	50
5.2	104	6	7	56	8	48

optimum level of output. This is where the value of the marginal product is just sufficient to cover the cost of the resource used (see Box 9.13.).

In Table 9.2, this occurs at a level that lies between 5 and 6 bags of fertilizer. The gross margin at this level of production is $50. Any application of fertilizer below this level results in a marginal value product that is greater than the cost of fertilizer. This means that the farmer would increase the gross margin if more fertilizer were applied. With 2 bags of fertilizer, the gross margin is $34. At fertilizer levels above 6 bags, the gross margin begins to decrease.

So long as the marginal value product exceeds the extra cost, the gross margin is increased by using more of the fertilizer input. As noted, there comes a point when it is not profitable to apply any more fertilizer to the crop. This is the point in the production process where the marginal gross income equals marginal costs.

The optimal fertilizer application rate is reached when: the *marginal value product* (additional revenue generated by applying one more unit of fertilizer) equals the *marginal cost* (additional cost of that unit of fertilizer).

However, because prices often fluctuate, the optimum level of output cannot usually be as easily predicted as in this example. Farmers therefore need to consider other factors as well when planning input use.

Will using this resource cost me more than the money received in return? This question is applied to many farm decisions. Will feeding poultry an additional 5 kg of feed add more to income (returns) than to costs? Will costs or income increase more if the output of milk per cow is increased to 100 l? If we add another hired labourer or expand the area under cultivation, will we add more to income than to costs?

The practical experience of farmers closely follows the concept of marginal analysis. Farmers who economize do not usually completely favour one product or resource over another. They normally decide whether to have a little bit more of one and not quite so much of the other. Farmers tend to apply inputs in small steps. They might test the application of fertilizer or seeds over a number of years, gradually trying out different combinations of inputs and outputs before deciding on an optimum level. As more experience is gained over the years through demonstrations or trials an optimal allocation of fertilizer can be reached. This often differs from recommendations that are given by agricultural extension workers. It is up to the farmers to test the combinations and arrive at their own decisions as to the most profitable application.

By conducting such detailed marginal and economic analyses, farmers can identify the optimal intensity of production that maximizes profitability rather than merely increasing production. This approach helps farmers strategically manage input use, reduce unnecessary expenses and achieve sustainable economic outcomes.

Box 9.13. The marginal concept.

The term 'marginal' has the same meaning as 'additional'. It can refer to either output or input. In both cases, it can be measured in either physical or financial terms. Thus, the marginal product per unit of input means the addition to total production achieved by adding one more unit of input. Similarly, *the marginal gross income* means the value of the marginal, or additional, product. If refers to the value added to the total gross income by adding one more unit of input. Other common marginal terms include marginal input and marginal cost, which refer, respectively, to added inputs and the cost of an added input.

9.2.6 Factor-factor relationships: resource substitution and combination

In farming, making decisions about the best combination of inputs is essential to achieving optimal production outcomes. These inputs, also known as factors of production, include resources such as labour, machinery, land, water and capital (we have, up to this point, referred to factors of production as livelihood assets or capital). This involves decisions about combining different inputs to achieve a production goal efficiently. Farmers seek the most cost-effective combination of inputs. Factor-factor relationships describe how different inputs interact within the production process, and how changes in the level of one input affect the use or productivity of another.

For instance, a farmer may need to choose between labour-intensive practices like hand weeding and capital-intensive options such as using a tractor or mechanical weeder. The choice of input combination depends on factors such as cost, availability and productivity. The objective is to find the most efficient mix of inputs that maximizes profit while minimizing cost.

Substitution

In farming, different methods can often achieve the same production goals. The concept of substitution comes into play when deciding between these methods. Substitution refers to replacing one input with another to maintain output while lowering costs or increasing efficiency. For instance, a farmer might consider substituting chemical fertilizers with compost. This decision depends on input costs, resource availability and the impact on productivity. If compost is cheaper, accessible and yields similar results, it may be more efficient. However, if compost requires more labour or takes longer to affect yields, those additional costs must be considered.

The principle of substitution helps farmers improve resource efficiency. By evaluating the costs and benefits of different input combinations, farmers can choose the approach that offers the greatest margin or profit. This enables better resource use, cost reduction and increased profitability.

What methods or techniques should farmers decide on under the different conditions of farm size, cropping patterns and capital availability? This is a genuine problem that farmers face. For example, choices may need to be made between two feeds such as maize and sorghum; two kinds of labour (hired and family labour); and chemical fertilizer vs organic manure. Each can substitute for the other. Machinery can also substitute for labour as well as draught power.

The question here is: Which of the alternative methods should the farmer use? The answer lies in determining:

1. The cheapest combination of resources needed to produce a product; and
2. The best combination of resources and technologies that result in maximum profits.

Optimizing resource use: substitution and cost minimization

Some decisions are straightforward. For example, the method of fertilizing a crop. The farmer could decide on the most suitable technology by comparing whether it would be cheaper to buy chemical fertilizer or apply compost. In other cases, the choice of inputs is more complicated, particularly when a combination is involved. For example, livestock may be fed various combinations of feed (roughage and concentrates). The challenge facing the farmer is what combination of resources is the best to apply? Farmers have to decide between different practices and technologies that require combinations of resources. The solution to this challenge draws on the principles of diminishing returns, marginal analysis and substitution.

In the production of maize, for example, we can apply fertilizer and compost. To produce 80 kg of maize, we can apply a small amount of fertilizer and a large amount of compost, a medium amount of both or any combination of the two. This is illustrated in Table 9.3.

The first two columns show numerous combinations of fertilizer and compost which produce an equivalent of 80 kg of maize. We can use 1 bag of compost and 7 kg of fertilizer, 2 bags of compost and 2.8 kg of fertilizer, and so forth.

The third column shows the amount of fertilizer replaced by a bag of compost (the rate of substitution); as we go from 1 bag of compost and 7 kg of fertilizer to 2 bags of compost and 2.8 kg of fertilizer, each bag of compost substitutes for 4.2 kg of fertilizer. A jump to the next combination allows one bag of compost to substitute for 1 kg of fertilizer. Here we have diminishing rates of substitution of compost for fertilizer; each added bag of compost substitutes for, or replaces, less fertilizer than the previous one.

Decisions on different practices that substitute for each other require finding the practice which results in the lowest cost of production. In this example, the farmer has 3 ha of maize under cultivation and the possibility of applying fertilizer or compost to the fields.

Which combination of fertilizer and compost will allow the farmer to farm at the lowest cost and produce 80 kg of maize?

The answer to the question of the best substitute depends partly on the costs of the resources which make up the practice. If fertilizer is high in price relative to compost, the costs may be lessened by substituting compost for fertilizer. But the costs of substitute inputs only show part of the picture. We also need to know the number of units of an input replaced by the substitute.

The total cost of each combination of compost and fertilizer is given in the last column of Table 9.3, calculated using compost costs at $10/bag and fertilizer costs at $20/kg. With a level of production of 80 bags of maize, a least cost combination occurs where 3 or 4 bags of compost and between 1.3 and 1.8 kg of fertilizer are applied. The cost calculated is $66. Cost is at a minimum because the extra cost that comes from substituting compost for fertilizer is less than any savings that can be made to the cost of fertilizer.

This is seen by observing that with 2 bags of compost, the extra cost of 1 more bag of compost is $10. Savings in the cost of fertilizer is $20 (i.e. 2.8–1.8 kg of fertilizer × $20). It therefore pays the farmer to substitute compost for fertilizer. Similarly, if more than 4 bags of compost are used, the extra cost of substituting compost for fertilizer is greater than the saving in fertilizer cost. For instance, the cost of the fifth bag of compost is still only $10 but the saving in the cost of fertilizer is only $2 (1.1–1.0 units of fertilizer

Table 9.3. A farmer growing maize; fertilizer and compost.

Substitution of fertilizer and compost in the production of maize			
Bags of improved compost (C1)*	Kg of fertilizer (F1)*	Rate of technical substitution (1 kg of fertilizer replaced by 1 bag of compost)	Total cost ($) (C1 + F1)
1	7.0	0	150
2	2.8	4.2	76
3	1.8	1.0	66
4	1.3	0.5	66
5	1.0	0.3	70
6	1.1	–0.1	82
7	1.3	–0.2	96

Compost costs $10 per bag; fertilizer costs $20 per kg.

× \$20). In this situation it is no longer profitable for the farmer to substitute fertilizer for compost.

The principle is that, in substituting one method for another, farmers must be certain that savings in the method of production will be greater than the cost of the resources added. The cheapest combination of factors to produce an output is found when the cost of any factor replaced is just greater than the cost of the factor introduced. By following the conditions shown previously, the lowest cost method of production can be selected.

Determining the optimal allocation of resources requires careful analysis of the costs and the returns or income associated with different levels of input use. Farmers must continuously assess the performance of their enterprises, adjusting their resource allocation strategies as conditions change, whether due to market fluctuations, changes in input prices or evolving environmental factors.

9.2.7 Enterprise combination and comparative advantage: product-product relationships

Smallholder farmers often have limited land, labour and capital, but multiple options for what to produce. The decision of how to combine different farming enterprises – crops, livestock or mixed systems – is a core part of farm planning and directly influences profitability and sustainability. This type of decision making falls under the product-product relationship, where the focus is on choosing the right combination of outputs to maximize returns.

Comparative advantage

Comparative advantage refers to selecting farming enterprises that offer the greatest economic benefit given a farm's specific conditions and resources. This economic principle helps farmers decide which crops or livestock activities to focus on by evaluating where their resources and conditions can generate the highest relative returns. Utilizing comparative advantage is crucial because it ensures that resources such as land, labour and capital are allocated efficiently, thereby optimizing farm profitability and sustainability.

SELECTION OF ENTERPRISES. One of the most critical decisions a farmer must make is selecting which crops or livestock enterprises to pursue. This choice is influenced by several factors, including the farm's natural conditions, the farmer's resources and market demand. The concept of *comparative advantage* plays a central role in this decision-making process. Comparative advantage means choosing enterprises that offer the highest profit given the farm's specific conditions (see Box 9.14.).

For instance, a farm located in a region with fertile soil and adequate rainfall might have a comparative advantage in growing high-value crops such as vegetables or fruits, while a farm in a drier area might focus on drought-resistant crops or livestock. Market demand also guides these decisions; producing what the market wants (at the right time and in the right quantity) can significantly boost profitability. Farmers must continually assess their comparative advantage, adapting their enterprise mix as market conditions, technologies and environmental factors change.

Even where there are such advantages, farmers still face choices. For example, a lowland farmer may have to decide between growing cereals for home consumption or growing cash crops. Similarly, upland farmers may have to decide between rearing livestock and cultivating horticultural crops. Farmers often decide by using some of the economic principles and concepts discussed here without even knowing it. Farmers must have good knowledge of their farms, including the type of soils and their quality, water sources and topography as well as an idea of the type of crops that can be grown. They need to know whether they have sufficient capital and labour to produce the enterprises and whether they might require more labour and machinery at particular times of the year, for example, during harvest. They must also have an idea of the gross margins that they can earn from the different enterprises.

WHICH CROP? Consider a farmer who is thinking of planting three crops – maize, millet and tomatoes – on three plots of land that are the same size but with varying soil fertility and climate. Which crop should be selected for each plot? The farmer needs information on yields, prices, inputs, costs and gross margins to help

Box 9.14. Practical examples of comparative advantage.

Crop choice based on ecological conditions: a farm situated in a region with fertile soils and abundant rainfall may have a comparative advantage in growing high-value crops such as vegetables or fruits, which typically yield higher profits per hectare. Conversely, farms in areas with limited rainfall may find comparative advantage in drought-resistant grains, legumes or livestock rearing.

Cash crops versus subsistence crops: farmers often face the decision to cultivate crops either primarily for market sale (cash crops) or for household consumption (subsistence crops). A lowland farmer may choose cereals for family food security or opt for higher-return cash crops to boost household income. Similarly, upland farmers might choose between horticulture for cash income or livestock for subsistence and market purposes.

make the crop selection. Table 9.4 shows the projected gross margin (yield multiplied by the price minus the variable costs) for the three crops when grown on each of the three plots.

The figures in the table show that Plot A has an absolute advantage over the other plots for all three crops. The gross margin generated from Plot A is higher than that of the others, perhaps because of higher yields as a result of better soil fertility. Out of the three crops, tomatoes have the greatest advantage (i.e. the highest gross margin). On Plot B, maize generates the highest gross margin. Plot C generates lower gross margins than both A and B for all three crops, with tomatoes being the most profitable crop.

The farmer has a choice of specializing by growing the same crop on all three plots or choosing a crop mixture (diversifying). In this example, the size of each plot is the same and the gross margin that can be earned via specialization or diversification can be seen.

SPECIALIZATION. If one crop is grown on all three plots the farmer could earn $156 from growing tomatoes, $145 from millet and $150 from maize. Focusing on a single enterprise (such as tomatoes across all plots) simplifies management, enables economies of scale and can maximize returns if the chosen enterprise consistently yields high profits. However, it can also expose farmers to greater risk due to market price fluctuations or climatic factors affecting that single crop.

DIVERSIFICATION. By diversifying with a different crop planted on each of the plots, the farmer could earn $168 from a combination of the three crops: $71 for tomatoes from Plot A; $57 for millet from Plot B; and $40 for maize from Plot C. In this case the farmer would do best to diversify. Diversifying crop production reduces risks by spreading investments across multiple enterprises, smoothing income flows and stabilizing overall returns. Diversification can also better utilize varying plot conditions, seasonal labour availability and can mitigate market volatility.

Thus, understanding comparative advantage and strategically balancing specialization and diversification, ensures efficient resource utilization, optimized profitability and enhanced resilience for farming households.

Optimizing the mix of farming enterprises

Most farmers are faced with a range of alternative products or enterprises that they could produce. Selecting an optimal mix of enterprises involves analysing various production possibilities to identify which combination generates the highest overall profits.

Farmers often manage multiple enterprises simultaneously, such as different crops or livestock. The product-product relationship addresses how these different farming enterprises interact, particularly in terms of resource allocation and profitability. The object of the decision is to identify the optimal enterprise mix considering available resources, market opportunities and risk diversification. This decision addresses the mix of different farming activities or enterprises to maximize profitability and resource efficiency. Choosing the optimal combination of enterprises is essential to maximize profitability, resource efficiency and risk management. For example, a farmer has limited land and must decide whether to plant maize for grain, maize for fodder or produce milk from

Table 9.4. Gross margin in $ per ha.

Plot	Tomatoes	Millet	Maize
A	71	68	59
B	54	57	51
C	31	20	40
Total	156	145	150

Table 9.5. Labour productivity in two alternative products.

Units of labour used	Total physical product of maize (bags)	Marginal physical product of maize (bags)	Total physical product of livestock (units)	Marginal physical product of livestock (units)
0	0	0	0	0
1	14	14	12	12
2	40	26	24	12
3	62	22	36	12
4	74	12	48	12
5	80	6	58	10
6	78	−2	60	2
7	70	−8	60	0

dairy cows. Maize grain might yield high returns in good market years but could also pose market risks. Fodder production supports dairy cows, ensuring consistent milk production, which stabilizes income and reduces market risk.

It has been shown that the goal of farmers as far as the operation of the farm is concerned is to make efficient use of all resources – land, water, labour, tools, capital, etc. The principle deals with the problem of allocation of these resources to those enterprises and activities that will maximize profit to the farm. It means putting to use those resources that are idle part of the year and making more intensive use of those already employed. This, in fact, is a key problem in many low-income countries, especially where underemployment and under- or unitilized land, labour and capital resources exist.

The farmer, in addressing this question, will first have to make the most of the comparative advantage that they may enjoy for particular lines of production. To exploit comparative advantage, the farmer must concentrate on those enterprises where the relative profits are greatest, considering the yields and current costs and prices. For most farmers, the availability of land, labour and capital is limited, and after having selected the enterprises for which the farm is best suited, they will need to consider allocating the scarce resources among the various enterprises in order to produce the greatest profit from the farm as a whole.

This is called a product-product relationship and is illustrated in the case of competing products by a hypothetical example in which just two alternatives are considered: maize production and livestock. This is a more complex problem than the previous example but uses the same principles. Maize and livestock compete for the farm labour resources available to the farmer, which amount to 7 units in total. It is assumed that other resources specific to each enterprise are fixed so that as labour inputs are increased, the marginal returns diminish. These can be seen in Table 9.5.

In Table 9.5 we can see that, if all 7 units of labour are used in livestock production, the total product is 60 units of livestock. In this situation, no labour is available for maize production, so the output of maize is 0. If one unit of labour is transferred from the livestock to maize production, there will still be an output of 60 units of livestock but in addition 14 bags of maize will be produced. If a second unit of labour is transferred to maize production so that only 5 units of labour are used in livestock production, this second unit has a 'cost' in terms of a reduction in livestock from 60 to 58 tons. This cost is measured in terms of the alternative foregone, i.e. the opportunity cost. In this case, the opportunity cost of increasing maize output from 14 bags to 40 bags by transferring 1 unit of labour out of livestock production is 1 unit of livestock. As more labour is transferred from livestock to maize production, the marginal product per unit of labour in maize diminishes. Similarly, if labour is transferred out of maize production into livestock, the opportunity cost, in terms of maize foregone, will increase.

To find the optimum combination of the two enterprises, the physical products are converted to value products (income) by multiplying the former with its per unit price. Similarly, the value product of labour can be estimated from knowing the wage rate of labour. In this example, maize is priced at $2 per bag, livestock at $7 per unit and the wage rate or factor costs is $10 per unit. The results are shown in Table 9.6.

The economic optimum level of labour use in maize production occurs at about 5 units where the marginal value product is $12. That is where the unit factor cost of labour is closest to the marginal value product of maize ($10 cost of labour and $12 marginal value product of maize). As 7 units of labour are available in total, the use of 5 units in maize production leaves 2 units free for livestock. This balance maximizes the farm's total profitability. Labour use in livestock remains below the economic optimum which occurs in the same way as for maize, at between 6 to 7 units of input, i.e. where the unit factor cost of labour is equal to the marginal value product of livestock. In this way, the two enterprises can be understood as competing for scarce labour products.

The same conclusion can be reached if we increase the labour input unit by unit. To maximize profits each unit of labour should be used where it will add most to returns. The first unit of labour should be used in livestock production where it can earn $84. The sixth unit of labour, however, would earn only $14 in livestock production, but if used in maize production it would earn $28 (the first unit of labour). The seventh unit of labour should also be used in maize production, where it can earn $52 (second unit of labour), as opposed to the $14 it would earn if used to produce livestock. If the labour inputs for both products are varied continuously then the most profitable allocation occurs where the marginal value product for each enterprise is equal. The principle involved in allocating resources between alternative uses is sometimes known as the *equi-marginal returns principle* (see Box 9.15.).

It should be pointed out that, if all the resources were allocated to the enterprise that gives the highest average value product, the profits of the farmer are not maximized. This can be understood from the earlier example. The average value product is higher in livestock production at $60 per unit of labour than under maize. If all 7 units of labour are used in livestock production a total return of $420 could be

Table 9.6. Marginal value product of labour used in maize and livestock production.

Units of labour	Marginal value product in maize production at $2 per bag	Marginal value product in livestock production at $7 per unit
1	28	84
2	52	84
3	44	84
4	24	84
5	12	70
6	–4	14
7	–16	-
Total value product $	140 (70 bags x $2)	420 (60 units x $7)
Average value product $/ unit of labour	20	60

Box 9.15. Equi-marginal returns principle.

The equi-marginal returns principle guides farmers to allocate resources so that the marginal returns (additional income generated by the last unit of input) are equalized across enterprises. If one enterprise's marginal returns are higher, resources should be shifted there until marginal returns equalize.

earned. If, on the other hand, 5 units of labour are used in livestock and 2 units in maize, the total return is higher at $486. In other words, profits will be greatest when the last $1 invested in an enterprise will earn the same as when invested in any of the other selected enterprises. And if a farmer allocates his resources in any other way, profits will fall.

The key to managing these product-product relationships is to balance the competing demands of each enterprise in a way that maximizes the overall profitability of the farm. This involves analysing the marginal returns of each enterprise – that is, the additional profit generated by allocating more resources to one enterprise vs another. Marginal analysis helps farmers determine the point at which investing additional resources in one enterprise yields diminishing returns compared to another enterprise.

In the maize and livestock example:

- Initially, livestock generates higher marginal returns so labour is first allocated there.
- As additional labour input into livestock decreases marginal returns, reallocating labour to maize eventually becomes more profitable.
- Optimal resource allocation occurs when marginal returns from maize and livestock enterprises are equal, maximizing total farm profitability.

By carefully analysing product-product relationships and applying these economic principles, farmers can optimize their enterprise combinations, maximize resource utilization and achieve sustainable profitability.

Enterprise relationships

As discussed in Chapter 4, different farm enterprises often compete for the same resources. For example, a farmer may lack sufficient labour to harvest two crops at the same time. In such cases, expanding one enterprise usually requires reducing another, as fixed resources must be reallocated. The output of one product may thus depend on the level of another. These conflicts can be addressed by adjusting the farm plan or modifying planting times, particularly when crop enterprises overlap.

Enterprise selection inherently involves *opportunity costs*. Using labour for maize production, for instance, limits its availability for livestock, resulting in forgone outputs. Farmers must assess these trade-offs and allocate resources to enterprises that generate the highest overall returns. Managing opportunity costs effectively is key to maximizing profitability.

Enterprises may also *supplement* one another when they use otherwise idle resources. They can be *complementary* when one provides benefits to another – for example, maize benefiting from the soil-improving effects of a prior green manure crop. Complementary relationships are especially important in mixed farming systems, where coordination between crops and livestock becomes more complex as livestock numbers grow. The goal is to keep resources productively engaged throughout the year, which requires wise selection of crop and livestock enterprises.

Farmers must strike a careful balance among competing, complementary and supplementary enterprises. Effective enterprise planning involves aligning activities with seasonal resource availability, seizing market opportunities and managing risk through diversification – all while ensuring efficient use of limited resources and maximizing farm profitability.

Enterprise combination decisions are among the most complex yet critical that farmers make. Using comparative analysis, opportunity cost evaluation and the equi-marginal returns principle, farmers can construct an enterprise mix that matches their conditions, maximizes their returns and ensures efficient use of their scarce resources. By integrating these concepts into regular farm planning and resource allocation, farmers can significantly enhance the profitability and sustainability of their operations.

9.2.8 Strategies for maximizing farm profitability

Profitability is the ultimate goal of most farming enterprises. To achieve this, farmers must adopt a set of strategic decision-making tools and practices that allow them to use resources effectively, minimize costs and respond proactively to changing market and environmental conditions. Three key strategies for enhancing farm profitability include (i) marginal returns analysis; (ii) cost-benefit analysis; and (iii) dynamic, responsive decision making.

Marginal returns and resource allocation

This is one of the applications of the concept of margins discussed earlier. The concept of marginal returns is crucial in making these decisions. Marginal returns refer to the additional profit or benefit gained from allocating an extra unit of a resource (e.g. land, labour, capital) to a particular enterprise. To allocate resources effectively, farmers should compare the marginal returns of different enterprises. For instance, if adding one more hectare to maize production increases profit by $200 but using that hectare for livestock grazing increases profit by $300, the farmer should allocate the land to livestock. However, this decision also depends on other factors such as market demand, input costs and risk. As resources are limited, the goal is to allocate them where they generate the greatest incremental benefit until the marginal returns from each enterprise are equal.

Cost-benefit analysis

Cost-benefit analysis (CBA) is a decision-making tool that helps farmers evaluate whether an investment or farming activity is financially worthwhile. It involves comparing the total expected costs of an action – such as purchasing new equipment, adopting improved seeds or expanding production – with the expected benefits, such as increased yields, higher market prices or reduced labour costs. If the benefits outweigh the costs, the investment is considered profitable. For small family farmers, CBA is crucial for making informed choices with limited resources. By systematically assessing costs (e.g. seeds, fertilizers, labour) and expected returns (e.g. crop sales, improved productivity), farmers can identify the most efficient use of their land, labour and capital. Even simple CBA calculations – such as estimating input costs against expected income – can help minimize financial risk and improve farm profitability. By applying CBA principles, smallholders can prioritize investments that enhance productivity and long-term sustainability.

Dynamic decision making

Product-product relationships are dynamic and can change over time due to shifts in market prices, changes in input costs or developments in farming technology. Therefore, farmers need to continually reassess their resource allocation decisions. Regularly monitoring the performance of each enterprise, staying informed about market trends and being flexible in adjusting the allocation of resources, are all critical components of successful farm management. For example, if the price of maize rises significantly, it might become more profitable to allocate additional resources to maize production, even if it means reducing the resources available for livestock. Conversely, if livestock prices surge, shifting resources towards animal production could be the better strategy. By carefully balancing competing enterprises, considering both complementary and competitive relationships, and focusing on maximizing marginal returns, farmers can optimize their resource use. This strategic approach not only enhances overall farm profitability but also contributes to the resilience and sustainability of their farming operations.

9.2.9 Importance of continuous monitoring and adaptation

Farming operates within a constantly changing environment – markets shift, input prices fluctuate, weather patterns vary and new technologies emerge. Therefore, economic decision making on the farm must be dynamic and responsive. Farmers should regularly evaluate their resource use, monitor market trends, reassess input efficiency and adjust enterprise mixes as needed. This continuous cycle of observation, analysis and adaptation allows farmers to: (i) maximize profits under changing conditions; (ii) reduce risks from market or environmental shocks; and (iii) enhance long-term sustainability and resilience. Monitoring of the farm and its enterprises is elaborated on further in Chapter 14.

9.2.10 Summary of economic decisions in farming

Economic decision making in farming can be summarized under three main categories, each linked to a specific type of input–output relationship, as shown in Box 9.16.

By applying and integrating these economic principles into daily farm management, farmers can make decisions that are both profitable and sustainable – ensuring that each input and enterprise contributes optimally to the overall success of the farm.

9.3 Economics Concepts for Financial Management Decisions

Effective financial management on a family farm involves more than just keeping track of income and expenses. It requires a deep understanding of economic principles that can guide decision making to enhance profitability and ensure sustainability. For small family farms, where resources are often limited and cash flow can be unpredictable, integrating these economic concepts into financial management decisions is crucial. This section explores key economic principles, such as cash flow management, investment evaluation, the time value of money and discounting and how they can be applied to the financial management of small family farms.

Integrating economic principles into financial management decisions is essential for the success of small family farms. By effectively managing cash flow, making informed investment decisions grounded in the time value of money and discounting, and understanding the time dimension in production and financial planning, farmers can improve decision making, enhance profitability and ensure the sustainability of their farming operations. These economic tools provide a robust framework for navigating the complexities of farm management, helping farmers build resilient and financially sound enterprises that can thrive in the face of uncertainty. This section explores these essential concepts, providing practical insights into how they can be leveraged to strengthen the financial health of the family farm.

9.3.1 Cash flow management

Cash flow refers to the movement of money into and out of the farm. It includes cash inflows, such as income from the sale of crops and livestock and cash outflows, such as payments for inputs, labour and other expenses. Understanding and managing cash flow is critical for ensuring that a farm has enough liquidity to meet its financial obligations throughout the year.

A positive cash flow indicates that the farm is generating enough money to cover its expenses, while a negative cash flow means that the farm might face difficulties in meeting its financial commitments. For example, a farmer might produce crops worth \$600 with variable costs of \$100, resulting in a profit of \$500. However, if the farmer only sells \$50 worth of produce

Box 9.16. Summary of economic decisions in farming.

Decisions	Factor-product relationship
1. Intensity of production – How much of an input should be used to maximize profit?	Factor-product example – fertilizer application for grain production: a farmer must determine the optimal amount of fertilizer where the marginal return from added yield just equals the cost of the fertilizer.
2. Factor combination – What is the most cost-effective combination of inputs?	Factor-factor example – labour vs machinery use: a farmer decides whether to use manual labour or hire machinery for land preparation, depending on availability and relative costs.
3. Enterprise combination – What combination of enterprises yields the highest total profit?	Product-product example – maize grain vs dairy production: a farmer evaluates which enterprise, or mix of both, provides the best return for available resources while managing risk and seasonal labour demands.

and the rest is consumed by the family, the cash inflow is insufficient to cover the $100 needed for variable costs. This situation highlights the difference between profitability and cash flow – a farm might be profitable on paper but still struggle with cash flow if sales are not timely or sufficient.

To effectively manage cash flow, farmers need to plan their income and expenses on a monthly or quarterly basis, forecasting periods when cash may be tight. For instance, a farmer planning to introduce a new crop, like tomatoes, must consider when the income from tomato sales will be realized and whether they will coincide with the timing of input purchases and other expenses. If cash inflows are expected to be insufficient to cover outflows at certain times of the year, the farmer might need to explore options such as adjusting the timing of sales, reducing other expenditures or securing a short-term loan.

9.3.2 Investment decisions and the time value of money

Investment decisions in farming require careful consideration of immediate costs vs future benefits. For example, when a farmer considers planting an almond orchard, the decision involves evaluating the initial investment needed for planting and maintaining the orchard against the expected future returns from almond sales. This evaluation is conducted by translating future earnings into their present value using a process called *discounting*.

Discounting is rooted in the principle of the *time value of money*. The time value of money means a dollar today is worth more than a dollar in the future because the present dollar can be invested to generate returns over time. This principle is crucial for making sound long-term investment decisions. For instance, if a farmer were individually offered $1000 today or $1000 a year from now, which would they choose? There are many reasons for choosing cash now instead of later. One might place the $1000 in a savings account earning 5% interest. This would generate $50 in extra income, an advantage that would be missed by receiving the $1000 1 year later. Alternative uses could earn the farmer even more money. Investing now means the farmer's money has more time to grow, highlighting the benefit of starting early to capitalize on compounding.

On the other hand, one may want the money for immediate needs, for example, why wait a year for new farm machinery when it can be purchased and used now? Alternatively, one may worry about the uncertainties of life. Either the giver or the receiver of the $1000 might not be alive in a year to fulfil their part of the transaction. These different reasons demonstrate the importance of the time value of money.

To determine the present value of future earnings, farmers apply a *discount rate* which is typically based on current interest rates or the expected rate of return on investments. For example, if a farmer expects to receive $6438 in 9 years and the discount rate is 15%, the present value of that amount would be $1828. This calculation helps the farmer assess whether the investment is worth pursuing in today's terms.

When making investment decisions, farmers must also consider the impact of *inflation*, which reduces the purchasing power of money over time. As the costs of inputs and services rise, the real value of future income may diminish, making it essential to factor in inflation when planning long-term investments and setting prices for agricultural products.

9.3.3 The time dimension in economic decisions

The time dimension plays a critical role in both production and financial planning within farm management. Agricultural production is inherently seasonal, with crops needing specific growing periods and livestock requiring time to mature. During these periods, various risks, such as weather changes or market fluctuations, can significantly impact farm operations and financial outcomes.

The time value of money underscores the importance of timing in financial management. Immediate expenses, such as purchasing seeds or fertilizers, have different financial implications compared to income that may be realized later, such as from the sale of harvested crops. For example, investing in an efficient irrigation system may entail significant upfront costs, but the potential increase in crop yields and subsequent income in future seasons could justify the expense. Conversely, delaying the sale of crops in anticipation of higher prices might be risky if the market declines or if storage costs outweigh the potential price gains.

Understanding these temporal aspects of economic decisions enables farmers to optimize resource allocation, synchronize production with market opportunities and manage risks more effectively. It also highlights the importance of planning for inflation, as rising costs over time can erode profitability if not adequately accounted for in pricing and financial projections.

9.4 Economic Concepts for Market Decisions

As discussed previously, making informed market decisions is essential for the success of small family farms operating in competitive and often unpredictable agricultural markets. Economic principles such as supply and demand, price fluctuations and market functions, influence profitability and long-term sustainability. By understanding these concepts, farmers can better anticipate changes, manage risks and make strategic choices about production, pricing and sales. This chapter explores key economic principles that shape agricultural markets and provides practical insights into how farmers can apply them to strengthen their market position and improve farm business performance.

9.4.1 Introduction to economic concepts for market decisions

Farmers operating in market-oriented systems must understand key economic concepts to make informed decisions about production, pricing and sales. Markets are dynamic, influenced by supply and demand, price fluctuations and consumer preferences. Without a strong grasp of these principles, farmers may struggle to maximize profitability, manage risks or adapt to changing conditions.

This section explores essential economic concepts that impact farm business decisions, including market functions, price dynamics, supply and demand interactions, and price elasticity. Understanding these factors helps farmers anticipate market trends, determine optimal pricing strategies and improve overall farm efficiency. By applying these economic principles, small family farmers can strengthen their position in agricultural value chains, enhance financial stability and make better decisions about when and where to sell their products.

9.4.2 The market and price dynamics

The workings of the market

In the simplest situation, farmers take their products to markets to exchange them for money; consumers take their money to markets to exchange it for products; the market sets the price at which those who want a product (buyers) can obtain it from those who have it to sell (farmers).

In a market-oriented system the price of a product is determined by supply and demand. Supply is the amount that farmers are prepared to sell. Demand is the amount buyers are able to buy. The market price is the price that both the buyer and the seller are willing to accept. If prices are too high consumers will not buy the product. If the price is too low the producers may not be willing to sell.

If supply increases but demand does not increase, the price will fall. This is often what happens in a market at the time of harvest, when there is an abundance of produce for sale and prices are low. Later in the season when there is less for sale, the price often increases. If demand increases but the supply does not, the price is likely to rise. Where a product is scarce, consumers are willing to pay more for it, thus pushing up the price. Fruits and vegetables out of season are examples of scarce products. When the price of a product increases, profits tend to increase, which encourages farmers to produce more of the product. On the other hand, if consumers do not want the product, its price falls and farmers make a loss. Such a situation may lead to farmers producing less of the product.

Supply and demand

Supply and demand determine market prices and, consequently, the profitability of farm operations. Supply refers to the amount of a product that farmers are prepared to offer for sale at various price points. It is influenced by factors such as production levels, input costs and the availability of resources. On the other side of the equation is demand, which represents the quantity of a product that consumers are willing and able to purchase at different prices. Demand is shaped by consumer preferences, income levels and the availability of substitute products.

The interaction between supply and demand is what sets the equilibrium price – the price at which the quantity supplied matches the quantity demanded. At this point, the market is in balance, with no excess supply or unmet demand. However, this balance can shift due to changes in either supply or demand. For instance, a bumper harvest may increase supply, potentially lowering prices if demand does not rise correspondingly. Conversely, a sudden increase in consumer demand can drive up prices if supply remains constant.

For farmers, understanding these market dynamics is crucial. It enables them to anticipate price movements, make informed decisions about when to sell their products, and set realistic expectations for their income. By timing their market entry to coincide with periods of high demand or limited supply, farmers can maximize their profits. Conversely, during times of oversupply or low demand, they might explore alternative markets or consider storage options to avoid selling at a loss. In essence, a deep understanding of supply and demand dynamics is key to effective farm management and long-term financial sustainability (see Box 9.17.).

Price changes

As we have discussed, prices change depending on changes in supply and demand. As a result of such changes, equilibrium prices and production levels also change. The farmer needs to recognize such changes and adjust the farm business to them. Supply and demand are influenced by price, as well as by the following factors:

- **Seasonality of production:** some produce ripens at a particular time of year, or during a season. This will create in that season an increase of supply. In the non-harvest season, there will be less supply of the produce. For example, tomatoes at harvest time will be abundant in supply and prices will usually fall. In times of the year where tomatoes are not harvested, supply will be less, and prices will rise.
- **Changes resulting from climate:** supply also varies from year to year. Changes in production frequently occur because of weather (including rainfall, floods and drought). Bad or good climatic conditions can decrease or increase the quantity supplied. Bumper harvests usually decrease prices; poor harvests usually increase prices. In these cases, the level of production is not affected by the price of the product but by climatic changes.
- **Production problems:** the incidence of pests, diseases and other hazards, such as fire, can decrease the supply.
- **Change in the cost of production:** a decrease in the cost of inputs or the introduction of more efficient technologies would decrease the cost of production. This makes it possible for the farmer to buy more inputs and produce more for the same cost of production. As a result, more tomatoes could be supplied at the old price. Conversely, an increase in the cost of labour employed on the farm would have the opposite effect.
- **Improved techniques:** improvements in technologies, such as the introduction of high yielding varieties, could increase the level of production resulting in a decrease in product price.
- **Expansion of a crop under cultivation:** farmers may decide to plant more of a crop because of the prospect of greater profits.
- **Changes in profitability of competitive products:** changes in the profitability of competing products can also cause a change in supply. If spinach is more profitable than tomatoes, for example, farmers will gradually shift their system towards it. This change would lead to a decrease in the supply of tomatoes.

There are also changes that occur that make consumers demand either more or less of a product at each price. In other words, the level of demand could change. For example, farmers could join to promote their sales of tomatoes. They might advertise the good quality, freshness and availability at times of the year when tomatoes are not usually found in the market. This may appeal to consumers leading to increased demand.

For fresh produce, such as vegetables, there are often short-term fluctuations in supply,

Box 9.17. Supply, demand and market price.

Supply: normally, the higher the price of a product, the greater the supply of that product. For example, farmers considering producing tomatoes would be encouraged by high prices. If they do produce tomatoes and the price increases, they would be encouraged to extend the area of land under the crop. They would also try to grow tomatoes with more or better quality inputs so that a higher yield could be produced. The following table shows the amounts of tomatoes supplied at different prices by all producers in the market. It shows that the tomatoes supplied change with the price. The higher the price the more abundant the supply; the lower the price the less abundant the supply.

Supply of tomatoes at different prices	
Price ($/kg)	Supply (kg)
12	40,000
10	32,000
8	25,000
6	**20,000**
4	13,000
2	7,000

Demand: the demand for a product normally rises when its price is lower. If the market price is high, consumers reduce their purchases. The following table shows the amounts of tomatoes in demand at different prices. If tomatoes are very expensive, consumers may substitute with other vegetables. If they are cheap, consumers will buy more.

Demand for tomatoes at different prices	
Price ($/kg)	Demand (kg)
12	3,000
10	9,000
8	15,000
6	**20,000**
4	25,000
2	35,000

The lower the price the higher the demand; the higher the price the lower the demand.

The market price is set at a point where supply and demand are in balance. Economists call this the equilibrium price. If we look at the tables on the demand and supply for tomatoes, the equilibrium price is $6 per kg (marked in bold). This price is the only one where there is balance between buyers and sellers. At this price, 20,000 kg of tomatoes is both in demand and offered for sale. The figure below demonstrates this concept.

(*Continued*)

Box 9.17. Continued

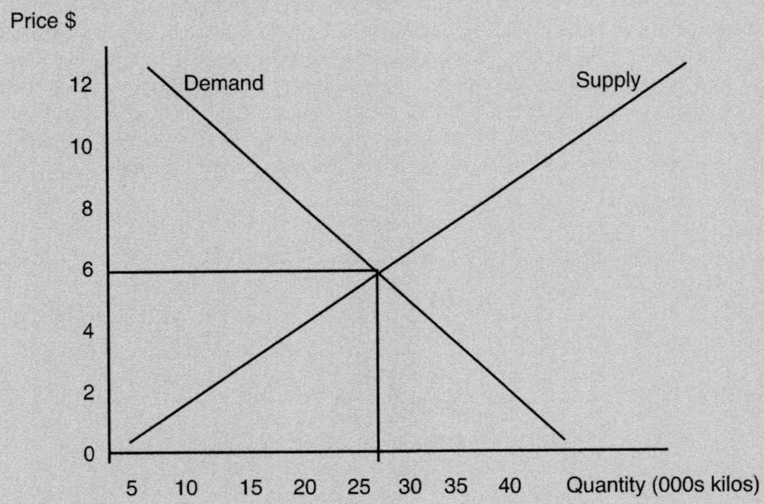

Suppose that, initially, the price of tomatoes is set at $8 per kg. At that price 15,000 kg of tomatoes would be in demand and 25,000 kg for sale. There would be an excess supply of 10,000 kg of tomatoes over the demand. This means that farmers would be left with a surplus that could only be cleared once the price was reduced. As the price falls, demand gradually expands, and supply slowly contracts until a price of $6 per kg is reached. There is then no need for the price to fall further.

Similarly, if the initial price is $4 per kg, 25,000 kg of tomatoes will be in demand but only 13,000 kg offered for sale. Consumers might even form queues to buy tomatoes. The price would then rise, giving an incentive for farmers to plant more. At $6 per kg there is no need for prices to rise further. This is called the *price mechanism*. It indicates the wishes of consumers and allocates productive resources accordingly.

sometimes a result of bad weather which can affect the harvest or of transport disruptions.

The following factors affect the demand for a product:

- **A change in the number of consumers:** extra people coming into a market can increase demand. Development projects, housing developments and other programmes often cause people to move to a new area.
- **An increase in income:** as countries get richer there will be a general increase in income and people will be able to afford new products. As people become richer, they tend to reduce their consumption of staples, such as rice and maize, and increase consumption of dairy products, fruits and vegetables.
- **A change in the prices of products that are close substitutes:** in some diets, spinach may be a substitute for tomatoes. If the price of spinach decreases, consumers may prefer to purchase it instead of tomatoes. Thus, even though there may have been no change in the price of tomatoes, the demand could decrease.
- **Availability of competing products:** if a wider range of products competing with tomatoes comes onto the market, consumers have a wider choice. This may lead to a lower demand for tomatoes. However, a lower demand may lead to lower prices, which could cause demand to increase again.
- **A change in taste:** a promotional campaign aimed at persuading consumers to purchase more of a product could also increase the demand. Similarly, changes in demand may come from changes in consumer tastes that take place gradually because of changes in diet.

- **Safety concerns:** Fears of contamination can spread rapidly among consumers, often leading to an immediate drop in demand – as was the case during the avian flu outbreak in the early 2000s.
- **Expectations of future price changes or shortages:** the fear that the price of a product could rise considerably the following week, for example, because of a planned transport strike, would motivate people to increase their demand and stock up. Clearly, this would not happen so much with perishable products.

The relationship between supply and demand and prices of products is thus extremely complex. A change in the price of one product can affect the demand and, in turn, the price of an entirely different product. Supply is likely to fluctuate much more than demand. Market prices are affected more by changes in production than by changes in demand.

Prices are also affected by what is called *inflation*. Inflation is the general increase of all prices throughout the country. Farmers cannot do anything directly about inflation. High inflation means that prices rise rapidly. Inflation affects both the price of inputs (seed, fertilizer, labour, etc.) and the price of products. For example, a bag of fertilizer which cost $10 last year may cost $12 a year later. Similarly, a kilogram of tomatoes that sold for $4 last year can now be sold for $5. Farmers must be aware of inflation and its impact on input and market prices.

9.4.3 Elasticity and price cycles

Normally, demand increases as prices fall and supply increases as prices rise. More people are interested in buying products that have a low price. Farmers are interested in increasing their production to follow higher prices. However, the supply of some products can be adjusted far more quickly than the supply of others. For example, the supply of mushrooms and tomatoes can be increased or decreased quickly in response to price changes. But other items such as fruit, coffee, tea and milk take longer to adjust. If the prices of such commodities fall, the decrease in supply does not come about immediately. First, changes in production would require significant changes, such as uprooting tree crops. A farmer would need to think through such changes very carefully. Second, even if the farmer were to make the required changes, it would take a long time for overall production to adjust.

Elasticity in farm management

Elasticity is a key economic concept that measures the responsiveness of supply and demand to changes in price. It plays a crucial role in farm management, as it influences how farmers make decisions about production, pricing and marketing strategies. When a product is elastic, even a small change in price can lead to significant shifts in the quantity supplied or demanded. For example, many vegetables are highly elastic; when prices drop, consumers may purchase significantly more, and when prices rise, they might reduce their purchases or switch to other products. This high sensitivity to price changes means that farmers producing elastic goods must be particularly attuned to market fluctuations, adjusting their production levels to capitalize on favourable prices or to avoid losses when prices fall.

On the other hand, when a product is inelastic, supply or demand changes very little in response to price fluctuations. Salt is a classic example of an inelastic product. The demand for salt is relatively unresponsive to price changes because it represents a small expense for most households, and its consumption is not easily reduced. Additionally, there are no significant substitutes for salt, so even large price changes result in only minor adjustments in demand.

For farmers, understanding the elasticity of the products they grow is essential for making informed decisions. In a market where a product is highly elastic, a farmer might decide to quickly ramp up production when prices are high, anticipating that demand will rise accordingly. Conversely, in a market with inelastic products, where price changes have little effect on demand, the focus may shift more towards cost management rather than adjusting production levels. By grasping the concept of elasticity, farmers can better predict how market prices will impact their sales and production, enabling them to maximize profitability while minimizing risk.

Price cycles

Price cycles are recurring patterns in the market prices of agricultural products, often driven by a combination of biological, economic and market factors. A classic example is the pig cycle, where pig production and prices fluctuate over several years due to the time it takes for changes in production to affect supply. Initially, high prices may encourage farmers to increase pig production, but since it takes time to raise pigs to market weight, the market might eventually experience an oversupply, leading to lower prices. As prices drop, farmers reduce production, which eventually leads to a shortage and higher prices again, thus continuing the cycle (see Box 9.18.).

Recognizing these cyclical patterns is vital for farmers as it allows them to anticipate future price movements and adjust their strategies accordingly. For instance, during periods of anticipated low prices, farmers might scale back production or explore alternative markets to avoid losses. Conversely, during periods of high prices, they might ramp up production or invest in expanding their operations to capitalize on the favourable market conditions. Understanding price cycles also helps in planning long-term investments, managing cash flow and reducing the impact of price volatility on farm income.

By staying attuned to elasticity and price cycles, farmers can navigate the complexities of agricultural markets more effectively, making decisions that not only enhance short-term profits but also contribute to the long-term sustainability of their farming enterprises.

Understanding these cyclical patterns is crucial for farmers in making strategic decisions and managing risks associated with price volatility. Farmers can use their knowledge of the pig cycle to anticipate and respond to

Box 9.18. An example of economic fluctuations.

The pig or hog cycle is a classic example of economic fluctuations observed in agricultural markets, particularly in livestock production. This cyclical phenomenon is driven by the responses of producers to price changes, which in turn are influenced by changes in supply and demand.

- **Period 1:** initially, let us assume the market is in equilibrium with pigs priced at $10 each and a total quantity of 10 units. A disturbance, such as a higher mortality rate among piglets, reduces the quantity available to 9 units. The scarcity of pigs leads to a price increase to $11. This higher price incentivizes farmers to increase production, anticipating higher future profits.
- **Period 2:** due to the biological and operational time lags in pig farming (ranging from the gestation period to the time it takes to raise a pig to market weight) the response to higher prices does not impact the market immediately. However, once this increased production enters the market, it causes an oversupply, leading to a price drop to $9. Reacting to this new lower price, farmers then plan for reduced production.
- **Period 3:** after another time lag, this decreased production results in a reduced supply of pigs, with quantities falling to 9 units again, which pushes the price back up to $11. This reinitiates the cycle as farmers respond to higher prices by planning to increase production once more.

If the elasticities of supply and demand are not identical, the dynamics of the cycle alter as follows:

- **Less elastic supply:** if the supply is less elastic than the demand, the fluctuations in quantity and price are less pronounced. For example, in Period 2, supply might only increase slightly to 9.5 instead of fully adjusting to 10, and in Period 3, both price and quantity adjust to 10.3 and 9.7, respectively. The cycle tends to stabilize and converge towards the equilibrium over time.
- **More inelastic demand:** conversely, if demand is more inelastic relative to supply, the fluctuations become more extreme, potentially leading to larger cycles or market crashes if the demand decreases significantly.

The pig cycle reflects how farmers react to market signals. Pig production, which can adjust relatively quickly to price changes due to shorter production periods compared to crops or cattle, shows more pronounced cycles. In contrast, crops, with lower supply elasticities, and cattle, which require longer periods to reach market weight, exhibit less defined cyclical patterns.

market fluctuations. For example, during periods of high prices, farmers may consider expanding their operations or taking advantage of favourable market conditions to maximize profits. Conversely, during periods of low prices, farmers may implement cost-saving measures, adjust their production levels or explore alternative markets to mitigate losses.

In conclusion, understanding these cycles enables farmers to make informed decisions, adapt their production strategies, and manage the risks associated with price volatility. By staying attuned to market dynamics and employing effective risk management strategies, farmers can navigate the cyclical nature of prices and strive for long-term profitability.

9.5 A Behavioural Economics Perspective

As noted previously in this book, farmer decisions are not always rational or, based on the economic principles that we discussed previously in this chapter, aimed at resource optimization. In practice, the decision-making process of farmers, in particular small family farmers, is more complex. This has been addressed by the new economics field 'behavioural economics', which offers a nuanced understanding of decision making that contrasts with the classical economic models described in this chapter (Kahneman, 2011).

When looking at the main decisions that smallholder farmers make – what to grow, how much to produce, how to use their resources and where to sell – behavioural economics helps explain why their choices may not always be fully logical or based on perfect information.

The behavioural economics response to the core farm management decisions that smallholder farmers make are described as follows:

1. **What to produce:** as discussed previously in this chapter, traditional economics suggests farmers will choose crops that maximize expected returns based on market prices and production costs. However, behavioural economics considers the impact of factors like farmers' aversion to risk, mental accounting, and the influence of social norms. For instance, a farmer may choose to plant traditional crops over more profitable alternatives because she or he is familiar with the crop or because of social pressures that may favour traditional farming practices in their community. Overcoming these biases might require extension workers to not only provide information but to convince the community before introducing new crops.
2. **How much to produce:** decisions on the scale of production are often influenced by farmers fearing loss. Farmers might under-produce out of fear of wasting money already invested on crops that may not yield well. This decision might be driven by memories of past failures. These effects can be mitigated by providing ways to reduce risks such as crop insurance. Behavioural changes can also take place by encouraging trial and error through pilot or demonstration plots where farmers can experiment with input applications that lower risk.
3. **How to allocate resources optimally:** as shown in this chapter, optimal resource allocation is crucial for maximizing farm productivity. However, farmers are often biased in their decision making and this can lead to less-than-optimal resource use. For example, if a farmer spends money on the first available input instead of shopping around or planning for the future, they may not have enough money left to buy critical inputs later in the season. Assistance may be needed to help farmers to budget and plan better. The tools needed to do this are described in forthcoming chapters.
4. **Where to market produce:** marketing decisions of farmers are often affected by what is called 'path dependence and conformity bias'. Farmers may choose to sell their produce at local markets they are familiar with, rather than exploring potentially more profitable options. This may be due to a strong reliance on familiar routines or copying the behaviour of other farmers in their community. Information campaigns that highlight the advantages of alternative markets can be helpful, along with platforms that link farmers directly to buyers, allowing them to replace traditional and often less profitable channels.

While the principles of economics are essential, agricultural advisors still need to be aware and knowledgeable of the broader decision-making processes of farmers when supporting them to make efficient use of resources and improve the profitability and sustainability of their farm businesses.

9.6 Use of Economic Principles in Agricultural Advisory Services

Understanding economic principles is vital for extension workers who support smallholder farmers. These principles are needed to guide farmers through complex decision-making processes that are essential for achieving sustainable and profitable agriculture. However, applying economic concepts in the smallholder context can be challenging due to the dynamic and often unpredictable environment in which these farmers operate.

Smallholder farmers face frequent market price fluctuations, rapid changes in production techniques due to innovation and time lags between making production decisions and selling their products. They also deal with inherent farming risks and operational complexities, such as joint products (e.g. grain and straw) and shared costs (e.g. a tractor used for multiple tasks). The indivisibility of many inputs, which cannot easily be scaled down, further complicates resource management.

Despite these challenges, a strong grasp of economic principles enables agricultural advisers to offer clear, effective guidance. It helps farmers understand market dynamics and the financial implications of their decisions. For example, choosing the best time to sell crops or livestock requires insight into price trends and consumer demand. Advisors with economic training can help farmers develop strategies to reduce the risks associated with such fluctuations.

Economic thinking also supports better production planning in response to both market conditions and environmental pressures. It encourages cost-benefit analysis, risk management and long-term planning. Extension workers play a key role in translating these ideas into practical advice tailored to farmers' specific needs – whether it is improving input efficiency, aligning production with market timing or adopting new technologies to increase productivity and sustainability.

Furthermore, economic principles inform broader decisions, such as whether to engage with formal markets, when to take on credit and how to respond to emerging opportunities. Extension workers with a solid economic foundation are better positioned to offer broad-based advice that addresses immediate challenges while supporting long-term resilience and growth.

Incorporating economic principles into advisory services is not just beneficial, it is indispensable. It transforms farm management advice into a powerful tool that empowers smallholders to make informed, strategic decisions, strengthening their livelihoods and contributing to the sustainable development of their communities.

9.7 Conclusion

In the complex and ever-changing world of farming, successful farm management hinges on the ability to integrate economic principles into daily operations. These principles are not just theoretical concepts; they are practical tools that empower farmers to make informed decisions, optimize resource use and navigate the myriad of challenges presented by modern agriculture. By grounding their practices in sound economic reasoning, farmers can effectively assess the profitability of various enterprises, manage financial risks and plan for long-term sustainability. This integration is essential not only for achieving immediate profitability but also for ensuring the resilience and enduring success of their farming operations.

Understanding the principles of substitution, efficiency and diminishing marginal returns, as well as opportunity cost and comparative advantage, to name a few of the concepts covered in this chapter, is invaluable for smallholder farmers willing to transition towards market-oriented farming and an essential part of the repertoire of agricultural advisors. This knowledge is particularly crucial for smallholder farmers who must make the most of their limited resources to remain viable in a competitive market. From the selection of enterprises based on comparative advantage and opportunity cost to the strategic allocation of resources considering the law of diminishing returns, these principles guide farmers in maximizing the efficiency and effectiveness of their operations. Moreover, understanding market dynamics, such as supply and demand elasticity and price cycles, equips farmers to engage with markets more strategically, enhancing their bargaining power and securing better prices for their products.

Notwithstanding the application of economic principles to their farm operations agricultural advisors need to be aware of the limits of rational decision making and the non-economic decisions that are often taken by farmers. Some of these decisions may be less than optimal when considering risk and uncertainty, tradition, intuition and decisions influenced by social pressure. All these factors in the decision-making process form the basis for effective farm management.

Additional Reading

Boehlje, M. D., & Eidman, V. R. (1984). *Farm Management*. Wiley.

Castle E., M. Becker and A. Nelson (1987) Farm Business Management: The decision-making process. Macmillan Publishing Company

Dillon, J. L., & Hardaker, J. B. (1980). *Farm Management Research for Small Farmer Development*. FAO.

Ellis, F. (1993). *Peasant Economics: Farm Households and Agrarian Development* (2nd ed.). Cambridge University Press.

Heady, E. O., & Jensen, H. R. (1954). *Farm Management Economics*. Prentice-Hall.

Heady, E. O., & Dillon, J. L. (1961). *Agricultural Production Functions*. Iowa State University Press.

Kahneman D. (2011) *Thinking Fast and Slow*, Penguin Random House.

Kahan, D. (2013). *Economics for Farm Management Extension*. Farm Management Extension Guide No. 2, FAO.

Kay, R. D., Edwards, W. M., & Duffy, P. A. (2015). *Farm Management: Planning, Enterprise, and Risk* (8th ed.). McGraw-Hill Education.

Makeham J., L.R. Malcolm (1986) The Economics of Tropical Farm Management. Cambridge University Press.

Nuthall, P. M. (2011). *Farm Business Management: Analysis of Farming Systems*. CABI.

Upton, M. (1976). *Agriculture Production Economics and Resource-Use*. Oxford University Press.

Upton, M. (1987). *The Economics of Tropical Farming Systems*. Cambridge University Press

10 Data and Information for Farm Management Decision Making: A Challenging Frontier

This chapter explores the significance of information in farm management, making a distinction between the various types of data that are essential for informed decision making and use. It will also look at the transformative role of information and communication technologies (ICTs) and 'big data' in small-scale family farm development, offering an in-depth analysis of their potential to enhance agricultural productivity, market access and climate resilience. The chapter will delve into the various digital tools that provide farmers with essential resources and information, from real-time weather updates to market trends and agricultural best practices. Following this, the narrative moves to big data and its capacity to offer deeper insights into farming operations through the analysis of varied data sets and thereby enabling more informed decision making and efficient resource use by both farmers and agricultural advisors. Furthermore, the chapter discusses the significant challenges and opportunities that these technologies present and concludes with an understanding of the role of agricultural advisors in facilitating the flow of information to farmers and its analysis.

10.1 Data and Information

It is necessary to distinguish between data and information to understand their differences regarding the decision-making process. Data refers to raw, unprocessed facts and figures, while information is the result of organizing, analysing and interpreting the data to derive meaning and relevance. Data alone lacks context and may be overwhelming to farmers. However, when transformed into meaningful information, it becomes a valuable resource that drives decision making.

- **Data:** the raw numbers and facts such as prices, costs, quantities, etc.
- **Information:** data that is processed in a way that is useful for decision making.

The relationship between data, information and decision making is reflected in the flow chart shown in Fig. 10.1.

Data can be categorized as: primary and secondary data; and quantitative and qualitative data. Primary data refers to data that has been collected first-hand by the farmer for a specific purpose, for example, crop yields. Secondary data refers to existing information that has been collected by others for purposes other than the decision at hand. It may include data from weather stations, market reports, financial reports, research publications, government reports or industry statistics. Quantitative data is data that can be counted and measured such as rainfall and temperatures; it is objective. Qualitative data is subjective; it consists of words and observations, not numbers.

To make decisions, farmers need information from different existing data sources. Farmers must develop the capacity to gather and analyse data and convert it into information that is useful in making farm management decisions. In addition to helping build that capacity, it is also one of the tasks of the agricultural advisor to turn data into information farmers can use. Turning data into information means essentially to give the data meaning and context. To do this the data

DOI: 10.1079/9781836990536.0010

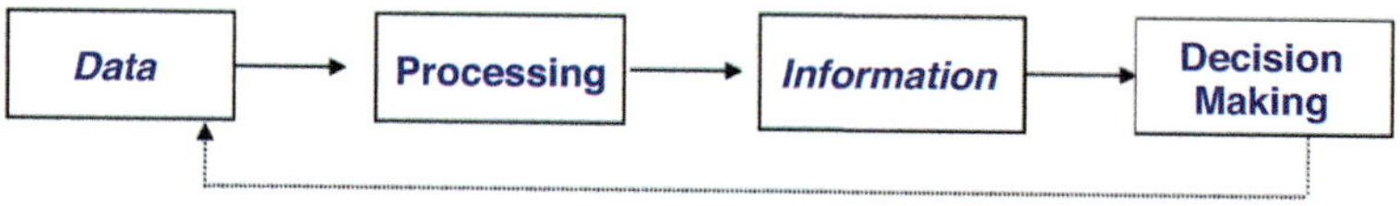

Fig. 10.1. The data–information–decision-making cycle.

needs to be analysed and sometimes combined with other sources of data (see Box 10.1.).

10.1.1 Primary and secondary data

As noted earlier, there are two main sources of data that can be used to generate farm management information: (i) primary data; and (ii) secondary data. Primary (first-hand) data is often used to describe the key problems confronting farmers with respect to their farm business. Examples of primary and secondary data sources are given below.

Primary data sources

Primary data is data collected by the farmer first-hand or personally, sometimes with the help of other farmers or an extension advisor. Sources of primary data include:

- **Farm records:** farm records are the farmer's foremost source of data. They provide recent and historical production and marketing data (and information). Crop yields, livestock production and cost information generated from farm records tell the farmer what can be achieved. Production records might tell the farmer how successful they were at managing the farm in the past.
- **Personal observation:** one of the most powerful sources of primary data is the farmer's personal first-hand observation of the state of production, infrastructure, equipment, marketing, etc. or some specific situation.
- **Handset (e.g. mobile phone) apps:** there are numerous apps available to assist with farm management. These can provide a range of data including field mapping, temperature monitoring, humidity monitoring, and pest and disease detection.
- **Weather stations and laboratories:** these can provide regular information about rainfall, soils, vegetation, erosion, etc.
- **Farm management surveys:** information is collected from individual farmers for a specific period, usually a season or year, for the purpose of determining the financial returns of the farm business and the factors affecting those returns. Data collected include both quantities and monetary values of crop and livestock production, income, expenses and capital assets, amongst others. Survey data is typically collected from large numbers of farms located in a homogeneous farming system area. Farm management surveys are based on interviews with farmers or key informants (knowledgeable experts such as researchers and extension workers). Farm survey data

Box 10.1. Communicating data and information to farmers.

Communicating data and information to farmers can be facilitated by following these points :

- The definitions of terms and presentation methods must be developed from the perspective of the farmer. This requires making sure that the data and information are relevant. Communicating the data must consider the capacity of the farmer who may not always have either a numeric background or technical knowledge of the extension message.
- The level of complexity of message must be suited to the farmer and their farming system; and the information adjusted accordingly.
- Visual aids like graphics and diagrams, along with interactive activities, are highly effective for engaging farmers and enhancing their understanding.

is often an essential prerequisite for farm and enterprise investigation and planning.

- **Other actors in the value chain:** while data from other actors (including other farmers) in the value chain is more likely to be secondary data, farmers and agricultural advisers can collect data about what these other actors are doing and their issues and concerns. This data helps create a context for decision making.

Secondary data sources

Secondary data is data that already exists or is documented in one form or another. The data could have been collected for some other purpose different from that of the secondary user – in this case, the farmer. There are a number of these sources available to farmers. These include data provided by:

- **Other farmers:** successful farmers learn from one another, and this is best done by sharing data (and information). Data on farm inputs, farm gate prices, yield, etc. can be gathered in this way, as can data on how other farmers are being affected by climate change, government or private sector policies, economic processes and market changes.
- **Lending institutions**: covering data on lending for agricultural enterprises and ventures.
- **Publications/journals/popular press:** farmers can also collect data from publications within and outside the ministry of agriculture and other government departments, from scientific journals and popular press such as farming magazines.
- **Television, radio and social media:** through mass media and social media, data is provided on a regular basis on a wide range of issues that affect farm management decisions. In addition to television and radio, farmers now have access to podcasts and webinar programmes and similar electronic information platforms.
- **Non-governmental organizations (NGOs):** international and local NGOs can provide data on a wide range of issues, depending on their areas of focus.
- **Agricultural input dealers:** private sector input and mechanization dealers often provide data and information from their records. In some cases, farm input suppliers, and in particular seed production companies might run well-managed demonstration farms to promote sales of their products to farmers. These suppliers also generate data on prices of various agricultural inputs and equipment.
- **National statistics:** historical yield and price information is often available from national statistical services. This information is often also useful. However, national data is often an average of the information collected from many farms and as such it does not tell the farmer exactly what he or she can get. Comparing historical farm yields to that of similar farms in the same area is an additional source of information on how the farmer can improve farm performance.
- **Agricultural extension advisory services:** farmers need to develop close working relationships with the agricultural advisors or extension workers serving their area. Advisors have access to data and information which may be difficult for the farmer to access on his own. Similarly, farmers can share data and information with their extension workers which can be used to advise other farmers.

Understanding the dynamics of these data sources is crucial for farmers. Agricultural advisors can play a pivotal role in bridging the gap between farmers and the wealth of available agricultural knowledge as well as in building the capacity of farmers to source, gather, interpret and use information. While direct information from other farmers and personal experiences forms the bedrock of a farmer's decision-making process, market information from traders is particularly vital for economic success. Agricultural advisors, therefore, need not focus solely on being the direct conduits of this information but should emphasize facilitating connections between farmers and these various sources. This involves organizing training sessions, workshops and field days where farmers can interact with experts, traders and peers. Additionally, agricultural advisors can leverage digital technologies to create online platforms and networks that enable farmers to access and share information more readily. By enhancing these connections, agricultural advisors empower farmers to tap into a broader knowledge base, making agriculture more productive, sustainable and profitable.

10.1.2 Qualitative and quantitative data

Qualitative and quantitative data are cornerstone elements in farm management. Each offers unique insights for decision making. Qualitative data delves into the subjective and often intangible aspects of farming operations, providing depth and context to the numbers that quantitative data offers. It includes farmers' experiences, perceptions and the quality of soil or plant health, which are not easily quantified but crucial for a comprehensive understanding of farm management. For instance, a farmer's narrative on the history of their land, such as previous crop choices and their effects on soil health, offers invaluable insights that numbers alone cannot provide. These narratives can guide decisions on crop rotation strategies or soil amendment needs.

Quantitative data, on the other hand, presents the measurable aspects of farming through numerical values, enabling precise and objective analysis. It encompasses yield rates, input costs, market prices and other variables that can be statistically analysed. For example, a farmer tracking the yield of different crop varieties in various plots over several seasons gathers quantitative data that can inform future planting decisions, potentially leading to higher productivity and profitability.

In the area of farm management, the integration of both data types enhances decision-making processes. Consider the case of pest management. Qualitative observations might include a farmer's notes on the appearance of pests, the conditions under which they thrive and the effectiveness of past control measures, providing a nuanced understanding of the problem. Quantitative data, such as the number of pests per plant or the percentage decrease in yield due to pest activity, offers concrete evidence on the magnitude of the issue and the efficacy of different treatments. Together, these data forms enable a farmer to choose the most effective, cost-efficient pest control strategy that is tailored to their specific circumstances.

Another example involves market analysis and product development. Qualitative feedback from consumers at a farmer's market on the taste, texture and overall appeal of a new vegetable variety can be invaluable for understanding market demand. This, combined with quantitative data on sales volumes, prices and customer demographics, can guide a farmer in deciding whether to increase the production of that vegetable variety or adjust marketing strategies to better meet consumer preferences.

Water management on a farm also benefits from a mixed-methods approach. Qualitative data might include observations on water runoff patterns, changes in water availability throughout the year and the effectiveness of different irrigation methods used on the farm. Quantitative data could involve measurements of water usage per crop, rainfall amounts and evaporation rates. Analysing both types of data helps in developing an efficient water management plan that optimizes water use, minimizes waste and ensures adequate hydration for crops.

Moreover, when considering expansion or diversification, farmers must evaluate both the qualitative aspects, such as their own expertise, labour availability and the potential fit of a new crop or livestock within their existing operations, alongside quantitative analyses of market trends, cost implications and expected return on investment. This comprehensive evaluation aids in making informed decisions that align with both the farm's capacities and market opportunities.

In essence, effective farm management is underpinned by the ability to synthesize qualitative and quantitative data. Qualitative insights offer a nuanced understanding of the farming system operations and market dynamics, while quantitative analysis provides the empirical evidence needed to make informed, strategic decisions. Together, they equip farmers with a holistic view of their operations, enabling them to navigate the complexities of modern agriculture with greater confidence and foresight.

10.2 Information Needs for Farm Management Decision Making

Farmers require timely and appropriate information at every stage in the farm production process: from the farm to the market and from the market to the farm. Information is needed in setting objectives, planning, implementing and controlling farm activities, and making more efficient use of their limited resources. The better skilled farmers are at using data and information, the better their farm decisions will be.

To be engaged successfully in market-oriented farming, small family farmers require the following kinds of information:

- from whom and where they can get low cost, good quality inputs;
- who the buyers of their produce are and where they are located;
- market access requirement and/or restrictions;
- the market potential in terms of quantity;
- consumer demands and legislation requirements in terms of quality of produce;
- what prices they can get for their products;
- what the price trends have been over time;
- what new enterprises and products they can introduce; and
- what new technology and practices to apply.

Ultimately this information will be used to make those critical key decisions:

- what to produce;
- how to produce;
- how much to produce; and
- where to market their produce.

Assisting farmers in obtaining the information they require is an important function of agricultural advisors. Advisors not only provide advice on production issues but also provide advice on market conditions and opportunities, as well as on all the other issues that farmers must consider when making farm management decisions. As farmers become more market oriented, agricultural advisors must also become more market oriented. If farmers cannot sell what they produce, then much of the extension advice on production techniques will have been wasted.

It is important to realize that farmers' circumstances are not static. There are often changes occurring that influence the farming operations. Changes in prices of products might lead to changes in the combination of enterprises, or changes in the pattern of resource use or changes in management practices. Changes in the needs of the farm-family household might lead to changes in the availability of family labour or in the food vs cash requirements of the family which in turn may lead to changes in farm enterprises or production methods. Therefore, whenever there is a change in the circumstances of a farmer, they need to review their situation carefully. Likewise, agricultural advisors must also review these changes and provide relevant advice including the suitability of new technologies and the introduction of new enterprises.

10.2.1 Uses of farm management information

Information is essential for farmers who wish to become fully market oriented and ensure that their production is in line with market demand.

Reducing risks

Farmers in low-income countries face numerous risks, including the risks of bad weather and variable market prices. The more knowledge a farmer has about price changes and the past profitability of enterprises, the better position they are likely to be in predicting the future. Accessible and reliable information on, for example, seasonal variations in prices, and changes in prices and in the weather over the years can be used to plan what to produce and when produce should be marketed. A farmer can reduce risk by collecting information about market prices and costs and returns of the alternatives open to them. They can also reduce risk by collecting information on weather and the likelihood of significant changes in temperature and rainfall. Farmers who keep records of these kinds of information are more likely to maximize their farm business efficiency.

Good information cannot, however, remove the risk altogether. Seasonal prices for perishable produce can fluctuate rapidly and significantly, depending on how much of the product arrives in a market on a particular day. Weather predictions may be inaccurate. But having the most accurate and relevant information possible is better than having none. Information empowers farmers to better assess their risks and to plan risk management strategies.

Deciding where to sell

It is important that farmers should be able to sell their produce at a feasible place. As discussed in Chapter 6, farmers may have several options:

- Sell at farm gate.
- Sell at a local market with a stall.
- Produce markets.
- Stock sales.
- Contact marketing.
- Community supported markets.
- Collective marketing.
- E-marketing.
- Sell directly to urban consumers.
- Sell directly to individual retailers.

The further along the marketing chain farmers go, the higher the price they will typically get. However, a high selling price may not mean that farmers will maximize their income. This depends on the costs involved when farmers decide to sell closer to the final consumer. It can be a lot more costly for a farmer to transport small quantities over a long distance than for a trader to transport large quantities over the same distance. Similarly selling directly to consumers is costly in terms of the farmer's time. The availability of information on market conditions at different locations or different points in the marketing chain is necessary for choosing where to market, as is an understanding of the marketing costs involved. The introduction of e-marketing has opened a whole new approach to marketing, including offering the opportunity to connect more easily with the final consumer.

Checking on prices received

With good information farmers can compare the prices they are receiving with the general level of prices in the market. Price information is now available in many forms. In addition to the traditional sources like the radio, there are many online services and websites that publish and even analyse prices. This has greatly improved the situation for farmers. If farmers receive prices lower than those broadcast or published, they may decide to find other traders, negotiate more assertively or try to improve the quality and presentation of their produce.

Deciding whether to grow produce out of season

A further advantage of having historical data on seasonal price trends is that farmers can make informed decisions about when to produce. With modern farming techniques and new varieties, it is now often possible to extend the harvest season and produce crops both before and after most other farmers.

Seasonal price data will give farmers a good idea of the prices they are likely to get from out-of-season production. However, farmers who do produce out-of-season also face higher investment costs (e.g. plastic tunnels) and higher production costs (e.g. for pest control), which may result in lower incomes. Agricultural advisors, therefore, need to be able to help farmers to calculate their likely production costs and relate them to expected selling prices.

There is also the possibility of producing during the counter-season. The countries in northern hemisphere represent a huge agricultural market, and when the northern hemisphere is experiencing winter, the southern hemisphere is experiencing summer. This means that when the farmers in the northern hemisphere cannot produce for their own markets, farmers in the southern hemisphere can try to meet the demand. For example, Peru has a healthy international counter-seasonal trade in products like avocados, citrus, grapes, blueberries and mangos. Peruvian farmers sell to markets in the USA, the European Union and Asia. Of course, there are many costs and logistics involved in counter-seasonal international trade but, if managed correctly, it can be very profitable.

Deciding whether to grow different crops

If new crops can increase their profits, farmers are often keen to try to grow them. More market-oriented production has encouraged farmers to diversify into new crops. Farmers need to be able to calculate whether growing other crops may be more profitable (agricultural advisors should help with these calculations). Information on past market prices is essential for them to make such decisions. In other cases, farmers may be considering diversifying into new crops to take advantage of market opportunities.

10.2.2 Data from on-farm demonstrations

A useful way for farmers to learn about analysing data and information is through demonstrations run by agricultural advisory services especially with the involvement of farm management specialists. In many countries, agricultural advisers are responsible for the design, execution and analysis of the results of such demonstrations. These demonstrations are particularly effective when the emphasis is placed on helping farmers understand the potential and risks of producing new products or using different production methods or technologies. As more farmers produce for the market, demonstrations are becoming increasingly important to test new enterprises as a way of making the most of market opportunities. This involves comparing one demonstration with another on comparable land types and with comparable technologies. Two sets of data are collected to enable a comparative analysis.

The active participation of farmers in the process of technology development and transfer is crucial. Feedback from farmers during the demonstration minimizes the possibility that technically and financially inappropriate practices would be promoted. The farmers' assessment of the results of demonstrations would complement the quantitative analysis which would normally be a part of the demonstration. When farmers participate, they can get early identification of potentially profitable technologies. Farmers add value to the process by using their accumulated experience to assess the extent to which new practices being tested are adaptable – thereby generating information that is more likely to be relevant to other farmers with similar circumstances.

A well-planned and executed on-farm demonstration will ensure that the data is collected as follows.

Background data

Background data refers to the history of the use of the land or field. The history may go back to two, three or more seasons. It will usually include information about the farming practices used to produce the particular crop or livestock. The data could include methods of land preparation, yields, fertility management, crop rotations, residue management, soil type, texture, slope, terrain, vegetation, system of water control, fallow periods, etc.

Technical input data

Data on the inputs applied to the demonstration should also be collected. These include the type of inputs used, the rate of application, the method of production and family and hired labour requirements. Due to the variability in the quality of labour, it is always advisable to distinguish between gender, age, family, non-family labour and to record the data accordingly. Input data is often underestimated, particularly when family labour is not accounted for. There is also a need to record the inputs of farmers' practices within the current farming system, as well as changes in inputs as a result of conducting comparative analyses.

Input–output response data

Data on the performance of the crop under the different technologies should also be recorded. This implies developing a response relationship between inputs and outputs. Assessments of performance are made at intervals over the season (crop stand establishment, live births, incidence of diseases) and at the end of the season once crop yields or weight gain of livestock have been assessed.

Product price and input cost data

Data on input and output prices are needed for farmers to make informed decisions. There are many sources of this data. They include individual farmers, groups of farmers, input suppliers and buyers, amongst others. Farmers need to know about the cost of the inputs that they will purchase to decide on new practices. Data on what inputs to apply, their unit and total cost and where to obtain them would be useful for farmers ready to innovate. Similarly, having access to up-to-date information on the prices of agricultural products enables farmers to assess the profitability of different enterprises and choose markets that offer the best returns.

Farmers' assessment data

Each farmer participating in the demonstration should actively engage in assessing the results of the demonstration. The data from farmer assessments is subjective, based on the farmer's own observations. There are at least two main ways that farmers and their families can make these assessments. Firstly, by making observations of crop growth and performance. Secondly, by getting their farm households to test the quality of food produced for the market. Such qualitative data will be very helpful in making decisions about changes to their production programmes.

10.2.3 Data from the market

Market information is often collected by government-run market information services. However, this is not the only, or even main, source of market information for farmers. Often, NGOs, local farmer organizations and commodity organizations also collect, analyse and publish market information. Additionally, small family farmers frequently rely on information by 'word of mouth' from other farmers and traders. As telecommunications improve many more commercially minded farmers are beginning to seek information via telephone, digital apps and online, directly from the major markets and other sources.

Relevant data includes the number of buyers each with their prices, promptness in payment and how the products should be packaged. Farmers producing for the market will require information from alternative markets for their products to decide which one(s) have better prospects.

Agricultural advisors and, in particular, specialists in farm management, have a role in identifying sources of data, assisting farmers in price data analysis and assessment of the cost of marketing produce. They should advise on the use of online sources and especially of digital apps to ensure that the farmers are not misled by unscrupulous businesses. They should also be able to advise farmers on reliable sources of input supplies and equipment. They should help farmers improve their ability to locate, access and interpret market information that will help them get better prices and support their farm planning activities.

The agricultural advisor could also facilitate direct meetings between buyers and groups of farmers for exchange of information to help farmers protect themselves from being taken advantage of by middlemen. This may lead to establishment of sale contracts between the parties. The farmers will learn first-hand the concerns that traders have about their products (quality, packaging, grading, etc.). Such feedback will help tailor production to the needs of the markets. This data will also guide the agricultural advisors in advising farmers on the most opportune time to sell.

10.3 ICT for Farm Management

The rapid advancement of ICT and digital technologies has transformed farm management, offering innovative tools to enhance productivity, decision making and market access for farmers worldwide.

Mobile phones, apps and digital platforms have become indispensable in bridging information gaps, particularly for smallholder farmers in rural areas. These technologies provide real-time data on weather, market prices and agricultural best practices, empowering farmers to optimize their operations and improve livelihoods.

Among the staple digital tools, spreadsheets have long played a crucial role in farm management, offering a structured approach to data entry, financial tracking, yield forecasting and resource planning. With enhanced capabilities such as automation, cloud integration and advanced analytics, spreadsheets remain an essential tool for farm record-keeping and decision support.

Beyond spreadsheets, *high-tech solutions* are reshaping agricultural management. Farm management software, decision support systems (DSS), digital marketplaces and Internet of Things (IoT) sensors allow farmers to monitor soil conditions, automate irrigation and optimize input use. Drones are increasingly deployed for precision agriculture, offering aerial imaging, crop health assessment and even pesticide application. These innovations enable data-driven farming, improving efficiency and sustainability.

Looking ahead, artificial intelligence (AI) represents the new frontier in digital agriculture, enabling predictive analytics, intelligent automation and data-driven decision making. AI-powered tools can analyse vast datasets, optimize resource allocation, detect diseases in crops and livestock and provide personalized recommendations to farmers. As these technologies continue to evolve, they promise to further revolutionize farm management, making agricultural operations more efficient, sustainable and resilient in the face of climate and market uncertainties.

Despite the clear opportunities these technologies create, challenges such as limited internet access, digital literacy and the cost of technology persist, especially in low-income regions (see Box 10.2.).

This section explores the role of ICT in agriculture, highlighting the use of mobile phones, apps and digital platforms, while addressing the barriers to their adoption and the potential for public-private partnerships to drive inclusive growth in the agricultural sector.

10.3.1 Mobile phones and apps

The use of mobile phones has revolutionized connectivity and information access within agriculture. Modern mobile phones are not just tools for communication; they are powerful computing devices that offer a wide range of functionalities similar to a computer. This technological advancement holds significant promise for smallholder farmers, particularly in rural areas where the penetration of traditional computing technology might be limited.

Globally, the ownership of mobile phones in rural regions is expanding rapidly, presenting a valuable opportunity for the dissemination of crucial agricultural information. Mobile phones are especially pivotal where literacy levels are lower and in communities speaking less widely spoken languages. Although text-based information dissemination via SMS presents challenges in such contexts, it remains a critical medium for transmitting data like market prices. Recognizing these constraints, efforts are being made to develop voice-based services and applications that can deliver information in local languages, thereby making content accessible to a broader audience. The development of mobile applications for smartphones has significantly transformed access to online services for farmers, enabling them to tap into a wide array of resources right from their cell phones. These apps not only provide crucial agricultural information but also facilitate market and farm management decision making, making them indispensable tools for modern farming.

Mobile phones facilitate a multitude of functions that empower farmers: they provide real-time weather updates, enable access to fluctuating market prices and offer platforms for mobile banking. These services collectively enhance a farmer's ability to make informed decisions, optimize farm output and improve financial management. Internet-enabled

Box 10.2. Percentage of rural households with internet access. From International Telecommunication Union (ITU), 2023.

Worldwide, 81% of urban dwellers used the internet in 2023, compared with only 50% of the population in rural areas. The urban–rural gap, measured as the ratio of the two percentages, has barely improved in recent years, from 1:7 in 2020 to 1:6 in 2023. Unsurprisingly, the urban–rural gap is the smallest in regions with high penetration: the gap has almost been bridged in Europe. The ratio stands at 1:2 in both the Americas and the Commonwealth of Independent States (CIS) regions, at 1:5 in the Asia-Pacific region and 1:6 in the Near East and North Africa region. In all these regions, progress has been modest since the start of the decade. In Africa, the gap remains wide: 57% of urban dwellers used the internet in 2023 compared with just 23% of people in rural areas, reflecting a ratio of 2:5 (2:9 in 2020). The gap between urban and rural areas varies significantly across income groups. The gap has almost been bridged in high-income countries (ratio of 1:1). In contrast, the divide remains deep in low-income countries: less than one in five people (17%) living in rural areas in these countries use the internet. People living in urban areas are nearly three times more likely to use the internet. This shows that the digital divide across income groups is magnified in rural areas. The internet use gap between urban areas of low-income countries and urban areas of high-income countries is 48 percentage points. The internet use gap is 71 percentage points between the rural areas of the two groups.

smartphones further extend these capabilities, connecting farmers to online training programmes, expert agricultural advice and virtual marketplaces. This connectivity eliminates the middleman, potentially increasing the farmers' profit margins significantly. Moreover, mobile banking not only offers secure financial management tools but also opens avenues for farmers to access credit and savings options, boosting their economic stability and resilience.

Interestingly, the proliferation of mobile phones, in general, did not require considerable public funding; private mobile phone providers invested in the systems and farmers and traders used the technology to search more widely and quickly for price information through their own contacts. The challenge now, especially in light of the vast expansion of mobile phone ownership among smallholder farmers over the past six years or so, is to find ways, when possible, through public–private collaboration, to broaden the scope of information provided in this way. Some examples are as follows:

- **India:** an SMS-based system automatically provides bid price and contact information for products in markets nationwide; the system is inexpensive because buyers perceive the advantage in providing the information at no cost.
- **Sri Lanka:** an SMS system provides gherkin farmers with daily information on issues related to produce in the market which assists them in immediately avoiding problems on their farms. Another project aims to lower information search costs throughout fresh produce supply chains via SMS, e-bulletin boards, price-reporting screens in markets and links with banks and extension services (De Silva and Ratnadiwakara, 2005). Since 2004, work has been ongoing in Sri Lanka to develop a cyber extension network as the country's foremost ICT initiative in agriculture. To date, over 50 cyber extension units have been established in various districts.
- **Kenya:** The M-Farm platform allows farmers to receive real-time price information and directly connect with buyers via SMS, eliminating the need for middlemen. This has enabled farmers to negotiate better prices and access wider markets.
- **Uganda:** The EzyAgric app provides farmers with access to information on inputs, weather forecasts, agronomic advice and connections to buyers. It also offers financial services like loans and insurance tailored for smallholder farmers.
- **Ghana:** The Esoko platform uses SMS and voice messages to provide farmers with market prices, weather updates and agricultural tips. It also facilitates communication between farmers, traders and buyers.
- **Ethiopia:** The 8028 Farmer Hotline allows farmers to call in and receive automated advice or speak to agricultural experts about pest control, farming techniques and weather predictions in their local languages.
- **South Africa:** The Vodacom Farmers' Club offers a mobile-based service providing agricultural information, market prices and financial tips, helping smallholders make informed decisions.
- **Latin America:** In Colombia, the Agrosavia platform integrates ICT tools to provide farmers with guidance on crops, pests and farming techniques, while also promoting sustainable practices. In Brazil, the Aegro app supports farm management by helping with financial planning, input tracking and productivity analysis.
- **Nigeria:** The National e-Agriculture Project sends farmers real-time price updates and agricultural information via SMS and mobile apps, helping to streamline supply chains and improve market efficiency.
- **Malawi:** Farm Radio International collaborates with local radio stations to broadcast farming tips and market prices, using mobile phones to enable two-way communication for questions and feedback.

Mobile applications in agriculture typically fall into several categories, each designed to address specific aspects of farm and market management.

Agricultural advisory apps

These provide real-time, location-specific advice on farming practices. These apps can deliver timely information about weather conditions, pest infestations and crop cycles, assisting farmers in making informed decisions about planting, irrigating and harvesting. An example of such an app is Plantix, which diagnoses plant diseases based on photographs farmers upload and offers management solutions.

Market price apps

This is another critical category offering farmers updates on current market prices for various crops and livestock. This information helps farmers decide the best times to sell their produce to maximize profits. Apps like AgriMarket allow farmers to compare prices across different markets, empowering them with the knowledge to bypass intermediaries and directly engage with buyers, thus improving their bargaining power.

Most of the extension effort using mobile phone apps has been focused on market information systems that include the collection and dissemination of freely available information on prices and market outlets. This places farmers in a better position to negotiate meaningfully with commission agents and intermediaries to attain premium prices. While market information is crucial, it is only useful if the system can guarantee the real-time collection and sharing of data on prices and transaction quantities; otherwise, it becomes ineffective. Additionally, developing grades and standards for farm products in collaboration with trade will also enhance this information and, therefore, analysing this information should be prioritized early on. Traders often identify different qualities within a broader category of standard produce, so capturing this informally recognized diversity (i.e. quality grades) and incorporating it into reporting of market prices and quantities can be an effective action. Providing real-time feedback to farmers on quality problems in the market that can vary seasonally (based on pest outbreaks on the farm) and technical solutions at the farm and post-harvest levels can be a very useful approach. Typically, this kind of information is collected from wholesale markets, where prices and quality-related discounts are set. However, less headway has been made in providing farmers with information about agribusiness opportunities, and this requires more focus on identifying the specific data needs and how to collect them effectively, including the collection methods used.

But, while market information is widely provided through mobile apps in the form of market price information, there is considerable potential to broaden the service and provide more comprehensive market intelligence that includes trend analysis, identification of key players, provision of trade information and so forth. Moreover, market information services need to be tailored to specific target groups, which this is an area that to date has been inadequately addressed.

Supply chain management apps

These apps streamline the process from farm to table, helping farmers connect directly with suppliers and buyers, and manage their supply chains more efficiently. FarmLogs, for example, helps farmers plan, manage and analyse their agricultural activities in one place, enhancing their ability to make better decisions and maintain records effortlessly.

Financial service apps

These apps have made significant strides in integrating financial inclusion for rural farmers who previously had limited access to banking services. Apps like M-Pesa have revolutionized mobile money services, allowing farmers to receive payments for their produce, pay for goods and services and access micro-loans and insurance products directly from their smartphones.

Each of these applications relies on a combination of GIS technology, real-time data analytics and user-friendly interfaces to provide tailored information and services. By leveraging such technologies, mobile apps not only enhance the decision-making capabilities of farmers but also contribute significantly to their operational efficiency and access to critical market information.

The development and proliferation of these apps signify a major shift in how agricultural and support services are delivered to farmers. They represent an integration of information technologies in agriculture that can lead to increased agricultural productivity and sustainability, especially when paired with traditional extension services and training on how to effectively use these digital tools. This holistic approach to digital agriculture supports a more informed, connected and financially inclusive farming community.

10.3.2 Computer spreadsheets

Computer spreadsheets have been used extensively for many years; they continue to have a significant role to play in managing data. Spreadsheets are computer programmes that utilize a grid layout to organize, calculate and analyse data, making them indispensable in farm management. Users can perform basic operations like addition, subtraction, multiplication and division, alongside advanced functions such as averaging, summing, calculating percentages and creating visual charts and graphs. This versatility is crucial for managing farm business aspects like budgets, production reports, sales, and income and expense records efficiently, without the need for costly specialized software.

In farm planning, spreadsheets enable 'what if' analyses, allowing farmers to model different scenarios and visualize potential effects of variables like crop prices, yield rates or input costs. For example, they can evaluate the financial outcomes of using different seeds or adopting new irrigation techniques to make informed decisions. Spreadsheets also facilitate the assessment of technological changes, policy shifts and the economic feasibility of new investments. Techniques, such as break-even budgeting and sensitivity testing, help in analysing how outcomes vary with changes in inputs, which is crucial for comprehensive risk assessments.

Despite these advantages, the effective deployment of spreadsheets in farm management often demands specialized skills that go beyond basic computer use, including a deeper understanding of economics and model building. Specialized farm economics advisory support is frequently necessary. Extension workers with an economic background can help farmers set up and utilize these models effectively or provide the results of complex analyses, ensuring that the unique aspects of a farm's operations are accurately represented and understood. This is explored further in Chapter 16.

10.3.3 Advanced digital tools for agricultural management

Building upon foundational ICTs, a suite of advanced digital tools is helping farmers harness data for smarter decision making. These technologies – ranging from secure supply chain systems to AI-driven analytics – enable smallholders to optimize inputs, reduce waste and improve market access. While their adoption in low-income contexts faces hurdles (including cost and digital literacy), successful case studies demonstrate their potential when appropriately adapted. This section examines key innovations and their practical applications in smallholder farming systems.

Artificial Intelligence

Artificial Intelligence (AI) has the potential to revolutionize farm business management by providing farmers with advanced tools for decision making, resource management and productivity optimization, and drawing together information from satellite-based technologies. Systems driven by AI can analyse vast amounts of data from various sources, such as weather forecasts, soil conditions, crop health and market trends, to offer real-time insights and recommendations. This enables farmers to make more informed decisions, reduce risks and enhance the efficiency of their operations. For instance, AI can help in determining the optimal time for planting and harvesting, predicting yield outcomes, and identifying areas where resources like water and fertilizers can be used more efficiently (see Boxes 10.3. and 10.4.).

Moreover, AI can enhance precision agriculture by enabling the detailed monitoring and management of crop and livestock production. Through machine learning algorithms and IoT devices, AI can analyse data from drones, sensors and satellite imagery, to detect issues such as pest infestations, disease outbreaks or nutrient deficiencies early on. By providing timely alerts and actionable insights, AI helps farmers address these problems before they escalate, thereby reducing losses and increasing overall productivity. This level of precision also allows for more sustainable farming practices,

Box 10.3. Use of Taranis. From Taranis, 2021.

An example of AI in farm management decision making can be found in the use of Taranis, an AI-powered precision agriculture platform. Taranis uses AI to process high-resolution images from drones, satellites and aircraft to identify crop issues at an early stage, such as pest infestations, nutrient deficiencies and disease outbreaks. For instance, a farm in Brazil utilized Taranis to monitor its soybean fields. The platform's AI capabilities detected signs of a fungal disease early on, which allowed the farm managers to take targeted action, applying fungicides only where needed. This early intervention not only prevented the spread of the disease but also reduced chemical usage, resulting in cost savings and minimized environmental impact. The successful application of AI in this case improved the farm's decision-making process, leading to better crop management and higher yields.

as resources are applied exactly where and when they are needed, minimizing waste and environmental impact.

However, the potential of AI is contingent on its accessibility and usability – both of which are currently limited for widespread use among advisory service providers and smallholder farmers in low-income countries. The same limitations that constrain the widespread use of ICT apply here, along with the additional cost of AI tools themselves. For AI to be a viable solution for smallholder farmers, these costs must be carefully managed to ensure that services remain affordable and accessible to the farmers who need them most. While AI holds significant promise for improving farm business management, addressing the infrastructural, educational and economic challenges is essential in realizing its full potential.

The need for smartphones and reliable internet connectivity poses challenges, as these resources are often limited in rural areas. Additionally, many smallholder farmers may lack the education or technical skills required to effectively use AI tools.

By focusing on affordable and accessible technologies, such as SMS-based services, mobile apps and community-based platforms, AI can support smallholder family farmers in developing countries. These tools improve their ability to make data-driven decisions, enhance profitability and reduce reliance on traditional systems that may disadvantage them.

Internet of Things (IoT) for farm management

As briefly introduced in Chapter 7, the Internet of Things (IoT) is revolutionizing farm management by integrating smart sensors, automated systems and real-time data analytics. IoT devices, such as soil moisture sensors, weather stations and GIS-enabled machinery, provide continuous monitoring of key variables like temperature, humidity, soil conditions and crop health. This data enables precision agriculture, where farmers can optimize irrigation, fertilization and pest control with minimal waste.

Beyond field management, IoT also enhances supply chain efficiency, livestock tracking and equipment maintenance by providing predictive insights and remote-control capabilities. Connected systems allow farmers to make informed decisions, reducing costs while increasing productivity and sustainability. When combined with AI and cloud computing, IoT unlocks advanced farm automation, enabling smarter, data-driven agriculture.

Drones and aerial imaging

As also briefly introduced in Chapter 7, these technologies provide aerial views of farmland for monitoring crop health, detecting pests or diseases and assessing soil conditions. They enable early intervention and targeted treatment. Beyond crop monitoring, drones assist in precision spraying, reducing chemical use and labour costs. Thermal and multispectral imaging helps detect irrigation issues and nutrient deficiencies. Autonomous drones can survey large areas quickly, aiding in yield estimation and land mapping. When integrated with AI and farm management software, they offer predictive insights, enhancing decision making. As costs decrease, drones are becoming an essential tool for precision agriculture and sustainable farm management.

Box 10.4. AI applications in farm business management for smallholder family farmers.

Financial planning and risk management

- Micro-credit scoring: AI-powered platforms use minimal data, such as mobile transaction history and crop records, to create credit scores, enabling smallholders to access micro-loans tailored to their needs.
- Community-based insurance: AI models predict risks like droughts or floods and help to design affordable insurance products that mitigate financial losses.

Market intelligence

- Price forecasting: AI tools predict prices for crops in local and regional markets, empowering farmers to time their sales for better returns.
- Demand identification: algorithms analyse nearby markets to identify high-demand crops, helping farmers choose what to grow based on potential profitability.

Supply chain optimization

- Transportation planning: AI optimizes local logistics, such as shared transport for farmer cooperatives, reducing costs and post-harvest losses.
- Community storage management: AI predicts storage needs for smallholders, preventing overstocking and ensuring produce is sold before spoilage.

Farm performance analytics

- Small enterprise profitability analysis: AI tools, accessible via mobile apps, track basic input costs and sales income, helping smallholders understand their profits.
- Low-cost cost-reduction insights: AI suggests affordable strategies, like bulk purchasing of inputs, to reduce expenses.

Decision support systems

- Crop and enterprise selection: AI recommends crops based on local conditions, market demand and available resources, guiding smallholders towards profitable choices.
- Simple scenario modelling: farmers can simulate decisions, such as investing in an irrigation system or growing a new crop, to evaluate the potential impact.

Customer relationship management (CRM)

- Local buyer matching: AI platforms match smallholders with local buyers or traders, enabling better negotiation of prices.
- Sales forecasting for smallholders: AI helps farmers anticipate sales volumes and manage relationships with cooperatives or buyers.

Policy and compliance

- Subsidy awareness: AI-based SMS systems inform farmers about available subsidies or grants in their area and guide them on how to apply.
- Simplified regulatory compliance: tools ensure that smallholder practices align with local farming laws and standards.

E-commerce integration

- Mobile marketplaces: AI-powered apps connect smallholders to local or regional digital marketplaces where they can sell produce directly to consumers or retailers.
- Price transparency: AI facilitates fair pricing by making market data accessible to smallholders, reducing exploitation by middlemen.

Continued

Box 10.4. Continued

Training and capacity building

- AI chatbots in local languages: virtual assistants provide real-time business advice, such as budgeting or pest control tips, in local dialects.
- Community learning platforms: AI-based training systems deliver interactive, mobile-friendly lessons on business planning and market engagement.

Sustainability and certification

- Simplified sustainability metrics: AI tracks sustainable practices like organic farming and links farmers to certification programmes, boosting market value.
- Carbon credit access: AI tools help smallholders participate in carbon credit markets, providing an additional income stream.

Decision Support Systems (DSS)

Software platforms that synthesize weather, soil and crop data to generate tailored recommendations (e.g. ideal sowing dates, fertilizer quantities). Tools, such as Digital Green's voice-enabled DSS in India, combine satellite data with local knowledge to address climate variability. Challenges include ensuring data accuracy and designing interfaces for low-literacy users. These tools can also be used for climate change adaptation. DSS tools can be used to integrate climate forecasts with agronomic advice, enabling farmers to pivot planting dates or crop choices in response to erratic rainfall patterns – a critical adaptation strategy.

Digital marketplaces/e-marketing

Online platforms (e.g. Ghana's Esoko; Indonesia's TaniHub) directly link farmers to buyers via mobile devices. These systems improve price transparency, reduce dependency on middlemen and minimize food waste through demand-driven harvesting. Their effectiveness depends on affordable mobile data and integrated payment solutions like mobile money. This tool can also be applied to climate change adaptation. Reducing food waste through efficient surplus redistribution avoids methane emissions from rotting produce, while shorter supply chains lower transport-related carbon footprints.

Farm management information systems

Farm management information systems (FMIS) have become a central component of modern farm management, providing critical support for effective decision making. These digital tools include a wide range of applications – such as AgriWebb and FarmLogs – designed to assist with planning, record-keeping, financial analysis and tracking key performance indicators. By using state-of-the-art technologies such as IoT, drones, remote sensing and data analytics, FMIS enable farmers to monitor input costs, labour efficiency, yield trends and resource use in real time. They support strategic operations like crop planning, financial management and risk assessment, ultimately improving productivity and sustainability.

New developments are making FMIS more accessible, especially for farmers with limited formal education. User-friendly interfaces, including voice commands and pictorial dashboards, are being introduced to simplify use. Many systems are now cloud based, allowing multi-device access and seamless updates from the field. Integration with IoT devices, satellite imaging and AI-driven analytics further enhances their functionality – providing insights into soil health, weather patterns, crop performance and yield forecasts. These tools are also being linked to digital marketplaces and government programmes, improving access to subsidies and expanding marketing opportunities.

Despite their significant potential, FMIS adoption still faces challenges. High costs, the need for technical expertise and issues of data compatibility and privacy remain barriers, particularly in low-income countries. Nonetheless, as systems continue to evolve with AI integration and better cloud functionality, they offer a powerful means of transforming farming practices and strengthening smallholder resilience (Fountas *et al*., 2015; Giua *et al*., 2021).

Blockchain

Blockchain is a decentralized digital ledger that creates tamper-proof records of transactions across supply chains. For smallholders, it enhances traceability (e.g. certifying organic produce), reduces post-harvest losses through real-time logistics tracking and ensures fair payments by minimizing fraud. Initiatives such as Kenya's FarmTrace coffee project demonstrate these benefits, using blockchain to verify ethical farming practices and improve market access. While not its primary application, blockchain's ability to streamline supply chains can indirectly address climate concerns: cutting post-harvest losses (a contributor to roughly 8% of global greenhouse gas emissions; FAO, 2022) and enabling transparent certification for climate-conscious markets. Challenges persist, however, as reliance on internet connectivity limits broader adoption among remote smallholders.

Table 7.1 in Chapter 7, outlined the primary functions and key climate adaptation benefits of a few apps designed primarily to assist farmers adapt to the effects of climate change. In addition, the digital tools and the technologies listed in Table 10.1 serve farm management more generally. They are designed to enhance farm productivity, profitability and operational efficiency. These technologies either specialize in optimizing resource use and market access or their primary value lies in reducing costs and streamlining decision making for smallholder businesses.

10.3.4 Challenges in the use of ICT, big data and AI

The integration of ICT, big data and AI into agriculture presents significant opportunities for improving farm management, decision making and market access. However, the adoption of these technologies – particularly by smallholder farmers – faces several challenges, including connectivity limitations, digital literacy gaps, data privacy concerns and cost constraints (see Box 10.5.).

Despite the growing availability of mobile-based information services, traditional mass media, such as radio, remains the most effective tool for reaching rural farmers, especially where literacy and mobile phone access are limited. Basic SMS-based services often provide narrow information, while more advanced digital platforms require reliable internet and smartphone access – both of which remain costly for many smallholder farmers. Additionally, cyber extension services and private information providers struggle with cost recovery, as smallholders may be unable to pay for these services directly. However, public-private partnerships and sponsorship models offer potential solutions for sustaining these services.

The lack of telecommunications infrastructure in rural areas further hampers the adoption of digital tools. Many remote regions lack reliable mobile networks and broadband access, limiting farmers' ability to access real-time market prices, weather forecasts and advisory services. Even when connectivity exists, it may be too expensive or insufficient for data-heavy applications such as online marketplaces, precision agriculture or remote sensing. Investments in digital infrastructure, affordable devices and local-language content are critical to ensuring inclusivity in digital agriculture.

While big data and AI offer powerful tools for improving productivity, resource efficiency and risk management, their effective use in smallholder farming is hindered by data quality issues, privacy risks and the need for technical capacity to analyse and interpret insights. Data-sharing networks and partnerships among governments, private sector actors and research institutions are essential to address these gaps. Ensuring data security through robust governance frameworks, encryption and anonymization techniques is key to building trust among farmers and stakeholders.

Nonetheless, the opportunities of big data and AI in farm management are immense. By leveraging diverse datasets, farmers can gain insights into crop performance, climate patterns and market trends, allowing for more precise decision making. AI-powered advisory services can provide personalized recommendations on planting times, fertilizer use and pest control, boosting productivity and resilience. Additionally, market intelligence generated from big data can enhance smallholder participation in value chains, enabling them to align production with demand, access premium markets and negotiate better prices.

Table 10.1. Digital technologies for general farm management.

Technology	Primary function	Key impacts
AI/Machine learning	Data analysis to extract patterns, predict outcomes and automate decisions	• Extract patterns (e.g. market trends, input-use inefficiencies), cutting costs • Predict outcomes (e.g. optimal sales timing, labour needs), identifying actional profit levers • Automate decisions (e.g. scheduling, resource allocation), reducing human guesswork to maximize resource efficiency
IoT (Internet of Things)	Connect physical devices (sensors, actuators, machinery) via the internet for real-time monitoring and automated control of farm operations	• Facilitates precision input management (irrigation, fertilization, feeding), reducing costs • Livestock monitoring (e.g. tracking animal health for early detection, improving yields) • Storage condition tracking (temperature & humidity in storage systems, reducing spoilage)
Drones	High-resolution aerial monitoring of crops, soil and infrastructure	• Targets input use by mapping nutrient/pest variability, reducing fertilizer/pesticide costs • Saves scouting time by covering 10x more area than manual checks, cutting labour hours • Improves record-keeping with geotagged field images, simplifying certification for premium markets
Decision Support Systems (DSS)	Integrates data for general advice	• Boosts yields through data-driven advice (e.g. optimal planting dates, pest thresholds) • Reduces input waste by tailoring fertilizer/water use to real-time field conditions • Mitigates risks with early warnings (e.g. disease outbreaks, adverse weather)
Digital Marketplaces	Connects farmers to buyers	• Increases farmer profits by directly connecting to buyers (bypassing middlemen markups) • Minimizes unsold produce through demand-driven sales and surplus redistribution • Provides price transparency to strengthen negotiation power
Farm Management Information Systems	Tracks inputs/yields/ finances	• Lowers input costs by identifying inefficiencies (e.g. overuse of fertilizers, labour) • Improves financial planning with digital records of expenses, yields and sales • Simplifies compliance for certifications (e.g. export standards) via automated documentation
Blockchain	Provides a decentralized and secure ledger for transactions and data in agriculture	• Enhances transparency and traceability in supply chains, ensuring food safety and reducing fraud • Secures financial transactions, enabling trust in digital payments and smart contracts • Strengthens record-keeping for compliance and certifications, reducing administrative burdens • Improves market access by verifying the authenticity of organic and fair-trade products

Box 10.5. Lessons from case studies.

The lessons drawn from the case study evidence suggest that:

- Investments in ICT infrastructure and hardware alone will not guarantee appropriate information for sustainable livelihoods. AI and digital tools should be seen as part of a broader support system, combining hardware with relevant, localized content and predictive insights.
- Solutions should be developed that leverage the numeric literacy that even those who cannot read and write usually possess. AI-driven visual and voice-based advisory tools can enhance accessibility for low-literacy users.
- ICT and AI solutions should be carefully planned, ensuring that deployment aligns with available infrastructure, including reliable power and internet access, to avoid common problems of erratic electricity supply and limited connectivity.
- The literacy level of users and the level of technical competence of the support teams are crucial for an effective system to work. AI-driven training tools and adaptive learning platforms can help build both farmer and extension worker capacity.
- Efforts should be taken to incorporate ICT into existing information services. This requires careful planning, training and a long-term perspective to ensure seamless integration with radio, extension networks and community knowledge systems.
- To ensure that the full potential of this technology can be realized in benefiting smallholder farmers, a cadre of local intermediaries should be supported in delivering appropriate information to farmers and rural entrepreneurs using ICTs in conjunction with traditional channels, on a cost-recovery basis.
- Existing schemes that use ICTs and AI successfully and sustainably should be identified and, if appropriate, replicated on a national and regional basis, ensuring that scaling efforts are informed by real-world successes.

Capacity building remains crucial. Training programmes must focus on data literacy, management and analytics, equipping farmers, extension workers and agribusinesses with the necessary skills to utilize data-driven tools effectively. Collaborative efforts between agricultural organizations, research institutions and governments can facilitate knowledge exchange, ensuring that digital innovations translate into tangible benefits for smallholder farmers.

By addressing these challenges and fostering inclusive access to digital tools, ICT, big data and AI can unlock transformative opportunities for smallholder agriculture, enhancing efficiency, resilience and market competitiveness.

A multifaceted approach to leveraging big data and AI for smallholder farmers

The integration of big data and AI in agricultural advisory services has the potential to transform smallholder farming. However, achieving widespread adoption requires overcoming infrastructural, economic and educational barriers. While mobile phones and digital platforms provide access to valuable agricultural information, traditional mass media, such as radio, remains the most effective tool for reaching broad farmer audiences due to literacy and cost constraints. Additionally, private-sector engagement, sustainable financing models and community-driven approaches play a critical role in ensuring the success of digital advisory services.

To fully harness the potential of big data and AI, the following multifaceted approach is necessary.

Enhancing media access and infrastructure

Governments and private stakeholders should collaborate to improve digital infrastructure, ensuring that mobile and internet access reaches even the most remote agricultural communities. Investments in telecommunications infrastructure – such as expanding broadband and mobile network coverage, deploying solar-powered internet kiosks and supporting community radio stations – can bridge connectivity gaps. AI-powered chatbots and voice-activated advisory services can further help reach low-literacy farmers by providing real-time, local-language insights via simple mobile phones.

TRAINING FARMERS AND EXTENSION WORKERS IN DIGITAL SKILLS. Digital literacy programmes should cover fundamental ICT skills, data interpretation and AI-enhanced decision-making tools. Training should be designed with a stepwise approach, starting with:

- Basic mobile literacy (using SMS, USSD and voice-based systems).
- Understanding satellite imagery and weather data for farming decisions.
- Using AI-powered recommendation systems that offer adaptive learning, tailoring educational content based on the farmer's location, crop type and prior experience.

Extension workers play a crucial role in supporting farmers; therefore, AI-driven knowledge platforms should be developed to provide real-time updates, best-practice guides and remote troubleshooting assistance. Voice-based AI assistants can further democratize access, enabling farmers to receive advisory services without requiring advanced literacy.

STRENGTHENING LOCALIZED AND INCLUSIVE LEARNING MODELS. For digital literacy initiatives to be effective, they must align with cultural contexts, language preferences and the realities of smallholder farmers. Strategies include:

- AI-powered voice recognition and translation tools to provide advisory services in local dialects.
- Community-led digital training hubs, supported by universities, NGOs and agribusinesses, to ensure that digital learning is accessible and scalable.
- Peer-to-peer learning platforms that use interactive chatbots or farmer-driven WhatsApp groups to share insights, troubleshoot issues and enhance adoption of new technologies.

SCALING SUSTAINABLE DIGITAL LITERACY THROUGH PARTNERSHIPS. Public–private partnerships play a pivotal role in making digital literacy training scalable and sustainable. Governments, tech companies and financial institutions can provide subsidized access to AI-driven educational platforms. They can also support training-of-trainers models, where local lead farmers receive advanced AI training and mentor their peers. And they can facilitate the integration of AI-driven agricultural curricula in vocational and extension training programmes, ensuring that future farmers and advisors are well-equipped to leverage digital tools effectively.

By embedding AI-enhanced digital literacy and capacity building into agricultural extension frameworks, smallholder farmers can move beyond passive data consumption to active, informed decision making, ultimately improving farm productivity, climate resilience and market access.

Affordability of technology and services

To ensure equitable access, cost-effective technology solutions must be developed. One option is subsidizing smartphones and data plans for farmers, particularly through telecom partnerships. Another is developing offline AI-powered apps that store and process key agricultural data without requiring constant internet access. A third involves encouraging business models where private companies provide advisory services at minimal costs, offset by sponsorships, government incentives or premium advisory packages for commercial farmers.

Localized content and personalized advisory services

For data-driven agriculture to be truly effective, information must be relevant to farmers' specific needs. This requires AI-driven content adaptation, ensuring that advisory messages are translated into local dialects and adapted to regional cropping systems. It also requires providing real-time, location-specific recommendations, integrating weather forecasts, pest alerts and market price trends. Supporting peer-to-peer learning through farmer WhatsApp groups, community-led radio programmes and digital platforms where farmers share experiences, is also essential.

Monitoring, feedback and trust building

The success of ICTs, big data and AI solutions depends on farmer trust and active participation. Continuous monitoring and evaluation ensure that technologies remain relevant and effective. This can be facilitated by using real-time feedback loops where farmers rate the usefulness of digital advisory services, helping refine AI-driven recommendations. Trust and participation can also be fostered by establishing data governance frameworks that prioritize data privacy, ownership and security, preventing exploitation of smallholder farmers' data. Finally, it is vital that partnerships between governments, private firms and farmer organizations to co-design AI-driven advisory tools are transparent.

By integrating ICT, big data and AI within this multifaceted approach, smallholder farmers can gain actionable, localized insights while ensuring that digital services remain accessible, affordable and inclusive. This model not only enhances agricultural decision making and market access but also strengthens rural economies and long-term farm sustainability.

10.3.5 ICT platforms for information delivery

The integration of ICT platforms into agricultural extension services has also been part of the revolutionary way information is delivered to smallholder farmers. These platforms include a diverse array of tools and facilities such as social media platforms, village kiosks, infomediaries, information centres, telecentres and multimedia centres, all designed to bridge the gap between knowledge sources and rural farming communities (see Box 10.6.). By leveraging these technologies, farmers gain access to timely and relevant information on best practices, market opportunities and weather updates, empowering them to make informed decisions and improve their livelihoods. These are discussed in this section.

Social media platforms

Social media platforms, such as Facebook, WhatsApp and X, have become exceptionally impactful in extending the reach of agricultural information. These platforms enable the rapid dissemination of instructional videos, infographics and expert insights, equipping farmers with actionable knowledge. Social media influencers, including lead farmers, researchers and extension specialists, play a vital role in promoting the adoption of innovative technologies and practices. By sharing success stories, testimonials and expert recommendations, these trusted figures inspire their followers to embrace new methods and tools, bridging the gap between innovation and practical implementation.

Mobile ICT units

Mobile ICT units represent an innovative example tried and tested in some countries. In exploiting this opportunity, the key constraints are identifying the most useful information, supplying it sustainably and packaging it in the most suitable dissemination channels. Literacy and language barriers are also fundamental problems in many countries, especially where different ethnic groups are found. Work is ongoing in donor-funded projects to use SMS to promote functional literacy in this context (R. Erskine-Smith, personal communication). In India and Bhutan, interactive voice response (IVR) systems via mobile phones have been particularly innovative especially among the less literate farmers. Even in areas without electricity, farmers are using mobile phones and charging their batteries with small hand-driven generators and torch batteries. Learning from this and other initiatives that provide opportunities for scaling up should become a top priority.

Farmers and agribusiness stakeholders could greatly benefit from access to additional information, such as the prices and availability of inputs like seeds, fertilizers, chemicals, credit, irrigation tools and other equipment. Providing this information, alongside technical guidance on recommended seeds and inputs, would support both farmers and rural entrepreneurs in improving productivity and market access. Combining this approach with internet- based tools creates opportunities to develop entrepreneurship training programmes for small producers and rural entrepreneurs. These programmes could cover essential skills such as production planning, profitability analysis, cash flow forecasting and business plan development.

While these technologies offer immense opportunities for delivering such resources, challenges remain – particularly the need to expand access to technology in rural areas and reduce the costs of outreach. Mobile phones, however, provide a cost-effective solution for sharing timely information with farmers, traders and agro-entrepreneurs. Evidence from studies (Jensen, 2007; Aker, 2008) shows that mobile phones improve market efficiency by reducing the cost of accessing information, particularly for leveraging regional price differences. This highlights the potential of mobile technology to transform marketing and agribusiness,

Box 10.6. Sri Lanka – cyber extension. From Wijekoon and Rizwan, 2009.

The cyber extension programme uses interactive multimedia as an information database and to develop low-cost audiovisual aids. The programme pursues five strategies: (i) use of interactive multimedia CD-ROMs as crop-based information material; (ii) use of interactive multimedia to develop low-cost audiovisual aids; (iii) internet delivery mechanisms with CD-ROMs; (iv) developing digital training material for extension and training; and (v) distance learning mechanisms. The programme focuses mainly on the provision of production-related information and market-price information. In 2007, a farmer database was established through the cyber extension mechanism to solve marketing problems. However, update of the database was not sustainable owing to shortages of field personnel. Another challenge faced has been to expand the outreach to farmers from agrarian service centres and create a demand for these information services. One consideration was to establish cyber user groups and set up joint ventures with other rural knowledge centres.

making it more inclusive and accessible for rural communities.

Village kiosks

Village kiosks have emerged as pivotal community managed information hubs in rural regions of India, Nepal and Bhutan, primarily funded through public initiatives. These kiosks serve as local centres where community members can access a range of services, including internet connectivity and mobile phone information. The establishment of these kiosks represents a significant advancement in making digital resources accessible, providing local farmers and entrepreneurs with crucial data on agriculture, market prices, weather forecasts and government services. The kiosks often become gathering points where community members not only access information but also receive training on how to effectively utilize these technologies for personal and professional development.

Parallel to these public efforts, the private sector has made significant strides in enhancing rural connectivity and information access. Private companies have been instrumental in deploying innovative solutions that complement the services offered by village kiosks. These include the development of mobile applications tailored to the needs of rural users, providing updates on market trends, and enabling e-commerce activities that connect farmers directly to buyers, bypassing traditional supply chains. The synergy between publicly funded village kiosks and private sector initiatives has created a robust ecosystem that supports the socio-economic development of rural communities, driving forward the integration of rural areas into the broader digital economy. This collaboration illustrates a successful model of public–private partnerships enhancing life in rural areas, offering scalable and sustainable solutions that leverage both community management and private innovation.

Infomediaries

Public sector initiatives have also focused on supporting local private sector-run information businesses. Communities and potential rural entrepreneurs have been assisted to operate as informal information search and distribution agents, sometimes called 'infomediaries'. To participate in these programmes, information providers need to be familiar with the structure of the rural communities and be keen to set up business as information suppliers. Public assistance has also been available to provide training in accessing and searching market-related information via the internet as well as conventional sources. In some cases, infomediaries access websites themselves and sell the downloaded information to local enterprises and individuals in the local community.

Information centres

Besides public sector initiatives, there has also been considerable private sector investment, particularly in India. ITC – one of India's leading private companies initiated the e-Choupal effort that places computers with internet access in rural farming villages. Many of these private sector initiatives are ICT based but include not only internet access but the production and publishing of technical information and the provision of individual and group advisory services. The Indian Society for Agribusiness Professionals (ISAP), for example, provides both market and specific technical information through its farm publications. Similarly, Indian Agribusiness Systems Pvt. Ltd (ISAL) has also been able to establish its credibility as a neutral, relevant and accurate information provider and many farmers are directly buying its publications. The field evidence suggests that market information of a specialized nature is appreciated by commercially minded farmers and is more likely to be purchased as a private good. Some examples can also be found of collaboration between financial service providers and information hubs, providing farmers and rural entrepreneurs with a fully integrated support service package. These private sector-led initiatives have a wide appeal and outreach, reaching thousands of customers within a country.

These efforts have created a highly profitable distribution and product design channel for the company – an e-commerce platform that is also a low-cost service system focused on the needs of rural India. The system has also helped to alleviate rural isolation, create more transparency for farmers and improve the productivity and incomes of farmers.

Telecentres

Another institutional innovation is telecentres which serve as a communal hub where the public can access a variety of information and communication technologies, primarily through the internet. These centres provide critical connectivity and technological services to low-income populations that might otherwise face barriers such as cost and distance. The potential impact of telecentres is significant, as they offer not just internet access but also a suite of services that can include training in computer skills, access to voice over internet protocol (VoIP) and integration with other media, like radio, to disseminate information widely.

In regions with young demographics, telecentres not only offer services but also foster technological literacy among youth, who are quicker to adapt to new technologies and can use these skills for educational and economic advancements. However, the sustainability of telecentres hinges on their ability to generate enough income to cover operational costs and maintain or replace equipment, a challenge that is not always met, leading to the risk of closure if financial viability is not achieved.

The challenges facing telecentres extend beyond financial sustainability to include issues like regular updating of the technology and content to keep pace with rapid changes in IT and the specific needs of their communities. In developed countries, models, like those found in Australian libraries, show some level of success by offering restricted but free access, funded by government support. This model, however, proves difficult to replicate in less affluent countries where resources are scarce, and sustained government or donor funding is not always feasible. Despite these challenges, telecentres that manage to operate sustainably can have a profound development impact. They improve the living conditions of numerous low-income users by providing them with access to information and services, thus broadening their opportunities and fostering socio-economic growth. This dual focus on sustainability and development impact makes telecentres a crucial element in the broader strategy to enhance digital inclusion and economic development in underserved regions.

Multimedia centres

The establishment and use of multimedia centres hold significant potential to transform the dissemination of agricultural information, and extension and training programmes for advisory service providers and farmers. These centres provide dynamic and interactive ways to engage with smallholder farmers, making complex information more accessible and actionable. By integrating video, audio and text, multimedia serves as an effective medium for conveying market and farm business management concepts. For example, instructional videos visually demonstrate agricultural techniques step-by-step, enabling farmers to understand and apply practical methods. Similarly, audio content, such as podcasts and radio broadcasts, delivers timely updates on market trends and expert advice, while infographics and interactive presentations make data accessible to those with limited literacy skills.

Multimedia centres also offer farmers the opportunity to access valuable resources while in the field, facilitating real-time learning and decision making. Farmers can view videos on pest management, listen to market price updates, or access other multimedia resources, empowering them to make informed choices on the spot. The interactive nature of these platforms fosters engagement and feedback, creating a two-way flow of information. This allows farmers to ask questions and receive tailored advice, ensuring that the information provided is relevant and actionable. Agricultural organizations and agribusinesses leverage the capabilities of these centres to deliver critical market intelligence and farm management training, driving improvements in agricultural productivity and sustainability.

The use of multimedia centres in delivering information, extension and training ensures inclusivity and caters to diverse learning preferences. This approach enables a wider outreach, allowing more farmers to benefit from essential agricultural knowledge and tools. By addressing various learning styles and overcoming literacy barriers, multimedia centres have the potential to significantly enhance the impact and effectiveness of agricultural extension services, fostering a more informed and resilient farming community.

These examples highlight the effectiveness of interpersonal communication and interaction in complementing information and communication technologies by providing personalized support to smallholder farmers. By fostering farmer-to-farmer and farmer-advisor communication, stakeholders can create a supportive environment that encourages knowledge sharing, collaboration and the exchange of innovative ideas. This combined approach of technologies and interpersonal communication maximizes the impact of interventions and contributes to the overall development of smallholder farming communities.

10.4 Application of Big Data for Small Farm Development

The emergence of big data represents a transformative moment in the collection and utilization of information across industries, agriculture being no exception. Characterized by its scale, speed and diversity, big data provides a foundation for modernizing farm management systems, offering innovative ways to interpret and act on complex agricultural challenges. By drawing on data from varied sources, such as climate models, market analyses and field sensors, it becomes possible to create a more comprehensive understanding of the interconnected factors influencing farm performance.

Smallholder farmers, often constrained by limited resources and vulnerability to changing conditions, stand to benefit immensely from the precision and predictive capabilities of big data. These tools enable them to refine their practices, reduce inefficiencies and respond proactively to emerging risks. When integrated into advisory systems, big data serves as a catalyst for practical innovation, bridging knowledge gaps and aligning farming strategies with emerging opportunities in a dynamic agricultural landscape. This section explores how these advancements can reshape the trajectory of small-scale farming, fostering a more resilient and inclusive agricultural sector.

The use of big data holds immense potential in improving decision making and generating valuable insights for smallholder farm development. By analysing large datasets, patterns, trends and correlations can be identified, providing actionable insights for optimizing productivity and addressing specific challenges faced by smallholder farmers. The ability to derive targeted interventions from comprehensive data analysis allows smallholder farmers to address these challenges more effectively and enhance the farm and enterprise profitability and competitiveness.

For smallholder farmers, leveraging big data offers a transformative opportunity to enhance farm management practices such as weather prediction, soil management, pest control and market trends. By accessing localized and timely data, farmers can tailor their activities to mitigate risks and maximize yields. For instance, weather prediction powered by big data enables farmers to plan planting schedules, adjust irrigation practices and prepare for adverse conditions, reducing vulnerabilities associated with climate variability.

Big Data also enhances crop performance monitoring and yield forecasting by integrating information from sensors, satellite imagery and historical records. These tools provide farmers with a detailed understanding of crop health, growth stages and nutrient needs, enabling the adoption of precision farming techniques. By applying inputs such as fertilizers and pesticides with accuracy, farmers can minimize waste and enhance efficiency. Additionally, insights into market trends and consumer preferences derived from big data analytics, empower farmers to align their production with demand, negotiate better prices and improve their access to premium markets. This information allows them to make informed decisions about crop selection, production volumes and marketing strategies. With access to such market intelligence, smallholder farmers can position themselves more strategically in the value chain, negotiate better prices and improve their market access.

Smallholder farmers can access big data through SMS messages, community radio or mobile apps designed for low literacy and local languages. If extension agents have the relevant big data tools, they can provide farmers with region-specific advice and training. Farmer cooperatives and NGOs can aggregate data and disseminate practical recommendations to farmers. Government departments and research institutions can use big data to improve policymaking and develop programmes that support smallholders.

The benefits of leveraging big data for smallholder farming are far-reaching. Data-driven recommendations on planting schedules, fertilization and pest management enhance resource efficiency and increase crop yields. Early warnings about extreme weather events or pest outbreaks mitigate potential losses and bolster resilience. Real-time market data enables farmers to make strategic decisions, negotiate better terms and reduce reliance on exploitative intermediaries. Over time, insights into soil health and water management foster sustainable farming practices, ensuring long-term productivity and environmental stewardship. By strategically utilizing big data, smallholder farmers can achieve greater profitability, resilience and sustainability in their farm business operations.

10.5 The Role of Agricultural Advisors in Data and Information

Agricultural advisors and in particular farm management specialists, play a pivotal role in bridging the gap between the wealth of information available and its effective dissemination to farmers. They serve as intermediaries, leveraging their knowledge and expertise to facilitate the transfer of information from research institutions, governmental organizations and other stakeholders to the farming community. They act as advisors, educators and trainers, equipping farmers with the necessary information, skills and tools to make informed decisions and improve their farm business management practices.

Agricultural advisors act as catalysts for change by providing guidance and technical support in data collection, analysis and interpretation. They help farmers understand the importance of data-driven decision making and its potential impact on farm productivity, profitability and sustainability. Agricultural advisors also assist farmers in identifying appropriate data sources, accessing relevant information and utilizing data analysis tools effectively. Their role extends beyond mere information dissemination, as they actively engage with farmers, build trust and provide ongoing mentorship and feedback. This work has taken on new importance with the emergence of the plethora of online and digital app services giving farmers unprecedented access to huge amounts of data. A central task in this connection is not only to facilitate access, but to build capacity to engage intelligently and critically with these services and to interpret the data they find. Farmers need to build capacity in determining the reliability, credibility and relevance of the data and information they find.

Part of this involves tailoring information and data to the specific needs and capacities of farmers. They understand that each farmer is different, with varying levels of literacy, technological access and comprehension. Agricultural advisors adapt their communication methods and strategies to ensure that information is conveyed effectively and understood by farmers. They break down complex concepts, use local language, visuals and hands-on demonstrations to enhance comprehension and facilitate knowledge transfer.

To effectively support farmers in adopting data-driven decision making, agricultural advisors can employ various strategies tailored to the local context and farmers' needs. These include training and capacity building, access to technologies and tools, establishment of platforms and networks for stakeholder collaboration and the use of mass media and innovative communication methods to disseminate information. These strategies have the potential to empower farmers to harness the power of data, make evidence-based decisions and improve their farm management practices. Their role can be powerfully instrumental in bridging the gap between data and farmers, ensuring that information is accessible, reliable, credible, relevant and utilized for the benefit of the farmers they serve.

10.6 Conclusion

The importance of timely and relevant data in enhancing farm management and decision-making processes cannot be overstated. This chapter underscores how the transformation of raw data into meaningful information empowers farmers to make informed choices, thereby improving productivity and profitability. It emphasizes that the role of agricultural advisors is pivotal in bridging the gap between the availability of data and its effective utilization. Advisors play a critical role in helping farmers understand, analyse and apply data to their specific contexts, ensuring that information is accessible and actionable for a diverse range of farming needs and challenges.

The emergence of digital tools, ICTs and big data has opened up new avenues for smallholder farmers, providing access to real-time information on weather, market prices and best practices. These technologies not only enhance efficiency but also reduce risks associated with climate variability and market fluctuations. However, the chapter highlights that the integration of these tools requires capacity building, infrastructure development and tailored support systems. Ensuring farmers have the skills to interpret and leverage data effectively is critical for maximizing the benefits of these innovations.

Despite the advancements in technology, this chapter acknowledges the persistent challenges that farmers face, such as limited access to reliable internet, high costs of advanced tools and varying levels of digital literacy. Addressing these barriers requires concerted efforts from governments, the private sector and non-governmental organizations to invest in infrastructure and training programmes. Moreover, creating localized, context-specific solutions – such as village kiosks, mobile applications in local languages and farmer learning groups – has been identified as essential for making technology and information accessible to all.

Ultimately, this chapter reinforces that the effective use of data and information in farm management is a collaborative effort that involves farmers, advisors, technology providers and policymakers. By fostering partnerships and ensuring the relevance and reliability of data, stakeholders can create an ecosystem where smallholder farmers thrive. This not only contributes to individual farm success but also to broader goals of agricultural sustainability, food security and rural development.

References

Aker, J.C. (2008) Does Digital Divide or Provide? The Impact of Cell Phones on Grain Markets in Niger Center for Global Development Working Paper no.154. Center for Global Development. Washington, D.C., US. Available at: https://ssrn.com/abstract=1093374 (accessed 17 March 2025).

FAO (2022) Tackling food loss and waste: A triple win opportunity. FAO, Rome. Available at: https://www.fao.org/newsroom/detail/FAO-UNEP-agriculture-environment-food-loss-waste-day-2022/en?utm_source=chatgpt.com (accessed 8 August 2025).

Fountas, S., Carli, G., Sørensen, C.G., Tsiropoulos, Z., Cavalaris, C. *et al.* (2015) Farm management information systems: Current situation and future perspectives. *Computers and Electronics in Agriculture* 115, 40–50. DOI: 10.1016/j.compag.2015.05.011. (accessed 7 February 2025).

Giua, C., Materia, V.C. and Camanzi, L. (2021) Management information system adoption at the farm level: Evidence from the literature. *British Food Journal* 123–3, 884–909. DOI: 10.1108/bfj-05-2020-0420/full/html. (accessed 7 February 2025).

Jensen, R. (2007) The digital provide: Information (technology), market performance, and welfare in the South Indian fisheries sector. *Quarterly Journal of Economics* 122(3), 879–924. DOI: 10.1162/qjec.122.3.879. (accessed 17 March 2025).

Silva, H. and Ratnadiwakara, D. (2005) Using ICT to Reduce Transaction Costs in Agriculture Through Better Communication: A Case Study from Sri Lanka. LIRNEasia. Srilanka. Available at: http://lirneasia.net/wp-content/uploads/2008/11/transactioncosts.pdf (accessed 17 March 2025).

Taranis (2021) Taranis case study: Soybean field in Brazil. Taranis. Available at: https://www.futurefarming.com/smart-farming/tools-data/taranis-sees-huge-potential-in-brazilian-agriculture/

Wijekoon, R. and Rizwan, M.F.M. (2009) ICT initiative for agriculture extension of the department of agriculture. In: Sivayoganathan, C. (ed.) *Experiences and Challenges in Agricultural Extension: Meeting farmer needs, Proceedings of Agricultural Extension Conference 2009*, Sri Lanka Agricultural Extension Association, Peradeniya.

Additional Reading

Ballantyne, P.G. (2009) *Information for Agricultural Development in ACP Countries: A Scoping Study*. CTA.

Basso, B., and Antle, J.M. (2020) Digital Agriculture to Design Sustainable Agricultural Systems. *Nature Sustainability*.

Chambers, R. (1997) *Whose Reality Counts? Putting the First Last*. Intermediate Technology Publications.

Chambers, R., and Conway, G.R. (1992) *Sustainable Rural Livelihoods: Practical Concepts for the 21st Century*. IDS Discussion Paper 296, Institute of Development Studies, University of Sussex.

Chambers, R. (1994) *The Origins and Practice of Participatory Rural Appraisal. World Development*, Vol. 22, No. 7, 953–969.

Chambers, R., Pacey, A., and Thrupp, L.A. (Eds.). (1989) *Farmer First: Farmer Innovation and Agricultural Research*. Intermediate Technology Publications.

DCM's Harilayi, Godrej's Aadhar, and Mahindra's Subh Labh (2005) *Corporate Agricultural Ventures in India*. Various Publications

Deichmann, U., Goyal, A., and Mishra, D. (2016) Will Digital Technologies Transform Agriculture in Developing Countries? World Bank Research Observer.

Devonish, E., Pemberton C.A. and Ragbir, S. (2000) Record keeping among small farmers in Barbados. Department of Agricultural Economics and Extension, University of the West Indies, St Augustine, Trinidad and Tobago.

Dudafa, U. J. (2013) Record keeping among small farmers in Nigeria: Problems and prospects. *International Journal of Scientific Research in Education*, 6(2), 214–220.

Griffith, E.F., Schurer, J.M., Mawindo, B., Kwibuka, R., Turibyarive, T. *et al*. (2023) The use of drones to deliver rift valley fever vaccines in Rwanda: Perceptions and recommendations. *Vaccines* 11–3, 605. DOI: 10.3390/vaccines11030605. (accessed 17 March 2025).

Ibrahim, U. (2023) Artificial intelligence in agricultural extension for sustainable development. *International Journal of Applied and Scientific Research (IJASR)* 1(3), 259–268. DOI: 10.59890/ijasr.v1i3.740. (accessed 17 March 2025).

Lezberg, S. (2019) The Role of Telecenters in Supporting Rural Agricultural Development: Evidence from Sub-Saharan Africa. *Telecommunication Policy*, Vol. 43, 1021–1032.

Minae, S., Baker, D. and Dixon, J. (2003) Status of farm data systems and farmer decision: Support in Sub-Saharan Africa: FAO, Rome, 2008.

Njagi, K. (2019) Ghana's farmers are using drones to boost crop yields - here's how. Rural Economic Forum. Geneva, Switzerland. Available at: https://www.weforum.org/stories/2019/10/ghanas-farmers-drones-boost-crop-yields/?utm_source=chatgpt.com (accessed 17 March 2025).

Rose, D.C. *et al*. (2019) How Can We Make Digital Farming More Inclusive? A Bottom-Up Perspective on Digital Technologies and Data in Farming. *Journal of Rural Studies*, Vol. 68, 225–236.

Saravanan, R. (2010) *e-Agriculture: Perspectives and Approaches*. New Delhi: New India Publishing Agency.

We Robotics (2018) Tanzania Drone Pilots Team up with IFPRI and Local Smallholder Farms. Blog. We Robotics. Wilmington, DE, US. Available at: https://werobotics.org/blog/tanzania-drone-pilots-team-up-with-ifpri-and-local-smallholder-farms?utm_source=chatgpt.com (accessed 17 March 2025).

Wolfert, S. *et al*. (2017) Big Data in Smart Farming – A Review. *Agricultural Systems*, Vol. 153, 69–80.

11 Budgeting

Farmers face constant challenges in organizing and managing their farm businesses to align with their goals and objectives. Decisions about the best ways to utilize resources, such as land, capital and labour, must be continually made to maximize economic returns. Budgeting serves as an essential management tool, providing a structured approach to analyse various alternatives and make informed decisions. Through budgeting, farmers can efficiently combine inputs to produce desired outputs, allocate resources to different products and choose the most beneficial combinations of products.

This chapter aims to describe the different types of budgets available to farmers, introduce basic economic principles and relate them to these budgeting tools. We will explore how budgeting helps farmers address key questions about resource use and farm management. By simulating different scenarios, budgeting allows farmers to experiment with potential outcomes before committing resources, uncover hidden costs, refine current operations, seek credit from lending agencies and learn to organize and reorganize more effectively.

We will examine various budgeting techniques, including enterprise budgets, break-even budgeting, partial budgeting, whole-farm budgeting and financial budgeting, showing how each assists in making farm management effective in contributing to the overall financial stability of the farm business. Additionally, the chapter will refer to some of the economic principles discussed in Chapter 9, this volume, necessary for resource allocation, such as adding inputs as long as the returns exceed costs, substituting inputs and products to maintain output and minimize costs, and using resources where they yield the greatest returns. By utilizing different types of budgets, farmers can systematically analyse their operations, plan for the future and make decisions that align with their financial goals and the overall sustainability of their farm businesses. The budgets referred to in this chapter introduce the basics of budget application, which will be further applied in the following chapters in Part 3, to the farm management cycle.

11.1 The Challenge for Farmers: Budgeting and Resource Allocation

Farmers continually face the challenge of organizing and managing their farm businesses to align with the goals and objectives of their families. As was indicated in Chapter 4, when we discussed the livelihoods context, decision making in this setting involves evaluating whether the alternatives considered are consistent with these established goals and objectives. As previously discussed, farmers should also consider the *opportunity cost* – the value of the best alternative forgone – to arrive at a rational decision. Every resource allocation decision comes with a trade-off. For instance, choosing to allocate more land to maize production means forgoing potential income from alternative crops like soybeans or vegetables. Recognizing these trade-offs allows farmers to maximize returns by selecting the most profitable enterprises.

Budgeting emerges as a crucial management tool that helps in developing and analysing these alternatives. Properly utilized, budgeting can provide valuable information to address numerous questions. It enables farmers to combine inputs into products, allocate resources to alternative products and choose optimal combinations of different products.

DOI: 10.1079/9781836990536.0011

Budgeting helps to answer the important management questions discussed in previous chapters:

- What is the best use of the available or accessible resources?
- What enterprises (crops and/or livestock) can be produced and which will contribute most returns to resources?
- How much of the land should be devoted to each enterprise?
- What equipment and machinery will be needed to produce the potential enterprises?
- What production practices should be used to produce each of the enterprises?
- How much labour (both family and hired) will be needed on the farm?
- What are the capital requirements?

Developing successful farm businesses requires significant management skills. Many farmers may not fully consider the economic environment that dictates the profitability of feasible production possibilities. As the above questions illustrate, farm management skills and knowledge are integral to success.

The problems of resource allocation and use in farming involve applying the following economic principles, as discussed in Chapter 9.

1. Adding units of an input as long as the value of the resulting output or added returns is greater than the added cost – the concept of the margin.
2. Substituting one input for another as long as the cost of the added input is less than the cost of the input that is replaced while maintaining output.
3. Substituting one product for another as long as the value of the added output is greater than the value of the output that is replaced.
4. Using each unit of resource where it gives the greatest returns when resources are limited.
5. Basing comparisons upon discounted values when considering different time periods and/or elements of risk.

The first three principles apply to situations where unlimited resources are available and there is perfect knowledge. The last two principles are relevant when resources are limited, and perfect knowledge is unattainable. Most resource allocation management problems faced by farmers can be addressed by applying these basic economic principles through budgeting. The various types of budgets discussed here can assist in making management decisions, with the choice of budget tool depending on the intention of the analysis. By understanding and implementing these principles, farmers can make more informed and effective decisions to optimize their farm operations and achieve their financial goals (see Box 11.1.).

11.2 What is Budgeting?

Budgeting is the process of planning how financial resources will be allocated and spent over a specific period. In the context of farm management, budgeting involves estimating the expected income and expenses associated with the different farming activities so that farmers can make informed decisions, set financial goals and monitor performance.

Budgeting procedures are often the most used techniques in planning. Budgeting does not draw up a plan; it only evaluates the plan in physical and financial terms. In this sense, it is only a planning aid. The actual plan is drawn up using a combination of experience, judgment and intuition following the processes we discussed in previous chapters. Budgeting is now the 'bread and butter' of conventional farm management analysis. It is popular because it is easy to use and supports practical, experience-based decision making instead of forcing a strict analytical method.

The process entails estimating future income (revenue), costs and profits to guide decision making. Sound budgeting is crucial for the success and sustainability of farming operations. It provides farmers with a roadmap for achieving their objectives, enabling them to make informed decisions and allocate resources effectively.

Using budgets gives farmers a way to:

1. test different options and predict possible outcomes before making real changes;
2. identify costs they might otherwise miss;
3. improve their current farm setup;
4. support loan applications with clear financial plans; and
5. build skills in organizing and reorganizing their farm operations.

Box 11.1. What is a budget?

A budget in farming is a financial plan that estimates the costs, revenues and expected profits for a specific farming operation over a set period, usually a growing season or year. It outlines how much money will be spent on inputs (e.g. seeds, fertilizers, labour) and predicts income from the sale of crops or livestock. A budget helps farmers manage resources effectively, track spending and plan for profitability.

Budgeting refers to the process of creating and managing this financial plan. It involves:

1. **Estimating costs:** calculating expected expenses, including seeds, fertilizers, pesticides, labour, equipment maintenance, fuel and other operational costs.
2. **Estimating income:** predicting income from the sale of crops, livestock or other farm products based on expected yields and market prices.
3. **Profit calculation:** subtracting total costs from expected income to determine potential profits or losses.
4. **Monitoring and adjusting:** once the season starts, the budget is used to track actual spending and income. If things deviate from the plan (e.g. higher input costs or lower crop yields), the farmer may adjust the budget accordingly.

Budgeting is a critical tool for farm management, helping farmers stay financially organized, make informed decisions and improve overall profitability.

In the next sections, we will look deeper into enterprise budgets, partial budgeting and break-even budgeting, illustrating how each type of budget aids in effective farm management and contributes to overall financial stability.

11.3 Types of Budgets

There are five basic types of budgets that can be used in the farm business management process. Each type of budget provides different information to the farmer for use in the decision-making process. The common thread in each type is that, if properly defined and used, the budget format permits the farmer to use economic logic to answer questions of what, how much and when resources should be used to achieve the goals and objectives as established by the farm family. The five types of budgets are:

1. Enterprise budget
2. Partial budget
3. Break-even budget
4. Whole-farm budget
5. Cash flow budget

An *enterprise budget* outlines the expected income and costs from producing a specific product using a particular set of practices over a defined period, such as a season or a year. It includes both variable and allocatable costs and helps the farmer estimate how profitable the enterprise is likely to be.

A *partial budget* is useful in analysing the effects of a change from an existing plan. This budget only considers revenue and expense items that will change with a defined change in the plan.

A *break-even budget* shows the point where income equals costs, meaning there is no profit or loss. It helps estimate the highest cost that can be afforded or the lowest income that must be earned for the enterprise to break even.

A *whole-farm budget* is a detailed summary of the main physical and financial aspects of the entire farm. It shows the different parts of the farm business and how they relate to each other and to the overall operation.

A *cash flow budget* is a planning tool that shows the expected flow of money (cash) in and out of the farm over a specific period – usually monthly, quarterly or annually. It helps farmers see when they will receive income (e.g. from crop sales or milk production) and when they will need to spend money (on things like seeds, fertilizers, labour or loan repayments).

At this point, it is helpful to remind ourselves of the definitions for terms used in this chapter (Table 11.1), some of which are discussed in previous chapters of this book.

Table 11.1. Recap on concept definitions.

Gross income	The total income (revenue) generated from the sale of crops, livestock and other farm products before deducting any expenses such as input costs, labour, or taxes.
Variable costs	Expenses (costs) that fluctuate (vary) based on production levels, such as seeds, fertilizers, fuel and labour, which increase or decrease depending on the amount of farming activity.
Fixed costs	Expenses (costs) that remain constant (fixed) regardless of production levels, such as land rent, insurance and equipment depreciation.
Total costs	The sum of variable and fixed costs, providing a comprehensive view of the financial investment required.
Gross margin	The difference between gross income and variable costs, representing the income available to cover fixed costs and generate profit.
Farm profit/ Net farm income	The amount of money remaining after all operating expenses, taxes and costs of production have been subtracted from total income generated by the sale of farm products. It is calculated by subtracting total costs from gross income. It represents the potential profitability of the enterprise.
Break-even price	The price at which the total income from selling a product exactly equals the total costs of producing it, resulting in no profit or loss.
Break-even yield	The minimum yield needed to cover all production costs without making a profit or loss at the current market price.
Working capital	The money needed to keep daily operations running smoothly. It is the cash available to buy seeds, feed, fuel and other supplies, as well as to cover regular expenses like wages or repairs. Essentially, it's the money the farm relies on to bridge the gap between spending on production and receiving income from sales.
Cash flow	The movement of money (cash) in and out of the farm, reflecting the cash income from sales and other sources, minus cash expenditures like input costs, labour and loan payments, over a specific period.

11.4 Enterprise Budgets

An enterprise budget provides a structured format for classifying information, allowing farmers to consistently analyse the economics of different enterprises and production systems. It is one of the main tools for deciding what to produce.

Enterprise budgets are financial tools that estimate the costs and potential returns associated with specific farming enterprises. They provide a detailed breakdown of expected income, expenses and profitability for individual crops, livestock or other farming activities. Each type of crop or livestock is considered an enterprise, and budgets can be constructed for different production levels or technologies. By analysing the financial aspects of each enterprise, farmers can make informed decisions regarding resource allocation, production strategies and marketing plans.

Enterprise budgeting assesses the expected profitability of a proposed enterprise, including four sections: (i) gross income; (ii) variable costs; (iii) fixed costs; and (iv) measures of enterprise profitability. This method allows for comparing the profitability of different enterprises, aiding in developing a comprehensive farm plan. An enterprise budget can be calculated on a per-unit basis, such as per hectare of land or head of livestock, for 1 year or one production period. Using common units allows for easy comparison between different enterprises, helping farmers select the most profitable options. Completed enterprise budgets provide data to compute the average cost of production, the break-even price and the break-even yield.

It is crucial for farmers to know how to prepare enterprise budgets and use them to assess the relative contribution of individual enterprises to farm income. The accuracy and value of budgeting depend on the skills and experience of the farmers and extension workers in forecasting and planning for the future. A positive gross margin contributes to covering the fixed costs of the farm, and maximizing the gross margin is equivalent to maximizing profit (or minimizing losses) because fixed costs for small farmers are often low and relatively stable. The more experienced the farmer and the simpler the farm, the more precise the enterprise budget is likely to be.

Enterprise budgets play a critical role in financial planning for farming operations. They offer several benefits:

- **Decision making:** enterprise budgets provide farmers with essential information to make decisions regarding resource allocation, investment and expansion. By comparing the expected costs and returns of different enterprises, farmers can identify the most profitable ventures and allocate resources accordingly, ensuring efficient use of resources and increased productivity.
- **Risk management:** budgeting and planning enable farmers to anticipate and mitigate potential risks and uncertainties. By identifying risks such as adverse weather conditions, market fluctuations or pest outbreaks, farmers can use enterprise budgets to develop contingency plans, explore insurance options and implement risk management strategies to safeguard their operations.
- **Financial management:** budgeting helps farmers maintain a clear understanding of their financial position. It allows them to identify potential sources of income, plan for expenses and allocate funds wisely, leading to improved cash flow management, reduced financial risks and better profitability.
- **Cash flow management:** enterprise budgets help farmers plan and manage their cash flow effectively. By estimating revenues and expenses, farmers can anticipate periods of high or low cash flow and plan accordingly, ensuring that they have sufficient funds to cover expenses and investments.
- **Loan applications:** lenders often require enterprise budgets as part of the loan application process. These budgets provide lenders with a clear understanding of the farm's financial viability and the potential for loan repayment.

Budgets can be divided into physical and financial parts. Farmers should remember that a change in an enterprise budget can result from any combination of changes with respect to the:

- selling price of the produce;
- costs of production;
- level of production;
- level of input use; and
- variations in the product mix (e.g. other enterprises).

Enterprise budgets typically include components such as revenue, variable costs, fixed costs, total costs and net income.

11.4.1 Examples of enterprise budgets

Enterprise budgets can be developed for various farming enterprises, including crop production, livestock raising, poultry farming, dairy operations and more. Following are a few examples:

- **Crop production:** an enterprise budget for a specific crop, such as maize or wheat, would estimate costs and revenues associated with land preparation, seed, fertilizers, pesticides, labour, machinery use, harvesting, storage and marketing.
- **Livestock raising:** an enterprise budget for livestock, such as beef cattle or pigs, would estimate costs and revenues related to animal purchases, feed, veterinary services, labour, breeding, housing, equipment, marketing and sales.
- **Poultry farming:** an enterprise budget for a poultry operation, such as broilers or egg production, would estimate costs and revenues associated with chick purchases, feed, medications, bedding, labour, equipment, utilities, marketing and sales.

These examples illustrate the specific components and considerations involved in developing enterprise budgets for different farming enterprises. Farmers can customize and adapt these budgets based on their own production systems, local market conditions and cost structures. An example of an enterprise budget is given in Table 11.2.

In this example, we see that the rice is profitable because it has a positive gross margin. In terms of planning, the farmer may consider producing rice because it is expected to make a profit. If it had a negative gross margin, then the farmer should generally not consider producing rice on their farm because it is not expected to be profitable.

In this example, we also see that even when fixed costs are deducted, the enterprise is still profitable. Remember that fixed costs are costs that do not vary with production and that they apply to the whole farm. In a farm business that has only one enterprise, all fixed costs can be allocated to that enterprise. However, in a multi-enterprise farm business, fixed costs are typically shared across enterprises based on their relative use of resources. For example, a tractor's depreciation might be allocated based on the time it is used for each enterprise, or property taxes might be distributed based on the acreage dedicated to each enterprise. This allocation helps accurately reflect each enterprise's share of fixed costs in the overall profitability analysis. It is also possible that an enterprise is sufficiently large to warrant having its own fixed costs. Examples include a small-scale poultry operation with its own dedicated housing and equipment, or an irrigated vegetable plot requiring a separate pump and irrigation system.

Thus, when comparing enterprises, it is most common to look at the gross margins to determine potential profitability. Table 11.3 gives an example of gross margins for four enterprises on a single farm.

As previously stated, a positive gross margin contributes towards covering the fixed costs of the farm. Therefore, maximizing the gross margin is equivalent to maximizing profit (or minimizing losses) because fixed costs do not change.

Thus, in this example, all the enterprises are potentially profitable, but peppers is the most profitable enterprise because it has the highest gross margin. With this information, the farmer can decide which enterprises to grow.

Farmers should remember that a change in gross margin can result from any combination of changes with respect to:

- the selling price of the produce;
- the costs of production;
- the level of production;
- the level of input use; and
- variations in the product mix (e.g. other enterprises).

For more precise planning, it is essential for enterprise budgets to be composed of a detailed breakdown of costs and income into both physical and financial terms as shown in the example for rice above. It provides the farmer with a much clearer picture of how the costs are composed.

As straightforward as an enterprise budget is, it is difficult to produce a completely accurate one. It is based on what is expected, not what is known. Even if the farmer has a lot of information about their enterprises from past performance, the enterprise budget is still predicting an uncertain future.

Even when farmers have a great deal of knowledge about how an enterprise performs under different production systems, they never have complete information regarding production conditions, such as weather and output markets. Typically, more information is available regarding the prices of inputs than on products since inputs are purchased during one period and products are sold in a later period. Thus, it is important to remember that the enterprise budget is an estimate that can be used to make decisions about what to produce, how much to produce, how to produce it and where to sell it.

In the next sections, we will explore additional budgeting and planning methods that complement enterprise budgets and provide a comprehensive approach to financial planning in farming.

11.5 Break-even Budgeting

Break-even budgeting is a technique for studying the relationship between costs and income at different levels of production. A break-even budget estimates the maximum acceptable level of a cost item or, alternatively, the minimum acceptable level of a benefit given an estimated level of cost. In other words, it looks at the scale of the enterprise where income equals cost. At this point, gross margins are zero.

As previously mentioned, the results obtained from actual farm conditions are often very different from the values included in the enterprise budget. Yields and output prices often vary considerably in reality because of changes in many factors. It is, therefore, useful to determine the minimum yield and product price required that enables the farmer to recover at least the variable costs incurred in farm enterprise production. The information included

Table 11.2. Enterprise budget for rice (1 ha).

Item	Unit	Quantity	Price ($)	Amount ($)
Rice sales	Tons	2.50	80.00	200.00
Gross income				*200.00*
Variable Costs				
Seed	Kg	84.00	0.25	21.00
Fertilizer: Nitrogen	Kg	77.00	0.30	23.00
Fertilizer: Phosphorus	Kg	35.00	0.10	3.50
Chemicals				19.00
Labour (hired)				2.00
Miscellaneous costs				7.00
Interest on working capital @ 10% for 4 months				5.60
Total Variable Costs				*81.10*
Gross Margin				*118.90*
Fixed Costs				
Machinery depreciation				10.00
Fixed labour				13.00
Land charge (rent)				15.00
Storage costs				3.00
Total Fixed Costs				*41.00*
Total Costs				*122.10*
Enterprise Profit (Gross Income - Total Costs or Gross Margin - Total Fixed Costs)				*77.90*

Table 11.3. Gross margin example for the Philippines (1 ha).

	Carrots	Yams	Cabbages	Peppers
Gross Income ($)	180	120	210	280
Variable Costs ($)	83	40	105	109
Gross Margin ($)	97	80	105	171

in enterprise analyses can be used to perform break-even analysis for product prices and yields.

There are several uses for break-even budgets:

1. **Pricing decisions:** help determine the minimum price needed to cover costs, guiding pricing strategies for crops, livestock or other products.
2. **Production targets:** establish the minimum production level (yield or output) required to cover costs, informing decisions on input use, management practices and resource allocation.
3. **Sales targets:** establish the minimum level of sales required to cover costs which helps the farmer with marketing decisions.
4. **Risk assessment:** by identifying the break-even point, farmers can assess financial

risks and understand the resilience of their enterprise to changes in prices, yields or input costs.

5. **Profitability analysis:** allows producers to gauge whether an enterprise can be profitable, serving as a benchmark to evaluate the enterprise's economic viability before committing resources.
6. **Investment decisions:** for potential expansions or new enterprises, a break-even budget shows the production level or price needed to justify an investment, aiding in capital planning and loan applications.

Overall, a break-even budget is a powerful tool in planning, decision making and risk management for farm operations.

In each of the uses mentioned, in preparing a break-even budget, the values of all the variables except a selected variable are known. For example, a farmer might be interested in substituting one variety of tomato for another, although the production potential of the new variety is unknown. In this situation, the break-even budget is constructed to estimate the minimum yield that would have to be achieved to make the change worthwhile. Alternatively, if the expected yield is known but the price is unknown, the budget could indicate the minimum price that must be obtained to make the change economically feasible. Farmers should be aware of the point at which they start making a profit.

In another example, perhaps a farmer is considering selling into a new market. To do so will affect their variable costs. They intend to sell their fruit at $100/t. A market investigation indicates that there is a potential to sell 100t at that price. The variable costs of production are US$3800 and increase by US$200/t for every additional 10t sold. While there is potential to sell 100t, they may not achieve that level of sales. The farmer wants to calculate the break-even point for sales and find out how much profit can be earned and the minimum amount that must be sold to at least break even. Table 11.4 shows the gross margins for each 10t sold. It shows that the farmer must sell at least 60t to break even and will earn a gross margin of $2400 if all 100t is sold.

This represents the break-even point at which neither profit is made nor loss incurred. The variable costs of the farm enterprise would be the same as gross income. It is very useful for farmers to know the break-even point because all enterprises must remain profitable, and this should be closely watched.

Understanding the break-even point also requires recognizing the impact of diminishing returns. While increasing input use can initially boost yields, each additional input eventually produces smaller incremental gains. This means that even if a farmer reaches the break-even yield, further input use may not always lead to higher profitability, as costs may rise faster than revenue.

Table 11.4. Break-even sales.

Units Sold	Sales Price ($)	Gross Income ($)	Variable Costs ($)	Gross Margin ($)
0	100	0	3,800	–3,800
10	100	1,000	4,000	–3,000
20	100	2,000	4,400	–2,400
30	100	3,000	4,800	–1,800
40	100	4,000	5,200	–1,200
50	100	5,000	5,600	–600
60	**100**	**6,000**	**6,000**	**0**
70	100	7,000	6,400	600
80	100	8,000	6,800	1,200
90	100	9,000	7,200	1,800
100	100	10,000	7,600	2,400

The break-even point can be calculated for both yields and prices.

11.5.1 Break-even yield

The break-even yield provides the farmer with information on the minimum yield that is required to make the enterprise profitable. It refers to the yield required to cover the variable costs at given product prices and input costs. The break-even yield is calculated by using this simple formula:

$$\text{Break-even yield} = \frac{\text{variable costs}}{\text{product price}}$$

This is the yield necessary to cover all variable costs at a given product price.

Variable costs = \$157.5/ha
Product price = \$80/t

Break-even yield is:

$$\frac{\$157.5/\text{ha}}{\$80/\text{t}} = 1.97\text{t/ha}$$

The farmer must achieve a yield of 1.97t/ha to break even.

Since the product price is only an estimate, it is often useful to compute the break-even yield for a range of possible prices as shown in Box 11.2.

This often provides some insight into the sensitivity of the break-even yield to changes in the output price.

Interpretation of break-even yield

If the actual yield is higher than 1.97t/ha (the break-even yield), it will be profitable for the farmer to grow the crop. Conversely, if actual yield is lower than its break-even yield, the farmer will incur a loss if the crop is grown.

11.5.2 Break-even price

The break-even price is the product price needed to just cover all costs at a given production level, and can be found from the equation:

$$\text{Break-even yield} = \frac{\text{variable costs}}{\text{expected yield}}$$

Using the same example, with an expected yield of 2.5t/ha, in this case, we find the following:

$$\text{Break-even yield} = \frac{\$157.5/\text{ha}}{2.5\text{t/ha}} = \$63/\text{t}$$

Assuming a yield of 2.5t/ha, the farmer must obtain a market price of at least \$63/ton to break even. You may notice that the break-even price is the same as the variable cost per ton.

Again, because the actual yield may be uncertain, the break-even price can also be computed for a range of possible yields as in Box 11.3. Different yields cause different break-even prices, and these prices can vary widely depending on the yield level.

This means that, if the price of the product is above the break-even price, it will be profitable to grow the crop. Conversely, if the price of the product is below \$63/ton, the farmer will incur a loss if they decide to cultivate the crop.

Since both the costs and the yield and output price in an enterprise budget are estimated values rather than actual values, the calculation of break-even points, yields and

Box 11.2. The break-even yield for a range of prices. (Variable costs: \$157.50)

Output price/ton (\$)		Break-even yield (tons)
40.00	→	3.94
50.00	→	3.15
60.00	→	2.63
80.00	→	1.97
90.00	→	1.75

Box 11.3. Break-even prices for a range of yields. (Variable costs: \$157.50)

Yield (ton/ha)		Break-even price (\$/ton)
1.0	→	157.50
1.5	→	105.00
2.0	→	78.75
2.5	→	63.00
3.0	→	52.50
3.5	→	45.00

prices can aid managerial decision making. By studying the various combinations of the break-even point, yields and prices, farmers can form their own expectations about the probability of obtaining a price and a yield that would just cover total costs. Break-even prices and yields can also be calculated from total variable costs rather than total costs.

11.6 Partial Budgeting

This section explains the concept of partial budgeting and its importance (see Table 11.5; Boxes 11.4–11.6). It also introduces an example of a partial budgeting form.

Partial budgeting is an important planning tool used by farmers to estimate the effect of relatively minor adjustments on the farm and its effect on overall farm profit. It looks only at those income and expense items that are affected by a proposed change. It is an approach that looks at changes at the margin. It involves considering what the situation would look like if the change were not implemented; compared to what the situation would be if the change were implemented. This differs from a total budget that includes all income and expense items for the entire farm.

Many of the day-to-day decisions made by farmers are really adjustments to the farming system and the farm operations. It may happen that the farmer needs to make a change in the weeding operation. Perhaps the piece of equipment they were going to hire is not available and the farmer must now hire labour instead. How will this affect the profitability of the enterprise? A partial budget will give an indication of the impact of this change, enabling the farmer to make an informed decision about making this change. If the farm business looks likely to be more profitable with the change than without the change, then the change is likely to be worth making.

Table 11.5. Four basic items typically considered in partial budgets.

Costs	Benefits
(a) new or additional costs	(c) new or additional income
(b) income forgone or reduced	(d) costs reduced or saved

So-called marginal or partial changes to a farm can include expansion, introducing a new enterprise, purchasing machinery or equipment or even downsizing the farm. These adjustment decisions often affect income and expenses. Partial budgeting is consequently a valuable instrument that can assess the effect of such changes on overall profitability, and choosing between technologies and enterprises, without having to develop full enterprise budgets. It is a form of marginal analysis designed to show the net increase or decrease in net farm income resulting from the proposed changes rather than to show profit or loss for the farm as a whole.

In the case where a new activity replaces an existing activity, there is likely to be some net income from the 'old' activity that will no longer be generated. There will be some income forgone, some costs saved and some net income given up. These effects must also be counted in evaluating a change to a farm system.

The partial budget evaluates whether the proposed change would be more profitable than the current situation. The difference between (a) + (b) and (c) + (d) will indicate whether the change is profitable (see Table 11.5). If (c) + (d) exceeds (a) + (b), the proposed change would increase farm income, provided that the change is technically feasible.

11.6.1 When to use partial budgets

Many small reorganization problems involve partial budgeting. These are undertaken when the basic farm plan is not changed, and the farmer is concerned with the marginal costs and returns resulting from a small change. In many cases where the farm is already planned, partial budgeting for small adjustments may be all that is required.

It is not always easy, or even feasible, to collect information on all the inputs and outputs for an enterprise or technology. Partial budgeting does not require this to be done. The tool only requires information about those outputs and inputs (i.e. both expressed in monetary or value terms) that will change as a result of small adjustments in the farm business.

Many changes that do not require a complete reorganization of the farm can frequently be identified. Farmers can employ

Box 11.4. Information needed in partial budgeting.

- Prices of both inputs and outputs.
- Costs and returns of the old practice/resource/plan.
- Costs and returns of the new practice/resource/plan.

From this information, the added as well as the reduced costs and income of both the old and new practice/resource/plan can be calculated.

Box 11.5. Uses of partial budgeting.

To compare two different systems of farming, such as:

- Use of animal vs tractor in land preparation.
- Transplanting vs direct seeding.
- Hand weeding vs weedicide application.
- Pesticide application vs integrated pest management.
- Manual threshing vs use of mechanical thresher.
- Sun drying vs use of mechanical dryer.
- Seasonal vs early planting to take advantage of high prices.

To help determine the profitability of varying input applications, such as:

- Applying 3 vs 4 bags of fertilizer.
- Applying 2 vs 3 pints of weedicide.
- Working 10 vs 15 labour days on the farm.
- Sowing 2 kg vs 3 kg of seeds per hectare.
- Transplanting vs direct seeding in planting.
- Use of caribou vs hand tractor in land preparation.
- Applying 1 vs 2 bottles of insecticide.
- Using Azolla vs chemical fertilizer.

To enable a farmer to choose the plan which gives better results, such as:

- Buying vs renting a farm machine.
- Feeding poultry with homemade vs commercial feeds.
- Using pump vs gravity as the source of power in irrigation.

To help determine the best age to market crops, poultry and livestock, such as:

- Marketing 5-week-old vs 6-week-old broilers.
- Marketing 3-month-old vs 4-month-old hogs.
- Marketing young vs old coconuts.
- Marketing young coconuts vs 'tuba' gathering.
- Marketing fresh maize vs dried kernels.
- Pick-up vs delivered in marketing farm products.
- Selling fresh vs processed products.

their resources in more than a single way in response to changes in product prices, market demand and the cropping pattern. Partial budgets are useful to evaluate changes such as:

- expanding an enterprise;
- selecting alternative enterprises;
- selecting different production practices;
- deciding whether to purchase or hire equipment;
- making a capital improvement;
- buying new equipment to replace hand labour or maintaining older equipment.

Partial budgeting is based on the principle that a small change in the organization of a farm

Box 11.6. Example of partial budget.

An example of a partial budgeting form is provided below. It is important to note that the categories on the left-hand side of the form are the two that increase profit: additional income and saved costs. On the right-hand side are the two that reduce profit: decrease in income and increased costs. Entries on the two sides of the form are summed up and then compared to find the net change in income.

Costs	Benefits
New costs Things you will have to buy because of the change or new activity. Only include a share of the total cost of capital items that will last longer than the period of the budget.	New income New or extra things that you will be able to sell or eat if you make this change.
Income foregone Things you will no longer have to sell or eat if you make this change.	Cost saved Things you will stop buying because of the change. You will only save a share of the cost of equipment that lasts longer than the budget.

For example, the shift from raising local Malaysian ducks to hybrid ducks brings the advantage of additional income due to the increased weight of the birds. However, it also brings additional costs in terms of extra labour, feed, etc. as well as savings through a reduction in the level of expenses on medicines and veterinarian care, among others. Income is also reduced as a result of decreased sales of meat. The overall additional gross income together with the reduced costs is US$132.4. In contrast, the total additional costs and reduced gross income is US$165.7. This results in a net change in profit of –US$33.3. This negative difference shows that the introduction of hybrid ducks would decrease profit. The partial budget is produced in the table shown here.

Partial budgeting

Problem: whether to introduce hybrid ducks or continue raising local Malaysian ducks

Added annual costs:	$	Additional Income:	$
Fixed costs		Added weight 60 kg (1 kg at $1.25)	75.00
Bigger shelter	50.00	Manure 30% wt (1 kg at $0.05)	0.90
Variable costs		Feathers 5% wt (1 kg at $0.50)	1.50
Extra labour	10.00		
Extra water	5.00		
Extra feed	50.00		
Reduced income:		Reduced costs:	
Malaysian duck 40 kg at $1.25	50.00	Vet and health	20.00
Manure 15% wt (1 kg at $0.05)	0.30	Mortality	5.00
Feathers 2% wt (1 kg at $0.50)	0.40	Feed efficiency	30.00
A. Total additional costs and reduced gross income	165.70	B. Total additional gross income and reduced costs	132.40
			→ 165.70
		Net change in profit (B minus A)	–33.30

Continued

Box 11.6. Continued

The outline of the partial budget suggests that there are five ways to increase the profitability of the farm business. Farm profit increases as the result of the following changes:

1. Increase income and decrease costs.
2. Increase income by more than the increased costs.
3. Increase one source of income by more than the decrease in other sources, with no changes in costs.
4. Decrease costs by more than the decrease in income.
5. Decrease one item of costs by more than the decrease in other items, with no change in income.

Although these five ways to increase profitability seem obvious, the farm decision is not easy. Farmers must discover the changes that are possible and evaluate them within the partial budget framework.

will have one or more of the following effects. It will:

- eliminate or reduce some costs;
- eliminate or reduce some gross income;
- cause additional costs to be incurred; or
- cause additional gross income to be received.

Partial budgeting gives an idea whether the proposed change is better or worse in terms of profitability compared with the current situation. However, it cannot by itself indicate whether both the before and after situations are profitable. Gross margins are used to do this.

11.7 Whole-farm Budgeting

The whole-farm budget is a strategic plan for the future use of farm resources and sets the direction for the farm business. As such it must align with the farm-family's goals and objectives to be effective. The whole-farm budget serves as an essential tool for evaluating the farm business and understanding the impact of these goals and objectives. By ensuring that the budget reflects the farm-family's aspirations and priorities, farmers can create a cohesive strategy that integrates personal and business aims, thereby reducing conflicts over resource allocation and enhancing overall farm efficiency.

The first step in developing a whole-farm budget is to assess the inputs available for the farm business. This often includes evaluating the amount and quality of the land available, labour and working capital which are typically the most limiting factors amongst smallholder farmers. Additionally, other elements such as the farmer's managerial skills, and the availability of markets need to be considered. Recognizing these assets is crucial because they form the foundation upon which the entire budgeting process is built. An accurate assessment of these elements ensures that the budget is realistic and grounded in the actual capabilities of the farm.

As shown in Table 11.6, the farmer has a 3 ha cropping system with two categories of land: irrigated and rain-fed. There is 1.7 ha under rain-fed farming and 1.3 ha under irrigation. The irrigated crops include vegetables (0.5 ha) and field crops (0.8 ha). The rain-fed cropping area is 0.8 ha of maize, 0.5 ha of millet and 0.4 ha of sorghum. Resource requirements per unit of each enterprise, or technical coefficients, are shown in the table. For example, a hectare of vegetables requires 1 ha irrigated land, 4 hours labour and $25 working capital.

Once the inputs are assessed, the next phase involves integrating resources, constraints, technical information and price data into a comprehensive plan. This integration process should result in a realistic whole-farm budget that considers all aspects of the farm's operation.

The outcome should be a strategic plan that provides clear direction for the farmer and family, guiding them on how to best utilize their resources to maximize returns. By carefully planning and aligning the budget with the farm's goals, farmers can make informed decisions that promote sustainability and profitability, ensuring long-term success for the farm business.

Table 11.7 contains the estimated income, variable costs and gross margins for the five potential crop enterprises to be considered in the whole-farm budget. The whole-farm planning process is elaborated on further in Chapter 14.

Table 11.6. Crops per hectare.

Category	Quantity available	Irrigated land		Rain-fed land		
		Vegetables	*Field crops*	*Maize*	*Millet*	*Sorghum*
Irrigated land (ha)	1.3	1	1			
Rain-fed land (ha)	1.7			1	1	1
Labour (annual hours)	900	4	3	3	2	0.5
Working capital ($)		25	10	8	6	2

Table 11.7. Gross margin whole-farm budget for a 3-hectare cropping system.

Category	Irrigated land		Rain-fed land			Total
	Vegetables	*Field Crops*	*Maize*	*Millet*	*Sorghum*	
Area (ha)	0.5	0.8	0.8	0.5	0.4	3.0
Yield (kg/ha)	12,000	6,000	4,000	3,000	2,500	-
Price ($/kg)	2.00	1.50	0.80	1.00	0.70	-
Gross income ($)	12,000	7,200	3,200	1,500	700	24,600
Variable costs ($)	5000	3,000	1,200	700	500	10,400
Gross margin ($)	7000	4,200	2,000	800	200	14,200

Notes:
- Yield (kg/ha): The average yield per hectare for each crop.
- Price ($/kg): The selling price per kilogram for each crop.
- Gross income ($): Calculated as Yield X Area X Price.
- Variable costs ($): The costs that vary with the level of output (e.g. seeds, fertilizers, irrigation, labour).
- Gross margin ($): Calculated as Gross income – variable costs.

11.8 Discounted Cash Flow Budgeting

So far, changes to the farm business have been analysed by constructing a budget of the activities of the farm as it currently operates and then comparing how it would operate with a new activity or technology, once the change is fully operational. This is called 'static' budgeting. The budget tools so far have involved creating a budget for a single production period, such as a growing season or fiscal year. They focus on estimating costs and revenues based on current conditions, without accounting for changes over time. This form of static budgeting is useful for short-term planning and decision making, providing a snapshot of the expected financial performance of the farm over a specific period.

In contrast, temporal budgeting, or budgeting over time, considers multiple production periods, incorporating the timing of activities, costs and revenues. This approach is essential for long-term planning, as it accounts for factors like investment decisions, crop rotations and the impact of current actions on future outcomes. By evaluating financial performance across several periods, temporal budgeting helps in making informed decisions that align with long-term objectives.

The time effect is best captured by referring to the concept of time value of money, discussed in Chapter 9. This is used in cash flow budgeting as a tool for investment appraisal. As a reminder, the time value of money means that a dollar today does not have the same value as a year from now. Understanding the time value of money is important in appraising and making farm investment decisions.

Realizing that money is worth more to a person in the present rather than in the future may appear obvious, but quantifying the effect is more complex. Interest rates are used to compare present and future claims on investments. Different interest rates imply different time values. Lenders of capital receive interest, and borrowers pay interest because of the perceived benefits of time. For example, a lender who provides $100 today at 8% interest per year is paid back $108 in 1 year. The $8 compensates the lender for alternative investments not made, personal consumption foregone or for the risk that the money might not be repaid.

The borrower and lender agree that $100 today is worth $108 a year from now. The borrower agrees to receive $100 now and pay back $108 in 1 year; the lender gives $100 now to receive $108 in 1 year. However, the lender is not convinced that $8 (8%) is enough. It might be that 9% interest would be necessary to convince the lender. In this case, the time value of money over the 1 year period would be $9.

11.8.1 Compounding and discounting

The time dimension of money is taken into account through the use of discounting. It is simply a technique by which future benefits and costs can be reduced to a present value. This can be better understood by explaining the notion of compounding and discounting.

Example: Compounding

Consider a case where a farmer lends $1000 to a neighbour at an interest rate of 5% annually. The following year the neighbour will owe $1050, consisting of the $1000 principal and $50 in interest ($1000 × 0.05). Suppose the neighbour wants to keep the money for two years. He or she must pay 5% for the use of the money for the first year and in the second year an additional 5%. However, in the second year, instead of the 5% interest being paid on the $1000, interest will be paid on $1050. In this way the interest is compounded. Thus, when borrowing or lending money, it is crucial to know whether the loan will be repaid with simple interest (where the interest is applied only to the original amount of the loan) or with compound interest as in the example.

Example: Discounting

Now suppose we ask a different question. If a borrower promises to pay the farmer $1200 at the end of five years at an interest rate of 8%, how much is that promise worth to the farmer today? Put another way, what is the present value of $1200 five years in the future if the interest is assumed to be 8%? To answer the question the farmer would have to divide the amount due by 1.08 for each project year, as shown in Table 11.8.

Table 11.8. Discounting: an example.

Year	Amount at beginning of year ($)	One plus interest rate	Amount at end of year ($)
1	1200	1.08	1111
2	1111	1.08	1029
3	1029	1.08	953
4	953	1.08	882
5	882	1.08	817

Thus, the present value of $1200, five years in the future, is $817.

This process of finding the present value of a future value is called 'discounting'. Discounting looks backward from the future to the present. The interest rate assumed for discounting is the 'discount rate' (see Box 11.7.). In this table, the discount factor of 8% for 5 years is given as 0.681. This is the present value factor. To find the present value of $1200 received 5 years in the future, the amount due in the future is multiplied by the discount factor for the fifth year. This gives: $817 (1200 × 0.681 = 817).

Box 11.7. Calculating the discount factor.

The discount factor (DF) is calculated as follows:

$$DF = \frac{1}{(1+r)^n}$$

Where:

- r is the discount rate (in this case 8%)
- n is the number of periods (in this case 5 years)

The calculation is as follows:

$$DF = \frac{1}{(1+0.08)^5} \approx 0.681$$

Discounting techniques in the context of evaluating farm investments are applied in Chapter 13.

11.9 Sources of Budget Information

When creating budgets, it is crucial to base them on the best information available. The reliability of a budget depends on the quality of the data used. For small family farmers, this data includes the quantity, price, method and timing of the inputs. Here are some useful sources of information for preparing budgets:

- **Farm records:** keep detailed records of the farm's activities, inputs and outputs.
- **Area summary analysis:** look at summary reports and data specific to the farm's region.
- **Research production data:** use data from agricultural research relevant to the farm's crops or livestock.
- **Typical budgets:** refer to sample budgets available from agricultural extension or farming organizations.
- **Farm literature:** read books, magazines and articles related to farming practices and budgeting.
- **Information from meetings**: attend local farming meetings, workshops and seminars to gather insights.
- **Neighbours:** talk to other farmers in the area to share knowledge and experiences.

Using a combination of these sources will help ensure the data used in budget preparation is accurate and relevant. Good farmers always need to verify the reliability of the data they collect to make sure it applies to their specific situation. However, it should be pointed out that the experience from one year is just an indicator and may not guarantee the same results in the following years. Regularly updating and verifying data will help create more reliable budgets and better farming decisions.

11.10 Advantages and Disadvantages of Budgeting

Budgeting offers several advantages that are crucial for the success and sustainability of farming operations. One primary advantage is that it provides a structured approach to financial management. By creating detailed budgets farmers can gain a clear understanding of the potential costs and returns associated with different crops or livestock. This enables them to make informed decisions regarding resource allocation and investment, ultimately leading to improved profitability.

Enterprise budgets are useful tools for comparing different farm enterprises. They help farmers identify which activities are most profitable and support better planning and decision making. By comparing expected costs and returns, farmers can choose the enterprises that best match their goals and resources.

However, to be effective, these comparisons must be made carefully. It is important to ensure that the budgets being compared are based on similar conditions – such as land quality, labour availability and market access – to avoid misleading conclusions. Accurate and relevant comparisons allow farmers to optimize their strategies and improve overall farm profitability.

Budgeting, however, also has its disadvantages, primarily stemming from its reliance on estimates and projections. The accuracy of a budget depends on the precision of the assumptions made about future prices, yields and costs. This can be challenging to predict accurately due to the inherent uncertainties in agriculture, such as weather conditions and market fluctuations. This can sometimes lead to discrepancies between the budgeted figures and actual outcomes, potentially resulting in financial shortfalls if the projections were overly optimistic.

Break-even budgeting offers a specific advantage in understanding the relationship between costs and income at various production levels. This type of budgeting helps farmers determine the minimum yield or product price required to cover their costs, providing a clear target for profitability. Calculating break-even points helps farmers decide if a certain level of production is worth doing. It shows whether an enterprise can cover its costs. However, the downside is that break-even budgeting only looks at one point – where income equals costs. It does not consider what happens if prices change or costs go up, which can make the results less reliable.

Partial budgeting offers the advantage of flexibility in evaluating the financial impact of small changes within the farm. It is particularly useful for assessing the economic consequences of introducing new technologies, expanding existing enterprises, or making other minor adjustments. By focusing only on the costs and benefits directly affected by the proposed change, partial budgeting allows farmers to make incremental improvements without needing a complete overhaul of the farm plan. However, a potential disadvantage is that partial budgeting does not provide a comprehensive view of the entire farm's financial situation. It only considers the marginal impact of changes, which might overlook broader financial dynamics and interactions between different enterprises.

Whole-farm budgeting gives farmers a complete picture of their financial situation by looking at every part of their farm operation. It helps farmers see how all their activities contribute to the farm's income and expenses, making it a powerful tool for decision making. By identifying costs and returns, it allows farmers to make better choices about investments and production, ensuring their resources are used efficiently. Whole-farm budgeting also helps assess risks and explore different 'what-if' scenarios. Since the budget can be aligned with the farm's long-term goals, it's a valuable guide to stay on track for the future. However, building a whole-farm budget takes time and effort. Gathering accurate information about every aspect of the farm can be challenging, especially with the amount of detail required to make it effective. If the data is not precise, the budget might give misleading results, which could lead to poor decisions. The budget also assumes a certain level of predictability, so if conditions change suddenly – such as unexpected price drops or extreme weather – it might not be as useful. Additionally, the focus on the farm as a whole might overlook smaller decisions in individual parts of the farm, even if those smaller changes could also improve profitability.

Discounted cash flow (DCF) budgeting helps farmers assess the value of future projects or investments by calculating what future cash flows are worth in today's values. By focusing on the present value of future cash, DCF budgeting shows how much each investment is truly worth over time. It is particularly helpful for long-term projects because it considers the time value of money, allowing farmers to prioritize investments that will bring the best returns in the future. It also helps manage risk by allowing farmers to see how changes in prices, yields or interest rates might affect returns, making it easier to plan for uncertain conditions. However, DCF budgeting also has some drawbacks. One challenge is that DCF calculations also rely on predictions about future cash flows, which can be even harder to make accurately than the more static budgeting methods. If the future cash flow estimates or discount rates are imprecise, the DCF results can be misleading. DCF budgeting is also a complex method that requires a good understanding of the method and mathematics behind it. Smaller farms or those without detailed financial data may find it difficult to apply this technique. Additionally, because DCF is geared towards long-term returns, it may overlook smaller, short-term opportunities that could improve profits but are not as obvious in a DCF analysis.

Before drawing conclusions from budgets, it is essential to carefully evaluate the resource situation of the specific farmer. Different farms, even with the same basic budget information, may require different management plans due to variations in resources. For example, budget data for a 1.6 ha farm can be used to prepare a budget for a 3.2 ha farm, but differences in resources and organization must be considered and adequately accounted for to ensure the result is reliable and useful.

Budgets are generally constructed to reflect future actions and as we mentioned, it can be challenging to predict future prices and yields accurately. Historical data can provide a basis for establishing initial levels of budget yield, price and timing data. Techniques such as forward contracting can be used to estimate future prices, but production and marketing risks still limit budget reliability. While 'best estimates' are used to develop budgets for farm business analysis, high degrees of variability introduce risk and pressure on the reliability of these estimates. Farmers must consider and evaluate these risks when determining the solutions that best meet the goals and objectives of the farm family.

Preparing budgets is also a time-consuming process. It involves detailed calculations and searching for relevant data sources. Although software and spreadsheets are available to assist with budget calculations, users must ensure that the returns from budgeting justify the cost of gathering the necessary information.

11.11 Participatory Budgeting

Traditional budgeting tools often fail to meet the needs of smallholder farmers in low-income countries. The complexity of conventional budgets, the difficulty of obtaining reliable farm-level data and the exclusion of non-cash resources such as household labour make them impractical for

Box 11.8. Step-by-step process for participatory budgeting. From Dorward *et al.*, 1997.

1. Form groups of similar farmers: the community selects groups of smallholder farmers with similar resource levels and enterprise mixes, often using a wealth-ranking exercise or similar method.
2. Identify key constraints: within each group, farmers discuss and rank the biggest challenges affecting their production and profitability.
3. Explore possible solutions: farmers suggest and rank potential opportunities that address their constraints.
4. Quantify inputs and outputs: the group agrees on the estimated costs, labour requirements and expected outputs for different opportunities, using local measures and categories rather than external financial data.
5. Develop simple budgets: farmers create basic financial projections to compare potential costs and returns of different opportunities.
6. Compare and discuss results: farmers evaluate the projected outcomes, contrasting them with their own experiences to refine decision making.

Participatory budgeting ensures that financial planning is grounded in real-life farming conditions and incorporates the farmer's own knowledge and priorities. This method strengthens decision making and helps ensure that any farm plan or investment aligns with local realities and available resources.

many farmers. Additionally, these tools rarely account for seasonal variations in resource availability or the use of locally available materials.

To overcome these limitations, participatory budgeting (see Box 11.8.) has been used to offer a more accessible and farmer-driven approach. This method enables farmers and extension workers to quantify and analyse resource use over time, whether for a specific enterprise or a particular resource across the farm. Rather than relying on external financial data, participatory budgets use farmers' own rankings, scoring systems and seasonal diagrams to assess costs, returns, and resource needs. This allows for a more practical and inclusive approach to farm financial planning.

Unlike conventional budgeting, participatory budgets:

- Are simple and easy to use, even for farmers with limited literacy.
- Include non-cash resources, such as family labour and locally available materials.
- Consider seasonal variations in input availability and cash flow.
- Can be applied at both the individual and group level.

In group settings, participatory budgeting allows farmers to compare their experiences and develop an average budget for a given enterprise, ensuring the approach reflects real conditions rather than theoretical models.

Participatory budgets are also a valuable tool for both extension support and research. They can be used to:

- Assist farmers in making practical financial decisions.
- Conduct needs assessments to identify constraints in farm enterprises.
- Screen and evaluate new technologies, assessing their feasibility in smallholder settings.
- Monitor and evaluate farm trials, helping farmers and researchers analyse the resource implications of new practices.

By jointly assessing resource requirements, potential costs and expected profits, participatory budgeting helps farmers make better-informed decisions and improves the adoption of technologies that align with their needs.

11.12 Conclusion

This chapter has highlighted the central role of budgeting in effective farm management, demonstrating how different types of budgets support planning, resource allocation and decision making. Budgeting is fundamental to enhancing financial stability and sustainability, enabling farmers to align their operations with both short-term priorities and long-term goals. It serves not only as a tool for outlining expected income and expenses but also as a roadmap for improving technical and economic efficiency, optimizing productivity and managing risks across the farm. The chapter has emphasized that budgeting is not just a tool for financial planning but also a means of ensuring the long-term sustainability and profitability of the farm enterprises individually and as a whole.

The utility of enterprise budgeting was underscored, particularly in assessing the profitability of specific farm activities. Enterprise budgets provide a comprehensive breakdown of revenues, variable costs and those fixed costs that can be allocated to specific enterprises, enabling farmers to compare the overall enterprise profitability of different crops or livestock activities. Additionally, enterprise budgeting aids in the preparation for potential risks and uncertainties, facilitating the development of contingency plans that safeguard the farm's financial health.

The chapter also introduced partial budgeting as a valuable tool for evaluating the economic impact of proposed changes within the farm. By focusing on the marginal costs and benefits of specific adjustments, partial budgeting helps farmers assess whether changes, such as introducing new technologies or expanding existing enterprises, will enhance overall profitability. This approach is particularly useful for making incremental improvements without overhauling the entire farm plan.

The chapter also discussed whole-farm budgeting, a tool that enables decision makers to assess the profitability, risk and sustainability of their entire farming system, thereby supporting strategic decision making and long-term planning. By viewing the farm as an integrated system, whole-farm budgeting fosters more comprehensive and balanced management decisions.

The chapter also made a distinction between static budgeting, which provides immediate insights, and temporal budgeting, which offers a deeper understanding of the financial implications of management decisions over time. Both approaches are essential for effective farm management economics, helping to balance short-term needs with long-term goals.

Together, these budgeting techniques empower farmers to navigate the complexities of modern agriculture, enabling them to adapt to market fluctuations, optimize resource use and achieve sustainable growth. Through sound budgeting practices, farmers can strengthen their financial management skills, secure their livelihoods and contribute positively to their communities and the environment.

And, of course, agricultural advisors will also need to be skilled in these areas of budgeting. In addition to providing training to smallholder farmers, extension workers will also need to be familiar with the capacity of the farmers they support in budgeting. They will need to be prepared to walk the farmers through the various budgeting techniques and help them interpret the results to make informed decisions about their farms. It is also important to instil the habit of thinking in terms of budgets to increase the objectivity of their farm management decisions. The use of budgets and the role of the farm management extension worker in supporting small family farmers is elaborated on in Part 3 of this book.

References

Dorward, P.T., Shepherd, D.D. and Wolmer, W.L. (1997) Developing farm management type methods for participatory needs assessment. *Agricultural Systems* 55(2), 239–256. DOI: 10.1016/S0308-521X(97)00009-7. (accessed 18 March 2025).

Additional Reading

Jobes R. Budgets (2006) Their use in Farm Management, Oklahoma Cooperative Extension Service

Kay, Ronald D., Edwards, William M., and Duffy, Patricia A. (2016) *Farm Management*. 8th ed., McGraw-Hill Education.

Malcolm, Bill, Jack Makeham, and Vic Wright (2005) *The Farming Game: Agricultural Management and Marketing*. 2nd ed., Cambridge University Press.

Olson, Kent D. (2004) *Farm Management: Principles and Strategies*. Iowa State Press, Ames, Iowa.

Rae, Allan N. (1977) *Crop Management Economics*. London: Crosby Lockwood Staples.

12 Farm Management Methods: Diagnosis

Farm management methods play a crucial role in guiding decision-making processes for farmers. Effective farm management requires a systematic approach that incorporates various tools and techniques to diagnose problems, plan strategies and exercise control over farm operations. This chapter, along with Chapters 13 and 14, aims to explore the significance of farm management methods in facilitating informed decision making and optimizing farm performance. This chapter focuses on diagnosis.

Diagnosis of the farm business is essential to understand how the farm business is performing. Information needed for diagnosis is collected mostly from individual farm records. Information collected must be analysed to understand how the farm or its individual enterprises are performing economically. This analysis can help to locate weaknesses and to correct them. It can also highlight areas of satisfactory performance so they may be exploited more fully. Some simple diagnostic techniques can assist farmers to see where management of the entire farm as well as of the individual enterprises can be improved and made more efficient and effective in meeting the goals of the farm business and farming family.

12.1 Farm Management Methods or Techniques: The Decision-making Process

A management method or technique is any formal process or procedure used to generate and apply information for decision making. It helps to analyse and identify possible solutions to problems and to monitor and evaluate the progress and effectiveness of those solutions once implemented. From this definition, two key aspects of a management method or technique stand out: it provides necessary information and it aids in problem solving. Fig. 12.1 shows the elements of the decision-making process.

Farm management methods or techniques offer farmers a systematic and structured approach to assess the current state of their operations. These methods enable farmers to gain a comprehensive understanding of their resources, capabilities and constraints. By conducting a thorough diagnosis of their farm businesses, farmers can identify areas that require attention and improvement. This diagnostic process involves collecting and analysing data related to production processes, financial performance, market dynamics and resource availability and condition.

As discussed in Chapter 5, the farm management cycle (Fig. 12.2) serves as a fundamental framework for farm management, encompassing five key stages: (i) diagnosis; (ii) planning; (iii) implementing; (iv) monitoring; and (v) evaluation. Each stage is essential in guiding decision making and ensuring effective farm management. Diagnosis involves gathering and analysing data to identify challenges, opportunities and trends. Planning enables farmers to develop strategies and set goals. Implementing is putting the plans into action. Monitoring allows for tracking implementation and progress, and for making adjustments, while evaluation assesses outcomes and provides insights for improvement. By embracing the farm management cycle, farmers can foster sustainable farm management practices and optimize their operations for long-term success.

With a clear diagnosis in hand, farmers can then move on to the planning stage. Farm

DOI: 10.1079/9781836990536.0012

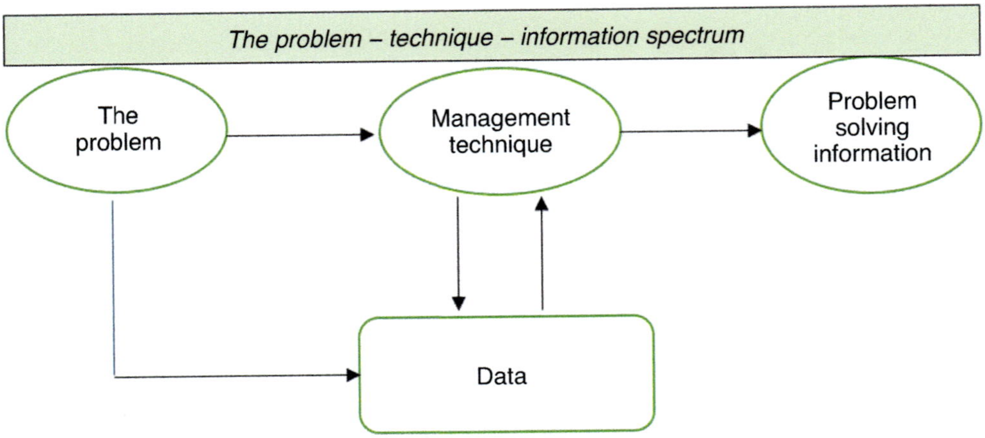

Fig. 12.1. The essential elements of the decision-making process.

management methods provide frameworks for strategic planning, allowing farmers to set realistic goals, develop action plans and allocate resources effectively. By considering factors such as market trends, input availability and production capacities, farmers can make informed decisions on enterprise choices, production methods, cropping patterns, livestock management and investment strategies. Through comprehensive planning, farmers can optimize resource allocation, minimize risks and seize opportunities in their operations.

Implementing is, perhaps, an obvious step, but it represents actions related to all the plans. Plans are put into action through concrete activities, such as planting schedules, input applications, labour deployment or investment in equipment and technology.

Monitoring is an integral part of the farm management process as it tracks the implementation process. By applying appropriate monitoring mechanisms, farmers can better control the progress of their plans and identify any deviations or bottlenecks. Regular monitoring allows farmers to assess the effectiveness of their strategies and make timely adjustments if necessary. It provides a feedback loop that helps farmers stay on track towards achieving their goals and ensures that their actions align with the desired outcomes.

Evaluation is the final step in the farm management cycle. Through evaluation, farmers assess the outcomes of their decisions and actions. By comparing actual results with the planned objectives, farmers gain valuable insights into the effectiveness of their management strategies. Evaluation provides an opportunity for reflection, learning and continuous improvement. It involves taking a longer look at what has been done and measuring it against the original expectations. How well did the plan work? To what extent were the goals achieved? It allows farmers to identify successful approaches, replicate them in future endeavours and learn from mistakes or areas that require further refinement.

Farm management methods not only enable farmers to optimize their resource allocation, enhance productivity and improve profitability but also contribute to the long-term sustainability of their operations. Sustainable farming practices are critical for preserving natural resources, minimizing environmental impact and ensuring the social and economic viability of agriculture. By integrating sustainability principles within the farm management framework, farmers can strike a balance between profitability, environmental stewardship and social responsibility.

Farm management methods are indispensable tools for farmers to navigate the challenges and complexities inherent in agricultural enterprises. These methods provide a systematic framework for diagnosing farm businesses, planning strategic interventions, monitoring progress and evaluating outcomes. By employing these methods, farmers can make informed decisions, allocate resources

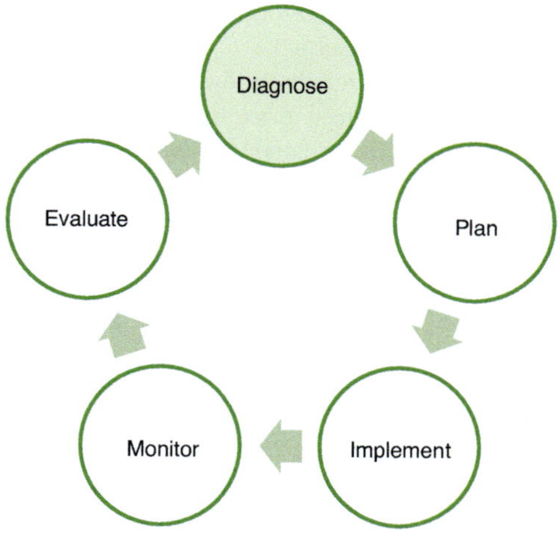

Fig. 12.2. The farm management cycle.

effectively and improve the overall sustainability and resilience of their operations. Ultimately, the utilization of farm management methods empowers farmers to navigate the ever-changing agricultural landscape and achieve long-term success.

12.2 Farm Business Diagnosis

In farm management, diagnosis plays a crucial role in assessing the current state of the farm business. It is a systematic process that involves gathering and analysing data from various sources to understand the different aspects that contribute to the farm's operations. The purpose of diagnosis is to provide farmers with a comprehensive understanding of their farm's strengths, weaknesses and areas for improvement.

One key aspect of diagnosis is the assessment of production systems. This involves examining the efficiency and effectiveness of the farm's production methods, including crop cultivation, livestock management and use of technology. By analysing the production systems, farmers can identify bottlenecks, inefficiencies, or areas where changes can be made to optimize resource allocation and improve productivity. It requires objectivity, especially a willingness to consider the effectiveness of comfortable, common, or long-standing practices and enterprise choices.

A critical factor in diagnosing farm production is the principle of *diminishing returns*. Increasing input use, such as fertilizer or labour, initially raises yields but, beyond a certain point, the additional output gained from extra inputs starts to decline. Understanding this helps farmers avoid excessive input use, which increases costs without proportional gains in output.

Financial performance is another critical component of diagnosis. It involves evaluating the farm's financial statements, including income, expenses and profitability. By analysing the financial performance, farmers can assess the overall profitability of the farm, identify areas of financial strain and make informed decisions regarding resource allocation and investment strategies. Financial performance assessment also requires understanding *marginal cost* and *marginal income*. Farmers should compare the additional revenue generated from expanding production with the additional costs incurred. If the *marginal cost* of increasing output exceeds the *marginal income*, it signals inefficiency and the need to reassess production strategies.

Market dynamics are also considered in the diagnostic process. This involves analysing market trends, demand-supply dynamics and price fluctuations. By understanding the market conditions, farmers can identify potential market opportunities, assess the competitiveness of their products and make informed decisions regarding marketing strategies and target markets.

Farmers should also consider *price elasticity of demand* when making marketing decisions. Some agricultural products, like staple grains, have *inelastic demand*, meaning that price fluctuations have minimal impact on quantity sold. However, perishable goods or high-value niche products tend to have *elastic demand*, where small price changes can significantly impact sales volume.

Resource availability is another important factor in diagnosis. This includes evaluating the availability and accessibility of key resources such as land, water, labour and inputs. Availability also includes the condition of those resources, such as the condition of the soil, the productive capacity of labour and the efficacy of inputs. By assessing resource availability and condition, farmers can identify any constraints or limitations that may impact their farm's operations and plan accordingly.

Diagnoses should also consider *opportunity cost* – the trade-off involved in using limited resources. For example, allocating land to cash crops may generate immediate income but could mean forgoing the long-term benefits of soil-enriching crops like legumes. Similarly, investing in one enterprise over another should be based on the potential returns of the next best alternative.

Thus, the purpose of diagnosis in farm management is to provide farmers with a comprehensive understanding of their farm's current situation. It helps farmers identify challenges and opportunities, enabling them to make informed decisions and develop strategies to address the identified issues. By conducting a thorough diagnosis, farmers can optimize resource allocation (including rethinking the resources themselves), improve efficiency, minimize risks and enhance overall farm performance.

Furthermore, diagnosis serves as a foundation for effective decision making. It provides farmers with accurate and relevant information, allowing them to evaluate different options, weigh trade-offs and choose the most appropriate course of action. By utilizing diagnostic methods, farmers can make strategic decisions that align with their farm's objectives and improve their chances of success. In the next sections the following diagnostic methods are discussed, after which the strengths and limitations of the methods are assessed.

- Market analysis.
- Market margins.
- Constraints and opportunities analysis.
- Gross margin analysis.
- Whole-farm income diagnosis.
- Efficiency measures.
- Trade-off analysis.
- Cash flow analysis for farm households.
- Risk analysis.

12.3 Market Analysis

Market analysis is an important method used in the diagnostic phase of farm management. It involves evaluating market dynamics, trends and demand-supply conditions to understand the opportunities and challenges within the agricultural marketplace. Market analysis techniques can include studying market size, consumer preferences, price trends, competitor analysis and market segmentation.

For example, farmers can analyse the market size for a specific product by studying data on the total demand and consumption patterns. They can also assess consumer preferences by conducting surveys or analysing consumer buying behaviour. This information helps farmers understand the potential market demand for their products and identify target customer segments.

Price trends analysis involves examining historical price data and identifying patterns and fluctuations in market prices. By understanding price trends, farmers can anticipate market movements and make informed decisions regarding the timing of production and sales.

Competitor analysis involves studying other farmers or agricultural businesses operating in the same market. Farmers can gather information about their competitors' products, pricing strategies, marketing techniques and market share. This analysis helps farmers identify their competitive advantages and develop strategies to differentiate themselves in the market.

Market segmentation involves dividing the market into distinct groups based on characteristics such as demographics, preferences or geographic location. By segmenting the market, farmers can tailor their products and marketing strategies to specific customer groups, allowing for more targeted and effective marketing efforts.

By conducting a thorough market analysis, farmers can identify potential market niches, assess the demand for specific products, and align their production and marketing strategies accordingly. This allows farmers to optimize their product offerings, target the right customers and enhance their competitive position in the market. Market analysis helps farmers make informed decisions regarding market entry, product diversification, pricing strategies and marketing approaches, ultimately improving their chances of success in the marketplace.

12.4 Market Margins

A marketing margin is the difference between the price paid by the consumer and the price received by the producer for an agricultural product. It represents the costs and profits of all marketing activities that occur as the product moves from the farm to the final consumer—such as transportation, storage, processing, packaging, wholesaling, and retailing.

Understanding marketing margins helps farmers assess how much of the final consumer price they receive, identify inefficiencies in the marketing chain, and explore ways to increase their share of the consumer price (for example, by selling directly to consumers or forming cooperatives).

By understanding marketing margins, the farmer can:

- See how much value is added after the product leaves the farm.
- Identify where profits are captured along the value chain.
- Explore options to reduce intermediaries or marketing costs—such as selling directly to local markets, joining a marketing cooperative, or improving product quality to capture a higher price.

In this way, knowledge of marketing margins helps farmers make better marketing decisions and improve their overall profitability.

12.4.1 How are marketing margins calculated?

Marketing margins are related to the prices received for produce. Costs must be related to these prices. Farmers selling their produce directly to the market are likely to get different prices at different times of the year. Farmers may also get different prices at different times of the day. In a fresh vegetable market, the price attained in the morning is usually higher than the prices attained at the close of day. Farmers need to understand how the markets they use operate, because this will affect the market margins.

As explained previously, the marketing margin is the difference between the prices the farmer receives for their produce and the costs incurred in marketing. An example is given in Table 12.1. In this case, the farmer sells their tomatoes in the nearest rural market.

For example, let us say that the farmer has a vehicle that can carry 200 kg of produce per trip. The running cost of the vehicle is $0.50/km. It is 10 km to the market.

This is the cost per kg or ton if transporting 200 kg. If transporting less than 200 kg, then the costs per kg or ton will be higher. Thus, one way to reduce marketing costs is to use transport optimally.

Table 12.2 presents a breakdown of marketing costs per kg for a product and shows the resulting market margin. Starting with a value of quantity sold at $0.81, we see a breakdown of the associated marketing costs per kg. These include market fees of $0.01, labour for packing, loading and unloading at $0.02, *en route* costs (such as roadtolls) at $0.01, transport to the wholesale market at $0.05, and packaging costs also at $0.05.

Table 12.1. Information to calculate market margin

Detail	Amount	Unit
A) Operating cost for transport	$0.50	Per km
B) Distance to market	10	km
C) Cost per trip (A x B)	$5.00	Per trip
D) Mass/weight per trip	200	kg
E) Cost per kg (C / D)	$0.025	Per kg
F) Cost per ton (E X1000)	$25.00	Per ton

Box 12.1. Example of Market Margins

Consider a tomato farmer who sells tomatoes at the farm gate for $3.00 per kg. By the time those same tomatoes reach the consumer in the city market, the retail price is $5.00 per kg.

The marketing margin is therefore:

$5.00 - $3.00 = $2.00 per kg.

This $2.00 covers the costs and profits of traders, transporters, and retailers involved in moving the tomatoes from the farm to the market.

Table 12.2. Determining the market margin.

Details		Per kg ($)
A. Value of quantity sold:		**0.81**
B. Marketing costs		
Market fees		0.01
Labour employed by farmer to pack, load, unload		0.02
Cost en route such as roadblocks		0.01
Transport to wholesale market		0.05
Packaging		0.05
	Total Marketing Costs (B)	**0.14**
C. Marketing margin (A – B)		**0.67**

The total marketing costs come to $0.14 per kg. After subtracting these costs from the value of quantity sold, the marketing margin is $0.67 per kg, indicating the net amount retained after covering all marketing expenses.

Based on this per kg breakdown, we can project the total income, marketing costs and marketing margin for larger quantities. Here is an example calculation for 1000 kg of product sold:

1. **Total value of quantity sold:** with a per kg value of $0.81, the total income for 1000 kg would be

   1000 kg × $0.81/kg = $810

2. **Total marketing costs**: at $0.14 per kg, marketing costs for 1000 kg would total

   1000 kg × $0.14 = $140

3. **Total marketing margin:** the margin per kg is $0.67, so for 1000 kg, the total marketing margin would be

   1000 kg × $0.67/kg = $670

A summary of projected totals for 1000 kg is as follows:

- Total value of quantity sold: $810
- Total marketing costs: $140
- Total marketing margin: $670

This projection shows that, for every additional 1000 kg sold, the farm could expect an additional $670 in net margin after marketing costs, assuming these costs and selling price remain constant per kg.

12.5 Constraints and Opportunities Analysis

Effective farm management requires identifying both obstacles and potential improvements. Constraints analysis helps pinpoint the specific challenges limiting farm productivity, whether they stem from resource shortages, technical inefficiencies or market conditions. By systematically assessing these issues, farmers can prioritize the most pressing constraints and develop targeted strategies to overcome them.

Once constraints are identified, opportunity analysis focuses on practical solutions. These may include improving production techniques, diversifying farm enterprises, or adopting better financial and marketing strategies. A structured approach ensures that opportunities are both feasible and beneficial to the overall farm system. By applying this methodical process, farmers enhance decision making, optimize resource use and improve long-term profitability and resilience in an ever-changing agricultural environment.

12.5.1 Constraints analysis

Understanding constraints analysis

Constraint analysis is a diagnostic method used to identify and analyse specific issues or challenges within a farm operation. It involves a systematic approach to examine the root causes of problems and their potential impact on farm performance. It is essential that both the farmer and extension worker understand the strengths and weaknesses of the farm business as a first step towards a more detailed diagnosis of the performance of individual enterprises. The results of the analysis determine the methods and solutions used to address the identified constraints.

For example, consider a farm that is experiencing a decline in crop yields. By conducting a problem analysis, the farmer can investigate various factors that may be causing the decline, such as soil fertility, pest and disease management, irrigation practices or inadequate crop nutrition. The problem analysis process involves gathering data, conducting field observations and analysing historical records to identify potential causes and their relationships.

Use of constraint analysis in diagnosis

Problem analysis is a valuable tool for diagnosing farm issues and formulating effective solutions. By utilizing problem-analysis techniques, farmers can break down complex problems into smaller components, identify cause-and-effect relationships and prioritize areas for intervention. Fig. 12.3 presents a constraints tree applied to understand what is causing low enterprise profitability.

Continuing with this example, once the farmer has identified potential causes through problem analysis, they can further investigate each factor to determine the most significant contributors. This may involve conducting soil tests to assess nutrient deficiencies, scouting for pests and diseases, evaluating irrigation systems for efficiency and reviewing crop management practices.

By systematically analysing the identified issues, farmers can develop targeted strategies to address the specific challenges. For instance, if soil fertility is found to be a significant factor affecting crop yields, the farmer can implement a nutrient management plan, including appropriate fertilization practices and soil amendments. If pest and disease management is identified as a key issue, the farmer can develop an integrated pest management programme to monitor and control pests effectively.

Constraint analysis helps farmers optimize resource utilization by directing their efforts towards the most critical areas. It enables them to develop tailored solutions that address the root causes of the problem, rather than just treating the symptoms. By addressing underlying issues, farmers can improve overall farm performance and mitigate potential risks or challenges.

A further step in analysing constraints is ranking the identified constraints. One way to do this is by constructing a constraints analysis matrix. The matrix helps the farmer analyse and prioritize the constraints using the following criteria:

- **Impact**: the level of effect the constraint has on performance.
- **Severity**: the urgency or criticality of the constraint.
- **Frequency**: how often the constraint occurs.

These can be rated on a scale of 1–5, with 1 being the lowest and 5 being the highest. This enables the farmer to consider the challenges methodically. Table 12.3 gives an example of how the farmer in our story might assess the results of their constraints tree.

Now the farmer is in a better position to identify opportunities to address these constraints. The ranking will help with deciding what to do first.

12.5.2 Opportunity analysis: Identification of opportunities

After the constraints have been identified and ranked, opportunities for improvement are identified and assessed. These should alleviate or remove existing constraints and may consist of husbandry improvements at the enterprise level and the introduction of alternative enterprises at the level of the whole farm. Improvements can relate to enterprises, the farm and the non-farm area. Once improvements are identified, the overall effects on the farm system should be appraised. In practice, many enterprises are interrelated technically as well as economically. For example, higher grain yields may increase the availability of straw as feed for livestock.

Table 12.4 expands the initial constraints analysis matrix to include opportunities, and reorders the constraints by their ranking which will help the farmer prioritize decisions.

There is no single right or wrong set of answers. Each analysis will differ depending

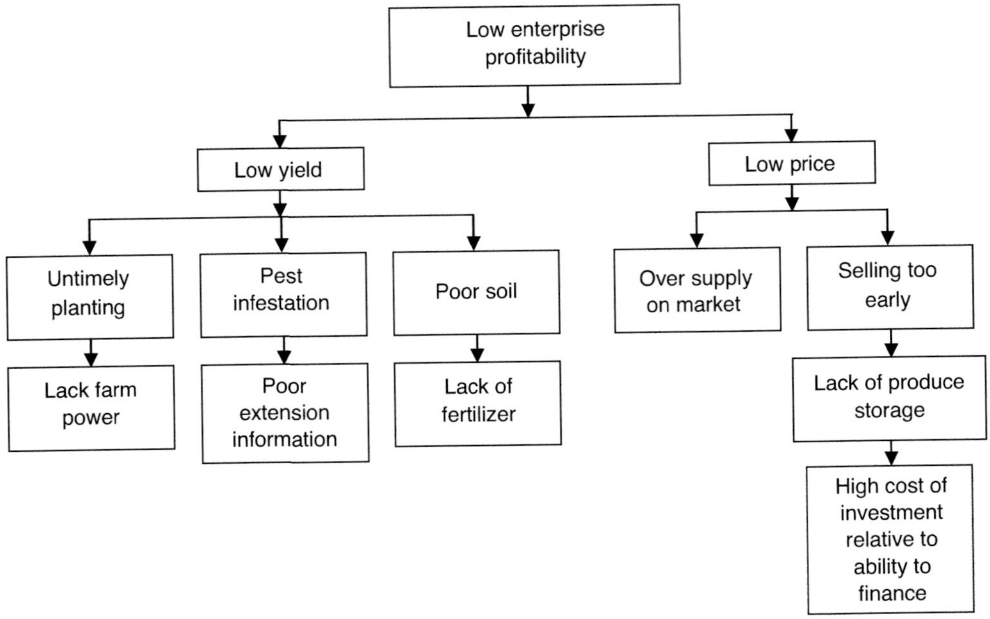

Fig. 12.3. Constraints tree.

Table 12.3. Ranking constraints.

Key constraint	Lowest level constraints	Impact (1–5)	Severity (1–5)	Frequency (1–5)	Total score	Rank (1–5)
Low yield						
Untimely planting	Lack of farm power	3	2	5	10	4
Pest infestation	Poor extension information	4	2	5	11	3
Poor soil	Lack of fertilizer	5	4	3	12	2
Low price						
Low price	Oversupply on market	5	5	4	14	1
Selling too early and lack of product storage	High cost of investment	5	4	5	14	1

on the resource base of the farm, the farmer's attitude towards risk, the reliability of information, etc. What is most important is that farmers begin to apply a systematic and methodical process of identifying constraints and opportunities. This will help the farmer improve his farm management skills, and will contribute to the sustainability and profitability of their farming activities.

When considering the opportunities for improvement, the farmer needs to follow the same systematic approach discussed previously. They need to investigate the feasibility of each identified opportunity. There are many questions to consider (see Box 12.2.), some of which require qualitative answers, and others which are answered using one or more of the budgets discussed in the Chapter 11.

Table 12.4. Constraints and opportunities analysis matrix.

Rank	Key constraint	Lowest level constraints	Opportunities	Changes to be made to current practice	Resources needed	Who is responsible?
1	Low price	Oversupply on market	Introduce a higher value crop; soybeans	Plant half the maize lands under soybeans	Extension information on how to grow soybeans	Farmer with support from extension service
1	Low price Selling too early and lack of product storage	High cost of investment relative to ability to finance	Investigate alternative options; seek loan	Train labour to pack and store for the market	Training support from extension service	Farmer with support from extension service
2	Low yield Poor soil	Lack of fertilizer	Buy fertilizer	Apply fertilizer on 50% of crops	Guidelines on using fertilizers	Farmer with support from extension service
3	Pest infestation	Poor extension information	Learn about integrated pest management at FFS	Apply ISP on fields	Training	Farmer with support from extension service
4	Untimely planting	Lack of farm power	Negotiate with brother-in-law to use tractor	Use tractor on at least 50% of the farm		Farmer

12.6 Gross Margin Analysis

12.6.1 Gross margin analysis and its application

Gross margin, which was discussed briefly in Chapter 11, is examined further in this chapter. As we learned previously, gross margin analysis is a diagnostic method used to assess the profitability of specific enterprises within a farm business. It involves calculating the difference between the total income generated by a particular enterprise and the variable costs associated with its production. By subtracting the variable costs from the income, farmers can determine the gross margin, which represents the financial return attributable to that specific enterprise.

Gross margin = gross income - variable costs

A farmer who grows crops worth \$60 at a variable cost of \$10 generates a gross margin of \$50 (\$60– \$10). The gross margin is a measure of what the enterprise is adding to (or taking away from) farm profits. As we know, variable costs rise and fall as an enterprise expands and contracts. The fixed costs of the farm are not affected by these changes. It is only the variable costs and value of production that change. If the extra variable cost is less than the value of extra production, the farmer will increase profits by growing that crop. Farm profits will be increased by the value of the gross margin for the enterprise.

Gross margin analysis enables farmers to compare the profitability of different enterprises and assess the viability of new technologies against existing practices. This analysis is particularly useful in evaluating the financial benefits of adopting improved methods by comparing the 'without new technology' and 'with new technology' scenarios. To ensure an accurate assessment, all variable costs, including non-purchased inputs such as family labour, manure and compost, should be accounted for. If the gross margin increases significantly with the introduction of new technologies, farmers will have a strong incentive for adoption. However, increasing productivity through technology must

Box 12.2. Questions to investigate the feasibility of an opportunity for improvement.

Intensification

- Which enterprises can be made more productive through improved production techniques?
- Which different production techniques are available but not yet considered?
- What will be the changes in costs of production per unit?
- What will be the changes in total production per unit?
- What is the effect of this change on the farm's cash flow?
- Is there adequate labour available to make the change?
- Are the necessary inputs and materials available?
- Are the suppliers of key inputs able to provide more inputs of the same quality on time?
- Will the extra production be sold at the same price?
- Will the extra production be available for sale or will the farm household use the extra production for home consumption?
- If the extra production is used for consumption, is the extra production substituting any other consumption and releasing cash?
- Is it possible for the farm household reduce the production of this other crop (releasing land for alternative production or fallow)?
- What will be the input requirements and output under conditions of low rainfall and other risks?

Diversification

- What market opportunities exist?
- Are there opportunities for increasing sales to the current market?
- Are there opportunities for selling produce in other markets in the vicinity?
- Are there opportunities to market produce further afield?
- Are there opportunities to diversify production and introduce new enterprises?
- What are the expected costs and incomes from doing so?
- What is the effect of this change on the farm's cash flow?
- Is there adequate labour available to make the change?
- Are the necessary inputs and materials available?

A small change may affect the whole farm household system, posing many questions to be answered by the farmer.

Alternatives

- What are the alternatives if the farm business does not have sufficient cash in the proposed period?
- Could the supplier deliver fertilizer on credit?
- If not, are there other institutions which could provide credit?
- If yes, what would be the extra cost?
- When must the debt be repaid?

This shows that each proposed change in the farm system must be evaluated on all its possible effects on the farm household system in the long run. It also shows the interrelationship between the farm and supporting systems.

also consider technical and economic efficiency. Technical efficiency focuses on maximizing output from available inputs, while economic efficiency ensures that these inputs generate the highest financial returns. Farmers should strike a balance between these factors to ensure that technological adoption enhances profitability without leading to excessive input costs.

The gross margins for each enterprise are combined to give the gross margin for the whole farm. Fixed costs are then deducted to arrive at whole-farm profit.

The key points are as follows:

- Gross margin measures only the relative profitability of the enterprise and not the actual profit. It will guide farmers on which enterprise to engage in or which technology to use given the resources available. Classification of costs into variable and fixed based on the cost and returns analysis is necessary.

- For crops, gross margin analysis is usually done on a per ha basis, if land is the most limiting resource. For livestock production the analysis is usually done on a per head basis.
- Gross margin analysis for crops is usually done at the end of the cropping season or year, after harvest of the crop as in the case of rice (two to three crops in the case of annual crops such as sugarcane and others).
- For perennial crops where harvests occur two to three times a year, the whole year harvest should be considered in the analysis for a given crop year since both yields and prices vary during the year.
- Inputs and output must refer to the same year being considered for analysis.
- In many cases, another resource such as family labour might be of greater concern. For example, some farmers might have extra land but few family members to do all the required work. In this case, gross margins can be calculated leaving labour out of the calculation until the end.

Table 12.5 sets out the example of a gross margin for 1 ha of maize in Guatemala. Using this information, the break-even information can be calculated (see Box 12.3.).

12.6.2 Gross margin for livestock enterprises and permanent (perennial) crops

Farming activities for perennial crops (e.g. fruit trees) and livestock enterprises extend over more than a single year. In these cases, gross income is defined as the difference between the closing valuation of produce stored, plus sales (including marketable produce and by-products consumed on the farm) and the opening valuation of produce stored plus purchases.

A gross income calculation for a livestock enterprise could be set up as shown in the example format shown in Table 12.6 (see Box 12.4.)

The gross income calculated for perennial crops would use the same calculation method.

Table 12.5. Gross margin for 1 ha of maize, Guatemala.

Item	Unit	Quantity	Unit price (quetzals)	Amount	Total (quetzals)
Income:					
- Maize yield	bag (90 kg)	20	1,000	20,000	
- Stover	load	1	400	400	
(A) Total Income					**20,400**
Variable costs					
- Seed	kg	10	70	700	
- Fertilizer	50 kg bag	1	1,000	1,000	
- Manure	tons	4	1,000	4,000	
- Pesticide	kg	5	100	500	
Labour					
- Land preparation	person days	18	50	900	
- Planting/manuring	person days	10	50	500	
- Weeding	person days	16	50	500	
- Harvesting/ threshing	person days	8	50	400	
(B) Total Variable Cost					8,500
(C) Gross Margin (A-B)					**11,900**

Box 12.3. Procedure for gross margin analysis.

The procedure for gross margin analysis is:

1. Calculate an average per farm yield or average amount of product for each activity.
2. Calculate average inputs per ha for each activity.
3. Calculate the gross income (return), which is the amount of produce times the farm-level price.
4. Calculate variable costs using market prices or imputed values for household resources, such as labour.
5. Calculate the gross margin (also called 'net return') by subtracting variable costs from the gross income (return).
6. Compare the gross margins among activities, to determine which is most profitable.

Table 12.6. Gross income for livestock enterprise.

	Item	$	
	Closing valuation (at end of the year)		
Less	Opening valuation (at beginning of the year)		
Equals	Increase/decrease in value of stock (inventory change)		(A)
	Income from sales (livestock)		
Plus	Income from sales (by-products)		
	Value of products used for home consumption		
Equals	Value of sales and consumption		
Less	Purchases of animals (during the year)		
Equals	Net sales		(B)
A+B	Gross income		

Changes in value of tree crops and the value of produce stored or consumed on the farm would be part of the gross income calculation.

Since it is possible to produce more than a single short-term crop from the same land area within a year, a distinction needs to be made between gross income for a particular season and gross income for a particular year. The gross income of a crop for the year may be the sum of the gross income for two or more crops grown during the year.

12.6.3 Gross margin analysis in decision making

Gross margin analysis plays a crucial role in decision making by providing farmers with valuable insights into the financial viability of different enterprises. By comparing the gross margins of various crops or livestock enterprises, farmers can identify those that are more profitable and consider expanding or optimizing their production. It is a key instrument for the farmer when deciding what to produce, how to produce, how much to produce and where to sell.

For instance, if the gross margin analysis reveals that crop A has a higher gross margin compared to crop B, the farmer may decide to allocate more resources, such as land and labour, to cultivate crop A. This decision is based on the understanding that crop A generates higher returns and contributes more significantly to the overall profitability of the farm.

Conversely, if certain enterprises show low or negative gross margins, farmers can assess the

Box 12.4. Gross margins for animal production.

In animal-based production, the unit used for calculating gross margin typically depends on the specific type of animal and production system. Here are common units used in various types of animal-based enterprises:

1. **Per animal (individual)**: for livestock such as cattle, sheep, goats or pigs, the gross margin is often calculated on a 'per animal' basis. This is useful for enterprises where each animal represents a significant investment and where individual animal productivity is a key driver of profitability.
2. **Per production cycle or batch:** in operations where animals are raised in batches, such as broiler chickens, meat rabbits or certain types of aquaculture, the gross margin may be calculated on a 'per production cycle or batch' basis. This unit is especially relevant when the entire group of animals is managed as a single unit over a defined period.
3. **Per weight of product:** for enterprises where production is largely measured in output weight (e.g. beef, pork or fish), gross margin is sometimes calculated 'per kilogram' or 'per ton of product sold'. This can provide insights into efficiency related to weight gain and feed conversion.
4. **Per hectare (in grazing systems)**: in grazing-based operations, such as beef or dairy cattle, sheep and goats, gross margin might be calculated on a 'per hectare' basis, particularly if land productivity is a primary concern and the operation's profitability is closely tied to land use.
5. **Per litre (for dairy)**: in dairy operations, gross margin is often calculated on a 'per litre' of milk produced. This unit is particularly relevant in evaluating efficiency and profitability related to milk yield per animal and per unit of feed or other inputs.

Choosing the right unit

The choice of unit depends on the primary drivers of income and cost in the specific production system, for example:

- **Per animal** is suitable for high-value livestock where individual animal health and productivity are key.
- **Per kilogram or per litre** works well in systems focused on weight or volume production, such as meat or milk.
- **Per hectare** is relevant in grazing-based systems where land use is essential to profitability.

Selecting the appropriate unit helps provide a clearer picture of profitability, efficiency and areas where improvements can enhance overall gross margins in animal-based production.

reasons behind it and explore alternative strategies or adjustments to improve profitability. For example, if an enterprise has a negative gross margin, indicating that the variable costs exceed the income, the farmer may need to reevaluate production practices, explore cost-saving measures, explore more lucrative markets or consider alternative crops or enterprises that offer better financial prospects.

Furthermore, gross margin analysis helps farmers assess the cost-effectiveness of production practices. By analysing the variable costs associated with each enterprise, farmers can identify areas where costs can be reduced or optimized without compromising profitability. This information enables farmers to make informed decisions about resource allocation, input procurement and production techniques. By optimizing variable costs, farmers can improve their overall profitability and financial sustainability.

Gross margins can also account for different marketing options. Different markets offer different market prices but also attract different marketing costs and risks. Gross margins can help farmers decide which is the most profitable market or combination of markets, including taking into account potential risks (see Box 12.5.).

12.7 Whole-Farm Income Diagnosis

This section discusses the concept of whole-farm income through a more detailed understanding of gross margin analysis. It explains the procedures for estimating annual whole-farm

Box 12.5. Pitfalls of gross margin analysis.

1. **Ignores fixed costs:** gross margin analysis only accounts for variable costs and incomes, leaving out fixed costs that relate, for example, to machinery depreciation, land or permanent labour. This can give an incomplete picture of the farm's overall profitability – especially on larger farms with higher fixed costs.
2. **Assumes constant prices:** it typically uses fixed prices for inputs and outputs, ignoring the fluctuations in market prices that can affect profitability.
3. **Does not consider risk:** gross margin analysis tends to ignore the risks and uncertainties associated with farming, such as weather variability, pests or diseases, which can significantly impact yields and profits.
4. **Short-term focus:** it often focuses on a single crop cycle or production period, ignoring long-term sustainability, soil health and changes in the market or technology that can influence the farm's future performance.
5. **Assumes fixed yields:** gross margin analysis usually assumes consistent yields, even though actual yields can vary due to environmental or management factors.
6. **Limited relevance for diversified systems:** for farms that operate with multiple enterprises or have complex, diversified production systems, gross margin analysis may oversimplify the interactions and dependencies between various components of the farm.

income. This discussion will help build capacity to analyse the farm business. While enterprise profitability analysis is limited to a particular enterprise, which is one part of the whole-farm business in a diversified farming system, whole-farm income analysis gives the overall picture of the farm performance. It shows the contribution of individual enterprises to the total income and net income of the farm business. Whole-farm net income is the year-by-year profitability of the farm as a whole.

There are two ways of calculating whole-farm net income, either by using gross margins or conducting enterprise profitability calculations. Whole-farm income is calculated by combining the gross margin of each of the farm enterprises and deducting fixed costs. Alternatively, it could be calculated by estimating the enterprise profit for each of the farm enterprises and aggregating to the level of the farm (Fig. 12.4.).

The final net income figure reflects the profit of the farm and is the reward for the capital and management contributed by the farm family during the year. Whole-farm net income is necessary to cover the family's living expenses and to pay taxes. The amount left over after accounting for living expenses and taxes can be reinvested into the farm activities.

Whole-farm net income is not the same as the actual cash earnings a farmer has on hand. To accurately determine cash earnings, it is important to adjust for non-cash costs, such as family labour, and depreciation. By separating these from whole-farm net income, we get a clearer picture of the true cash available to the farmer.

The example in Table 12.7 shows how to arrive at the whole-farm net income or profit; a farmer with 3.5 ha of land has gross margins from three economic activities, as shown in the table.

The whole-farm gross margin is $585. To calculate the whole-farm profit, the farmer needs to deduct the fixed costs for the whole farm, which in this example is $200 (see Table 12.8).

Since the fixed costs do not alter with changes in production, if the farmer can increase the gross margins on the farm, profits will automatically increase. Since most small-scale farmers have very few fixed costs, the gross margin is a useful indicator of overall farm profit. Planning can be carried out in terms of gross margins.

Table 12.9 gives an example calculation of the whole-farm gross margin for a farm with livestock and crops.

12.7.1 Farm diagnosis with the aid of gross margin

The section discusses the use of gross margins in diagnosing farm performance to provide greater insight into the use of gross margins.

Gross margin = gross income – variable costs

Whole-farm net income = total gross margin for all enterprises – fixed costs

Fig. 12.4. Gross margin and whole-farm net income calculations. Figure is Author's own.

Table 12.7. Comparative gross margins.

Economic activity	ha	Gross margin/ha ($)	Gross margin ($)
Millet	1.5	150	225
Maize	0.5	120	60
Groundnuts	1.5	200	300
Totals	3.5		585

Table 12.8. Calculating farm profit.

Item	$
Whole-farm gross margin	585
Fixed costs	200
Farm profit	385

As mentioned previously, gross margin is a key performance measure widely used to assess the performance of a farm business. As we know, many factors affect the output and the variable costs of an enterprise. To assess these factors, it is necessary to have a detailed breakdown of physical and financial information relating to the enterprise. At this stage, assumptions will be made, strengths and weaknesses will be identified and steps taken to improve performance.

The intention is to use the gross margin to identify weaknesses in farm organization. If farm income is too low, the faults that are likely to be found can be classified as follows:

- The gross margin per ha or per unit of output may be too low, due to low yields, expensive stock or excessive variable costs such as animal feed.
- Farming is not intensive enough, i.e. there are not enough high-value crops or livestock.
- Fixed costs may be too high (fixed labour costs, machinery and power, rent or other overheads).

The example in Table 12.10 shows how gross margin analysis can be applied to diagnose the farm. The example shows three farm situations reflecting low productivity, low intensity and high fixed costs.

When analysing the three situations, a number of strategies could be formulated for improvement. Some suggestions are given in Table 12.11.

A low total gross margin on a particular farm may be due to: a low ratio of land under high value crops, insufficient mixed cropping, too much fallow land, low yields, low prices or high variable costs. After the comparative analysis is conducted, it is useful to analyse the causes of the problem further, as very often the problems of a farm may have more than a single cause in which case a combination of solutions may be needed.

Further, there should be a realistic relationship between the gross margin and fixed costs. High fixed costs associated with labour, machinery, renting land, etc. should be matched by intensive farming (a high gross margin). As previously noted, the fixed costs of small-scale farms are usually low. Therefore, there is little opportunity to increase profits by reducing

Table 12.9. Whole-farm gross margin.

Economic activity	Unit	Gross margin/unit ($)	Gross margin ($)
Millet	1.5 ha	150	225
Maize	0.5 ha	120	60
Groundnuts	1.5 ha	200	300
Goats	2 LSU*	25	50
Totals	3.5 ha 2 goats		635

*LSU: Livestock units (in this case 2 goats).

fixed costs. Farmers would be more successful in increasing profits by focusing on ways to increase gross margins. That can be achieved only by increasing gross income (through improving productivity, producing higher value enterprises, selling to more profitable markets, etc.) or through reducing production costs (primarily by finding ways to reduce the cost of inputs). An analysis of the causes of low profits, as discussed in the section on constraints and opportunity analysis, will help the farmer identify what decisions to make to increase gross margins.

12.8 Measuring Farm Size and Volume of Business

Small farm income is often linked to the size of the farm business, but this does not necessarily reflect how efficiently the farm is managed. The size and volume of a farm business can be assessed using the indicators described in this section.

12.8.1 Total area of the farm/number of livestock

Farm size can be measured by:

- total land area or area under crops; and
- number of livestock (for livestock enterprises).

However, land size alone can be misleading. Land must be standardized to account for differences such as fertility, irrigation and quality. This can be done using measures like rental value or land value.

The area under major crops, especially standardized across farms, provides a more refined measure. Additionally, the concept of total cropped area is important. For instance, a farmer who grows two crops annually on 4 ha has 8 cropped ha, while one who grows just one crop has only 4.

This links to cropping intensity, defined as:

$$\text{Cropping intensity} = (\text{total cropped area/farm size}) \times 100$$

A 200% cropping intensity means two crops are grown on the same land in 1 year. However, comparing cropping intensity across farms can be misleading if the crop types differ in duration – e.g. short-duration vegetables vs. long-duration crops, like sugarcane or wheat.

In livestock farming, the number of animals can indicate business size, but differences in breed quality also affect output and profitability.

On mixed farms (both crops and livestock), using just one measure does not reflect the full scale of operations. A combination of both may be used, but comparisons across such farms are limited.

12.8.2 Total capital managed

This refers to the total value of all assets, including land, machinery and inventories. Capital can be calculated using opening or closing values. While useful when farms have similar outputs, capital comparisons are less meaningful when farms specialize in different enterprises.

Moreover, farms may be *over-* or *under-capitalized*, making it hard to determine if their capital levels are near the optimum. Capital comparisons are only helpful if the farms being compared are operating at similar efficiency levels.

Table 12.10. Applying gross margins to farm diagnosis.

Enterprise	Normal	Low gross margin (1)	Low intensity (2)	High fixed costs (3)
Rice	2.5 ha at $220/ha = $550	2.5 ha $110/ha = $275	3.0 ha at $180/ha = $540	2.5 ha at $220/ha = $550
Coffee	0.8 ha at $350/ha = $280	0.8 ha at $300/ha = $240	0.8 ha at $300/ha = $240	0.8 ha at $350/ha = $280
Beans	0.5 ha at $170/ha = $85	0.5 ha at $70/ha = $35	0.5 ha at $150/ha = $75	0.5 ha at $170/ha = $85
Maize	1.2 ha at 40/ha = $48	1.2 ha at $20 = $24	0.7 ha at $30/ha = $21	1.2 ha at 40/ha = $48
Total gross margin	5.0 ha = $963	5.0 ha = $574	5.0 ha = $876	5.0 ha = $963
Fixed costs	5.0 ha at $70/ha = $350	5.0 ha at $70/ha = $350	5.0 ha at $70/ha = $350	5.0 ha at $90/ha = $450
Profit	$613	$224	$526	$513

Table 12.11. Opportunities to address identified problems.

Problem	How to improve the present system
1. Low gross margin	Improve crop yields (by improving fertility and resolving drainage problems, diseases, etc.) Improve post-harvest operations (poor harvest, poor handling, etc.) Improve marketing conditions, improve inputs purchase system (by creating associations or farmer groups)
2. Low intensity	Change to a more intensive crop system (e.g. yams and hot peppers) Improved cultivation techniques (e.g. fertilization, irrigation) Grow more high-value crops (e.g. yams, hot peppers)
3. High fixed costs	Minimize buildings and fencing Streamline layout of fields to economize on labour and machinery Specialize to ensure full use of expensive equipment and/or buildings Check on other overhead expenses

12.8.3 Net worth

Net worth is calculated as:

Net worth = total assets - total liabilities

This indicator reflects the farm's financial position at a specific point in time. Comparing net worth across years shows financial progress, but comparisons across farms should be cautious. A higher net worth may simply mean one farmer owns more assets or has less debt; it doesn't necessarily mean they are more efficient or managing a larger or better farm business.

These indicators – land area, livestock numbers, capital managed and net worth – are useful for estimating the size of a farm business, but comparisons should be made only between similar types of farms and farming conditions and interpreted carefully, especially over time.

12.9 Efficiency Measures

This section describes some of the efficiency measures that can be used to analyse the causes of low enterprise profitability and productivity.

Farm and enterprise profitability is analysed by comparing gross income and costs. As noted in Chapter 9, profitability is a measure of how efficient the farm business is in using the resources available to generate income. If a problem of low profitability can be traced to a low level of production, the reasons given might be the unavailability of farm resources or, alternatively, the inefficient use of those resources available to the farmer. If production levels are satisfactory, low profitability may be traced to low prices attained for produce sold, or alternatively excessive costs for purchased inputs used. If inadequate resources are employed, the farmer would do well to extend the area of land under cultivation or alternatively, increase labour supply, expand livestock or obtain more capital. If the resource base cannot be increased, then fixed costs should be carefully evaluated. Steps should be taken to reduce those costs that do not significantly affect production. This calls for greater efficiency in production, better product prices and lower costs.

Efficiency can be measured in two ways: technical and economic efficiency. On a farm, balancing technical efficiency and economic efficiency is essential for profitability and sustainability.

To recap, *technical efficiency* means using resources – such as seeds, water, labour and equipment – in a way that gets the highest possible output. For example, it aims to produce the most milk from each cow or the highest yield per hectare. Technical efficiency is all about maximizing productivity with the resources available (Fig. 12.5.).

Economic efficiency, on the other hand, focuses on maximizing profit by balancing the cost of inputs with the income generated from outputs. It means not just producing a lot but doing so in a way that ensures the highest financial return. Sometimes, producing at maximum output can reduce profit if the extra cost of resources outweighs the income from the extra production (Fig. 12.6.).

$$\text{Technical efficiency} = \frac{\text{production (output)}}{\text{resources used}}$$

Fig. 12.5. Technical efficiency.

$$\text{Economic efficiency} = \frac{\text{total income}}{\text{total costs}}$$

Fig. 12.6. Economic efficiency.

So, farmers need to find a balance: pushing for high productivity (technical efficiency) while carefully managing input costs to ensure strong profits (economic efficiency). This balance helps them make the best use of their resources while achieving the highest possible return on investment.

Economic efficiency can be improved by attaining better prices for products sold and reducing the cost of inputs. This might be done by seeking better market outlets, searching for low-cost suppliers and substituting farm-produced inputs for purchased inputs. Excessive costs for inputs and materials can result in low economic efficiency (e.g. manure instead of fertilizer).

Technical efficiency can be improved in several ways: by adopting better production practices, using more productive inputs and by improving the productivity of labour. If enterprise profitability is low and market prices are good and technical efficiency cannot be improved, it is possible that the wrong enterprises are being produced. Alternatively, if technical efficiency measures are found to be satisfactory, another explanation for low profitability might be the receipt of below premium product prices. Further investigation might discover that this could be the result of low-price cycles, poor marketing practices, or inferior product quality. Alternative marketing outlets and marketing tools should then be considered.

The following measures are often useful in assessing the performance of farm enterprises:

- The level of yields and prices achieved.
- The quantities of variable inputs used.
- Total fixed costs.
- Various physical and financial performance and efficiency measures identified as relevant.

By using these indicators, and in particular measures of efficiency, farmers would be in a better position to assess their farms.

Low profitability can often be traced to poor efficiency in one or more areas of the farm business. Both technical and economic efficiency measures should be examined.

Measuring technical efficiency involves evaluating how well a farm uses its resources to produce output (see Fig. 12.7). Here are some common ways to measure technical efficiency:

1. **Yield per unit of land**: this is one of the simplest and most common ways to measure technical efficiency, particularly for crop farming. It measures the output (e.g. kilograms, tons, litres) produced per unit of land. The most common measures for specific crops are: cavans per hectare for rice and maize; kilograms or tons per hectare for rice, maize, copra and abaca; and piculs per hectare for sugarcane.
2. **Yield per tree:** tree crops are usually assessed in terms of average production per tree. For example, production of coconut is measured in terms of nuts or gallons of coconut toddy per tree.
3. **Livestock enterprises:** technical efficiency is usually expressed in physical terms such as pounds or gallons of milk per dairy cow, egg production per layer, kilograms per broiler and pigs per litter.
4. **Output per unit of input:** this measures the amount of output produced for each unit of a specific input (e.g. labour hours, litres of water or kilograms of fertilizer).

These measures help farmers assess how efficiently they are using resources to maximize output and where improvements can be made.

As noted above, the measures of economic efficiency differ from physical measures primarily because they are expressed either in monetary values or as a rate or percentage relating to capital use. In the next sections, some examples are given of indicators used to measure economic efficiency.

$$\text{Output per labour hour} = \frac{\text{total output}}{\text{total labour hours}}$$

$$\text{Output per litre of water} = \frac{\text{total output}}{\text{total water used}}$$

Fig. 12.7. Output per unit of input.

$$\text{Feed cost per kg of production} = \frac{\text{total feed cost}}{\text{total production}}$$

Fig. 12.8. Feed cost per kg of production.

12.9.1 Feed cost per unit of production

This measure indicates the cost of feed required to produce one unit of output in a livestock enterprise. It is determined by dividing the total feed cost for the enterprise by the total production output (Fig. 12.8).

The total production output (e.g. kilograms of meat or litres of milk) should account for:

- the quantity sold and consumed;
- adjustments for inventory changes; and
- the exclusion of any purchased livestock or milk not produced on the farm.

The result is expressed as a monetary value per unit of production (e.g. \$/kg of meat or \$/l of milk). This measure is influenced by both feed prices and livestock feed efficiency. Therefore, it should only be compared within the same type of livestock enterprise, as different species have varying feed conversion rates.

12.9.2 Value of crop per cultivated area

This value measures the income generated per unit of land cultivated (Fig. 12.9). It is calculated by dividing the total gross income of all crops produced during the season or year by the area under cultivation. The result will be a monetary value per unit of land (e.g. \$/ha).

This metric provides insight into the revenue productivity of cultivated land, but it does not

$$\text{Value of crop per cultivated area} = \frac{\text{total gross income of crops produced}}{\text{total area under cultivation}}$$

Fig. 12.9. Value of crop per cultivated area.

$$\text{Rate of capital turnover} = \frac{\text{total gross income}}{\text{total capital used}}$$

Fig. 12.10. Rate of capital turnover.

account for production costs. Therefore, it should not be interpreted as a measure of profit or financial efficiency. Instead, it is best used to compare the income potential of different crops or farming systems within similar conditions.

12.9.3 Rate of capital turnover

This rate of capital turnover indicates how efficiently capital is being used in production. It is determined by dividing the total gross income of farm production by the total capital used in the farm business (Fig. 12.10).

In this context, total capital includes fixed capital, working capital, livestock as productive capital, and cash and operating reserves. It does not include the value of land or personal assets used for the farm business.

The result is a ratio – a number without units – that tells us how many times the total capital is 'turned over' (i.e. how many times its value is earned back) in a given period. For example, a rate of capital turnover of 0.3 – or 30% if expressed as a percentage – means that the farm's total gross income is 30% of the total capital invested in the farm business in that period. A higher rate of capital turnover indicates that farm income is high relative to the capital invested, allowing capital to be recovered more quickly.

The rate of capital turnover varies by farm type. Dairy, pig and poultry farms typically have higher rates because of faster production cycles and higher revenue per unit of capital. Beef farms tend to have lower rates due to longer production cycles and high capital requirements. Predominantly crop-based farms usually have intermediate values.

Since farm enterprises differ in capital structure and revenue generation, the rate of capital turnover should only be compared between farms of the same general type.

Many other efficiency factors have been used, and others could be derived to suit specific needs. It is of little value comparing only one ratio and taking action to improve it, as this will affect other ratios. The only true criterion is overall efficiency for which no single, universally reliable ratio so far exists.

12.10 Trade-off Analysis

Trade-off analysis is a diagnostic method used to evaluate the costs and benefits of different options or decisions in farm management. It involves examining the trade-offs between various factors such as production costs, yield potential, resource availability and environmental impact. Trade-off analysis helps farmers understand the consequences and implications of choosing one option over another and supports decision-making processes that consider multiple objectives. Trade off analysis is closely related to the concept of opportunity cost discussed in Chapter 9.

For example, consider a farmer who must decide between two different crop varieties: variety A and variety B. The farmer can conduct a trade-off analysis by considering factors such as yield potential, input requirements, disease resistance and market demand. If variety A has higher yield potential but requires more inputs and has lower disease resistance compared to variety B, the farmer needs to assess the trade-offs between higher yield potential and the associated costs and risks. By understanding these trade-offs, the farmer can make an informed decision based on their specific goals, resource availability and risk tolerance.

Trade-off analysis is commonly used in agricultural decision making across various aspects of farm management. Here are a few examples:

- **Crop selection:** farmers often face the trade-off between different crop options when planning their planting strategy. They need to consider factors such as market demand, input requirements, disease resistance and environmental sustainability. For instance, a farmer may compare the trade-offs between growing high-value specialty crops that require intensive management and have higher profit potential vs staple crops with lower profitability but higher resilience to environmental stress. Trade off analysis is closely related to the concept of opportunity cost discussed in Chapter 9. When weighing trade-offs between enterprises, the value of the next-best alternative forgone – the opportunity cost – should be taken into account. The opportunity cost of choosing one crop over another is the potential profit from the crop selected over the crop that is foregone. Understanding opportunity costs helps farmers make informed decisions by comparing the benefits of different resource uses in their context.
- **Resource allocation:** trade-off analysis is crucial when allocating limited resources such as land, labour and capital. Farmers need to weigh the trade-offs between different enterprises or activities in terms of profitability, risk and resource requirements. For example, a farmer may evaluate the trade-offs between expanding an existing enterprise that has proven profitability and diversifying into a new enterprise with higher market potential but higher initial investment and associated risks.
- **Sustainable practices:** trade-off analysis is essential when making decisions that balance agricultural productivity with environmental and social sustainability. Farmers may assess the trade-offs between intensive farming practices that maximize short-term productivity but may deplete soil health or cause environmental degradation, and sustainable farming practices that prioritize long-term ecological stewardship. By considering these trade-offs, farmers can adopt practices that promote both profitability and environmental sustainability.

Trade-off analysis allows farmers to make informed decisions by considering the costs and benefits of different options and weighing the trade-offs associated with each choice. By understanding these trade-offs, farmers can align their decisions with their goals, optimize resource utilization and achieve a balance between multiple objectives for long-term farm success.

12.10.1 Procedure

The first step in carrying out a trade-off analysis consists in gathering information about the requirements of a new activity in terms of natural, human, physical, financial and social capital. This information can be gathered using tools such as the market assessment, constraints and opportunities analysis, and budgeting tools. Table 12.12 gives an example of a trade-off analysis framework.

The resources required in each category are listed in the second column of the worksheet shown in Table 12.13. When this information is complete, farmers and household members can begin to answer the questions shown in column 3, which will guide them through an analysis of trade-offs.

Finally, the household can rate the possibility of responding as high, medium or low. The final result will be a prioritization of activities to better match the proposed livelihood strategy. The results can be recorded in the last column of the worksheet.

The results of different economic activities can be compared to determine which is most feasible for the household and/or which best fits within the household livelihood strategy.

In this trade-off analysis, the farmer has evaluated the resources required to introduce soybeans and weighs them against the farm's capacity to respond. *Human capital, financial capital* and *physical capital* are likely the most critical factors influencing the decision, with the farmer needing to ensure they have the labour, financial resources and equipment needed to make the venture successful. Balancing these trade-offs will allow the farmer to make an informed decision about whether introducing soybeans is the right move.

12.11 Cash Flow Analysis for Farm Households

Cash flow analysis is a diagnostic method used to assess the inflow and outflow of cash within a farm household. It provides valuable insights into the financial health and stability of the household, helping farmers make informed decisions, plan for expenses and achieve financial sustainability. The concept of cash flow was introduced in Chapter 9. This section elaborates on its use and application for financial diagnosis of the smallholder farm business.

Cash flow analysis for farm households focuses on tracking the sources and uses of cash within the household. It involves monitoring and evaluating the timing and amount of cash inflows and outflows to ensure sufficient liquidity and effective financial management. Cash flow analysis allows farmers to:

- **Evaluate financial health:** cash flow analysis provides an overview of the financial health and stability of a farm business. It helps farmers understand whether there is a consistent inflow of cash to cover expenses, repay debts and invest in growth.
- **Identify cash shortfalls:** by analysing cash flow, farmers can identify periods or situations where cash inflows are insufficient to meet outflows. This helps them anticipate and address potential cash shortfalls, avoiding financial strain and ensuring business continuity.
- **Plan and budget effectively:** cash flow analysis enables farmers to develop realistic budgets and financial plans based on the timing and availability of cash. It helps align expenses with available funds and facilitates effective financial decision making.
- **Improve financial management:** by gaining insights into cash flow patterns, farmers can identify areas for improvement in financial management. They can assess the impact of different factors such as seasonality, market fluctuations, or payment delays on cash flow, and develop strategies to enhance financial stability.

12.11.1 Cash flow analysis: Use and application

Importance of cash flow analysis

Cash flow analysis helps farmers understand the financial dynamics of their household and assess their ability to meet financial obligations. It provides insights into the timing and availability of cash, allowing for effective budgeting and expense planning. It assists in identifying potential cash flow gaps, financial risks and opportunities for improvement.

Table 12.12. Trade-off Analysis framework.

Proposed Activity:			
Farm Resources	Resources required to introduce new activity	Trade-offs	Household's capacity to respond (high, medium, low)
Natural Capital		Would areas of production or crops have to be changed or replaced? Would preserved areas have to be put into production? Would the sustainability of the natural resources by compromized?	
Human Capital		What changes would be required in the use of labour?	
Social Capital		How much time would be taken away from social, organizational or cultural activities?	
Physical Capital		Would the use of equipment, machinery or installations have to be redistributed? Would new investments in machinery, equipment or installations be necessary?	
Financial Capital		Would certain expenditures need to be curtailed? Would new debts be necessary?	

Components of cash flow analysis

Cash Inflows include income from farm sales, off-farm employment, government payments, loans and other sources of cash. *Cash Outflows* comprise household expenses such as living costs, debt repayments, farm inputs, family withdrawals and savings.

Benefits of cash flow analysis

Cash flow analysis facilitates effective financial planning by ensuring adequate cash reserves for day-to-day expenses and future investments. It helps in identifying potential financial challenges or shortfalls and taking proactive measures to address them. It enables farmers to make informed decisions about resource allocation, debt management and investment opportunities.

Integrating cash flow analysis into the financial management of a farm household enhances financial stability and supports long-term sustainability. Cash flow analysis is particularly valuable when assessing cash shortfalls, which occur when cash outflows exceed cash inflows. Here are some ways to integrate cash flow analysis effectively:

- **Seasonal farming:** farmers engaged in seasonal agriculture may experience periods with limited cash inflows while facing ongoing expenses. By conducting a cash flow analysis, farmers can identify these cash shortfall periods and plan accordingly. They may explore options such as securing credit in advance, diversifying income sources or implementing cost-saving measures during periods of low cash inflows.

Table 12.13. Example trade-off analysis for soybeans.

Proposed Activity: Introduce soybeans as a new crop in the farm rotation			
Farm resources	Resources required to introduce new activity	Trade-offs	Household's capacity to respond (high, medium, low)
Natural capital	• Land area: soybeans would require arable land, which would involve shifting some areas currently dedicated to my maize crop • Soil health: Soybeans can improve soil nitrogen levels (due to their ability to fix nitrogen), but I may need to consider crop rotation practices	• Some of the areas of my maize production will need to be shifted or replaced to make space for soybeans • I may have to convert underutilized land to production, which could lead to higher environmental pressure (e.g. soil erosion risk, nutrient depletion) • The sustainability of natural resources might be impacted if the rotation is not properly managed, and overuse of soil might result in a reduction in fertility	**Medium**: My household could adapt by reallocating land or using existing farm practices for crop rotation. However, careful planning will be needed to manage land sustainability
Human capital	• I will need additional labour for planting, maintaining and harvesting soybeans • Existing labour will need retraining in the specifics of soybean farming, because they are new to this crop	• Labour time required for planting and harvesting is going to compete with my other farm activities (e.g. managing current crops, livestock) • Existing workers will need training, which will be an additional cost in terms of time and resources	**Medium**: My household is able to adjust to the additional labour requirements, but I may require temporary seasonal help
Social capital	• Introducing soybeans is going to involve time attending agricultural meetings, networking with soybean buyers and learning from other farmers • The community or farm social activities might be interrupted during peak planting or harvest times	• Family time and social, organizational or cultural activities might be sacrificed during key times (e.g. harvest season) • This could potentially affect relationships with neighbours or social groups	**Low**: My family is very involved in community activities, and I am heavily involved in crop management (especially at harvest). The introduction of soybeans might take significant time away from these obligations
Physical capital	• Soybeans will require specialized equipment for planting, maintenance and harvesting, such as planters, seeders or harvesters. • I will have to redistribute and adapt some of my equipment to handle soybeans	• My current machinery could be used for soybeans but I will need specialized attachments • There will be increased maintenance costs	**Medium**: My family will be able to invest in necessary equipment

Continued

Table 12.13. Continued

Proposed Activity: Introduce soybeans as a new crop in the farm rotation			
Farm resources	Resources required to introduce new activity	Trade-offs	Household's capacity to respond (high, medium, low)
Financial capital	• Initial investment in seeds, fertilizers and new equipment is needed for soybean production • Operating costs for new crops will increase, especially in the first few years as I adjust to soybean production	• Time spent on my other crops may need to be reduced to reallocate funds for the new crop • My cash flow is tight; I will need to take on new debt to finance the start-up costs • The return on investment from soybeans will not be immediate, which will affect short-term profitability	**High**: My farm is financially stable. We will be able to reallocate resources and take on debt to fund the soybeans.

- **Market fluctuations:** changes in market prices or demand can impact cash flow for farmers. If cash inflows decrease due to lower sales income, farmers need to assess their ability to cover expenses. By analysing cash flow, they can identify potential cash shortfalls and consider strategies such as adjusting pricing, exploring new markets or improving marketing efforts to mitigate the impact of market fluctuations.
- **Capital investments:** cash flow analysis is essential when considering significant investments, such as purchasing new equipment or expanding infrastructure. Farmers must evaluate the cash outflows associated with these investments and assess whether they have sufficient cash inflows to cover them. This analysis helps prevent cash shortfalls that could arise from overextending financial resources.
- **Seasonal labour expenses:** farms relying on seasonal labour may face increased cash outflows during peak labour periods. Cash flow analysis helps farmers assess the timing and magnitude of these expenses, allowing them to plan for the additional cash requirements. They may consider options such as securing pre-season financing or adjusting labour schedules to align with cash inflows.

By conducting cash flow analyses in situations where cash shortfalls are likely, farmers can proactively manage their finances, address potential gaps and ensure the financial stability and sustainability of their farm businesses. See Box 12.6. for the steps to prepare a cash flow.

12.12 Risk Analysis

Risk analysis is a diagnostic method used to assess the potential uncertainties and vulnerabilities that may affect farm performance and sustainability. It involves identifying and analysing risks, evaluating their likelihood and potential impact, and developing strategies to mitigate or manage them.

In diagnosis, risk analysis helps farmers anticipate and understand the potential risks they may face in their farm operations. It allows them to assess the vulnerabilities of their production systems, market dynamics and external factors such as climate change, pests or policy changes. By conducting risk analysis, farmers can make informed decisions about risk management strategies, develop contingency plans and enhance the resilience of their farm.

For example, consider a vegetable farm conducting a risk analysis to assess the potential impact of pest infestations on their crop yield. The farmer identifies the risk of pest outbreaks as a significant concern based on historical data and regional pest trends. The farmer further analyses the likelihood of pest infestations based

Box 12.6: The steps involved in preparing the Cash Flow are as follows:

1. Identify Inflow and Outflow

List the income and expenditure items and when they occur in the year, as shown in the table below. This is a time plan schedule.

Cash Inflow			Cash Outflow		
Description	Month	Income ($)	Description	Month	Expenditure ($)
Sale of maize	January	250	Money spent on farm inputs (Maize)	March	300
	March	300			
	September	300			
	October	100			
	November	100		September	300
Sale of Cassava	Apr Aug	340	Money spent on farm inputs (Cassava)	April	37
Sale of milk	March- Sept	420 120	Money spent on farm inputs (livestock)	January	30
				Feb to September	400
				October to December	90
Sale of chicken	January	130	Brooding cost and feeding (chickens)	Sept	60
	September	60			
Planned sale of beans	July	450	Money spent on inputs (Beans)	April	200
				July	40
	December	400		October	180
				December	40
		0	Money to cover living expenses	Jan to Dec	360
		0	Money to cover school expenses	February	200
				May	140
				September	100
		0	Money to cover health expenses	January- December	180
Total		2970	Total		2657

Continued

Box 12.6. Continued

2. Prepare a cash flow chart

From the inflow outflow table, we can work out the monthly balance. This can be done by entering all of the information in a cash flow chart as shown.

CASH COMING IN	Jan	Feb	Mar	Apr	May	Jun	Jul	Aug	Sep	Oct	Nov	Dec
Sales of farm products:												
Maize	250		300						300	100	100	
Cassava				340				120				
Milk			60	60	60	60	60	60	60			
Chicken	130								60			
Planned sale of French beans							450					400
Total Cash Inflow	380	0	360	400	60	60	510	180	420	100	100	400

CASH GOING OUT	Jan	Feb	Mar	Apr	May	Jun	Jul	Aug	Sep	Oct	Nov	Dec
Payments & Purchase of inputs:												
Maize inputs			300						300			
Cassava inputs				37								
Farm inputs livestock	30	50	50	50	50	50	50	50	50	30	30	30
Chicken feeding expenses									60			
Beans inputs				200			40			180		40
Household expenses:				0			0	0	0	0	0	0
Living expenses	30	30	30	30	30	30	30	30	30	30	30	30
School fees		200			140				100			
Health Expenses	15	15	15	15	15	15	15	15	15	15	15	15
Total Cash Outflow	75	295	395	332	235	95	135	95	555	255	75	115

Continued

Box 12.6. Continued

3. Calculate the monthly net Cash Flow

This is done by subtracting the expenses from the income for each month. It will be positive if income is greater than expenses; and negative if income is less than expenses. The results are added to the bottom of the Cash Flow table as shown in the table below.

	Jan	Feb	Mar	Apr	May	Jun	Jul	Aug	Sep	Oct	Nov	Dec
Total Cash Inflow	380	0	360	400	60	60	510	180	420	100	100	400
Total Cash Outflow	75	295	395	332	235	95	135	95	555	255	75	115
Monthly Net Cash Flow	305	-295	-35	68	-175	-35	375	85	-135	-155	25	285

4. Calculate the cumulative net Cash Flow

In order to assess whether the family have enough cash over the year to cover the introduction of beans we have to construct a cumulative Cash Flow. The cumulative net Cash Flow is calculated by adding the monthly net Cash Flow with the cumulative net Cash Flow of the previous month.

For example, in January, the monthly net Cash Flow was $305. This also the cumulative net Cash Flow because it is the first month in the season. In February, the monthly net Cash Flow is -$295. Adding these two together gives tells us that by the end of February, the cumulative net Cash Flow was $10.

	Jan	Feb	Mar	Apr	May	Jun	Jul	Aug	Sep	Oct	Nov	Dec
Monthly net Cash Flow	305	-295	-35	68	-175	-35	375	85	-135	-155	25	285
Cumulative balance	305	10	-25	43	-132	-167	208	293	158	3	28	313

Note: All of these calculations (Steps 2, 3 and 4) are meant to appear in a single table.

5. Analyse the net Cash Flow

This example shows that the family has a shortfall of cash in March, May and June. This means that even though beans is a profitable enterprise the family does not have the cash available to cover the expenditures expected to occur in those months. What can the family do? There are a number of possibilities:

- The farmer could decide not to introduce beans
- The farmer could try to save some money in order to cover the financial shortfall
- The farmer might decide to cut back on some of the inputs used for growing beans
- The farmer might decide to reduce some of the area under maize and cassava in order to reduce costs.
- The farmer might sell some of his livestock to cover the financial gap
- The farmer might decide to take a loan to cover the shortfall

on weather conditions, pest life cycles and pest management practices. They also assess the potential impact of pest infestations on crop yield and financial losses.

Based on the risk analysis, the farmer identifies strategies to mitigate the risks, such as implementing integrated pest management practices, enhancing crop monitoring and scouting, and investing in pest-resistant varieties. The farmer may also develop contingency plans, such as securing insurance coverage for crop losses due to pest damage.

As mentioned in Chapter 9, farming is risky. Risk occurs when the outcomes of a decision cannot always be predicted. Farmers need to make decisions that take these risks into account. Appreciation of risk and its management is aided by gathering information about a particular risk. For family farmers to understand risk it needs to be effectively communicated to them. All too often risk is not well communicated to farmers, which creates a barrier to understanding.

In scientific terms, we talk about probabilities. The probability of something happening 10 times in 100 is 0.1. The probability of an occurrence is often better understood as a 'natural frequency'. This approach, especially when applied to risk analysis, helps in clarifying probabilistic events, aiding better decision making.

Farmers generally prefer more certainty and less risk, but they understand that potentially returns are often associated with higher risk levels. When making decisions involving risk, knowing the probability of outcomes is essential. Probabilities offer a way to quantify belief in an event's occurrence. For example, an event with a 0.4 probability implies it might occur 4 times out of 10 and not 6 times, which could also be described as odds of 6:4 against. If the likelihood of an event is 0.6, this is 4:6 in favour, showing that a greater than 50% probability makes it a more likely outcome. Understanding probabilities is fundamental to analysing risk (see Box 12.7.).

12.12.1 Calculating expected value: example for smallholder crops

A valuable technique in assessing risk is calculating the expected value of potential outcomes. For smallholder farmers, this might involve considering different yields and prices for a crop like maize. Suppose a smallholder farm estimates three yield scenarios: a poor season (yielding 1 t/ha with a 30% probability (0.3)), an average season (yielding 2 t/ha with a 50% probability (0.5)), and a good season (yielding 3 t/ha with a 20% probability (0.2)). Given a stable price of \$150/t, the expected yield per hectare can be calculated as follows:

$$\text{Expected yield} = (1 \text{ ton} \times 0.3) + (2 \text{ tons} \times 0.5) + (3 \text{ tons} \times 0.2) = 0.3 + 1 + 0.6 = 1.9 \text{ tons/ha}$$

$$\text{Expected income per ha} = 1.9 \text{ tons} \times \$150 = \$285$$

Using expected values provides a clearer financial outlook than simply using an average. While historical averages are based on past patterns that may not recur, expected values reflect a more realistic scenario based on current probabilities and conditions, allowing farmers to make informed decisions about crop investments based on potential financial outcomes.

12.12.2 Sensitivity analysis

Sensitivity analysis looks at the sensitivity of an enterprise's profitability to changes in key variables and assumptions. It involves examining how changes in different variables, such as input costs, market prices or yield levels, impact enterprise profitability. Sensitivity analysis can be used with any of the farm business planning and management techniques. It simply quantifies the outcome of a change in selected key variables such as yield or price of outputs or usage level or price of inputs. By conducting sensitivity analysis, farmers can identify key risk

Box 12.7. Probability.

Probability refers to the likelihood that an event will occur. It is typically expressed as a number between 0 and 1 or as a percentage. A probability of 1 (or 100%) means the event is certain to happen, while a probability of 0 (or 0%) means it is impossible.

When considering multiple possible outcomes, their total probability must total 1. For example, if there is a 50% chance of rain, then the probability of no rain must be 50% (0.5), ensuring the total remains 1.

factors and understand their potential effects on the farm's profitability.

The tool helps to answer the questions, 'what if...?'

- What if our production decreases, or increases?
- What if the price of our product goes up, or down?
- What if the cost of an input changes?
- What if the family increases the amount of crop it consumes?

The preparation of farm budgets often involves judgements about the probability of an event occurring. The basic elements of the budgets, such as input and output prices and quantities, seldom represent certain events that can be adequately represented by single values. It is desirable therefore that in preparing budgets some consideration is made to the possible variations in some of the values used.

In preparing these budgets, the effect of risks impacting on price, cost, yield and input requirements can be mapped out and the calculations adjusted accordingly.

For example, the background data given in Table 12.14 illustrates how yields, on-farm consumption and prices influence the farmer's cash income. It further compares the situation in the normal year with the situation in the bad year. Yields and prices tend to move in opposite directions for food crops. When yields fall there is a scarcity of the crop, and prices tend to rise. When yields rise there is an abundance of the crop, and prices tend to decline. The gross income in a normal year is 320/ha, as given in the calculation shown in Table 12.14.

In the bad year, output falls by 50%. On-farm consumption requirements remain the same, and the quantity of produce which is marketed decreases. Prices rise somewhat but the farmer's cash receipts are still greatly reduced. This can be translated into the gross margin calculation in Table 12.15.

From this example, one can see that in a bad year the decrease in total farm income is much larger than the decrease in total variable costs. As a result, the bad year gross margin of 67.5 is much less than the normal year gross margin.

The importance of this calculation is that the farmer can be expected to have a gross margin of at least 67.5 every year: good, normal and bad.

Where maize and other food crops are concerned, the amount of produce marketed by the farmer may be even more variable than yields. This is because the farmer markets only that portion of the food crop that is not required for on-farm consumption. If yields decrease, on-farm consumption may remain almost the same because the farmer will first reduce the amount of produce marketed before reducing on-farm consumption. When low rainfall is noted before the crop is planted and continues throughout the crop season, a farmer may anticipate the effects shown in Table 12.16.

12.12.3 Risk analysis in assessing farm vulnerabilities

Risk analysis can be applied in various contexts to assess farm vulnerabilities. Here are some examples:

- **Financial risk analysis:** farmers may conduct a financial risk analysis to evaluate the potential impact of fluctuating market prices, input costs or interest rates on their profitability. For instance, a livestock farmer may assess the risk of feed price volatility and analyse the potential financial impact on their operation. This analysis helps farmers develop strategies to manage

Table 12.14. Background budget data (cash only).

Future state	Good year	Normal year	Bad year
Produce yield in tons	15	10	5
Produce consumed on farm (tons)	2	2	2
Marketed produce (tons)	13	8	3
Farm gate price per ton	35	40	55
Total farm income (cash from sales)	455	320	165

Table 12.15. Gross margin estimates.

Future State	Good year	Normal year	Bad year
Income			
Income from sales	455	320	165
Variable costs			
Wages of hired labour:			
- Land preparation and planting	30	30	30
- Harvesting	22.5	15	7.5
- Seed purchases	10	10	10
- Fertilizers and chemicals	50	50	50
Total variable costs	112.5	105	97.5
Gross margin	342.5	215	67.5

Table 12.16. Impact of a decrease in the yield of food crops on costs and prices.

Area of effect	Effect
Yield	Decrease
On-farm consumption of crop	No change
Product price	Increase
Land preparation cost	No change
Application of materials used in planting	No change
Application of materials used after planting	No change
Cost of purchased materials purchased used in planting	No change
Cost of purchased materials purchased used after planting	No change
Harvest costs	Decrease

price risks, such as entering into forward contracts, diversifying income sources or implementing cost-saving measures.

- **Climate risk analysis:** farmers can conduct a climate risk analysis to assess the vulnerability of their farm to climate change impacts. They evaluate potential risks associated with extreme weather events, changing precipitation patterns or temperature variations. For example, a fruit orchard owner may analyse the risk of late frosts damaging blossoms and develop strategies such as installing frost protection systems or planting frost-tolerant varieties.
- **Investment risk analysis:** farmers may perform a risk analysis when considering new investments or changes in farm operations. For instance, a farmer contemplating the purchase of expensive machinery may conduct an investment risk analysis to assess the potential risks, benefits and uncertainties associated with the investment. This analysis helps farmers evaluate factors such as equipment reliability, maintenance costs, financing options and market demand to make informed investment decisions.

By applying risk analysis methods in these examples, farmers can identify potential vulnerabilities, evaluate the likelihood and impact of risks and develop strategies to mitigate or manage them effectively. This proactive approach supports farmers in minimizing potential losses, optimizing their decision-making processes and enhancing the overall resilience and sustainability of their farm operations.

12.13 Advantages and Disadvantages of Farm Management Diagnostic Methods

Farm management diagnostic methods help assess farm performance, identify challenges and explore opportunities for improvement. They support informed decision making, optimize resource use and enhance productivity while also aiding in risk management. However, their effectiveness depends on data availability, and some methods can be complex, time-consuming or subject to bias. Additionally, they may not always capture the full complexity of farm systems. Despite these challenges, when applied with awareness of their strengths and limitations, diagnostic methods remain essential tools for improving farm efficiency, sustainability and resilience in an increasingly uncertain agricultural environment.

12.13.1 Benefits of utilizing diagnostic methods

Utilizing diagnostic methods in farm management offers several advantages that can greatly benefit farmers and enhance their decision-making processes. These methods provide a systematic and structured approach to assess farm businesses, identify challenges and capitalize on opportunities. Some key benefits of utilizing diagnostic methods include:

- **Comprehensive understanding:** diagnostic methods enable farmers to gain a comprehensive understanding of their farm operations by collecting and analysing relevant data. This in-depth analysis allows farmers to identify strengths, weaknesses and areas for improvement, leading to informed decision making.
- **Informed decision making:** diagnostic methods provide farmers with valuable insights and information, allowing them to make informed decisions. By understanding the current state of their farm, farmers can develop strategies, allocate resources effectively and optimize their operations for better outcomes.
- **Efficiency and productivity:** diagnostic methods help identify inefficiencies and bottlenecks in farm operations. By addressing these issues, farmers can enhance efficiency, optimize resource allocation and improve overall productivity.
- **Risk management:** diagnostic methods, such as risk analysis, enable farmers to assess potential risks and vulnerabilities. This information allows them to develop risk management strategies, such as diversification, insurance or contingency plans, to mitigate the impact of uncertainties.
- **Sustainability and resilience:** diagnostic methods help farmers assess the social, economic and environmental aspects of their farm operations. By integrating sustainability principles, such as livelihood analysis or gross margin analysis, farmers can enhance the long-term sustainability and resilience of their farm businesses.

12.13.2 Limitations and challenges associated with diagnostic methods

While diagnostic methods offer numerous benefits, it is important to acknowledge their limitations and challenges. Each method has its own set of limitations that may impact their effectiveness. Some common limitations and challenges associated with diagnostic methods include:

- **Data availability and quality:** the effectiveness of diagnostic methods relies on the availability and quality of data. Limited access to data or unreliable data sources can hinder the accuracy and reliability of the analysis.
- **Subjectivity and bias:** diagnostic methods involve subjective judgments and interpretations. Different individuals may have varying perspectives, leading to potential biases in the analysis and decision-making processes.
- **Complexity and time-intensiveness:** some diagnostic methods, such as livelihood analysis or trade-off analysis, can be complex and time-consuming. They require a significant amount of data collection, analysis and expertise, which may pose challenges for farmers with limited resources or technical knowledge.
- **Limited scope:** diagnostic methods may have a limited scope, focusing on specific aspects of farm management. While they provide valuable insights into those specific areas, they may not capture the entirety of

a farm's complexity or address all potential issues.

- **Contextual variations:** the effectiveness of diagnostic methods can vary depending on the specific context and circumstances of each farm. What works well for one farm may not be directly applicable to another due to variations in resources, geographical location or market conditions.

Despite these limitations and challenges, the benefits of utilizing diagnostic methods in farm management outweigh the drawbacks. By understanding the limitations and adapting the methods to their specific context, farmers can effectively leverage these tools to make informed decisions, enhance their farm's performance and achieve long-term success.

12.14 Advisory Service Support for Farm Management Investigation

Extension workers play a crucial role in assisting small family farmers to better understand and manage their farms. They achieve this by gathering detailed farm management data through comprehensive farm surveys or targeted case studies. This information provides the foundation for effectively applying practical farm management diagnostic tools, enabling advisers to assess the current state of the farm, identify key constraints, and provide guidance on improving productivity, profitability and sustainability. Farm management diagnosis is not only a technical process, it strengthens decision-making capacity and supports farmers in managing their farms as businesses.

As discussed, there are several tools available for diagnosing the performance of farms and their enterprises. Selecting an appropriate diagnostic method depends on the specific objectives of the analysis and local farming conditions. Extension workers who conduct enterprise and farm studies or who help farmers maintain farm records typically aim to:

- Assess the relative profitability of various farm enterprises.
- Identify underlying causes of profitability variations.
- Determine best practices and effective techniques for farm operations.
- Establish optimal input requirements for each farm enterprise.

To achieve these objectives, the diagnostic studies must be carefully planned and implemented, using the most appropriate tools to support smallholders in improving farm performance. These methods enable extension workers and farmers to collaboratively analyse production and financial outcomes, assess enterprise viability and make informed decisions that lead to more efficient, profitable and resilient farm operations.

In many low-income country contexts, farm-level investigations have proven particularly valuable for evaluating enterprise profitability and helping farmers determine which aspects of their farm businesses contribute most to returns on investment. This information directly informs decisions on resource allocation. Investigations also support financial analyses to assess small farmers' capacities to repay loans. Extension workers often assist farmers in preparing seasonal cash flow budgets, highlighting the need for savings or short-term credit during periods of production gaps.

Investigative tools applied in participatory settings help uncover underlying causes of poor farm performance – such as declining soil fertility or limited access to inputs – and assist farmers in prioritizing technologies and improved practices. These tools can also facilitate trade-off and risk analyses, allowing farmers to compare strategic options (e.g. investing in irrigation vs expanding livestock) by weighing expected returns against risks such as climate variability or market instability.

Effective use of these diagnostic tools requires continuous training for extension workers, both in technical skills and participatory facilitation methods. Simple software programmes can further assist extension workers in farm enterprise analyses. Equally important is the advisers' ability to translate diagnostic findings into practical, context-specific recommendations.

With strong facilitation and support, farm diagnostic studies form an essential prerequisite for planning, monitoring and evaluation, thus completing the farm management cycle. These investigations also provide invaluable information for applied agricultural economic research, supporting the design of extension programmes and policy formulation. Investing in the capacity of agricultural advisers to apply these tools effectively is critical to enhancing farm performance and ensuring long-term viability.

12.15 Conclusion

The farm management cycle serves as a fundamental framework for effective farm management, guiding decision-making processes and ensuring the optimization of farm performance. The cycle, consisting of diagnosis, planning, implementing, monitoring and evaluating, provides a systematic approach that allows farmers to navigate the complexities and challenges of agricultural enterprises.

Throughout this chapter, the significance of diagnostic methods within the farm management cycle has been explored. These methods play a crucial role in assessing the current state of farm operations, identifying challenges and capitalizing on opportunities. By employing diagnostic methods, farmers gain a comprehensive understanding of their strengths and weaknesses, enabling them to make informed decisions that drive their farm's success.

The farm management cycle emphasizes the importance of utilizing diagnostic methods at each stage. The diagnosis stage requires farmers to gather and analyse data, providing a foundation for strategic planning. The planning stage allows farmers to develop strategies, set goals and allocate resources effectively. The implementing and monitoring stages ensure that plans are implemented and progress is tracked, enabling timely adjustments. Finally, the evaluation stage assesses the outcomes of decisions, facilitating continuous improvement.

It is important to recognize the need for sustainable farm management within the planning and control framework. As stewards of the land, farmers must consider the long-term environmental, social and economic impacts of their decisions. Integrating sustainability principles within the farm management framework ensures the preservation of natural resources, minimizes environmental impact and promotes the overall well-being of agricultural systems.

By embracing diagnostic methods and incorporating sustainability considerations, farmers can achieve optimal decision making and farm performance. These methods provide the tools and techniques necessary to diagnose farm issues, plan strategies and exercise control over farm operations. They enable farmers to navigate challenges, optimize resource allocation, enhance productivity and improve profitability.

Additional Reading

C.S. and J.S. Nix (1973). *Farm Planning and Control*, Cambridge University Press. Barnard.

Dillon, J.L. and J.B. Hardaker (1993). *Farm Management Research for Small Farmer Development, FAO Farm Systems Management Series No. 6*. Food and Agriculture Organisation of the United Nations, Rome.

Food and Agriculture Organisation of the United Nations (1989). *Farming Systems Development: Concept, Methods, Applications*. Rome.

Makeham, J.P. and L.R. Malcolm (1986). *The Economics of Tropical Farm Management*. Cambridge University Press.

Nuthall, P. L. (2016). *Farm Business Management: The Fundamentals of Good Practice*. CABI, ISBN: 978-1780646572

Upton, M. (1973). *Farm Management in Africa*. Oxford University Press, London.

Upton, M. and J.M. Dixon (eds) (1994). Methods of Micro-level Analysis for Agricultural Programmes and Policies, FAO Farm Systems Management Series No. 9, Food and Agriculture Organisation of the United Nations, Rome.

13 Planning

Planning is the backbone of successful farm management, ensuring that farmers make informed decisions, to optimize resource allocation, manage risks and improve profitability. This chapter explores the critical role of planning in farm management, presenting a range of methodologies and tools that guide farmers in making informed and strategic decisions. The chapter begins by outlining the importance of planning within the farm management cycle, emphasizing how structured approaches to production, financial management and market engagement contribute to long-term sustainability. It highlights the benefits of planning, including improved resource efficiency, better risk management and enhanced decision-making capabilities. Different types of planning are explored, including strategic planning, which focuses on long-term goals and market positioning; farm planning, which integrates production decisions with resource constraints; and market planning, which aligns production strategies with consumer demand. This chapter also examines labour planning, detailing how farmers can optimize their workforce while balancing family and hired labour. Financial cash flow planning is addressed, guiding farmers on how to forecast income and expenses to ensure financial stability. Investment planning follows, explaining the appraisal of capital investments using techniques like net present value (NPV) and internal rate of return (IRR) to ensure long-term profitability. Advanced tools such as linear programming are introduced, illustrating how farmers can maximize returns through mathematical optimization of land, labour and capital. The concept of whole-farm planning is emphasized as a comprehensive approach that integrates multiple planning components for enhanced productivity and resilience.

Through practical examples, this chapter equips farmers, extension workers and policymakers with structured methodologies to enhance farm planning, ensuring that agricultural enterprises remain competitive, profitable and sustainable.

13.1 Planning and the Farm Management Cycle

Planning is the second stage in the farm management cycle and involves setting goals, determining strategies and outlining the steps needed to achieve desired outcomes (Fig. 13.1).

In the context of farming, planning encompasses various aspects such as crop production, livestock management, resource allocation, marketing and risk management. Sound planning is crucial for the success and sustainability of farming operations. They provide farmers with a roadmap for achieving their objectives, enabling them to make informed decisions and allocate resources effectively. Some key benefits of planning in farming include:

- **Resource allocation:** planning helps farmers optimize the allocation of resources such as land, labour, capital and machinery. It ensures that resources are utilized efficiently and that the right inputs are allocated to each production activity. This leads to increased productivity and cost-effectiveness.
- **Risk management:** planning enables farmers to anticipate and mitigate potential risks and uncertainties. By identifying risks such as adverse weather conditions, market fluctuations or pest outbreaks, farmers can develop contingency plans, explore

DOI: 10.1079/9781836990536.0013

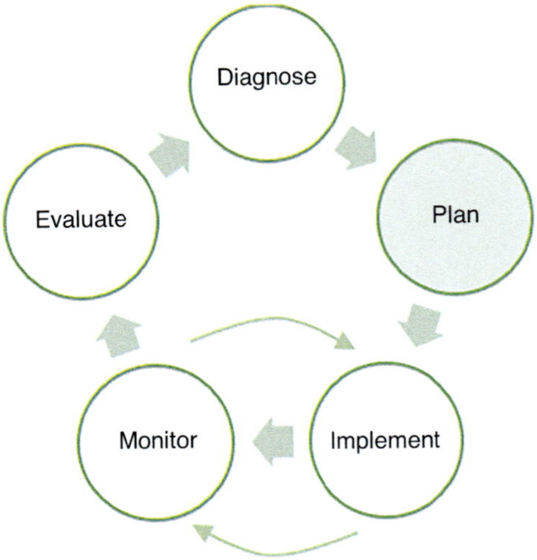

Fig. 13.1. The farm management cycle: planning.

insurance options and implement risk management strategies, to safeguard their operations.

- **Decision making:** planning provides farmers with a solid foundation for decision-making. It helps assess the feasibility of new ventures, evaluate the profitability of different crops or livestock enterprises, and make informed choices based on financial projections and market trends.

13.2 The Planning Process

The planning process involves several sequential steps that help farmers develop comprehensive and effective plans. Although the specific steps may vary depending on the farm's size, type and goals, the general planning process typically includes the following stages:

1. **Goal setting:** farmers define their long- and short-term goals, considering factors such as production targets, profitability objectives, sustainability goals and personal aspirations.
2. **Analysis:** farmers conduct a thorough analysis of their farm's strengths, weaknesses, opportunities, and threats (SWOT analysis). They assess internal factors such as available resources, infrastructure and capabilities, as well as external factors such as market trends, competition and regulatory frameworks (See Chapter 12.).
3. **Action planning:** based on the goals and analysis, farmers develop action plans outlining the specific steps required to achieve their objectives. These plans may include production schedules, marketing strategies, financial projections and risk management measures.
4. **Implementation:** farmers execute the action plans, putting them into practice through activities such as planting, managing livestock, marketing products and monitoring performance (See Chapter 14.).
5. **Monitoring and evaluation:** farmers regularly monitor and evaluate their progress towards the goals set in the planning phase. They collect data, compare actual results with projected outcomes and make adjustments when necessary to ensure continuous improvement (See Chapter 14.).

13.3 Overview of Different Planning Approaches

There are various planning approaches that farmers can adopt depending on their specific circumstances and preferences. Some common types of planning include:

- **Strategic planning:** strategic planning focuses on long-term goals and the overall direction of the farm. It involves analysing market trends, identifying growth opportunities and formulating strategies to achieve sustainable competitive advantage.
- **Market planning:** as we discussed in Chapter 6, market planning refers to the process of analysing market opportunities, understanding customer needs and preferences, and developing strategies to effectively promote and sell farm products. The purpose of market planning is to align farm production and marketing efforts with consumer demand, optimize sales potential and achieve a competitive edge in the marketplace.
- **Farm planning:** farm planning involves the systematic process of setting goals,

making decisions and allocating resources to achieve desired outcomes in agricultural operations. A farm plan can cover a single enterprise or the whole farm (all the enterprises within the farm). It helps assess the profitability and viability of each enterprise and guides resource allocation decisions for the farm as a whole.

- **Planning labour:** planning labour involves determining how much labour is needed to effectively carry out their operations. A labour plan also outlines when labour is needed at different times and whether family labour or hired labour is required.
- **Financial cash flow planning:** cash flow planning is a vital component of farm management that focuses on managing and projecting the movement of cash into and out of the farm business. It involves forecasting the inflows and outflows of cash over a specific period, typically on a monthly or annual basis. Cash flow planning is crucial for farmers as it helps ensure sufficient liquidity, facilitates timely payments and supports effective financial decision making.
- **Investment planning:** investment planning in agriculture involves making informed decisions regarding capital investments to enhance farm productivity, profitability and long-term sustainability. It focuses on identifying, evaluating and prioritizing investment opportunities that align with the farm goals and objectives. Investment planning considers factors such as available resources, market conditions, technological advancements and risk assessment, to guide farmers in making sound investment decisions.

While small family farmers are not likely to be in a position to carry out all these types of planning on their own, it is important that they take at least the concepts and ideas into account when planning their farm businesses. Ultimately, the farmer will need to develop some kind of plan to know what is going to be produced, what resources (including inputs, labour and finance) will be required and when, and where the products will be sold. The planning discussed in this chapter is intended primarily for the planning stage in the farm business cycle, but can also be used at other times in the cycle.

By understanding and utilizing these different types of planning, farmers and agricultural extension workers can select the most suitable type or combine multiple types of planning to develop comprehensive and tailored plans for their farming operations.

In the upcoming sections, we will explore each type of planning in more detail, providing practical guidance and examples to assist farmers in implementing these methods effectively.

13.4 Strategic Planning

As noted previously – and in Chapter 5, – strategic planning is a process that involves defining the long-term goals and objectives of a farming operation and determining the strategies and actions needed to achieve them. It goes beyond day-to-day operations and focuses on the overall direction and sustainability of the farm. Strategic planning helps farmers make informed decisions, allocate resources effectively, and adapt to changing market conditions and environmental factors.

In farming, strategic planning involves analysing internal and external factors that can impact the farm's success. This includes assessing strengths, weaknesses, opportunities and threats (SWOT analysis), considering market trends, consumer demands, technological advancements, environmental regulations and other relevant factors. By understanding these factors, farmers can develop strategies that leverage their strengths, address weaknesses, seize opportunities and mitigate threats.

Strategic planning is closely linked to a farm's long-term goals and objectives. It provides a roadmap for achieving those goals by identifying the necessary actions and resources. Farmers must have a clear vision of what they want to achieve in the future and align their strategic planning efforts accordingly.

For example, long-term goals might include expanding the farm's production capacity, diversifying into new markets or products, improving sustainability practices, or passing on the farm to the next generation. Strategic planning helps identify the steps, investments and strategies needed to accomplish these goals. It ensures that short-term decisions and activities are aligned with the broader vision for the farm's future.

A strategic plan for a farm typically includes the following key components:

- **Mission statement:** a mission statement articulates the farm's purpose, values and identity. It defines what the farm aims to achieve and the principles that guide its operations. The mission statement serves as a foundation for strategic decision making.
- **Vision:** the vision outlines the desired future state of the farm. It describes what the farm aspires to become and the overarching goals it seeks to accomplish. The vision provides a clear direction and inspires stakeholders to work towards a common objective.
- **Goals and objectives:** goals and objectives are specific targets that the farm aims to achieve within a defined timeframe. These targets may include financial objectives, production targets, market expansion goals, sustainability benchmarks or any other areas that align with the farm's vision and mission.
- **Strategies and action plans:** strategies outline the broad approaches or methods the farm will adopt to achieve its goals. Action plans break down these strategies into actionable steps, specifying who is responsible for each task, timelines and required resources. Strategies and action plans provide a roadmap for implementation.
- **Resource allocation:** strategic planning involves identifying the resources needed to execute the strategies. This includes financial resources, human capital, technology, infrastructure and other inputs required to achieve the farm's goals. (Also see the discussion on livelihood assets in Chapter 4.)

By developing a comprehensive strategic plan, farmers can align their operations with long-term objectives, anticipate challenges, seize opportunities and adapt to changes in the agricultural landscape. Strategic planning provides a framework for decision making, fosters innovation and enhances the farm's competitive position in the market.

13.5 Market Planning

As discussed in Chapter 6, marketing plays a crucial role in farm business success; the chapter explored the concepts of markets and marketing, providing a foundation for understanding how agricultural products reach consumers. Additionally, Chapter 12 introduced market margins as a diagnostic tool for assessing profitability along the marketing chain. This section builds on those discussions, using these insights as key inputs for market strategy formulation and planning.

Market planning refers to the process of analysing market opportunities, understanding customer needs and preferences, and developing strategies to effectively promote and sell farm products. It involves a systematic approach to gathering market intelligence, identifying target markets, creating value propositions, and developing marketing strategies and tactics that resonate with the intended customers. It enables farmers to make informed decisions about product development, pricing, distribution channels, promotion and customer relationship management.

13.5.1 Identifying target markets and customer needs

Effective market planning begins with identifying target markets and understanding customer needs. This involves conducting market research, gathering demographic data and analysing trends to identify potential customer segments that align with the farm's products and capabilities.

By understanding customer needs, preferences and buying behaviour, farmers can tailor their products, packaging, pricing and marketing messages to attract and retain target customers. This customer-centric approach helps build brand loyalty, foster positive customer experiences and create a competitive advantage in the marketplace.

Chapter 6, provided a detailed outline for conducting a market analysis. This should be carried out as a part of developing a marketing plan.

13.5.2 Developing marketing strategies and tactics

Once target markets and customer needs are identified, farmers can develop marketing strategies and tactics to reach and engage with their target audience. This involves considering the elements outlined here.

Product positioning

Farmers must define the unique selling propositions and competitive advantages of their products. This helps differentiate their products from competitors and create value in the eyes of customers.

Pricing strategies

Determining the optimal pricing strategies is crucial for maximizing profitability while remaining competitive in the market. Farmers should consider factors such as production costs, market demand, competition and perceived value by customers, when setting prices. Pricing decisions may involve strategies such as cost-based pricing, market-based pricing, value- based pricing or dynamic pricing (see Box 13.1). By finding the right balance between income generation and market competitiveness, farmers can enhance profitability and customer satisfaction.

Pricing considerations could also involve considering elasticity of demand. If a farmer produces a highly elastic product, market fluctuations could significantly impact income. In contrast, inelastic products, such as staple grains, may help stabilize earnings.

Selling strategy and distribution channels

Farmers have several options in selling their produce. One way is to sell at the farm gate. Alternatively, the farmer could sell directly in the market. The choice of strategy, however, depends on the location of the farm relative to the market, the costs involved in marketing and the volume of produce to be sold. Farmers also need to decide on the most effective distribution channels to reach their target customers. This could involve direct sales through farm stands or farmers' markets, partnerships with retailers, online sales or other distribution methods based on customer preferences and convenience.

Promotion and communication

Farmers should develop strategies to promote their products effectively. This may include advertising, public relations, social media marketing, content creation, influencer collaborations or other communication channels that resonate with the target market.

Box 13.1. Pricing strategies.

- **Cost-based pricing**: setting prices based on the total cost of production (e.g. materials, labour, overheads) plus a desired profit margin. For example, if producing a crop costs \$5 per unit and the farmer wants a 20% profit, the price would be \$6.
- **Market-based pricing**: Setting prices based on what competitors are charging for similar products in the market. This approach ensures the farm remains competitive while meeting market expectations.
- **Value-based pricing**: Setting prices based on the perceived value of the product to the customer, rather than just costs or competition. For example, organic produce might command a higher price because customers value its health and environmental benefits.
- **Dynamic pricing**: Adjusting prices in real-time based on factors like demand, seasonality or market conditions. For instance, prices for fresh vegetables might increase during a shortage or decrease during a surplus.

Customer relationship management

Maintaining strong relationships with customers is crucial for repeat business and brand loyalty. Farmers should consider strategies for customer engagement, feedback collection and personalized experiences, to build long-term relationships and enhance customer satisfaction.

13.5.3 Incorporating market planning into overall farm planning

Market planning should be integrated into the overall farm planning process to ensure alignment between production and marketing efforts. By considering market factors and customer demands from the early stages of farm planning, farmers can make informed decisions about enterprise selection, product diversification, production volumes and resource allocation.

Market planning also requires farmers to anticipate market trends, adapt to changes in consumer preferences and respond to competition. By regularly evaluating market conditions, customer feedback and sales performance, farmers can adjust their strategies, refine their product offerings and optimize their marketing efforts, to maximize profitability and long-term success. Incorporating market planning into overall farm planning fosters a customer focused mindset, enhances market responsiveness and enables farmers to leverage opportunities for growth and profitability in a competitive marketplace (see Box 13.2.).

Planning for the market could cover the steps outlined in Fig. 13.2.

Agricultural advisors can assist farmers in analysing the possible strategies by developing the constraints opportunities step to include solutions and actions and presenting them into a matrix, as given in Table 13.1. The matrix should enable the farmer, with the assistance of the advisor, to appraise whether a farm enterprise is worth continuing.

The solutions to marketing problems can often be very simple and may not require major changes to be made in production or for new technologies to be introduced. The marketing strategy should, consequently, be as simple as possible. As part of the strategy, the farmer should examine the combination of factors needed to satisfy the needs of the consumers and, through them, increase farm profits.

The market plan format in Table 13.2 provides a structured approach to planning the sale of farm products by outlining key marketing strategies. It can help farmers determine when and how much to sell, identify target markets, select distribution methods and define pricing strategies while anticipating potential challenges. By using this format, farmers can make informed marketing decisions, ensuring better market positioning.

The overall marketing strategy could include several specific methods relating to sales and pricing.

Regardless of whether the farm is small or large, the market-oriented farmer should formulate a marketing strategy and develop a market plan that responds to consumers' needs. Farmers aim at maximizing profits, while buyers aim at maximizing their satisfaction. Profits can be maximized by selling large quantities of a small number of products at the highest possible price. Buyer satisfaction can be maximized by increasing the range of products available for consumption and purchasing these products at the lowest possible price consistent with their income and the quality they require.

13.6 Farm Planning

Farm planning is the process of translating a farmer's goals and available resources into a structured plan for agricultural production. It involves creating detailed resource use plans and financial plans for the farm business. As mentioned previously, it involves the systematic process of setting goals and objectives based on the assessment of resources done in the diagnosis stage, making decisions and allocating resources to achieve desired outcomes in agricultural operations. A well-designed farm plan ensures that resources are used optimally, risks are managed and the farm's long-term sustainability is secured.

This section starts with a review and consolidation of the key elements of the farm planning process and then discusses linear programming and whole-farm planning.

Box 13.2. Developing a market plan.

When developing a market plan, a farmer should consider several key questions to align their production and sales strategies with market demands and opportunities. These questions can be grouped into categories.

1. Understanding the market
   - Who are my target customers (e.g. individuals, wholesalers, retailers, processors or institutions)?
   - What are the key needs or preferences of my target market (e.g. organic produce, local sourcing, specific varieties)?
   - What is the demand for my products (e.g. seasonal demand, price sensitivity, quality expectations)?
   - Who are my competitors, and what are their strengths and weaknesses?
2. Product strategy
   - What products should I grow or produce to meet market demand (e.g., staple crops, niche products, value-added items)?
   - What quality or grading standards are required (e.g., size, colour, freshness, packaging standards)?
   - Should I diversify my product line or specialize?
3. Pricing strategy
   - What prices are customers willing to pay for my products? (Consider market research, surveys and customer feedback.)
   - How do my prices compare to competitors?
   - Can I set prices to cover costs and achieve profitability? (Include production costs, transportation and marketing.)
4. Sales and distribution
   - Where should I sell my products (e.g. farmers' markets, direct-to-consumer, local stores, online)?
   - What distribution channels shall I use (e.g. delivery, pickup points, partnerships with retailers)?
   - How will I handle transportation and logistics?
   - What is the cost and reliability of my distribution options?
5. Promotion and communication
   - How will I let customers know about my products (e.g. advertising, social media, word-of-mouth)?
   - What unique value proposition will I communicate to customers (e.g. sustainable practices, local production, affordability)?
   - How can I build trust and loyalty with customers?
6. Risk management
   - What risks could affect my ability to market my products (e.g. weather, price fluctuations, competition)?
   - What contingency plans do I have for unexpected changes in market conditions?
   - Do I have access to alternative markets if demand shifts?
7. Monitoring and evaluation
   - How will I track sales and customer feedback?
   - How can I measure the success of my market plan?
   - What adjustments can I make to improve results over time?

13.6.1 Developing the farm plan: key steps and considerations

This section focuses on the specific steps and considerations involved in creating a farm plan, building on the general planning process outlined in Section 13.2 on the planning process. While Section 13.2 provides an overview of the planning cycle (goal setting, analysis, action planning, implementation and monitoring), this section delves deeper into

The current market situation: the general background on the market in which the product will be sold. It begins with a general idea of who the buyers are and what they want, followed by anything else that describes the market in which the products would be sold (e.g. existing supplies, packaging preferences, etc.)

↓

Constraints and opportunities analysis: based on an assessment of the market opportunities, the farmer or farmer group identifies the opportunities and constraints that the farm faces and realistically evaluates the farm's internal strengths and weaknesses in dealing with its market situation.

↓

The marketing strategy: based on the analysis carried out above, the farmer draws up a plan to address the marketing objectives of the farm. The strategy should include a clear definition of buyers, buyer needs and the prices attained for produce sold.

Fig. 13.2. Market strategy formulation.

Table 13.1. Analysis of constraints and opportunities.

Constraints	Opportunities
No local market Poor transport services	Potential exists for early crop production when supplies are short
Solutions	**Actions**
Organize a local farmers' market Encourage buyers to use their own transport	Encourage growing early crops and develop appropriate production techniques

Table 13.2. Format for the preparation of a market plan.

Product	Marketing strategies						
	When?	How much?	Where?	How?	To whom?	Farm gate price	Likely problems

the practical application of these steps within the context of farm planning, covering:

- Setting farm-specific goals and objectives.
- Assessing available resources (land, labour, capital, etc.).
- Selecting enterprises and diversifying production.
- Planning crop rotations and land allocation.
- Using budgeting and gross margin analysis to evaluate profitability.

Setting farm goals and objectives

The first step in farm planning is to define the farmer's goals and objectives. These may include increasing income, improving food security, diversifying production or achieving environmental sustainability. Goals should be specific, measurable, achievable, relevant and time-bound (SMART). For example, a farmer might aim to increase maize yields by 20% over the next 2 years or diversify into vegetable production to reduce reliance on a single crop. Clear goals provide direction and help farmers prioritize their efforts.

Assessing available resources

Once goals are established, the next step is to assess the farm's available resources. This includes evaluating the quantity and quality of land, labour, capital, water and other inputs. Farmers should consider:

- **Land:** how much land is available, and what is its suitability for different crops or livestock?
- **Labour:** what is the availability of family and hired labour, and how does it vary seasonally?
- **Capital:** what financial resources are available for purchasing inputs, equipment or expanding operations?
- **Water:** is there reliable access to water for irrigation, and how can it be managed efficiently?
- **Infrastructure:** are there adequate storage facilities, tools or machinery to support production?

This assessment helps farmers identify constraints and opportunities, ensuring that their plans are realistic and feasible.

Selecting enterprises and diversifying production

With a clear understanding of their goals and resources, farmers can select the most suitable enterprises (crops or livestock) for their farm. Enterprise selection should consider:

- **Market demand:** what products are in demand locally or regionally, and what prices can be expected.
- **Resource requirements:** how much land, labour and capital are needed for each enterprise.
- **Risk tolerance:** how vulnerable the enterprise is to risks such as drought, pests or price fluctuations.
- **Diversification:** whether the farm should focus on a single enterprise (specialization) or multiple enterprises (diversification) to spread risk and stabilize income.

Planning crop rotations and land allocation

Crop rotation is a critical component of farm planning, as it helps maintain soil fertility, control pests and reduce the risk of crop failure. Farmers should plan rotations based on the nutrient needs and growth cycles of different crops. For instance, legumes can be rotated with cereals to fix nitrogen in the soil, while root crops can follow leafy vegetables to break pest cycles.

Farmers should also evaluate marginal cost and marginal income in planning crop rotations. Expanding a crop enterprise should only continue as long as the marginal income exceeds the marginal cost. If the marginal cost surpasses marginal income, further expansion reduces profitability. (See Chapter 9.)

Land allocation involves deciding how much land to dedicate to each enterprise. Farmers should consider:

- **Yield potential:** which crops, or livestock are likely to produce the highest yields given the available resources.
- **Market opportunities**: which enterprises are most profitable based on current market conditions.
- **Risk management:** how land can be allocated to minimize the impact of adverse weather or market fluctuations.

Budgeting and gross margin analysis

Budgeting is an essential tool for evaluating the financial viability of different enterprises. Farmers can use gross margin analysis to compare the profitability of various options. As discussed previously, the gross margin is calculated by subtracting variable costs from gross income. By comparing gross margins, farmers can identify the most profitable enterprises and allocate resources accordingly. Budgeting also helps farmers anticipate cash flow needs and plan for expenses such as input purchases or labour hiring.

Integrating elements into a whole-farm plan

The final step in farm planning is to integrate all the above elements into a cohesive whole-farm plan. This plan should outline:

- **Production targets:** how much will be produced of each crop or livestock product.
- **Resource allocation**: how land, labour and capital will be distributed across enterprises.
- **Timeline:** when key activities such as planting, weeding and harvesting will take place.
- **Contingency plans:** how the farm will respond to risks such as drought, pests or price drops.

A well-structured farm plan provides a roadmap for achieving the farmer's goals while ensuring efficient use of resources and long-term sustainability.

13.6.2 Linear programming

Linear programming is a powerful mathematical optimization technique that plays a crucial role in farm planning. It enables farmers to make optimal decisions regarding resource allocation, production levels and profit maximization. For example, a farmer with 5 ha of land must decide how much to allocate to different crops based on expected income and costs. By using linear programming, they can determine the most profitable crop mix while staying within resource constraints. By formulating linear programming models, farmers can address complex resource allocation problems and find efficient solutions.

At its core, linear programming involves formulating an objective function and a set of constraints. The objective function represents the goal to be achieved, typically maximizing profits or minimizing costs. The constraints represent limitations on resources, such as land availability, labour capacity, capital investment and input quantities.

In farm planning, linear programming models consider various factors to determine the optimal use of resources. These models also incorporate *marginal cost analysis*, ensuring that production decisions align with profitability. By optimizing land, labour and capital allocation, farmers can maximize returns while maintaining efficiency.

Production constraints

Farmers must account for production limitations, such as the available hectarage, labour availability, and constraints on inputs like water, fertilizers and machinery. Linear programming helps determine the optimal allocation of these resources across different crops, enterprises or activities.

Market prices

Market prices for agricultural commodities are subject to fluctuations. Linear programming models incorporate market prices as variables to optimize crop selection and production levels. By considering the profitability of different crops based on market prices, farmers can determine the optimal mix of crops to maximize overall farm revenue.

Resource availability

Scarce resources like land, labour and capital need to be allocated efficiently. Linear programming models help farmers determine the optimal use of these resources to achieve the desired production levels and maximize profits. For example, the models can guide decisions on how much land should be allocated to specific crops, how to allocate labour across different tasks, or how to allocate capital investments in machinery and infrastructure (see Box 13.3.).

Input optimization

Linear programming also aids in optimizing input usage. Farmers can determine the optimal quantities of inputs like fertilizers, pesticides and water for each crop to achieve maximum yields while minimizing costs. This ensures that inputs are used efficiently, reducing waste and environmental impact.

Once a linear programming model is formulated with the objective function and constraints, specialized algorithms are used to find the optimal solution. These algorithms systematically explore different combinations of resource allocations to identify the best outcome that maximizes profits or minimizes costs while satisfying all constraints.

A simplified linear programming matrix for farm planning is given in Table 13.3.

In this example, the objective function is to maximize profit and there are five constraints. The matrix is presented in the form of equations.

To maximize profit:

Maximize $= Z = P_1X_1 + P_2X_2 + P_3X_3 + P_4X_4$
Where:

- X_1, X_2, X_3, X_4 represent the hectares allocated to each crop.
- L_1, L_2, L_3, L_4 represent the labour required per hectare for each crop.

CONSTRAINTS

1. Land constraint: $X_1 + X_2 + X_3 + X_4 \leq 5$ ha
2. Labour constraint: $L_1X_1 + L_2X_2 + L_3X_3 + L_4X_4$ $\leq$ total available labour
3. Capital constraint: $C_1X_1 + C_2X_2 + C_3X_3 + C4_{X4}$ $\leq$ available capital
4. Water constraint: $W_1X_1 + W_2X_2 + W_3X_3 + W4_{X4}$ $\leq$ water supply
5. Fertilizer constraint: $F_1X_1 + F_2X_2 + F_3X_3 + F_4X_4$ $\leq$ fertilizer supply

INTERPRETATION

- Each crop (e.g. maize, beans, tomatoes, wheat) competes for limited farm resources.
- The farmer must decide the optimal allocation (X_1, X_2, X_3, X_4) that maximizes profit.
- Constraints ensure that land, labour, capital, water and inputs are not exceeded.
- The solution to this model provides the best combination of crop enterprises given the resource limits.

This simplified format helps farmers determine how to allocate their land and inputs optimally for maximum returns.

By utilizing linear programming in farm planning, farmers can make informed decisions about resource allocation, production levels and crop selection. The models help optimize resource utilization, reduce inefficiencies and maximize overall farm profitability. Moreover, linear programming can account for various scenarios and allow farmers to conduct sensitivity analyses, enabling them to understand how changes in market conditions or resource availability impact optimal decisions.

It is worth noting that while linear programming provides valuable insights, it simplifies the real-world complexities of farming systems. Factors such as weather variability, market uncertainties and biological interactions are not explicitly considered in linear programming models. Therefore, it would be essential for farmers to supplement linear programming with their experience, judgment and local knowledge to make well-rounded decisions.

It is important to note that while mathematical techniques such as linear programming can be highly effective for optimizing farm plans, they can also be complex. However, Microsoft Excel and other simple software tools offer built-in functions that facilitate these calculations. In practice, small-scale farmers may require specialized assistance from extension workers to apply these methods effectively. This topic is further explored in Part 4 of this book.

On a practical level, many smallholder farmers may have limited access to computers

Table 13.3. Linear programming matrix for farm planning.

Variables	Crop 1 maize	Crop 2 beans	Crop 3 tomatoes	Crop 4 wheat	Total available
Land (ha)	X_1	X_2	X_3	X_4	$\leq$ 5 ha
Labour (hours)	L_1X_1	L_2X_2	L_3X_3	L_4X_4	$\leq$ Total labour
Capital ($)	C_1X_1	C_2X_2	C_3X_3	C_4X_4	$\leq$ Available capital
Water (litres)	W_1X_1	W_2X_2	W_3X_3	W_4X_4	$\leq$ Water supply
Fertilizer (kg)	F_1X_1	F_2X_2	F_3X_3	F_4X_4	$\leq$ Fertilizer supply
Profit ($/ha)	P_1X_1	P_2X_2	P_3X_3	P_4X_4	Maximize $\Sigma(PX)$

Box 13.3. Not exceeding constraints.

Each resource has a finite limit, and the farmer must allocate it efficiently across competing crops. For example, the total land allocated to maize, beans, tomatoes and wheat cannot exceed the available 5 ha. Similarly, the labour hours, capital, water and fertilizer used for all crops must stay within the farm's capacity. These constraints ensure that the farm operates within its resource limits while maximizing profitability.

and the necessary software to implement linear programming independently. As a result, the widespread application of these techniques relies on support from specialized agricultural advisors in farm management, or private sector business service providers to bridge the technological gap and make data-driven decision making more accessible. An example of an application of linear programming is given in Box 13.4.

13.6.3 Whole-farm planning

Whole-farm planning is a core approach in farm management also focusing on the optimal allocation of resources and the selection of farm enterprises to maximize efficiency and profitability. It similarly ensures that land, labour and capital are distributed strategically across competing farm enterprises to achieve the highest economic returns. This approach, sometimes referred to as farm programme planning, serves as a simplified form of linear programming, utilizing basic budgeting techniques and stepwise allocation methods to help farmers explore different production plans, evaluate trade-offs and determine the most feasible and profitable strategy.

A whole-farm plan provides an overview of available resources, enterprise options and the objectives of the farm family. Farmers may need to adjust their farming systems, modify enterprise selection or scale, and adapt to market changes, production techniques or marketing strategies. The plan outlines expected income from chosen enterprises, detailing methods and levels of production, cost estimates and projected revenue. It also considers market outlets, input sources, financial requirements, labour availability and risk management. Additionally, a spatial representation of the farm may be included, indicating field locations, enterprise distribution and crop rotation plans.

Essentially, developing a whole-farm plan involves two interconnected stages. The first is the preparation stage, where the farmer evaluates options for enterprises, production systems, markets and other factors. The second is the finalization stage, where the farmer consolidates decisions into a coherent plan for implementation. The farm plan represents all the decisions the farmer has made regarding what to produce, how to produce it, how much to produce and where to sell the products.

To develop an effective farm plan, the following data is essential:

- Inventory of available resources (land, labour, capital, infrastructure).
- Potential crop and livestock enterprises that align with market demand.
- Resource requirements per unit of each enterprise (e.g. input needs, labour requirements).
- Enterprise gross margins to assess profitability.

Once enterprises are identified and resources assessed, farmers can evaluate enterprise combinations to maximize total income while ensuring feasibility in terms of land, labour, time constraints and financial viability. A well-structured whole-farm plan supports optimal decision making, helping farmers achieve greater productivity, profitability and sustainability.

When preparing the plan, the farmer needs to take into account the objectives, the possible enterprises that can be grown and the resources that are at his or her disposal. The plan preparation should include consideration of both existing and new enterprises that have market potential and align with the farmer's goals. The whole-farm planning layout is similar to that

Box 13.4. Ameena uses linear programming.

Ameena, a farmer with 5 hectares of land, planned her planting season carefully. Her primary crop is rice, and she cannot abandon that. But she also considered growing vegetables for higher profit margins. However, she faced constraints: only 120 hours of labour were available, and labour-intensive vegetables required more time per hectare than rice.

Using a linear programming approach, Ameena with the assistance of her extension worker defined her objective to maximize profit:

($Z = \$500x_1 + \$800x_2$)

They calculated her constraints as:

$x_1 + x_2 \leq 5$ha for land (the amount of land planted to rice plus the amount of land planted to vegetables must be less than or equal to 5 ha.)

($20x_1 + 30x_2 \leq 120$ hours) for labour (the amount of labour used on the two crops cannot be more than 120 hours)

They then solved the problem.

The solution revealed that Ameena should plant 3 ha of rice and 2 ha of vegetables, optimizing her limited resources for maximum returns. At harvest, her profit soared, reaffirming the power of mathematical planning in farming.

Here are the steps followed by Ameena to prepare her farm plan.

Step 1: What is Ameena Trying to Do?

She wants to make the most money from her 5 hectares of land.

- She must plant some rice (x_1)
- She is considering planting vegetables (x_2)

Goal (Profit):
Make as much profit as possible
Profit = \$500 × rice hectares + \$800 × vegetable hectares
So: $Z = 500x_1 + 800x_2$

Step 2: What Are Her Limits? (Constraints)

1. Land Limit:
She has only 5 hectares total
$\rightarrow x_1 + x_2 \leq 5$

2. Labour Limit:
She has only 120 hours of labour

- Rice takes 20 hours/ha
- Vegetables take 30 hours/ha

$\rightarrow 20x_1 + 30x_2 \leq 120$

Step 3: What Are Her Options?

Ameena (with her extension worker) checks which combinations of rice and vegetables fit her limits and give the most money.

She tries a few options:

Continued

Box 13.4. Continued

Option (x_1, x_2)	Lamd Used	Labour Used	Profit (Z)
(0, 0)	0	0	$0
(5, 0)	5	100	$2,500
(0, 4)	4	120	$3,200
(3, 2)	5	120	$3,100

But (0,4) means no rice, which is not acceptable.
So the best practical solution is:
3 hectares rice and 2 hectares vegetables

Step 4: What Should Ameena Do?

- Plant 2 ha of vegetables
- Plant 3 ha of rice
- Use all 5 ha of land and all 120 hours of labour
- Make a total profit of $3,100

Why This Worked

- She used maths to test her options
- She stayed within her land and labour limits
- She respected the need to grow rice
- She chose the option with the highest profit and practical balance

This is how linear programming helps small farmers make better decisions.

presented in the linear programming matrix. A typical whole-farm planning format is given in Box 13.5.

To develop the farm plan, the farmer requires the following basic data:

- inventory of resources;
- potential crops and livestock enterprises;
- resource requirements per unit of each enterprise; and
- enterprise gross margins.

Once the potential enterprises have been identified and the available resources have been assessed, the farmer can evaluate combinations of enterprises. The goal is to maximize total income while ensuring technical feasibility (e.g. land, labour and time constraints) and financial viability (e.g. cash flow and profit margins). An example of a linear programming matrix is given in Box 3.6.

How should the farm plan be prepared?

The farm plan involves several steps:

1. The plan should first be looked at in physical terms. An assessment will be needed of the resources available to the farmer and the area of each crop and number of each type of livestock available. This stage also involves identifying the potential crops that can be grown from the farmer's resource base.
2. The next step of the plan is to prepare budgets of the different enterprises that can be produced. The enterprise budgets or gross margins are brought to a per unit basis and in this way are compared. The range of enterprises should also include new enterprises identified through the market assessment as previously mentioned. The farmer should consider such decisions as whether to specialize in a few enterprises

Box 13.5. A planning format.

The typical format of a whole-farm plan and budget depends on the number and kind of enterprises involved and the details desired by the planner. A complete plan and budget, however, usually contains the following elements:

- Farm inventory.
- Production plan showing the choices of enterprises including their extent.
- Calendar of operations.
- Marketing plan.
- Labour plan.
- Financial plan including a cash flow schedule.
- Risk management plan.
- Summary budget.

or alternatively, to diversify production and spread risks. Supplementary enterprises such as small-scale livestock can also be included especially if they are held around the homestead and do not compete for scarce resources. These enterprises will add to the overall farm profit.

3. The next step is to use the enterprise budgets to determine which of the alternative crops and livestock enterprises would yield the highest gross margin for the whole farm. This is done while considering the constraints of the limited supplies of labour and sometimes capital.

   While farmers have a number of resources under their management control, there may be one or two resources that are really in short supply and prevent the farmer from expanding profits on the farm. These are called the most limiting resource, the resource that limits production. Ways are used to identify which resource it is.

   Once the limiting resource is found, the problem facing the farmer is to know how much of an enterprise to produce and with what resources to maximize profit. Enterprises would then be selected in order from the highest gross margin per unit of limiting resource to the lowest. A selected enterprise would be expanded until the limiting resource is all used up.

   A problem often encountered is to find which resource it is that is the most limiting. The farmers' judgement should be used to assess this, as they are most likely to know. Most often land is treated as the limiting resource and planning is conducted on the basis of gross margin per hectare, but this may not always be suitable, especially among small farms where family labour is often in scarce supply. Very often the limiting resource depends on local circumstances.

4. The farm plan is initially constructed on the basis of per unit (land, livestock) gross margin data. In order to estimate the net farm income, the total gross margin generated from the farm plan must take into account the fixed costs. This generates the whole-farm income and the annual profitability of the farm as a whole.

   It should be remembered that whole-farm income and profitability are not the same as the cash the farm family earns. To know precisely the amount of cash available, the non-costs such as family labour and depreciation must be deducted from the farm income.

5. The next stage is to test the sensitivity of the plan to key variables. This can be done in relation to the goals of the farm family (profit, food, leisure), availability of cash, labour supply and food requirements. A sensitivity analysis is required that repeats some of the enterprise budget calculations for different values of the key variables, indicating to the farmer the risks involved.

6. The final stage is to compile the farm plan.

13.7 Planning Labour

As discussed in Chapter 6, labour management is a crucial aspect of farm management, particularly for small family farmers who rely

Box 13.6. Example of optimizing income and maximizing available resources.

The basic objective of individual farmers is to optimize farm income given the available resources. For this, the farmer should choose the most profitable combination of enterprises suitable for a given agro-ecological location along with available resources for preparing the farm plan. This example involves a few enterprises and limited resources and as such can be resolved by applying a simple farm planning technique as described in the following table.

Budgeted gross margin by enterprise.

Enterprise	Total cost ($)	Cash cost ($)	Yield (mt/ha)	Gross income ($)	Gross margin ($)
Rice	21,735	12,730	2.8	30,800	9,065
Cauliflower	44,546	29,740	15.0	120,000	75,454
Cabbage	44,820	29,820	25.0	100,000	55,180

Basic information for farm plan:

1. **Determine the resource availability and requirement**
   - **Land availability:** in our example the farmer operates 0.76 ha of irrigated land.
   - **Capital:** the farmer has $16,000 cash in hand for crops grown in the summer season.
   - **Labour:** the farmer has two full-time adult family members. The hired cost of labour is $60/day.

In the following table, rice, cabbage and cauliflower are considered as the competitive crops.

Resource requirement by each crop

Resource	Available	Resource requirement by each crop/ha		
		Rice	Cauliflower	Cabbage
(1)	(2)	(3)	(4)	(5)
Rice land	0.76 ha	1		
Cauliflower land	0.40 ha		1	
Cabbage land	0.40 ha			1
Family members in September	30 days	28	50	50
Cash available/ requirements ($)	16,000	12,730	29,740	29,820
Gross margin per ha ($)		9,065	75,454	55,180

The table lays out a matrix to enable calculations. It gives three crop enterprises: rice, cauliflower, and cabbage. 30 family member days are available in September. The cash available is $16,000. All of the figures under resource requirements are for 1 ha. For example, 1 ha of rice requires 28 days from family members in September and $12,730 cash. The same applies for the other crops.

The total land available is 0.76 ha. All 0.76 ha is suitable for rice. Only 0.40 ha is suitable for cauliflower and/or cabbage. This means that out of 0.76 ha, 0.40 ha is suitable for vegetable production. If this is used to produce vegetables, the remaining 0.36 ha is available for rice.

The labour requirement for the month of September is then estimated for each crop. The total available family labour is 30 days for this month. Cash requirements and cash available to the farmer are also shown in the table.

Now that we have information on requirement and availability of resources such as land, labour and capital resources, the relevant question is: how can the resources be allocated so that profit can be maximized? For this, we divide the available resources by requirement and take the smallest number as the determinant. In this case, there is $16,000 cash available and 30 days family labour.

Continued

Box 13.6. Continued

2. **Obtain the crops with the highest gross margin**

Looking at the table, the crop with the highest gross margin is cauliflower. The area allocated for cauliflower is 0.40 ha out of the total of 0.76 ha. To determine the family labour required to cultivate cauliflower, we multiply 50 person days (the requirement as shown in the table) with 0.40 ha (the area allocated for cauliflower). Why is this done? Remember that the resource requirement is for one hectare of the crop, thus, we need to determine the proportion for 0.40 ha. Based on this, the family labour required for 0.40 ha is 20 (0.40 × 50) days out of 30 days available, leaving 10 days unallocated.

We now consider the crop with second highest gross margin, in this case cabbage. Due to land constraint, because the 0.40 ha suitable for vegetable growing has already been allocated to cauliflower, no cabbage can be grown. The remaining 0.36 ha of land, therefore, can now be planted with rice. Similarly, the remaining 10 days of family labour, which is the required labour for 0.36 ha rice, can now be used.

To calculate the capital requirements for cauliflower, we multiply 0.40 ha by $29,740/ha which gives us $11,896. Subtracting $11,896 from $16,000 (cash available), we have $4,104 left. Again, we cannot grow cabbage due to cash limits (we need $11,896 also for this). We can allocate the remaining $4104 for rice. However, we will have to limit our rice area to only 0.32 ha. Remember, 1 ha of rice requires $12,730. Therefore, 0.36 ha would require $4583. We have only $4104 left which is approximately 90% of $4583. Therefore, we can plant 90% of the remaining 0.36 ha which is 0.32 ha.

The combination of cauliflower and rice generates the highest income compared to alternatives, i.e. $30,182 ($75,454 × 0.40) from cauliflower and $2,900 ($9.065 × 0.32) from rice generates a total farm income of $33,082.

From this example, it is clear that budget analysis does not follow strict rules. It is quite simple but could be a little tedious if more resource constraints and activities exist. (In such cases computer packages such as linear programming are available for optimization of income or minimization of cost of resources.)

primarily on family labour, supplemented by casual workers during peak periods such as planting and harvesting. In that chapter we covered key labour management concepts, whereas this chapter focuses on the practical aspects of planning labour for the farm, with an emphasis on differentiating labour roles and responsibilities based on gender dynamics.

Effective labour planning allows farmers to balance family and hired labour with mechanization, ensuring that workloads are managed efficiently without overwhelming the workforce. Recognizing gender roles in farm labour is critical, as men and women often have distinct responsibilities in agricultural production. By integrating gender-sensitive labour planning, farmers can improve productivity, promote equity and ensure that all labour contributions are valued.

13.7.1 Assessing labour needs and requirements

Assessing labour needs is an essential step in planning labour, helping farmers determine the quantity and type of labour required for different farming activities. This assessment must consider farm size, crop or livestock production, seasonal variations and technological advancements, ensuring that tasks are allocated efficiently and fairly.

When assessing labour needs, farmers should take into account:

- Skill levels and experience required for various tasks, recognizing that men and women may have different skill sets and knowledge.
- Physical capabilities, considering tasks that require strength vs those that require precision or specialized knowledge.
- Availability of labour, including family members, hired workers and mechanization options, while acknowledging the role of women in both productive and reproductive labour (e.g. childcare responsibilities).

By conducting a gender-inclusive assessment, farmers can ensure equitable distribution of work, better use of labour resources, and fair compensation for all workers. This also informs labour recruitment and training strategies to enhance skill development among both male and female workers.

13.7.2 Developing a labour profile

Developing a labour profile involves creating a detailed overview of farm labour requirements, ensuring that gender differences in tasks and roles are taken into account for each position or task (see Box 13.7.). Farmers should identify specific responsibilities and outline the skills, experience and any required certifications or training for each role.

The labour profile should include information on:

- Tasks performed and the estimated time required, ensuring equitable task distribution between men and women.
- Frequency of tasks and their physical demands, recognizing work that may disproportionately affect one gender.
- Seasonal variations in labour demand, identifying when additional labour (casual workers) may be needed.

By developing a labour profile, farmers can better understand their labour needs, ensure fair workload distribution and equal access to opportunities for men and women, and make informed decisions regarding labour allocation, recruitment and training. It also helps farmers identify any skill gaps or training needs within their labour force and develop strategies to address them, such as providing training programmes or seeking external expertise when required.

In addition, the labour profile serves as a valuable resource for workforce management, allowing farmers to plan work schedules, allocate tasks efficiently and ensure that the right people with the appropriate skills are assigned to specific jobs. It also supports effective communication and coordination among the labour force, leading to improved productivity and overall farm performance.

More specifically, the labour profile shows graphically the seasonal labour requirements of each enterprise and the total demand of all enterprises for each month of the year (see Box 13.8.). The procedure for constructing a labour profile is as follows:

1. Calculate the person days required for each enterprise;
2. Calculate the monthly requirements in person-days for each enterprise;
3. Construct a labour profile for all farm enterprises as a whole;
4. Assess the person-days available to the farmer from the family labour supply; and

Box 13.7. The importance of planning and managing labour.

Labour costs are often a high percentage of the total production costs. It is essential, therefore, to carefully plan the use of family and hired labour. More efficient use of labour can be planned on two levels: (i) the individual enterprise; and (ii) the whole farm.

At the individual enterprise level, labour planning is used to improve the performance of the different operations associated with the enterprise. Here the problems are essentially practical and deal with the way operations are conducted. At the whole-farm level, best use of labour throughout the year is assessed, focusing on securing and deploying labour most efficiently and cost effectively. These two levels, however, are closely linked. Any change in the type of farm enterprises selected and farm operations requires farmers to examine the labour resources and requirements. Labour requirements in peak periods (e.g. transplanting, weeding or harvesting period) could be met partially through the employment of part-time or casual labour.

A farm will rarely carry sufficient full-time labour to cover all its demands throughout the year. If it does, it is likely to have excess labour at various times of the year or season. It is necessary to consider the peaks and troughs of seasonal labour availability in relation to the farm labour requirements. To do so, it is necessary to calculate the standard figures of labour requirements for the different crops and livestock within the farm.

Box 13.8. Example: constructing a labour profile for Aaron's farm.

The first step in creating a labour profile involves calculating the person days required for each enterprise, as shown in the following table.

Enterprise	Available person days	Required person days
Cabbage	80	90
Peppers	130	140
Yams	130	140
Carrots	70	80
General farm work	62	68
Total	472	518

The following table gives an example of person day requirements for some agricultural enterprises by smallholder farmers in Asia.

Enterprise	Labour availability (person days per ha or head)	Jan	Feb	Mar	Apr	May	Jun	July	Aug	Sep	Oct	Nov	Dec
Cabbage	80	0	0	3	5	0	0	0	18	24	25	4	0
Pineapple	120	0	0	4	8	0	0	28	37	24	13	7	0
Peppers	130	0	0	4	9	0	0	30	40	33	8	7	0
Yams	130	0	0	40	0	4	0	0	32	33	5	12	5
Carrots	70	0	0	19	2	2	0	0	26	9	3	6	3
Tannia	240	0	0	74	0	0	0	0	72	54	10	21	10
Sweet potato	120	0	0	11	22	3	6	6	1	12	45	14	2
Ginger	120	0	0	6	0	90	18	3	3	0	0	0	0
Sugar cane	120	11	11	11	10	9	9	9	9	9	10	11	11
Dairy cows	60	5	5	5	5	5	5	5	5	5	5	5	5
Beef cattle	45	4	4	4	4	3	3	3	4	4	4	4	4
	Total	20	20	181	65	116	41	84	247	207	128	91	40
General farm work (approx. 15%)		3	3	27	10	17	6	12	37	31	19	14	6
Total general		23	23	208	75	133	47	96	284	238	147	105	46

The following graph offers a representation of the labour profile.

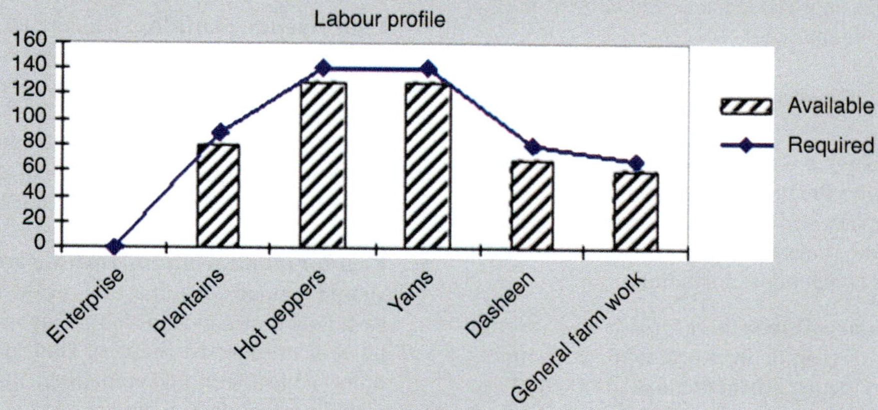

5. Examine the labour supply and demand profiles and formulate a strategy for dealing with labour shortfalls and surpluses.

As noted above, many of the farm operations are carried out specifically by women and the labour profiles should be disaggregated and broken down by gender.

It is necessary to consider the highs and lows of seasonal labour availability in relation to the farm labour requirements. By modifying the cropping pattern and making changes to the enterprise operations it is possible to achieve a better allocation of labour and ensure its more efficient use. Low periods, for example, may be used for conducting more general farm maintenance operations. And labour requirements during high periods (e.g. transplanting, weeding, harvesting, etc.) could be met through hiring part-time or casual labour or alternatively by the introduction or more efficient use of mechanization, or a combination of both.

An example seasonal calendar framework is given in Table 13.4.

13.8 Financial Cash Flow Planning

Cash flow planning provides farmers with a clear understanding of their financial position and helps them anticipate and manage cash flow challenges. It allows farmers to plan for capital investments, operational expenses, debt servicing and unforeseen contingencies. By having a solid grasp of their cash inflows and outflows, farmers can make informed decisions, maintain financial stability and improve overall farm profitability.

13.8.1 Developing a cash flow projection

Developing a cash flow projection involves estimating the timing and amounts of expected cash inflows and outflows. To create an accurate cash flow projection, farmers need to consider various components, including:

- **Cash inflows:** these include cash revenue from crop or livestock sales, government payments, rental income, grants and any other sources of cash inflows. Farmers should estimate the timing and amounts of these inflows based on market prices, marketing plans (including contracts), production forecasts and historical data.
- **Cash outflows:** these encompass all farm-related cash expenses, such as input costs, labour expenses, utility bills, loan repayments, taxes, insurance premiums, equipment maintenance and other operational costs. Farmers should estimate these expenses based on past records, projected production levels, market conditions, contractual obligations and upcoming financial commitments.
- **Timing considerations:** it is essential to consider the timing of cash inflows and outflows accurately. Some expenses may be seasonal or have specific due dates, while revenue may be received at different intervals throughout the year. Understanding the timing of cash flows helps farmers anticipate periods of surplus or deficit and plan accordingly.

13.8.2 Managing cash flow challenges and uncertainties

Cash flow was touched on in Chapter 6, when discussing the need for finance. Cash flow challenges and uncertainties are common in agriculture due to factors such as market volatility, weather variability, price fluctuations and production risks. Effective cash flow planning can help farmers manage and mitigate these challenges. Here are some strategies to consider:

- **Contingency planning:** building contingency plans into the cash flow projection allows farmers to anticipate and address potential cash flow shortfalls. This may involve creating reserves or exploring financial instruments such as lines of credit or operating loans to bridge temporary gaps.
- **Regular monitoring and review:** farmers should regularly monitor and review their cash flow projections to identify any deviations or unexpected changes. This enables timely adjustments and proactive measures to address cash flow challenges.

Table 13.4. Example seasonal labour calendar: labourers required (person days).

1 Ha Maize, 1 ha Sweet potato and 10 livestock

Month Activity	Jan	Feb	Mar	Apr	May	Jun	Jul	Aug	Sep	Oct	Nov	Dec	Total
FARM ACTIVITIES													
Buying inputs		1					1						2
Ploughing										1	1		2
Planting											2	2	4
Weeding	3	3		2									8
Fertilizing											1	1	2
Herding cattle	1	1	1	1	1	1	1	1	1	1	1	1	12
Harvesting					3	3	3						9
Remove stover									2				2
Marketing					1	1	1						3
Farm maintenance								2	2				4
Total Farm Labour	4	5	1	3	5	5	6	3	5	2	5	4	48
HOUSEHOLD ACTIVITIES													
Fetching water	0.5	0.5	0.5	0.5	0.5	0.5	0.5	0.5	0.5	0.5	0.5	0.5	6
Fetching firewood & fuel	0.5	0.5	0.5	0.5	0.5	0.5	0.5	0.5	0.5	0.5	0.5	0.5	6
Food preparation	0.5	0.5	0.5	0.5	0.5	0.5	0.5	0.5	0.5	0.5	0.5	0.5	6
Cleaning	0.5	0.5	0.5	0.5	0.5	0.5	0.5	0.5	0.5	0.5	0.5	0.5	6
Tending to children	0.5	0.5	0.5	0.5	0.5	0.5	0.5	0.5	0.5	0.5	0.5	0.5	6
House repairs				1			1				1		3
Total Household Labour	2.5	2.5	2.5	3.5	2.5	2.5	3.5	2.5	2.5	2.5	3.5	2.5	33
Total Labour Required	6.5	7.5	3.5	6.5	7.5	7.5	9.5	7.5	7.5	4.5	8.5	6.5	83
Family Members Available	5	3	3	4	4	3	4	3	4	4	4	5	46
Additional Labour Required	1.5	4.5	0.5	2.5	3.5	4.5	5.5	4.5	3.5	0.5	4.5	1.5	37

- **Diversification and risk management:** diversifying income sources, implementing risk management strategies, and adopting practices that reduce the impact of external factors can help minimize cash flow uncertainties. For example, crop diversification, forward contracting and crop insurance can mitigate the risk of price volatility and production losses.
- **Effective working capital management:** efficient management of working capital, including inventory control, timely invoicing, credit management, and negotiation of payment terms with suppliers and customers, can optimize cash flow and liquidity.

By adopting these strategies and incorporating them into the cash flow planning process, farmers can better manage cash flow challenges, reduce financial stress, and maintain a healthy and sustainable farm operation.

13.9 Investment Planning

Assessing investment opportunities involves evaluating potential projects or initiatives that require capital investment. Farmers need to consider various factors, including market demand, production feasibility, technological requirements, resource availability and financial viability. By conducting a thorough analysis of investment opportunities, farmers can identify projects that align with their farm's strategic direction and have the potential for favourable returns.

Alongside assessing opportunities, it is crucial to evaluate the risks associated with investments. As noted in previous chapters of this book, risks can include market fluctuations, weather variability, technological uncertainties and production risks, as well as policy changes. Farmers need to carefully analyse and understand the potential risks involved in each investment opportunity to make informed decisions and develop appropriate risk mitigation strategies. Risk in farm planning is elaborated on further in Section 13.10, this chapter.

13.9.1 Financial analysis and evaluation of investment projects

Financial analysis is essential for investment planning, allowing farmers to assess the feasibility and profitability of potential investments. A key factor in decision making is *economies of scale* – while purchasing larger machinery may reduce per-unit costs, it may not be financially viable if the scale of operations is too small. Farmers must determine their break-even point to ensure the investment generates sufficient returns.

Another crucial aspect is *opportunity cost* – how allocating resources to one investment affects alternative opportunities. For instance, investing in irrigation may enhance yields but could delay spending on farm mechanization. Farmers need to weigh these trade-offs to prioritize investments with the highest returns.

To assess financial viability, key metrics such as the payback period, simple rate of return, NPV and IRR provide insights into potential returns and risks. A comprehensive evaluation should consider capital costs, expected revenues, operating expenses and cash flows over time. By applying financial analysis, farmers can make informed investment decisions, optimize resource allocation and enhance long-term profitability.

13.9.2 Simple methods of investment appraisal

The farmer can use different methods to appraise investments that have different timings of costs, returns and benefits. Investing in capital items typically means that a large expense (the initial purchase) occurs in one period while the benefits are distributed over several future periods. Investment appraisal requires background information that includes an estimate of the annual net income from the investment, the initial cost of the investment, the salvage value and the interest or discount rate to be used. The simplest methods for assessing the profitability of an investment are:

- Payback period.
- Rate of return.

Each of these methods has advantages and disadvantages which will be discussed. Since investment decisions can be quite complicated, extension workers may need to assist farmers in selecting the appropriate method and in making the correct decisions.

The payback period method

The payback period method calculates the number of years it would take for an investment to return its original cost through the generation of income. This method assesses the time that is required until the cumulative income arising from the investment equals its initial cost. The method is also used to highlight those investments that are not viable (e.g. ones that never achieve payback). It can also be used to select the most appropriate source of finance. For instance, those investments having a short payback period would only require short-term finance.

To calculate the payback period, the farmer needs to assess the annual cash flow[1] that is expected after making the investment. This is the difference between the cash inflows and cash outflows for each year in the future. In comparing any two investments, one with a shorter payback period would be more attractive than one with a long payback period (see Box 13.9.).

The payback method is attractive because it is easy to calculate and is a simple way of comparing alternative investments. It is also easily understood by farmers. If the farm is short of cash, it is essential that cash income is generated. However, the payback method also has deficiencies. It ignores the cash flows arising after the payback has occurred as well as the timing of the income flows. Money received earlier in the life of the investment is more valuable than money received later. For example, by selecting investment A, the method ignores the higher cash flows from investment B in year 4 and year 5 as well as the greater net cash flow. The payback period does not really measure profitability but is more a measure of how quickly the investment will contribute the farm's liquidity.

Rate of return method

The rate of return recognizes that it is not only income that is important to the farm but also the amount of capital used to produce it. Income is regarded not by itself but as a return on the capital used. More precisely, the rate of return expresses the average annual net income as a percentage of the investment. The concept of net income is used and is calculated by subtracting the average annual depreciation of the investment from the average annual net cash.

The rate of return on capital is then calculated as follows:

$$\textit{Rate of return} = \frac{\textit{average annual net income} \times 100}{\textit{cost of investment}}$$

A return on capital is compared to the cost of capital and is used in comparing alternative investments.

For the example cited previously see Fig. 13.3.

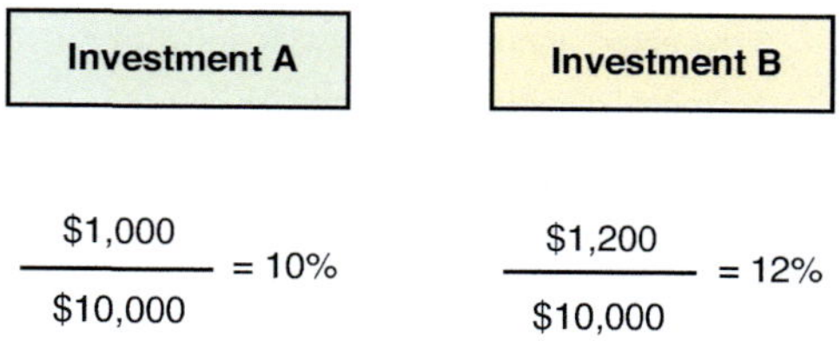

Fig. 13.3. Comparative return on capital.

This method ranks investment B higher than investment A.

This result is different from using the payback period method. The rate of return is preferred to the payback period because it considers the earnings of the investment over its entire life. The advantage of this methodology is that it is simple to calculate and to rank competing investment opportunities. Investments with higher rates of return are naturally preferred. However, it uses average annual income and thereby fails to consider the size and timing of the annual income flow possibly making errors in investment selection. This is particularly true when there is increasing or decreasing net income. For example, investment A would have the same 10% rate of return if there was no net cash flow for the first 4 years but $15,000 in year 5. The average cash flow would remain the same at $3000 per year. Furthermore,

Box 13.9. Example annual cash flow for two investments.

For the two investments outlined here, both need an initial capital outlay of $10,000 but give rise to different cash flow patterns. For simplicity, the salvage values are assumed to be zero.[2] It is convention to treat the initial cost of the investment as if it occurs in year 0. When the annual cash flow is the same for each year of the investment, the payback period is calculated by dividing the amount of the investment by the expected annual cash flow. When the cash flows are irregular and unequal they are accumulated, and the payback period is assessed by taking the year where the cash flow is equal to the cost of the investment.

- **Investment A:** the payback period is 3.3 years. The cost of the investment ($10,000) is divided by the annual cash flow ($3,000).
- **Investment B:** the payback period is 4 years. The cumulative net cash flow covers the cost of the investment ($10,000) in year four.

Therefore, investment A is preferred over investment B because it has the shorter payback period.

Year	Investment A	Investment B
0	$10,000	$10,000
1 expected annual cash flow	$3,000	$1,000
2 expected annual cash flow	$3,000	$2,000
3 expected annual cash flow	$3,000	$3,000
4 expected annual cash flow	$3,000	$4,000
5 expected annual cash flow	$3,000	$6,000
Total	$15,000	$16,000
Payback period (years)	3.3	4.0
Average annual cash flow (5 years)	$3,000	$3,200
Less annual depreciation	–$2,000	–$2,000
Average annual net income	$1,000	$1,200

the method ignores the time value of money, discussed in Chapter 9, and elaborated on here.

Discounted investment appraisal

The previous section considered some simple methods of investment appraisal. This section of the book focuses on the application of the three most common discounted methods for assessing the profitability of an investment. These are:

- Net present value (NPV).
- Internal rate of return (IRR).
- Cost- benefit analysis

Refer back to Chapter 9 to review the concept of discounting and to Chapter 11 for its computation under section 11.8 - Cash Flow Budgeting.

Net present value (NPV)

The NPV of an investment is the sum of the present values for each year's net cash flow less the initial cost of the investment. It is also called the discounted cash flow (which was discussed earlier) method since it uses the discounting method of analysis. This method considers the time value of money and the stream of cash flows over the entire life of the investment.

A general rule regarding the number of years to include in the cash flow analysis is to choose a period of time that is comparable to the life of the major investment item. In the case of farm machinery and equipment, this is often taken as between 5 and 7 years and for buildings 30 to 40 years.

With this method, the farmer would accept an investment with a positive NPV, reject those with a negative NPV and be indifferent to a zero value. The rationale behind accepting investments with a positive NPV can be explained in two ways. First, it means the rate of return on the investment is greater than the discount rate used in the calculations. A second explanation is that the farmer can afford to pay more for the investment and still achieve a rate of return equal to the discount rate used in calculating the NPV (see Box 13.10.).

The selection of the discount rate consequently affects the result of the appraisal. When a higher discount rate is used, the NPV decreases and vice versa. At some higher discount rates, the NPV values would be zero, and at an even higher rate it could become negative.

An example of calculating the discount factor is given in Box 13.11.

An example of an NPV calculation using a discount rate of 8% is given in Box 13.12. In this example, the farmer could pay up to $11,979 ($10,000 + $1,979) for investment A and $12,048 ($10,000 + $2,048) for investment B and still receive an 8% return or more on invested capital. Both investments show a positive NPV using an 8% discount rate.

Box 13.10. NPV in Smallholder Farming

Net Present Value (NPV) is a financial calculation used to determine whether a farming project or investment is profitable over time by comparing the present value of expected benefits (e.g. income from crops or livestock) with the present value of expected costs (e.g., seeds, labour, and fertilisers). It accounts for the fact that money earned or spent in the future is worth less than money available today due to inflation, risks, and alternative uses of funds.

In simple terms:

- If the NPV is positive, the project is expected to make a profit and is worth pursuing.
- If the NPV is negative, the project is expected to result in a loss.

For example, a smallholder farmer considering investing in a new irrigation system would calculate the NPV to determine if the long-term increase in yields and income outweighs the upfront costs and annual maintenance expenses when considering the time value of money.

Box 13.11. Finding the present value

Find the present value of $6,438 received 9 years in the future if the discount rate is 15%?
The relevant factor in the discounting tables shows that at a 15% rate and for a period of 9 years, the coefficient* is 0.284. The future amount is multiplied by the discount factor to obtain a present value of $1,828 ($6 438 × 0.284 = $1,828).
The formula for calculating the discount factor is as follows:

$$\text{Discount Factor} = \frac{1}{(1+r)t}$$

Where:

- r = discount rate (expressed as a decimal, e.g., 10% = 0.10)
- t = time period (year) of the cash flow

*The coefficient for calculating Net Present Value (NPV) is the discount factor, which adjusts future cash flows to their present value. It is based on the chosen discount rate (reflecting the time value of money) and the year in which the cash flow occurs.

Box 13.12. Calculating NPVs using an 8% discount rate

Year	Investment A			Investment B		
	Net cash flow $	Discount Rate	Present Value $	Net cash flow $	Discount Rate	Present Value $
1	$3,000	.926	2,778	1.000	.926	926
2	$3,000	.857	2,571	2.000	.857	1,714
3	$3,000	.794	2,382	3,000	.794	2,382
4	$3,000	.735	2,205	4,000	.735	2,940
5	$3,000	.681	2,043	6,000	.681	4,086
		Total	11,979			12,048
		Less investment cost	10,000			10,000
		Net present value	1,979			2,048

Internal rate of return (IRR)

The Internal Rate of Return (IRR) is a financial measure used to evaluate the profitability of a farming project or investment by calculating the discount rate at which the net present value (NPV) of all future cash flows (inflows and outflows) equals zero. This is the maximum rate of interest that the farmer can afford to pay for the resources used in order to recover the original investment and its operating costs and still break even. In simple terms, it represents the percentage return the farmer can expect to earn on their investment. From another angle the IRR can be understood as the earning capacity of the investment.

The formal selection decision for the internal rate of return is to accept all investments with a return equal to or greater than the discount rate. If the IRR of the investment as a whole is lower than the discount rate, the investment is not profitable. If the IRR is higher than the farmer's rate of return from other investments, the project is worth pursuing. If the IRR exceeds the interest rate, the farmer may try to borrow as much as possible to increase the return on his or her resources. For smallholder farmers, the IRR helps answer the question: "*How efficient is this investment at generating returns compared to its costs?*"

An example of an IRR calculation for an investment is given in Box 13.13. Since there is no formula for finding the IRR, a trial and error procedure is needed to equate the net present value of the cash flow with the zero break-even point. An interest rate of 14% was arbitrarily chosen as a first estimate of the IRR. The calculation shows a positive NPV of $296, indicating that the IRR is still high and in excess of zero, the break-even point. An even higher discount rate is used next (16%) assuming that this will result in a lower NPV. The result shows that there is a negative IRR of -$178. The actual IRR lies somewhere between 14% and 16%.

The most difficult aspect of this trial-and-error process is making an initial estimate. If the estimate is too far from the final result, then several trials will have to be made to find two rates that are sufficiently close enough to allow the breakeven point to be calculated. This is done by interpolation, i.e. finding a desired value between two other values.

The rule for interpolating the value of the IRR between two discount rates (with a negative and positive NPV) is as follows:

$$\text{Low discount rate} + \frac{\text{High NPV}}{\text{High NPV} - \text{Low NPV}} \times (\text{High discount rate} - \text{Low discount rate})\ \%$$

Box 13.13. An example of calculating the IRR for an investment

Year	Net cash flow $	14%		16%	
		Present Value Factor	Present Value $	Present Value Factor	Present Value $
1	3,000	.877	2,631	.862	2,586
2	3,000	.769	2,307	.743	2,229
3	3,000	.675	2,025	.641	1,923
4	3,000	.592	1,776	.552	1,656
5	3,000	.519	1,557	.476	1,428
		Total	10,296		9,882
		Less Investment Costs	-10,000		-10,000
		NPV	296		-178

This procedure is applied to the above example. The lower discount rate is 14%. The difference between the two rates is 2%. The present worth of the lower discount rate cash flow is $296 and the present worth of the higher rate is -$178. The sum of the present worth of the cash flows at the two discount rates, ignoring the signs, is $474.

Through interpolation the IRR is calculated as follows:

$$14 + \frac{296}{296 - (-178)} \times (16 - 14)\,\% = 15.20\%$$

At a discount rate of 15.2% the investment just breaks even.

Finally, it should be pointed out that an IRR can only be calculated when at least one value in the net cash flow is negative. If all the values are positive, no discount rate can make the net present value of the cash flow equal to zero. In this event the IRR would be infinite.

Each of these methods has advantages and disadvantages and will be discussed in the course of this section. Since investment decisions can be quite complicated, extension workers will likely need to assist farmers in selecting the appropriate method and making the correct decisions. But before looking at these methods, it is important to understand the time value of money and take it into account when discounting.

Selection of discount rate

To use discounted measures of investment appraisal, it is essential to choose an appropriate discount rate. This rate usually reflects the cost of capital for the farmer - the expense of acquiring funds to finance the investment.

For most small family farmers, the cost of capital is often measured as the interest rate on loans they can access. However, if the investment is funded through a combination of borrowed capital (loans) and equity (the farmer's own funds), the discount rate should reflect both sources. In such cases, the rate should be weighted to account for:

- The borrowing cost (interest rate) for the loaned portion.
- The expected return needed to justify the use of the farmer's own money (equity capital).

By balancing these factors, the discount rate represents the true cost of financing the investment.

A weighted average cost of capital (WACC) calculation is needed, estimating the rate of return for both forms of capital tied up in the investment: debt and equity. The weighted average cost of capital then reflects the cost of all forms of financing that the farmer may use. The WACC can be calculated as follows:

$$\text{WACC} = \frac{\text{Equity capital}}{\text{Total capital}} \times \text{Return needed to attract equity capital} + \frac{\text{Borrowed capital}}{\text{Total capital}} \times \text{Borrowing rate}$$

This requires information on the:

- Cost of capital funds from each source employed by the farm; and
- Weights to be given each source of capital in computing a weighted average cost.

The proper set of weights to use is the relative proportion of each type of capital in the farmer's desired financing of the investment. The assumption is that the farmer will, in fact, fund the capital investment in the proportions assumed in the weighting system. If bank loans represent 75% of a farmer's funds, it should be assigned a weight of 0.75. If equity capital represents 25% of the investment, it would be given a weight of 0.25.

While the interest rate charged on long-term bank loans is used as the cost of capital from lending, placing a meaningful cost on equity capital is much more difficult. This is because equity has no clear cost. There is no interest paid. However, there is an opportunity cost. If the farmer puts his or her money into an investment he or she is denied the use of these funds for other purposes. If by investing in his or her own farm, the farmer foregoes making an investment that will earn 10%, he or she should view equity capital as costing the farm the alternative 10%.

The opportunity cost of farmers' equity differs between farmers since individuals differ in their abilities to find alternative investment possibilities for their capital and in their management capacity and attitude to risk. With this in mind farmers must use their judgement in assessing their cost of equity. The important point is for the farmer to recognise that equity capital is not "costless" and, in fact, should be regarded as a more expensive source of capital than bank loans.

An example of calculating the weighted cost of capital is given in Box 13.14.

In this example, the weighted average cost of capital is 8%. Investments promising a return of less than 8% (i.e. investments with a negative net present value using 8% as the discount factor or investments with an IRR of less than 10%) would not be desirable. On the other hand, investments promising a rate of return of something over 8% would be acceptable.

Comparison of NPV and IRR

The two discounting measures give the same answer to the simple question of whether the investment pays. Many extension workers or farmers might prefer the NPV criterion for its simplicity, unambiguous quality and straightforward way of selecting among mutually exclusive investments.

Some, however, may prefer the IRR criterion because it is understood easily by those unfamiliar with the discounted measures. The IRR can be explained readily as the maximum rate of interest that a farmer could pay if all resources used in the investment were borrowed. But if the investment alternatives considered by the farmer are mutually exclusive, the IRR is not a useful measure and the NPV would be more appropriate. The NPV better serves this purpose because it measures the absolute surplus of the benefits over the costs, discounted. Furthermore, it escapes the problem of having to estimate the opportunity cost of capital to the farmer.

Cost-benefit analysis

Cost-benefit analysis (CBA) is a tool used to evaluate the economic feasibility of an investment decision by comparing its anticipated discounted benefits and costs. The economic worth of the investment is expressed as the cost-benefit ratio

Box 13.14. Calculating the weighted cost of capital

Source	Amount	Weighted cost of capital
Loan (7% IR)	5,000 (67%)	7% of 67% = 4.7%
Equity capital (10% IR)	2,500 (33%)	10% of 33% = 3.3%
Total	$7,500 (100%)	8.0%

(CBR) – the ratio of the present value of costs to the present value of benefits.

The cost and benefit are discounted separately to their present values, and the resulting CBR provides a discounted measure of the investment's overall value. A CBR greater than or equal to 0 indicates that the investment is economically viable.

The CBR can be calculated from the following formula.

$$CBR = \frac{\sum_{t=1}^{n} \frac{C_t}{(1+r)^t} + I_0}{\sum_{t=1}^{n} \frac{B_t}{(1+r)^t}} = \frac{PV(C)}{PV(B)}$$

Where:

- CBR: cost-benefit ratio
- B_t: benefits for year t
- C_t: operating and maintenance costs in year t
- I_0: initial investment cost
- r: discount rate
- n: number of years in the analysis period
- PV (C): present value of costs (including initial investment)
- PV (B): present value of benefits
- Σ: sum of

The figure describes the calculation – taking all the future costs (C) and adjusting them to present value PV (using the discount rate) and initial investments added – totalling them up (numerator). The same process is taken for the benefits stream (denominator). All future benefits are discounted in the same way. Finally, the total benefits are divided by the total costs.

Decision rule:

- If CBR > 1, the investment is accepted → because benefits exceed costs.
- If CBR < 1, the investment is rejected → because costs exceed benefits.
- If CBR = 1, the investment breaks even → benefits are equal to costs.

The formal selection criterion for the cost-benefit ratio measure of investment viability is to accept all investments with a cost-benefit ratio of greater than 1 when the cost and benefit streams are discounted at the opportunity cost of capital.

An example is given in Box 13.15.

13.9.3 Incorporating investment planning into overall farm planning

Investment planning should be integrated into the overall farm planning process if new capital investments are required. By considering investment opportunities alongside other planning components, farmers can align their capital investments with the farm's strategic goals, financial capabilities and resource allocation plans to ensure a more comprehensive coverage of the farm plan.

Farmers should then consider how investment projects fit into the farm's long-term vision and objectives. Additionally, investment planning should be synchronized with other planning aspects such as market planning, production planning, financial planning and risk management. This integration ensures that investments are aligned with market demands, production capabilities and financial feasibility.

Furthermore, incorporating investment planning into overall farm planning, when needed, allows farmers to prioritize and allocate resources effectively. It helps them determine the optimal timing and sequencing of investments, considering factors such as cash flow availability, resource requirements and interdependencies between the different aspects of the plan.

13.10 Risk in Farm Planning

As we have noted throughout this book, farming is inherently uncertain. Farmers face the challenges of climate change, fluctuating market prices, pest or disease outbreaks and many others posing significant risks (see Box 13.16.). A farmer's attitude towards risk plays a crucial role in shaping their decision-making process and how they respond to these uncertainties. Each farmer has a distinct risk profile, influenced by personal preferences and their tolerance for uncertainty. For instance, some farmers are risk-averse, prioritizing stability and security by avoiding uncertain outcomes, even if it means sacrificing potential gains. In contrast, others are more risk-tolerant, willing to embrace higher levels of uncertainty in pursuit of greater rewards.

Box 13.15. Example: Cost-benefit analysis of an irrigation project.

Suppose a farmer group wants to invest in a small irrigation system.

Background data:

- Initial investment (Year 0): $1,000
- Annual benefits (increased yields, etc.):
- Year 1: $500
- Year 2: $500
- Year 3: $500
- Annual operating/maintenance costs:
- Year 1: $100
- Year 2: $100
- Year 3: $100
- Discount rate: 10% (or 0.10)
- Project duration: 3 years

Steps:

1. Create a discount factor table

   The discount factor (DF) for each year is:

   $DF_t = \frac{1}{(1+r)^t}$

Year	DF at 10%	Cost ($100 × DF)	Benefit ($500 × DF)
1	0.909	90.90	454.50
2	0.826	82.60	413.00
3	0.751	75.10	375.50
Total	–	248.60	1,243.00

2. Add initial investment to total costs
   Total PV of costs = 248.60 + 1000 = 1248.60
3. Calculate CBR
   CBR = total PV of costs/total PV of benefits = 1248.6/ 1243.0 = 1.0045

Interpretation:

CBR = 1.004

The project barely breaks even; benefits are slightly more than total costs.
Small increases in benefit or cost reduction could make the project viable.

Box 13.16. Factors influencing farmer's attitudes towards risk.

Several factors influence a farmer's attitudes towards risk. These factors can be categorized into internal and external factors. Internal factors include personal characteristics, experience, risk perception and financial position. External factors encompass the farm's location, market conditions, government policies, and social and cultural norms. Understanding these factors helps to comprehend why farmers may differ in their risk attitudes.

Recognizing and understanding these diverse risk preferences is essential for developing risk management strategies that align with a farmer's comfort level, goals and capacity to absorb potential setbacks. Tailored approaches can help farmers make informed decisions, balancing risks and rewards in ways that support their long-term resilience and success.

In farming, *game theory* – an economic decision-making approach – can serve as a valuable tool to guide smarter choices (Dillon, 1962; Agrawal and Heady, 1968). It provides farmers with analytical tools to evaluate their options under climate uncertainty, helping them align their decisions with their risk preferences. By considering all possible outcomes and how varying conditions, such as weather, might influence these outcomes, game theory enables farmers to make more informed and strategic decisions.

The key decision criteria in game theory capture different attitudes that farmers may have towards risk and uncertainty. These criteria are outlined further on, accompanied by examples using simplified payoff matrices for crop choices under diverse weather scenarios.

Farm planning under conditions of risk requires tools to account for the multiple variables involved such as prices, production levels for different products and the cost of inputs. These variables make computations complex, often necessitating computer assistance.

To illustrate, we consider the consequences of different farm plans formulated at the beginning of a season. This involves the following steps:

1. **Estimate probabilities of rainfall patterns**: define probabilities for three types of rainfall, i.e. wet, average and dry years. Determine rainfall ranges for each type to structure the analysis.
2. **Calculate enterprise margins**: compute the gross or net margins for each enterprise based on yields, prices and costs under the different states of weather (e.g. wet, average and dry conditions). Separate margins are prepared for each enterprise and for each weather state.
3. **Develop optimum farm plans**: using the calculated gross margins, create separate farm plans, one optimized for each rainfall scenario.
4. **Construct a payoff matrix**: for each planned area under cultivation, prepare a matrix of outcomes that reflects the variations in weather conditions and their impacts. The payoff matrix summarizes the financial returns for each scenario and helps evaluate risk and uncertainty in decision making.

13.10.1 The payoff matrix

A payoff matrix (or payoff table) is a tool used to evaluate potential outcomes under different scenarios. It records financial returns (payoffs) for each combination of actions (farm plans) and uncertain events (weather conditions). This tool allows farmers to assess the risks and rewards of various strategies. The components of a payoff matrix are as follows:

- **Actions**: decisions available to the farmer, such as which crops to grow or production plans to implement.
- **Events**: external conditions like weather variability, market prices or input costs.
- **Payoffs**: financial outcomes (e.g. gross margins) for each action–event combination.
- **Probabilities**: likelihood of each event occurring, if estimable.

Example: crop selection

A farmer must decide which crop to grow – maize, beans, cotton or vegetables – based on probabilities of five weather types. Table 13.5 shows expected income per hectare for each crop under varying conditions. The probabilities for the weather types are shown in Table 13.6.

The farmer's decision depends on their approach to risk, as detailed in the next section.

13.10.2 Decision-making criteria

When making farm planning decisions under uncertainty, farmers need systematic ways to compare different options and assess potential risks and rewards. Various decision-making criteria help farmers choose the best crop or enterprise based on factors such as expected income, risk tolerance and potential regret.

Table 13.5. Expected income/ha for each crop depending on various weather conditions.

Crop	Very dry	Dry	Average	Wet	Very wet
Maize	100	130	180	200	220
Beans	250	210	170	120	80
Cotton	120	170	230	170	110
Vegetables	120	130	170	190	210

Table 13.6. Probabilities for various weather types.

Weather type	Very dry	Dry	Average	Wet	Very wet
Probability	0.10	0.20	0.30	0.20	0.20

Some methods, like the expected value criterion, rely on probability-based calculations to estimate the most profitable choice. Others, such as the maximin criterion, prioritize security by selecting the option with the least risk of loss. Optimistic farmers may use the maximax criterion to pursue the highest possible return, while those concerned about missed opportunities might apply the regret criterion.

By understanding these different approaches, farmers can align their decisions with their personal risk preferences and available market information, ultimately improving their ability to plan for the future. The following subsections outline the six key decision-making criteria and how they can be applied in farm planning. The figures used are from the previously mentioned crop example (see Tables 13.5 and 13.6).

Expected value criterion

This method assumes that the outcomes are not equally likely and is used when information about the probabilities regarding weather conditions is available.

This approach calculates the expected income for each crop by weighing payoffs against their probabilities. For example, the expected income for maize is:

$$0.10(100) + 0.20(130) + 0.30(180) + 0.20(200) + 0.20(220) = 174$$

The expected incomes for all crops are shown in Table 13.7; based on this criterion, maize offers the highest expected income.

Table 13.7. Expected incomes for all crops.

Crop	Expected income
Maize	174
Beans	158
Cotton	171
Vegetables	169

Equal likelihood criterion

This method assumes that all possible outcomes are equally likely and is often used when no information about probabilities is available. If probabilities are unknown, each state of weather is assigned equal weight (20% in this case). The farmer calculates average income for each crop. For example:

$$\text{Maize: } 0.2(100 + 130 + 180 + 200 + 220) = 166$$

The expected incomes under equal likelihood are shown in Table 13.8.

A farmer without any risk preference would grow either maize or beans, as both yield the highest expected income.

Table 13.8. Expected incomes under equal likelihood.

Crop	Expected income
Maize	166
Beans	166
Cotton	160
Vegetables	164

Partial optimist criterion

This method allows farmers to balance their degree of optimism and pessimism by using a coefficient to weigh the best and worst outcomes. Each farmer could assess their level of optimism on a scale of 0–1, where 0 is a very pessimistic value and one is very optimistic.

The decision rule for the expected income (EI) is:

$$EI = (\alpha \times \text{best outcome}) + [(1 - \alpha) \times \text{worst outcome}]$$

If we assume that the farmer gives themself a subjective value of 0.7, the analysis produces the result given in Table 13.9.

Table 13.9. Expected outcomes.

Crop	Expected income
Maize	184
Beans	199
Cotton	194
Vegetables	169

A moderately optimistic farmer would grow beans, which provides the highest expected value.

Maximin criterion

This criterion seeks to minimize the worst-case scenario, suitable for risk-averse farmers. The decision rule is to choose the crop with the highest minimum income from Table 13.10. A risk-averse farmer would grow vegetables to ensure the highest minimum income.

Table 13.10. Estimated minimum income of crops.

Crop	Minimum income
Maize	100
Beans	80
Cotton	110
Vegetables	120

Maximax criterion

Optimists or risk-takers focus on maximizing potential gains. The crop with the highest maximum income is chosen from Table 13.11. A risk-tolerant farmer would grow beans, which offer the highest possible income.

Table 13.11. Estimated highest maximum income of crops.

Crop	Maximum income
Maize	220
Beans	250
Cotton	230
Vegetables	190

Regret criterion

This method minimizes the regret (opportunity cost of missed better outcomes) by comparing each decision's payoff with the best possible payoff for a weather type and the payoff of the chosen crop.

The decision rule is to choose the crop with the lowest maximum regret. For example: Regret for maize in very dry weather: 220 (very wet) − 100 (very dry) = 120

The maximum regrets for all crops are given in Table 13.12. A farmer minimizing regret would grow vegetables.

Table 13.12. Estimated maximum regrets for all crops.

Crop	Maximum regret
Maize	120
Beans	170
Cotton	120
Vegetables	90

By applying these criteria, farmers can align their decision-making processes with their attitudes towards risk and uncertainty. This structured approach ensures more informed and strategic choices, tailored to their individual circumstances. For a summary of decision criteria and results see Table 13.13.

For small-scale farmers, these techniques could help them to navigate the unpredictability of climate change – weather variability, price fluctuations and input availability. Risk-averse farmers may choose crops with stable minimum returns, while risk-tolerant farmers may focus on maximizing potential gains.

Table 13.13. Summary of decision criteria and results.

Decision criterion		Farmer's attitude	Crop selected
1	Expected value	Neutral	Maize
2	Equal likelihood	Neutral	Maize/Beans
3	Partial optimist	Mixed	Beans
4	Maximin	Risk-averse	Vegetables
5	Maximax	Risk-tolerant	Beans
6	Regret	Concerned about regret	Vegetables

Ultimately, integrating decision-making frameworks into farm planning has the potential to empower smallholder farmers to make strategic, data-driven choices but this will require the support of agricultural extension workers to guide them through these processes.

13.11 Advisory Support for Planning

Effective farm planning is a critical component of successful farm management, yet many small family farmers may lack the technical expertise, tools or resources to develop comprehensive plans independently. This is where agricultural extension and advisory services play a pivotal role. Building on the diagnostic phase, where extension workers help farmers assess their current situation, advisory support for planning focuses on guiding farmers through the process of setting goals, allocating resources and developing actionable strategies to achieve their objectives.

Extension workers can assist farmers in translating their goals into practical plans by facilitating structured planning processes. This includes helping farmers define clear, measurable and achievable objectives, whether they relate to increasing productivity, diversifying income sources or improving sustainability. Extension workers with farm management skills can help farmers use simple tools like budgeting and gross margin analysis to plan their farms, compare the profitability of different crops or activities, and make better use of their land, labour and cash resources. By providing hands-on training and support, extension workers can empower farmers to make informed decisions that align with their resources, market opportunities and risk tolerance.

Moreover, advisory services can help farmers integrate various planning components – such as market planning, labour planning and investment planning – into a cohesive whole-farm plan. Extension workers can assist in identifying potential risks and developing contingency strategies, ensuring that farmers are better prepared to navigate uncertainties such as fluctuating market prices or adverse weather conditions. By fostering a collaborative approach, extension workers can help farmers build confidence in their planning processes and ensure that their plans are both realistic and adaptable to changing circumstances.

Ultimately, the role of advisory services in farm planning is to bridge the gap between theoretical planning concepts and practical application. By providing tailored guidance, tools and ongoing support, extension workers can help small family farmers develop robust plans that enhance productivity, profitability and long-term sustainability. This collaborative effort not only strengthens the capacity of individual farmers but also contributes to the resilience and growth of the broader agricultural sector. The role of extension workers in farm planning is further visited in Part 4 of this book.

13.12 Conclusion

Planning plays a vital role in the success and sustainability of farming operations. Throughout this chapter, we have explored the significance of sound budgeting and planning in agriculture, and the various approaches that can be employed to develop effective farm plans. Planning provides farmers with a structured framework to make informed decisions, allocate resources efficiently and achieve their goals. By implementing formal planning approaches, farmers can benefit from a systematic and strategic approach to managing their farms. These formal approaches, such as budgeting, strategic planning, market planning, investment planning and farm planning, allow farmers to analyse their operations comprehensively, identify potential risks and capitalize on opportunities.

By embracing formal planning approaches and particularly with support from their local extension worker, farmers can unlock numerous benefits. These include enhanced decision making, improved resource allocation, increased stakeholder engagement and greater adaptability to changing market conditions and risks.

Therefore, it is essential for farmers to recognize the importance of budgeting and planning in farming, and the advantages of formal planning approaches. By utilizing these approaches, farmers can chart a clear path to success, optimize their resource utilization and contribute to the long-term sustainability of their farm businesses.

Throughout this chapter we have frequently mentioned extension workers and agricultural advisors. Many of the concepts and tools presented in this chapter may be beyond the capacity of many small family farmers to apply on their own and will need the help of extension workers. To be able to respond effectively, extension workers will need to familiarize themselves with the concepts and tools presented in this chapter (along with Chapter 6) and practice the tools until they are proficient and able to engage farmers with them.

Endnotes

[1] Annual cash flow, and not annual net income, is used because the farmer must repay the investment loan in cash, and it is often the case with small family farms that they will consume some portion of their production. While this counts towards profitability, it does not contribute towards the actual cash required to repay an investment loan. Using cash flow ensures the assessment reflects the farmer's ability to cover loan repayments and other cash obligations.

[2] Whenever salvage values exist, they should be added to the net cash income since they represent an additional cash receipt.

[3] The coefficient for calculating NPV is the discount factor, which adjusts future cash flows to their present value. It is based on the chosen discount rate (reflecting the time value of money) and the year in which the cash flow occurs.

References

Agrawal, R.C. and Heady, E.O. (1968) Application of game theory models in agriculture. *Journal of Agricultural Economics* 19–2, 207–218. DOI: 10.1111/j.1477-9552.1968.tb01326.x.

Dillon, J.L. (1962) Applications of game theory in agricultural economics: Review and requiem. *Australian Journal of Agricultural Economics* 6–2, 20–35. DOI: 10.1111/j.1467-8489.1962.tb00299.x.

Additional Reading

C.S. and Nix J.S. (1973). Farm Planning and Control. Barnard, Cambridge University Press.

Brown, M.L. (1979). Farm Budgets: From Farm Income Analysis to Agricultural Project Analysis. Published for the World Bank by Johns Hopkins University Press, Baltimore.

J.L. and J.B. Hardaker (1993). Farm Management Research for Small Farmer Development, FAO Farm Systems Management Series No. 6. Dillon, Food and Agriculture Organisation of the United Nations, Rome.

Malcolm B., Makeham J. and Wright V. (2012) The Farming Game. Cambridge University Press.

Makeham, J.P. and L.R. Malcolm (1986). The Economics of Tropical Farm Management. Cambridge University Press.

Upton, M. (1987). African Farm Management. Cambridge University Press.

Upton, M. (1973). Farm Management in Africa, Oxford University Press, London.

William C. Found (1970). A Theoretical Approach to Rural Land-use Patterns. Edward Arnold Press.

14 Monitoring, Evaluation and Learning

This chapter develops the theme that farming must be profitable, with a focus on the third, fourth and fifth stages of the farm management cycle, namely implementing, monitoring and evaluating (see Fig. 14.1).

The chapter underscores the necessity for farmers to actively monitor the implementation of their plans and evaluate their costs and returns throughout their operations. In today's competitive agricultural landscape, maintaining a sustainable farming business hinges on understanding and controlling costs while maximizing returns. To this end, the chapter underscores the necessity of monitoring farm performance and evaluating costs and returns. It introduces record-keeping, benchmarking and comparative analysis to help farmers optimize operations and make informed decisions.

A key element of this approach is the use of the budgeting tools explained in Chapter 11, which can also be used for controlling and monitoring farm businesses. By comparing the performance of the farm against industry standards or peer benchmarks, farmers can identify opportunities for improvement, optimize operations and adopt best practices that enhance productivity and profitability. Comparative analysis and benchmarking help to empower farmers to make well-informed decisions about resource allocation, assess investment opportunities and manage costs strategically.

The chapter will conclude by tying back to Chapter 12 Farm Management Methods: Diagnosis, reinforcing the idea that planning and decision making in farming is an iterative process. This cyclical approach ensures continuous improvement and adaptation, allowing farmers to respond proactively to changing conditions and new challenges. By understanding that budgeting and financial analysis are ongoing activities, farmers can maintain a dynamic and responsive management strategy that supports long-term success and sustainability.

14.1 Importance of Monitoring, Evaluation and Learning

Monitoring, evaluation and learning (ME&L) are essential components of effective farm management, enabling farmers to maintain control over their operations and make well-informed decisions. Through continuous assessment, farmers can optimize efficiency, improve productivity and ensure sustainable farm practices.

ME&L in farm management refers to the systematic process of collecting, analysing and interpreting data to assess the performance and progress of farm operations. It involves tracking key parameters such as production output, resource utilization, financial performance and regulatory compliance. The primary objective is to provide accurate and timely information that helps farmers enhance efficiency and achieve their desired outcomes.

Monitoring is an ongoing process that involves gathering data related to specific activities, indicators and targets. It allows farmers to track their operations in real time, identify deviations from expected actions and results and implement corrective measures as needed. By closely observing crop growth, livestock health, labour performance, resource allocation and financial status, farmers can take proactive steps to optimize productivity and minimize risks.

Evaluation, on the other hand, is a systematic analysis of the collected data to assess overall farm performance, strengths, weaknesses

DOI: 10.1079/9781836990536.0014

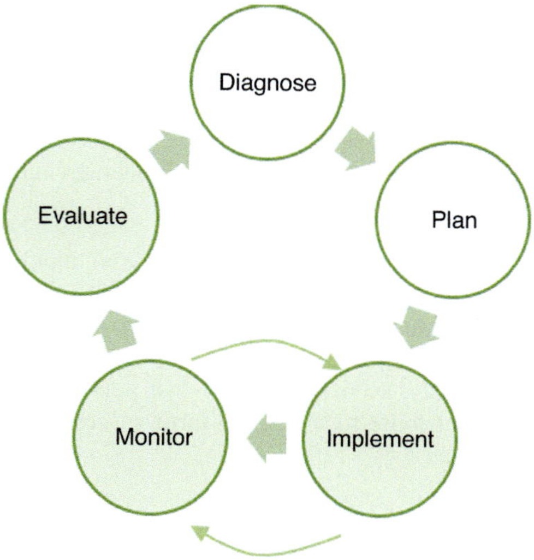

Fig. 14.1. The farm management cycle.

and areas for improvement. It provides farmers with a comprehensive understanding of their farm's operational effectiveness and helps refine management strategies to enhance long-term sustainability.

Learning is a critical component of the system, as it ensures that insights gained from monitoring and evaluation are used to improve farm performance. By drawing lessons from these activities, farmers are better equipped to adapt to a changing farming environment.

ME&L is not merely about passive data collection; it is a dynamic process that empowers farmers to actively manage and improve their operations. By integrating monitoring into daily farm activities, farmers can detect inefficiencies early, ensure better decision making and respond swiftly to changing conditions. Evaluation complements this by providing insights into long-term trends and guiding strategic planning.

A key aspect of ME&L is tracking the implementation of the farm business plan, as discussed in Chapter 13. Monitoring ensures that planned activities align with expected outcomes and that any gaps in execution are promptly addressed. This systematic approach helps farmers stay on course, optimize resource use and adapt to evolving challenges in farm management.

By adopting robust ME&L practices, farmers can enhance efficiency, improve profitability and ensure the sustainability of their agricultural enterprises. Ultimately, a well-structured ME&L system strengthens decision making, fosters resilience and contributes to the long-term success of farming operations.

14.2 Link between Profitability and Monitoring Costs and Returns (Income)

Profitability is the goal of any farm operating as a business. To achieve sustainable profits, farmers must closely monitor their costs and returns to maintain financial health. By tracking input usage against output levels, farmers can identify when additional inputs – such as fertilizer or labour – no longer yield proportional increases in production (the principle of diminishing returns). Recognizing this principle enables better resource allocation and cost management.

Effective cost monitoring helps farmers identify unnecessary expenses, reduce waste and implement cost-saving measures. It also supports budgeting and forecasting, facilitating stronger financial planning and informed long-term decision making.

Similarly, monitoring returns provides insights into the financial performance of different farm activities. Understanding marginal costs and marginal income (marginal value product) is essential; farmers should expand production only if the additional income exceeds the additional cost. This ensures investments remain profitable rather than leading to unnecessary expenditures.

Tracking returns also helps evaluate the relative performance of various crops, livestock or farm products, guiding farmers towards the most profitable enterprises. Identifying high-performing activities allows farmers to replicate successful practices across their operations. Moreover, monitoring aids in improving marketing strategies, pricing decisions and market selection, ultimately with the aim of optimizing profitability.

The connection between profitability and cost-return monitoring is clear: systematic tracking and evaluation enable farmers to optimize resource use, minimize costs and maximize returns. This process provides valuable insights into strengths, weaknesses, opportunities and threats within farm operations, facilitating strategic decision making.

The following sections of this chapter explore key components of monitoring and evaluation systems, the importance of record-keeping, comparative analysis and benchmarking, balancing data accuracy with practicality, and integrating monitoring processes into farm planning.

14.3 Key Components of Effective Monitoring and Evaluation Systems

Monitoring and evaluation practices include the following.

- **Data collection and recording:** effective monitoring begins with collecting relevant and reliable information about farm activities. Smallholder farmers should focus on key performance indicators (KPIs) such as yield, input use, costs and sales. Data can be gathered through simple methods like field observations, record-keeping, notebooks or tally sheets. Where possible, mobile phones or basic spreadsheets can improve data organization and retrieval. Consistent record-keeping helps farmers track progress, plan better and make informed decisions.
- **Analysis and interpretation of data:** once data is collected, analysis helps farmers identify patterns and trends. Simple methods such as comparing yields across seasons, tracking costs against income, or assessing the efficiency of inputs can offer valuable insights. Monitoring should also evaluate technical and economic efficiency. Recognizing strengths and areas needing improvement allows farmers to adjust practices to boost productivity and profitability.
- **Reporting and feedback:** sharing and discussing farm performance with family members, farmer groups, extension workers or cooperatives, encourages learning and improvement. Farmers can use simple reports, discussions, or even visual tools like charts, to understand their progress. Feedback from peers or advisors helps refine strategies, solve challenges and adapt to changing conditions.

By adopting basic but structured monitoring and evaluation practices, smallholder farmers can improve decision making, optimize resources and enhance farm productivity and profitability. Even simple, low-cost approaches can contribute to long-term farm success and resilience.

14.4 Recorded Information on Smallholder Family Farms

Recorded information is a powerful tool for smallholder family farmers, significantly enhancing their ability to manage their farms effectively. Given the challenges of fluctuating market prices, resource scarcity and climate risks, maintaining accurate records is particularly valuable. With structured and well-organized data, smallholder farmers can identify high-yield, profitable crops and make decisions that align with current market trends, ultimately securing a better return on investment.

Participation in value chains also becomes more feasible when smallholder farmers maintain recorded data. Accurate and well-maintained records allow farmers to demonstrate production consistency, meet quality standards and comply with buyer requirements. This information strengthens their bargaining power, enabling them to negotiate fair prices with buyers and traders. Furthermore, proper documentation facilitates access to government subsidies, agricultural grants and safety net support programmes specifically designed for smallholders. Additionally, with documented production data, farmers can more readily qualify for credit and investment opportunities, enhancing their financial sustainability and growth potential.

The impact of recorded information extends beyond individual farm operations, enhancing transparency and accountability in smallholder farming. Accurate records make it easier for farmers to collaborate with cooperatives, agribusinesses and NGOs, which often require documentation to validate support, including input supplies, training and technical assistance. Moreover, the ability to track progress and measure the outcomes of partnerships and interventions builds confidence and credibility within the agricultural sector. Over time, this documentation increases smallholders' chances of forming long-term, productive partnerships.

However, despite its benefits, maintaining comprehensive records poses challenges for many smallholder farmers, especially those in resource-constrained areas with limited infrastructure and services. Literacy barriers, lack of access to digital tools and time constraints often make record-keeping overwhelming. To address these barriers, simplified record-keeping methods are essential. Pictorial or colour-coded formats, as well as pre-designed paper templates, can improve accessibility for farmers with limited literacy.

Additionally, innovative digital tools, such as mobile apps with voice-input features or user-friendly interfaces, can further support smallholders by making data entry easier and more intuitive. These technologies help bridge the gap between traditional and modern record-keeping practices, allowing farmers to document essential data more regularly and efficiently.

14.4.1 Methods of record-keeping

Effective record-keeping requires appropriate methods and tools to ensure accuracy, accessibility and efficiency. Farmers can choose from various options based on their preferences, resources and technological capabilities. The most used methods include the following.

- **Manual record-keeping (records, notebooks)**: traditional paper methods, such as records, notebooks and logbooks remain widely used due to their accessibility and cost-effectiveness. Farmers can create customized templates or use pre-designed forms to record relevant information. This method is particularly suitable for farmers who prefer a hands-on approach or have limited access to technology.
- **Farm management handbooks**: farm management handbooks, previously mentioned in Chapter 9, can also serve as a practical tool for smallholder farmers to systematically monitor their farm performance and profitability. These handbooks typically include structured templates for recording production activities, input usage, expenses and revenues, and can serve as a reference for best practices, helping smallholders adopt improved farming techniques and business management strategies. When used effectively, these handbooks have the potential to empower farmers with the knowledge and tools to track their progress, optimize resource allocation and ensure long-term farm sustainability.
- **Farm management software and mobile applications**: dedicated farm management software and mobile applications offer comprehensive solutions for record-keeping and farm operations management. These tools integrate various record-keeping functions, including financials, production tracking, inventory management and labour management, into a centralized platform. They also provide real-time data access, automated calculations, reporting capabilities and advanced analytical tools. While these solutions can significantly streamline record-keeping processes and improve decision making, farmers must choose software that aligns with their specific needs, considering factors such as cost, compatibility and user-friendliness. This method is similarly more sophisticated and if used by smallholder farmers is likely to require the support of specialized farm management advisors.

In conclusion, effective record-keeping is a cornerstone of successful smallholder farm management. By utilizing appropriate methods and leveraging modern tools, farmers can enhance productivity, strengthen market participation and secure financial sustainability.

Addressing challenges related to accessibility, literacy and technology adoption is crucial in ensuring that all smallholders can benefit from improved record-keeping practices. As digital solutions continue to evolve, there are opportunities to further integrate smallholder farmers into modern agricultural value chains and, thereby, empowering them with data-driven monitoring capabilities for improved decision making.

14.4.2 Types of records to maintain

Recognizing the diversity of smallholder farming systems, levels of literacy, and technological access, it is useful to present a range of formats that cater to different contexts and preferences. These formats should be user-friendly, adaptable and capable of capturing the necessary information required for effective farm business management. As previously noted, this includes simple paper formats as well as digital formats – spreadsheets and mobile applications – which offer the advantage of automated calculations, data analysis and ease of sharing.

The following points describe the various types of records required for more effective farm management.

- **Production records**: document crop yields, livestock performance, breeding activities, feed consumption and other production-related metrics. These records help farmers assess productivity, identify trends and make informed decisions regarding crop selection, animal health and breeding strategies. They also help farmers identify strengths and weaknesses in their production process and optimize the use of inputs to increase efficiency and output.
- **Inventory records**: document the acquisition and use of inputs (e.g. seed, fertilizer, feed, chemicals, etc.). These records allow more commercially oriented smallholder farmers to assess use of inputs and their costs and make decisions about purchases, allocations and storage.
- **Labour records**: help smallholder farmers track daily farm activities of the smallholder family and hired labour. Using checklists, notebooks or basic logbooks, farmers can record who performed tasks, what was done, when it was completed and the time spent on each activity. Maintaining these records ensures that essential farm tasks are carried out as planned and helps identify any delays or issues that need attention. By tracking labour use, farmers can assess productivity, manage costs and decide how best to allocate work among family members or hired workers. These records also provide useful information when seeking advice from extension officers or planning for future farming activities. Regularly updating and reviewing labour records helps smallholder farmers make better decisions, reduce waste and improve farm operations for better productivity and income.
- **Sales records**: allow smallholder farmers to keep track of all sales transactions enabling them to understand their income. Simple records should include the type of product sold, quantity, price, buyer and date. This helps farmers monitor market trends and improve their selling strategies.
- **Cash inflow records**: recording all cash received from farm sales, loans or other sources gives a clear picture of total income. This helps in budgeting and planning for farm expenses.
- **Cash outflow records**: tracking all farm expenses, such as seeds, fertilizers, tools, labour, and loan repayments, allows farmers to control costs and maintain a stable cash flow.
- **Home consumption records**: recording the amount of farm produce used for household consumption helps farmers assess total production value, even when some crops are not sold.
- **Profit and loss records**: summarize income and expenses over a season or year and can assist farmers to see if they are making a profit. This allows them to adjust spending and improve financial management.
- **Loan and contract records**: track loans, credit agreements and contracts with

buyers or suppliers to ensure that farmers meet their financial commitments and avoid disputes.

- **Fixed asset records**: record information about long-term assets, such as tools, equipment and buildings, and helps farmers plan maintenance and replacements to keep their farms running smoothly.
- **Weather records**: track rainfall and temperature to assist farmers to understand weather patterns and how they affect crop yield. This information can aid in better planning for future seasons.

By maintaining these basic records, smallholder farmers can improve their financial management, plan effectively and make informed decisions for a more profitable and sustainable farm. See Tables 14.1 to 14.11 for examples.

14.4.3 Best practices for record-keeping

To make record-keeping practical and useful for smallholder farmers, it is important to follow simple and effective strategies that enhance accuracy, accessibility and usability. Farmers should establish a routine for recording essential farm activities, such as planting dates, input usage, costs, yields and sales. Using easy-to-understand formats, such as simple tables or pictorial representations, can help ensure consistency and reliability.

Maintaining accuracy is key to useful records. Farmers should use consistent units of measurement, clearly label entries, and double-check their information to avoid errors. Keeping records up to date by setting a regular time for data entry – such as after fieldwork or sales – ensures that information remains relevant and useful for decision making.

Protecting records is also important. Paper records should be kept in a dry and safe location to prevent damage from weather, pests or misplacement. Farmers using electronic records should save copies on external devices or cloud storage to prevent data loss. Simple security measures, such as keeping records in a locked container or using passwords for digital files, help safeguard valuable information.

Organizing records logically makes them easier to use. Farmers should group records into clear categories, such as financial transactions, crop production, livestock management and labour use, as referred to in the previous section. Using notebooks with dividers, folders or labelled envelopes for receipts can make retrieval simpler when needed.

By following these best practices, smallholder farmers can improve their ability to track farm performance, plan for the future and make informed decisions. Accurate and well-organized records enable farmers to monitor profitability, identify successful practices and adjust strategies to improve farm productivity and sustainability.

Table 14.1. Example crop production records.

Enterprise	Land size (ha)	Expected yield (kg/ha)	Total yield (tons/kgs/bags)
Maize	2	3,000	6,000 kg
Groundnuts	1	2,000	2,000 kg

Table 14.2. Example livestock production records.

Enterprise	Number of animals	Expected yield	Cycle	Total yield
Layers	200	216/layer/cycle	10 months	43,200 eggs
Dairy cows	5	20 litres/cow/day	1 year	36,500 litres

Table 14.3. Example inventory record.

Date	Item	Opening Stock	Quantity Added	Quantity Used	Closing Stock	Location	Remarks
01-Jan-24	Fertilizer (Urea)	100 kg	50 kg	30 kg	120 kg	Storage shed A	Used for Field 1
01-Jan-24	Seeds (Maize)	20 kg	10 kg	5 kg	25 kg	Storage shed B	Remaining for next season
01-Jan-24	Diesel	40 L	20 L	15 L	45 L	Fuel tank	Regular tractor use

Table 14.4. Example labour record.

Date	Activity	Number of people working	Number of hours taken	Total number of hours taken
10-Oct-24	Land preparation	2	2	4
20-Nov-24	Planting	1	6	6
21-Nov-24	Planting	2	5	10

Table 14.5. Example cash inflow record. (Enterprise: eggs.)

Date	Sales/Output	Quantity	Unit Price ($)	Total Income ($)	Comments
12/04/24	Sold eggs	20 trays	4	80	Sold to hospital
15/04/24	Sold eggs	10 trays	3	30	Sold at town market
20/04/24	Sold eggs	15 trays	3	45	Sold at town market
22/04/24	Sold eggs	12 trays	3	36	Sold at town market
25/04/24	Sold eggs	18 trays	2	36	Sold at farm gate
Total				247	

Table 14.6. Example cash outflow record. (Enterprise: Eggs.)

Date	Operation/ Input	Quantity	Unit Cost ($)	Total Cost ($)	Comments
25/1/24	Purchase point of lay hens	20 hens	15	300	Bought from Monare's
1/2/24	Feed for laying hens	1 × 50 kg sack	70	70	Bought from Shiluli's
18/2/24	Vaccines for laying hens	1 injection per hen; 0 hens	1	70	Bought from Monare's
25/2/24	Constructed chicken housing	2 rooms	35	70	Did some of the work myself
5/03/24	Purchased layers	ten hens	16	160	Bought from Monare's
5/03/24	Transport costs to take eggs to market	1	5	5	Hired from Nakuru
5/03/24	Packaging/cartons	100	100/10	10	Bought at Agricor
6/03/24	Paid workers	2x 10 days	3/worker /day	60	Hired from village programme
Total				745	

Table 14.7. Example home consumption record. (Enterprise: Eggs.)

Date	Consumed items	Quantity	Unit price ($)	Total value ($)	Comments
1/2/24	eggs	10	0.25/egg	2.50	Family
18/2/24	eggs	15	0.25/egg	3.75	Mrs Ndlovu
25/2/24	eggs	5	0.25/egg	1.25	Spoilt
5/03/24	eggs	20	0.25/egg	5.00	Family
5/04/08	eggs	12	0.25/egg	3.00	Family
Total		62		15.50	

Table 14.8. Borrowing, loan and contract record.

Date	Type (Loan/ Contract)	Details	Amount/ Terms	Due Date	Payment Schedule	Status	Remarks
YYYY-MM-DD	Loan	Loan from Cooperative Bank	$500 @ 5% interest	YYYY-MM-DD	Monthly instalments	Ongoing	3 instalments paid
YYYY-MM-DD	Contract	Sale of 100 kg maize to Vendor A	$200, delivery in Oct	YYYY-MM-DD	One-time payment	Awaiting delivery	Delivery scheduled

Table 14.9. Example profit and loss record. (Enterprise: maize. from March to May 2024 in an area of 5 ha.)

	Quantity	Unit price or cost ($/unit)	Total ($)
Income			
Sales			
Home consumption			
Other			
Total Income (a)			
Expenses			
Inputs and materials:			
Seed			
Fertilizer			
Pesticide			
Field operations:			
Land preparation			
Ploughing			
Planting			
Weeding			
Harvesting			
Total Costs (b)			
Profit/Loss (a-b)			

Table 14.10. Example fixed asset record.

Date of purchase	Item	Purchase price ($)	Life (years)	Comments
1/2/24	Hand tools	800	10	To be replaced during 2026
3/4/24	Sprayer	1000	10	Bought from Monare's
4/4/24	Irrigation pipes	2000	10	Bought from Agricor
6/7/24	Houses for laying hens	1500	40	Built it myself

Table 14.11. Weather records template.

Date	Weather Type	Temperature (°C)	Rainfall (mm)	Wind Conditions	Remarks
02/04/24	Sunny	30	0	Mild	Ideal day for weeding
03/04/24	Rainy	24	15	None	Delayed harvesting
04/04/24	Overcast	26	2	Moderate	Irrigation skipped

14.4.4 Challenges and alternatives to record-keeping

Introducing record-keeping for smallholder farmers in low-income countries poses challenges. Limited literacy rates, lack of awareness and cultural barriers among smallholder farmers can hinder their understanding of the benefits and importance of record-keeping. Additionally, inadequate access to technology, such as computers and smartphones, will limit the ability of farmers to maintain digital records. Moreover, the time and effort required to keep detailed records can be perceived as an additional burden for already resource-constrained smallholder farmers.

Examining record-keeping initiatives in low-income countries reveals a diverse range of outcomes, highlighting both successful and unsuccessful cases. These examples offer valuable insights into the potential benefits and challenges associated with implementing record-keeping practices. The field experiences with record-keeping are elaborated on in the case study examples given in Boxes 14.1. to 14.3.

These examples from different smallholder farming contexts highlight the varied outcomes of record-keeping initiatives. While successful cases demonstrate improvements in farm management, productivity and income generation, failures underscore challenges such as low literacy rates, limited resources and inadequate support systems. Lessons from these experiences can inform future initiatives and improve implementation strategies.

While record-keeping may not be widely relevant for all smallholder farmers, there are specific circumstances where it can be particularly beneficial. For example, farmers engaged in market-oriented production, those involved in contract farming arrangements, or those seeking to access credit and insurance services will find record-keeping essential.

Extension programme managers, private advisory service providers and policymakers play a critical role in promoting record-keeping by identifying when and how to implement these practices effectively. Their support can include training, development of user-friendly tools and facilitating access to financial and technical assistance.

By aligning resources with the specific needs of farmers, stakeholders – including government agencies, financial institutions, cooperatives and agribusinesses – can enhance adoption and impact. Accurate and accessible records enable smallholder farmers to make informed decisions, improve farm efficiency and strengthen their participation in agricultural markets.

14.4.5 Minimum data recording

An effective record-keeping system for smallholder farmers need not be comprehensive and complex. Instead, it should be practical, easy to maintain and focused on essential information that supports key farm management decisions. Given the limited time and capacity available to many smallholder farmers, record-keeping should concentrate on monitoring the most important farm enterprises and the critical factors influencing profitability. To simplify the demands of record-keeping, farmers should focus on the most critical information the farmer needs to monitor and evaluate the farm plan being implemented. To simplify the process, farmers should prioritize recording information that directly impacts their farm's financial and operational success, ensuring that the data collected is both manageable and useful for decision making. The following sub-section provides a simple guide to deciding on what information is needed, what records would be required and some ideas about the methods of keeping records.

Deciding What to Monitor and Evaluate

STEP 1: ALIGN RECORD-KEEPING WITH THE FARM PLAN. Since the farm plan has already identified key challenges, objectives and strategies, record-keeping should focus on tracking progress towards solving those challenges. The farmer should ask the following:

- What specific actions am I taking to implement my farm plan?
- What outcomes do I expect from these actions?
- What simple records will show me whether these actions are working?

For example, if the plan includes increasing rice yields by using improved seed varieties, the farmer does not need to track everything about rice production, just a few key indicators, such as seed variety used, fertilizer application, rainfall and final harvest per hectare.

STEP 2: DECIDE WHAT IS CRITICAL TO TRACK. To determine which information is essential for monitoring implementation and evaluating outcomes, farmers can use a simple filtering process.

1. Does this factor directly measure progress towards my farm plan's goal?
   If the goal is to increase rice yield, key factors might include total harvest per hectare, seed variety used and fertilizer application. Tracking factors that do not contribute to this goal – such as general weather conditions – may be unnecessary unless they directly affect the yield.
2. Does this information help me evaluate my plan's success?
   - If the plan aimed to increase yield by 10%, recording harvest quantities per hectare allows for a direct comparison with past seasons.
   - If the goal was to reduce input costs, tracking fertilizer and pesticide expenses will show if spending has decreased.
3. Is this information realistic to record?
   - If rainfall is a key factor but the farmer has no rain gauge, it may be more practical to record whether the season was 'wet', 'normal' or 'dry' rather than trying to track exact millimetres of rain.
   - Instead of daily tracking, recording harvest weight weekly or at the end of the season reduces effort while still providing useful data.

For example, a farmer's plan aimed to increase rice yield by 15% using an improved seed variety. To evaluate whether the plan worked, the farmer does not need to track every activity. Instead, the most useful records might include:

- Date of planting and harvest (to compare the growing period with past seasons).
- Seed variety used (to confirm that the improved variety was planted).
- Fertilizer and pesticide amounts applied (to see if inputs were used correctly).
- Final harvest weight per hectare (to measure if yield increased by the targeted 15%).

By focusing only on these few key records, the farmer can evaluate whether the plan met its intended goal without excessive paperwork.

STEP 3: USE SIMPLE AND PRACTICAL RECORD-KEEPING METHODS. Once the farmer decides what is essential, the records should be easy to maintain:

- Use a calendar or notebook – mark planting dates, fertilizer use or weekly sales.

Box 14.1. Examples of record-keeping initiatives.

Farm record-keeping practices of farmers in Gujarat, India

In 2015, a detailed study was conducted in Anand Taluka, Gujarat, India, to assess the state of farm record-keeping practices among local farmers. The area, known for its diverse agricultural activities, provided a representative setting for examining how well farmers manage their business records, which is crucial for effective planning and resource management in agriculture. The study revealed that 62.67% of the farmers in Anand Taluka were actively preparing budgets – crop-wise, year-wise or season-wise – with the majority (50%) opting for crop-wise budgeting. This reflects a robust inclination towards control and monitoring of their farms. Furthermore, it was found that while half of the farmers surveyed maintained some form of farm records, the vast majority (97.33%) documented expenses and revenues. However, only 38.67% kept comprehensive financial records, which are critical for informed farm management.

The investigation highlighted several key insights into the methods and impacts of record-keeping. Most farmers who kept records did so manually, which they found to be laborious and time-consuming. Despite these challenges, 56% of the farmers updated their records daily, while a smaller group (8%) did so monthly. Approximately 75% of these farmers maintained four to five different types of records, underscoring the depth of their engagement with record-keeping practices. Those who kept records identified significant benefits, such as the ability to pinpoint strengths and weaknesses in their farming operations and to lay a strong foundation for future farm planning. Conversely, farmers who did not maintain records cited illiteracy in record-keeping, lack of time and a general lack of awareness about the benefits of various record-keeping practices as major obstacles. The study also found a significant correlation between a farmer's educational level, landholding size and income; and their likelihood of maintaining farm records, highlighting the broader socio-economic factors influencing farm management practices in Anand Taluka (Prajapati *et al.*, 2015).

Farm record-keeping, Ghana

In 2021, a significant study was undertaken to assess record-keeping behaviours among small-scale pineapple farmers in the Nsawam Adoagyiri municipality, located in the eastern region of Ghana. This research revealed that record-keeping was a common practice within this farming community, with each farmer maintaining at least six different categories of records. These records varied in their frequency – daily, weekly and monthly – and were maintained through a mix of manual and computerized methods. The findings demonstrated a positive correlation between the practice of keeping records and various factors such as the farmers' level of education, their experience in farming and the size of their farms. Farmers reported multiple benefits from keeping detailed records, including the ability to make accurate productivity projections, manage improvements in their agribusiness operations, provide clear insights into farm operations, identify strengths and weaknesses, and facilitate responses to inquiries.

However, the study also highlighted several challenges that discouraged some farmers from maintaining records. A significant 41.7% of the farmers cited the lack of formal training in record-keeping as a major obstacle, echoing concerns about high levels of illiteracy and low numeracy skills previously noted in similar communities across Africa. Additionally, 26.7% of farmers felt that time constraints hindered their record-keeping efforts, while 17.5% mentioned stress associated with data entry as a deterrent. Another notable challenge, reported by 14.1% of the participants, was the fear of tax inspections by officials from the Ghana revenue authority, which some perceived as intrusive. These findings underscore the need for targeted educational programmes and awareness campaigns to enhance record-keeping practices among small-scale farmers. By promoting such practices through agricultural extension services and sharing the successes of pineapple farmers in the municipality, there is potential to improve agribusiness outcomes significantly, not only in Nsawam Adoagyiri but across the country. (Manteaw *et al.*, 2021).

Case study: Model Extension Learning and Doing Project, Philippines

In 1993–1994, the department of agriculture/agriculture training institute (DA/ATI) in the Philippines, in collaboration with the Japan international cooperation agency (JICA), initiated the Model Extension Learning and Doing Project. This pilot project was launched in three selected areas and aimed to

Continued

Box 14.1. Continued

explore new extension methodologies that would enhance farmer incomes through improved farm management and related techniques. Integral to this project was a farm recording component, which was designed to instil among farmers the practice of maintaining detailed farm records, as part of a broader strategy that included Integrated pest management, organic farming and water management.

The project demonstrated clear financial benefits from the integrated approach to extension services. Farmers who consistently kept farm records reported a significant increase in income, attributing this to better management decisions informed by their records. These decisions included choosing the most profitable farm enterprises, optimizing fertilizer use and improving water management practices. The continuous interaction with extension workers, who regularly visited and consulted with farmers, and the formation of farmer groups for collaborative activities, further bolstered the effectiveness of these practices. The project not only improved the farmers' technical skills but also fundamentally changed the attitudes and perceptions of agricultural extension workers towards farmer engagement and support.

Despite the successes documented in the pilot areas, the project faced challenges in sustainability and scalability. Although farmers acknowledged enhanced decision-making capabilities and improved management strategies due to the training and record-keeping activities, the continuation of these practices without ongoing support from extension services was uncertain. There was no subsequent replication of the project in other areas, raising concerns about the long-term viability and adoption of farm record-keeping practices in the absence of integrated support mechanisms. This case highlights the need for sustained commitment from local officials and extension staff to ensure that the benefits of farm record-keeping and associated agricultural practices are realized on a larger scale and maintained over time (Deomampo, 2003, unpublished material).

Farm data systems review study: a synthesis of country studies in Africa

In December 2001, a comprehensive study, commissioned by the Food and Agriculture Organization (FAO) was undertaken to assess farm data systems across several African countries. The research aimed to scrutinize the existing farm data systems to bolster decision making in farm management. The study reviewed a myriad of data collection methods across governmental and non-governmental organizations, encompassing a broad spectrum of agricultural data from crop production to farm management practices. The findings indicated a widespread engagement in record-keeping, particularly emphasizing the role of farm data in enhancing productivity and management strategies. It also highlighted a positive correlation between detailed record-keeping and improved decision-making capabilities among farmers, thereby underscoring the strategic importance of farm data.

However, the study also uncovered several challenges impeding the effective utilization of farm data. Among these were the lack of formal training in data management, the time-consuming nature of record-keeping, and a pervasive fear among farmers of tax-related repercussions from accurate financial reporting. Despite the clear benefits demonstrated by effective data systems, such as the ability to make informed management decisions and improved financial oversight, these barriers significantly hampered the broader adoption of record-keeping practices. The research called for enhanced educational programmes and increased awareness to encourage systematic data collection and utilization. Additionally, it recommended the integration of these practices within the broader agricultural extension services to ensure their sustainability and to maximize their impact on small-scale farming operations (Minae, 2002, unpublished material).

- Tally sheets – for tracking harvest weights or input use.
- Use simple symbols or colour codes – a green mark for a good yield and red for a poor yield, making patterns easy to spot.
- Mobile phone records – farmers with access to phones can store basic data via SMS or apps.

By aligning record-keeping with the farm plan, filtering for critical data and using simple tracking methods, farmers can monitor their farm's progress without unnecessary complexity. This ensures that decisions are based on useful insights rather than excessive paperwork.

Box 14.2. Dairy record-keeping tools: transforming smallholder farming in developing countries.

With global milk demand projected to grow significantly by 2030, smallholder dairy farmers in developing countries face the challenge of staying competitive in a rapidly evolving industry. A recent review conducted across sub-Saharan Africa, South Asia and South-east Asia explores the tools available for recording and managing dairy farm data. It emphasizes the importance of record-keeping for improving productivity, animal health and farm management. The study also highlights challenges, innovative technologies and opportunities to make these tools more accessible and impactful.

The review categorizes dairy record-keeping into four essential components: (i) animal identification and registration (I&R); (ii) traceability (AT); (iii) animal health information (AHI); and (iv) performance recording (PR). Tools range from traditional methods, like notebooks and spreadsheets, to advanced digital systems, such as mobile apps and IoT-integrated devices. In sub-Saharan Africa and South Asia, mobile phones have emerged as the most accessible technology for smallholder farmers, offering features like vaccination reminders, milk production tracking and breeding schedules. For instance, Kenya's iCow app has gained popularity for delivering practical solutions to common management challenges.

However, the review notes that while digital tools hold great promise, many farmers still rely on manual methods. Limited internet connectivity, low digital literacy and high costs of advanced technologies remain significant barriers in developing regions. Tools requiring manual data entry often miss the opportunity to provide real-time insights, which are crucial for decision making in areas like animal health and milk yield optimization. Bridging this gap requires targeted investments in training and infrastructure.

The study highlights the role of diverse stakeholders – governments, cooperatives, researchers and private companies – in designing and implementing record-keeping tools. Successful initiatives in East Africa, for example, show that involving farmers directly in the development process leads to better adoption and more practical solutions. These collaborations have resulted in tools tailored to smallholder farmers' specific needs, such as low-cost digital platforms and cooperative-led record-keeping systems.

As the dairy industry grows, the potential of record-keeping tools to transform smallholder farming is undeniable. By focusing on affordability, user-friendliness and farmer training, these tools can significantly improve productivity and efficiency. Empowering farmers with the ability to track, analyse, and optimize their operations is key to meeting future milk demand while ensuring sustainable livelihoods for millions of smallholders (Resti *et al*., 2024).

14.4.6 Incentives for farmers to keep records

To encourage farmers to engage in record-keeping, it is crucial to establish clear incentives that recognize the value and benefits associated with the practice. Incentives serve as powerful motivators for farmers to maintain accurate and comprehensive records, and they play a vital role in fostering a supportive environment conducive to the adoption of record-keeping in smallholder agriculture.

One key incentive for farmers is access to preferential credit terms. Financial institutions often require robust and reliable data to assess the creditworthiness of farmers. By maintaining records that showcase their production history, income and financial management, farmers can demonstrate their creditworthiness and secure loans with favourable interest rates and repayment terms. This could provide them with the additional capital to invest in their farms, expand their operations and improve their overall productivity and profitability.

Inclusion in value chain programmes is another important incentive for farmers to engage in record-keeping. Many value chain actors, such as processors, exporters and retailers, require traceability and quality assurance. By maintaining records that track the origin, inputs and production practices of their crops, farmers can meet the stringent requirements of these value chain actors. Inclusion in value

Box 14.3. Farm record-keeping: a key to empowering Nigerian youth farmers.

In Ekiti State, Nigeria, a study has shed light on the record-keeping behaviours of rural youth farmers, highlighting both its potential and challenges. With an average age of 28 and a mean of 12 years of schooling, these young farmers are better educated than many of their older counterparts, offering hope for modernizing agricultural practices. However, despite their relatively high literacy levels, the study found that their overall record-keeping behaviour was fair, with significant room for improvement. Most farmers kept only basic financial and production records, while neglecting important supplementary records like climate data and farm layouts.

The research revealed that poor attitudes towards record-keeping were a major barrier. Many of these young farmers viewed it as time-consuming or irrelevant, with little perceived benefit to their operations. Only 24% had a positive attitude towards record-keeping, while most cited forgetfulness and the fragmented nature of their farms as challenges. The study also found that farmers' knowledge of record-keeping was moderate, with a knowledge score of 66%. Encouragingly, the youth expressed strong intentions to adopt better practices, particularly for financial and production records, with 86% and 77%, respectively, indicating willingness to keep these records.

Education emerged as a critical factor influencing record-keeping behaviour. Farmers with higher levels of schooling showed better record-keeping practices, emphasizing the need for targeted literacy and training programmes. Extension agents also played a pivotal role in raising awareness about the importance of record-keeping. However, limited access to extension services and farming associations meant many youth farmers lacked the support needed to develop comprehensive record systems.

The study identified several practical barriers to record-keeping. The fragmented and scattered nature of farmlands made it difficult for farmers to manage records effectively. Mixed cropping systems further complicated the process, as tracking costs and outputs for multiple crops on the same plot posed significant challenges. Additionally, poor understanding of record-keeping's benefits and a lack of affordable tools discouraged adoption.

To address these challenges, the study recommended tailored training programmes and literacy initiatives for youth farmers, with a focus on managing records for fragmented plots and mixed cropping systems. Extension services should also work to improve farmers' attitudes by demonstrating the practical benefits of record-keeping, such as better farm management and access to credit. With these steps, Nigerian youth farmers could transform their record-keeping practices, paving the way for more productive and profitable agricultural enterprises (Chwawa, 2019).

chain programmes opens up opportunities for these farmers to access premium markets, secure long-term contracts and receive higher prices for their produce, thereby enhancing their income and market competitiveness.

Government support can also serve as a powerful incentive for record-keeping. Governments can offer various forms of support to farmers who maintain records, such as subsidies, grants and technical assistance. By demonstrating their commitment to record-keeping, farmers become eligible for these support programmes, which can provide them with additional resources, training and access to improved technologies. This support enhances their capacity to effectively manage their farms, adopt sustainable practices and overcome challenges, ultimately contributing to their overall success and resilience.

Participation in knowledge-sharing networks is another incentive that can motivate smallholder farmers to engage in record-keeping. Being part of networks, such as farmer groups, cooperatives or agricultural extension programmes, allows these farmers to exchange information, learn from one another's experiences and access valuable resources. By maintaining records, farmers can contribute to these networks by sharing their insights, successes and challenges. In return, they gain access to a wealth of knowledge, best practices and innovative approaches that can enhance their farm management skills and decision-making abilities.

In aligning the benefits of record-keeping with the needs and aspirations of smallholder farmers, advisory service providers play a critical role. It is essential to communicate the advantages of record-keeping in terms that resonate with farmers' priorities, such as increased productivity, enhanced profitability, risk mitigation and long-term sustainability. Farm management advisors should provide training, technical assistance and user-friendly record-keeping tools that simplify the process and minimize the burden on smallholder farmers. Additionally, raising awareness through educational campaigns, workshops and demonstrations can help farmers understand the potential benefits of record-keeping and instil a sense of value and importance in this practice.

14.5 Comparative Analysis and Benchmarking

Comparative analysis and benchmarking are essential tools that help farmers evaluate performance and make informed decisions. By comparing farm data, practices and performance indicators against industry standards or similar farms, farmers can identify strengths, pinpoint areas for improvement and uncover growth opportunities. For example, if benchmarking reveals that a farm's yield per hectare is below the regional average, the farmer can investigate and adopt more effective cultivation techniques. Conversely, lower input costs compared to peers may validate the efficiency of current practices and highlight opportunities for further optimization. Similarly with respect to pricing, benchmarking can be very useful. If competitors achieve higher revenues with similar products, monitoring price responsiveness (elasticity of demand) can inform adjustments to pricing strategies and market positioning.

These methods also enhance understanding of competitive positioning and market dynamics. By regularly assessing their performance against leading farms, farmers can stay updated on industry trends and innovations, allowing them to adjust strategies, optimize resource use, reduce costs and increase profitability. Additionally, comparative analysis provides a structured approach to setting realistic goals and measuring progress over time, fostering continuous improvement.

14.5.1 Benchmarking as a tool for better farm decision making

Benefits of comparative analysis and benchmarking

Benchmarking enables farmers to evaluate their production methods, financial performance and resource utilization, by comparing key metrics with similar operations. This helps identify cost-saving opportunities, improve productivity and enhance financial management. By analysing input costs, labour expenses and machinery efficiency, farmers can pinpoint inefficiencies and make data-driven adjustments. Identifying areas such as excessive input use or underutilized resources allows for better allocation and cost management. Regular benchmarking keeps farmers informed of industry trends and innovations, ensuring they remain competitive. It enables them to refine strategies, adopt best practices and effectively allocate resources to maximize profitability. By systematically measuring performance against benchmarks, farmers can implement gradual enhancements, ensuring long-term sustainability.

Use of benchmarking to compare farm performance

Benchmarking farm performance against industry standards or similar farms helps reveal areas where a farm has a competitive advantage. This advantage may arise from specialized production techniques, unique market opportunities or efficient resource management. Identifying these strengths allows farmers to differentiate their products, enhance market positioning and explore new market segments.

Additionally, benchmarking provides valuable insights into market trends and emerging opportunities. By comparing performance with industry benchmarks, farmers can recognize demand patterns, consumer preferences and economic shifts that influence decision making. This enables them to align production strategies, diversify offerings, and adapt to changing market conditions.

Performance comparisons should also consider *economies of scale*. Farms operating at a larger scale may have lower per-unit costs, allowing them to achieve higher margins. However, benchmarking can reveal whether smaller farms maintain advantages through more efficient input use or specialized market strategies.

Moreover, benchmarking of crop profitability also has the potential to help farmers identify key performance indicators that significantly impact farm profitability, allowing farmers to focus record-keeping efforts on the most critical areas. As previously discussed, by comparing profitability metrics across similar farms or production systems, farmers can pinpoint the main cost drivers and income determinants – such as yield levels, input costs and market prices. Benchmarking highlights variations in performance and identifies where efficiency gains can be made, reducing the need for extensive, unfocused data collection. Instead of tracking every possible variable, farmers can concentrate on recording data related to the most influential factors, such as fertilizer application rates, labour costs and selling prices. This targeted approach to record-keeping minimizes the administrative burden while ensuring that essential data is available for decision making, ultimately improving profitability and resource allocation.

Another key use of benchmarking is assessing resource allocation and management practices. Comparing resource utilization with industry benchmarks helps identify inefficiencies, wastage and areas where improvements can be made. This insight enables farmers to optimize resource use, improve productivity and make informed investment decisions in machinery, labour or technology. See Box 14.4. for applications of benchmarking.

Integrating comparative analysis and benchmarking in farm management

Incorporating comparative analysis and benchmarking into farm management enables farmers to make data-driven decisions, enhance efficiency and remain competitive in a dynamic agricultural landscape. Learning from peers and industry standards allows for continuous performance improvement, better resource allocation and strategic growth. By leveraging these tools, farmers can optimize operations, strengthen market positioning and achieve long-term sustainability.

14.5.2 The benchmarking process

The benchmarking process starts by identifying farms and farmers that are performing well and are successful at what they do. It requires a thorough understanding of their farming practices to identify strengths and weaknesses, and the steps needed to improve performance. The performance of these 'benchmark' farms is set as a standard for other farmers to compare against.

There are two main parts to benchmarking: (i) collecting data from farms; and (ii) comparing this data against that of a more successful farm. To make meaningful comparisons, it is necessary to collect similar data from each farm. Farmers often engage in informal benchmarking. For example, a farmer may notice another farmer with a larger harvest or one who gets a better price for the same product at the same market and wonders why. Or a farmer might hear of another reducing costs by introducing a new technology and consider adopting the same approach. By observing and talking to successful farmers, others can learn how to improve their own farm performance. These informal comparisons can lead to improved outcomes.

However, to maximize benefits, farmers need to learn how to benchmark in a formal manner, following a systematic process. Formal benchmarking involves a general strategy of:

- Examining your own farm and identifying areas for improvement.
- Identifying a similar farm that is performing better.
- Studying the benchmark farm in detail to understand what it does better.
- Comparing the performance of the two farms and understanding the reasons for differences.
- Planning and implementing changes based on what you have learned.

The sequence of steps is as follows:

1. **Identify key performance measures**: select suitable comparative farm data from a farming system closely related to the one under study. Key performance measures that reflect

Box 14.4. Applications of benchmarking.

Benchmarking provides a standard for comparison and can be applied in various ways:

- Comparing the performance of any farm to a more successful farm.
- Comparing the past performance of a farm to current performance.
- Comparing a farm plan with the actual outcome.
- Checking production levels to determine if the farm is technically efficient.
- Evaluating production costs to ensure economic efficiency.
- Examining production and marketing processes to determine their effectiveness.
- Learning from the experience of other farmers to generate new ideas.

Benchmarking can be internal, comparing different periods within the same farm, or external, comparing one farm with another.

Benchmarking can be conducted at all stages of the farm decision-making cycle, from diagnosis and planning to implementation. There is no need to wait for a problem to emerge before considering benchmarking.

Internal benchmarking

Internal benchmarking involves comparing the farm business's performance with its past results. This internal assessment helps identify weaknesses and opportunities for improvement. By analysing performance over time, farmers can make informed changes to improve their operations. Quick results can often be seen, but the challenge lies in knowing what specific actions to take for improvement. Solutions may often involve external insights, leading to external benchmarking.

Questions for farmers to consider include:

- How does your farm business compare with other farms of similar size and type?
- What can you learn from others, to improve your own farm's performance?

External benchmarking

External benchmarking involves comparing a farm's performance with that of other successful farms with similar enterprises. The benchmark may be a competing farmer or a successful one willing to share best practices. This process can involve studying multiple benchmark farms to get a comprehensive understanding of effective farming practices. By learning from these successful farms, others can adopt and implement practices that lead to improved performance.

farm performance must be identified, such as market-related measures (e.g. final market price achieved, quality of harvested produce, marketing costs) and output-input related measures (e.g. yield per hectare, cost per ton of packaging, milk produced per kilogram of feed). Other efficiency factors could include fertilizer purchases as a percentage of production costs, area cropped per man equivalent, livestock units per hectare of grazing land and total farm income per $100 of total input costs.

2. **Evaluate key performance measures by comparison with similar farms**: obtain comparative information from sources such as farm survey data, farm management information publications, extension services and farmers' associations. This comparative data is crucial for evaluating how a particular farm's performance measures up against similar farms.
3. **Identify the best performance for each key performance measure**: comparing the performance of a particular farm with other farms or a group of farms is essential. Performance is typically measured in terms of overall profitability, gross margin performance of enterprises, yields, selling prices, quantities of variable inputs used, total fixed costs and other relevant physical and financial performance measures.
4. **Identify the technological or marketing practices leading to the best performance**: farm business analysis is used to identify the indicators and their associated values in comparison with best-in-class cases. Extension workers can organize

meetings where farmers discuss the root causes of their performance success and failures. These discussions help identify whether production or marketing practices are responsible for high performance.

5. **Assess the transferability of the best practices to the particular farm**: analysis of key performance measures should suggest the extent to which best practices can be transferred. Factors preventing the transfer of techniques, such as unsuitable soils, insufficient rainfall or lack of required skills, must be considered.
6. **Investigate the potential benefits and implications of using the best practices**: once the best marketing and production practices are identified, a cost-benefit analysis must be conducted to evaluate their implementation. This step ensures that the benefits outweigh the costs and that the practices are feasible.
7. **Implement the practices and monitor their performance**: after identifying the appropriate production and marketing techniques, they need to be implemented on the farm. Monitoring the performance of the relevant enterprises ensures that the changes are improving performance in line with expectations.

By following these steps, farmers and extension workers can systematically analyse farm performance, identify areas for improvement and implement strategies that enhance productivity and profitability. Benchmarking thus becomes a powerful tool for continuous improvement, enabling farmers to stay competitive in a dynamic agricultural landscape.

Benchmarking procedures usually need to be led by extension workers, but they can also be effectively conducted by a farmer with strong farm management skills. Benchmark studies, however, may require additional training in farm management to set the performance measures needed for comparison. This training ensures that the benchmarking process is grounded in accurate and relevant performance metrics (see Box 14.5.).

Look for similarities when benchmarking

Benchmarking involves collecting data from individual farms within an area that have similar farm enterprises, scales of operation and levels of technology. These farms should experience similar impacts from changes in climate, prices and technologies. Selecting farmers with comparable abilities, objectives, and attitudes towards risk ensures that the analysis and comparisons are relevant and useful. By focusing on farms with these similarities, the benchmarking process can yield meaningful insights that are applicable across the group.

Benchmarks from data collection

For benchmarking to be effective, it is essential to establish measures (benchmarks) of farm business performance. Data can be collected from individual farms, groups of comparable farms, or through discussions with groups of farmers to reach a consensus on performance standards. Two key points about data collection in formal benchmarking are as follows.

1. Farmers need accurate information about their own farms to compare their business performance with others.
2. Benchmarking is usually conducted among farms in the same area to ensure that the farms being compared are similar in type and size.

Detailed records of farm activities are crucial for making benchmarking easier and more effective. For the dos and don'ts of benchmarking, see Box 14.6.

By following these guidelines, benchmarking can become a powerful tool for improving farm management practices, enhancing productivity, and fostering a culture of continuous improvement among farmers.

EXAMPLE: COMPARATIVE ANALYSIS OF FARM PROFITABILITY. A maize farmer, Steve, wants to evaluate the profitability of his farm compared to similar farms in his region. He gathers financial and operational data from his farm and compares it with benchmark data prepared by the agricultural economics department of the agricultural university and collected from farm surveys conducted annually.

Steve collects key financial and production indicators from his farm and compares it with

Box 14.5. Benchmarking: as an extension tool.

Benchmarking can be seamlessly integrated into extension work. Extension workers with a good understanding of local farm performance are well-positioned to facilitate benchmarking analyses. If this knowledge is lacking, it is advisable to survey the area to understand the farming system, identify the challenges farmers face and pinpoint well-performing farmers.

Most farmers aspire to improve the productivity and profitability of their farms but may lack the knowledge or discipline for self-assessment. Being part of an extension learning group supported by a skilled extension worker can foster this discipline. Effective benchmarking requires a range of complementary skills, including group formation, farm diagnosis, farm business management, technical farming knowledge and learning facilitation.

Identifying and engaging community leaders that are open to adopting new ideas and methods can help drive the benchmarking process. These leaders can set a positive example and assist in forming learning groups and facilitating benchmarking exercises.

Box 14.6. Dos and don'ts for benchmarking.

Dos for benchmarking

- Ensure the enterprise, activity or function selected is critical to the farm business.
- Clearly define the problem and address it quickly, preferably before the next crop or livestock cycle.
- Seek agreement on the focus of the benchmarking exercise.
- Inform other farmers in the area about the benchmarking activity.
- Keep the exercise simple and match the complexity to the farmers' interest and skills.

Don'ts for benchmarking

- Do not benchmark topics where performance is already good.
- Avoid benchmarking processes that are not important.
- Do not select topics that are too broad or poorly defined, leading to unfocused efforts.
- Avoid major strategic benchmarking initiatives if farmers are not committed to changing their current practices.

the average of the best performing farms in the region (Table 14.12).

From the analysis, it is clear Steve's farm yields 6.5t/ha, while the benchmark farms average 8.0t/ha. This indicates a potential yield improvement of 1.5t/ha (≈23%), which could significantly increase income. Due to lower yields, Steve earns $18,750 less in total income than the benchmark farms, despite having the same farm size and market price.

Steve's variable costs are slightly lower ($45,000 vs $50,000), but his cost per ton is higher ($9.23 vs $8.75), suggesting less efficient input use or lower economies of scale. Fixed costs are lower, which may indicate fewer investments in infrastructure, machinery or technology.

Steve's net profit is $8750 lower than the benchmark farms, with a profit per hectare gap of $175/ha. This suggests that optimizing productivity and cost-efficiency could improve farm profitability.

To increase yields, Steve could improve soil fertility management (better fertilization and crop rotation), adopt improved seed varieties and implement precision farming techniques. This may require investment in farm machinery to enhance operational efficiency. Alternatively, to reduce overall production costs he could consider increasing labour productivity and improving pest and disease management efficiency.

Since Steve's total income is lower than that of the benchmark farms, he could consider exploring opportunities for value addition through processing and identify premium markets that offer higher prices.

Table 14.12. Comparative analysis.

Indicator	Steve's farm	Benchmark farms (average)
Farm size (ha)	50	50
Yield (tons/ha)	6.5	8.0
Selling price ($/ton)	250	250
Total income ($)	81,250	100,000
Variable costs ($)	45,000	50,000
Fixed costs ($)	15,000	20,000
Net profit ($)	21,250	30,000
Profit per ha ($)	425	600
Cost per ton ($)	9.23	8.75

By adopting best practices in yield improvement, cost management and investment, he can enhance profitability and competitiveness in the market.

14.5.3 Gross margins in comparative analysis

Assessing the profitability of a farm requires more than knowing its total income or expenses. Farmers and advisors need tools that reveal *where* profits are being generated and *why* differences occur between farms.

Two simple yet powerful measures—the system index and the index of performance—use gross-margin data to evaluate both the potential profitability of a farm's enterprise mix and the efficiency with which that potential is realised. Together, these indices help diagnose whether variations in farm income arise from the choice of enterprises or from how well those enterprises are managed.

Gross margin data can be collected for each farm enterprise and for the whole farm.

The system index

The system index (also called the index of potential) shows how intensive and profitable a farm's chosen mix of crops or enterprises could be. It tells us whether the farm is focusing on high or low-margin activities compared to similar farms. Using gross margin per hectare as the key measure, it allows comparison of a farm's enterprise mix with the group average.

To calculate the system index:

1. Multiply the area of each crop by its standard gross margin per hectare.
2. Add up these values to get the farm's total standard gross margin.
3. Divide by the total cultivated area to get the standard gross margin per hectare.
4. Express this as a percentage of the group's average gross margin per hectare.

This percentage is the system index. Farmers could use this index to assess whether their farming system has the potential of producing a high gross margin per hectare.

The index of performance

The index of performance shows how well a farm is doing compared to its potential. It compares the farm's actual gross margin with what it could earn using standard margins. Performance is influenced by yields, prices and variable costs and the index measures how efficiently the farm turns its potential into real profit.

These indices can show whether there are weaknesses in the farm system or in the production performance. If the system index or the index of potential is low, the farmer should consider modifying his system and producing more high gross margin enterprises. If the index of performance is low the farmer should aim at improving the efficiency of production of individual enterprises.

Term	Measures
Index of Potential (System Index)	How good the farm's enterprise choices are
Index of Performance	How well the farm achieves results from those choices

Thus, together, these two indices help identify problems:

- A low system index suggests the farm should shift to more profitable enterprises.
- A low performance index suggests the farm should improve efficiency within existing enterprises.

Example

An example farm from a sample of farms in Musanze District, Rwanda, has a total area of 5.5 hectares. The crops cultivated are yams, maize, cowpeas, sorghum, cassava and vegetables and some of the land remains fallow. The land use breakdown is given in the second column of Table 14.13.

The total gross margin for this example farm is \$282 or \$52 per hectare. This is slightly higher than the average gross margin per hectare of \$45.8 for similar farms within the group.

To determine the system index and the index of performance, standard (average) gross margins per hectare and for the farm must be calculated. These figures are provided in columns 3 and 4 of the Table 14.13. Using these standard gross margins, the calculated standard gross margin per hectare for this farm's mix of crops is \$60.

Index of Potential

- Average standard gross margin for similar farms: \$45.8 per hectare
- Standard gross margin for the example farm: \$60 per hectare
- Calculation of Index of Potential: $(60 \div 45.8) \times 100 = 131$
- *Interpretation*: The potential of this farming system (131) is considerably higher than the group average (100).

Factors affecting the index of potential

A more detailed look at how land is used can help guide this decision. In this example, the system index is affected by:

- The amount of land left fallow
- The extent of mixed cropping
- The proportion of high-margin crops grown

Although the example farm has a lower-than-average proportion of high-margin crops, its lower than average percentage of fallow land and high proportion of mixed cropping result in a high system index (see Table 14.14).

Index of Performance

- Actual gross margin per hectare for the example farm: \$52
- Standard gross margin per hectare (based on actual crops and cultivated areas, multiplied by standard group average margins): \$60
- Calculation of Index of Performance: $(52 \div 60) \times 100 = 87$
- Interpretation: The performance of this farm (87) is below the group average (100).

A lower performance index shows that some farm enterprises are not performing efficiently. The final column of Table 14.13 shows the actual gross margins per hectare for each crop, which can be compared to the average (standard) margins. In this case, yams, cowpeas, sorghum and cotton have lower- than- average gross margins. This suggests that the yields, prices, and variable costs for these crops should be reviewed and compared with group standards.

The analysis can also be extended by examining how different production techniques influence the gross margin per hectare for individual crops. Evidence from farms in Musanze District, Rwanda, shows a clear and consistent pattern: profitability increases with a higher system index, and a strong performance index further enhances returns. The most profitable farms are those that score highly on both indices, combining efficient management with a productive and well-balanced enterprise mix.

Table 14.13. Calculation of Standard Gross Margin for Example Farm

Crop	Hectares (A)	Standard or average gross margin per hectare for the group ($) (B)	Total ($) (A x B)	Actual gross margin per hectare on example farm ($) (C)	Actual GM ($) (AxC) (D)
Yams	0.59	377	222.4	284	167.6
Maize	1.17	54	63.5	59	69.0
Cowpeas	0.48	46	21.8	42	20.2
Sorghum	0.025	30	0.8	24	0.6
Cassava	0.06	118	7.1	150	9.0
Cotton	0.05	116	5.8	110	5.5
Vegetables	0.06	156	8.6	168	10.1
Total including fallow	5.5		330.0		282.0
Standard gross margin per hectare		60 ($330/5.5ha)			
Total Actual Gross Margin($)					$282.0
Actual gross margin per hectare($)				52 (282/5.5)	

Source: adapted from Upton (1973)

Table 14.14. Factors affecting the index of potential.

	Example farm	Group average
% fallow	55.8	77
% mixed cropping	162	152
% high margin crops	26.5	28.8

Footnote: % fallow for the example farm is derived from Table 14.13 (2.4 ha cultivated area out of total farm area of 5.5 ha., gives 3.06/5.5 = 55.8% fallow).

In contrast, farms that operate intensive systems but with weak performance tend to be the least profitable. This suggests that improving yields, prices, and production efficiency should take priority before attempting to intensify the farming system or expand the enterprise mix.

The system index and performance index provide the greatest insight when they are analysed together. Each captures a different dimension of profitability—one reflects the potential of the farming system (enterprise choice), while the other reflects the efficiency with which that potential is realised.

When the two are compared, farms can be classified into four broad categories:

- **High system index and high performance index** — These are the *best-performing farms*. They are both efficient and well-diversified, achieving high profits through sound enterprise selection and strong management.
- **High system index and low performance index** — These farms operate potentially profitable systems but are *intensive yet inefficient*. Their high potential is undermined by weak management or

technical inefficiency, indicating a need for better use of resources and improved practices.

- **Low system index and high performance index** — These farms are *efficient but low in potential*. They make good use of their resources but are limited by an unprofitable or poorly chosen enterprise mix. Increasing returns would require restructuring the farming system towards higher-margin activities.
- **Low system index and low performance index** — These are the *least profitable farms*. Both the enterprise structure and management efficiency are weak, pointing to the need for substantial changes in both farm planning and operational performance.

This combined interpretation helps both farmers and extension agents identify whether low profitability results from poor enterprise selection, inefficient production, or a combination of the two. It also provides a practical framework for determining whether improvements should focus first on restructuring the farming system or on enhancing technical and managerial efficiency within the existing system.

14.5.4 Benchmarking: some concerns

While benchmarking is a valuable tool in farm management, it is not without its weaknesses, which must be understood to maximize its effectiveness. Some key concerns include the challenge of finding comparable benchmarks, the construction of averages and standards, deficiencies in record-keeping, inadequate diagnostics and the sharing of information among farmers.

Finding comparable benchmarks

The diversity of farming systems can make it difficult to find comparable benchmarks. Each farm family is unique, with distinct goals, resources, abilities and attitudes towards risk. No two farms are the same, and differences in ecology and resource bases often complicate standard comparisons. For instance, farms grouped by their main enterprise may still vary significantly due to differences in soil quality and climate. In such cases, the extension worker becomes crucial in facilitating the process to find the best or most representative benchmark farm possible, ensuring that comparisons are meaningful and applicable.

Developing recommendation domains can help mitigate many of the pitfalls associated with comparative analysis in smallholder farm management. Recommendation domains are essentially subsets of farms or regions that share similar characteristics – such as climate, soil type, production goals and socio-economic conditions – that allow for more tailored analysis and recommendations. By grouping farms based on these criteria, recommendation domains create more relevant comparisons, reducing the likelihood of drawing inaccurate conclusions that can arise when smallholder farms are compared across dissimilar contexts.

Construction of averages and standards

In some situations, specialists are employed to construct benchmarks by collecting data from similar farms in the vicinity and calculating standards based on average performance. These averages provide the theoretical 'benchmark' that is used in comparative analysis. However, there are limitations to using averages as a standard for comparison. Averages are artificial and do not always represent reality, as there will always be farms that are more profitable and efficient than the average. Additionally, average costs calculated for the whole year do not account for seasonal and price variations. Extension workers must understand these limitations, especially when helping farmers develop action plans with projected outcomes that may vary, potentially discouraging farmers if not properly managed.

This critique arises because comparative analysis focuses on average relationships rather than marginal ones, with marginal analysis being a more accurate indicator of economic efficiency (Candler and Sargent, 1962). The economic analyses described in Chapter 6, further elaborate on the concept of margin analysis in relation to economic efficiency.

Deficiencies in record-keeping

In many cases, farmers do not record data in sufficient detail to allow meaningful comparisons. This lack of detailed record-keeping can hinder the effectiveness of benchmarking. Extension workers should be aware of this reality when setting up farmer learning groups and emphasize the importance of keeping accurate records. Positive results from benchmarking exercises can help demonstrate the value of record-keeping, encouraging farmers to maintain better records for future comparisons and decision making.

Inadequate diagnostics

Benchmarking may highlight a weakness but does not necessarily identify the cause. The identified weakness is a symptom of a gap in farm management rather than the root cause. There is often a tendency to stop the diagnosis once the weakness has been identified. Farmers need to dig deeper until the root cause is found, ensuring that corrective actions are effective and address the underlying issues. This may require encouragement from extension workers.

Sharing of information among farmers

Farmers are often reluctant to share their financial positions due to fears of competition. Even when information is recorded, there may be a lack of willingness to be open and honest about business problems. Poor results tend not to be discussed, which can hinder the collective learning process. Extension workers can play a crucial role in fostering a culture of openness and collaboration, helping farmers understand that sharing information can lead to mutual benefits and improved performance across the community.

14.6 Data and Information for Monitoring and Evaluation

As has been noted frequently, monitoring and evaluation depend on data and information. We covered this topic rather extensively in Chapter 10. However, it is useful to revisit the subject in the context of monitoring and evaluation.

14.6.1 The importance of data accuracy and detail

Ensuring data accuracy is essential, but it must be balanced with relevance and practicality to equal effective farm management decisions. The following explores how to prioritize key data points, avoid information overload and utilize estimation techniques to enhance monitoring and evaluation without unnecessary complexity.

Balancing the need for data accuracy with practicality

When it comes to monitoring and evaluation in farm management, data accuracy is crucial for making informed decisions. However, it is essential to strike a balance between accuracy and practicality. Collecting highly detailed and accurate data for every aspect of the farm can be time-consuming and resource intensive. Therefore, it is important to focus on collecting data that is both relevant and feasible to obtain. Farmers should identify the key data points that have a significant impact on their decision-making process and prioritize collecting accurate information for those aspects. This approach ensures that the data collected is reliable and useful for analysis and decision making, without overwhelming the farm management process.

Importance of collecting relevant data rather than excessive data

In addition to accuracy, the relevance of data is equally important. Collecting excessive data can lead to information overload and make it challenging to extract meaningful insights. It is crucial to identify the specific data points that align with the farm's goals, objectives and areas of focus. By collecting relevant data, farmers can focus their monitoring and evaluation efforts on the aspects that have the most significant impact on farm performance and profitability. This targeted approach ensures that the data collected provides actionable insights and contributes to effective decision making.

Utilizing estimation techniques and sample data for cost-benefit analysis

Obtaining highly accurate data for every single farm operation can be impractical. However, estimation techniques and sample data can help fill in the gaps and provide a reliable representation of the overall farm performance.

Farmers can utilize statistical methods and sample data to estimate costs, returns and other performance indicators. By employing estimation techniques, farmers can make informed decisions based on a reasonable approximation of the farm's financial and operational performance. This approach allows for efficient monitoring and evaluation, even in situations where collecting comprehensive data may be challenging.

14.6.2 Key data points for effective monitoring and evaluation

Certain key data points play a critical role in effective monitoring and evaluation in farm management. These data points may vary depending on the farm's specific objectives and activities, but they commonly include:

- **Financial data:** this includes income, expenses, investments, debt and cash flow. Financial data provides insights into the farm's profitability, efficiency and financial stability.
- **Production data:** crop yields, livestock performance, feed conversion rates and other production-related metrics help assess the farm's productivity and efficiency in utilizing resources.
- **Input data:** tracking inventory and input usage helps monitor resource utilization, identify waste and optimize input management.
- **Labour data:** recording hours worked, tasks completed and labour costs helps evaluate labour efficiency, allocate resources effectively and estimate labour requirements for future planning.
- **Compliance and regulatory data:** maintaining records related to regulatory requirements, permits, certifications and environmental compliance ensures the farm operates within legal and ethical guidelines.

Collecting and analysing these key data points enables farmers to gain insights into their farm's performance, identify trends, make informed decisions and track progress towards their goals. By understanding the importance of data accuracy and detail, focusing on relevant data, utilizing estimation techniques when necessary and monitoring key data points, farmers can enhance their monitoring and evaluation processes, make better-informed decisions and drive sustainable farm management practices.

14.7 Monitoring Implementation of a Farm Plan

In Chapter 13, we discussed the need for the farmer to develop a farm plan based on the diagnoses conducted in the first stage of the farm management cycle. Obviously, the plan is meant to be implemented. And, therefore, it is equally important to monitor its implementation. Monitoring the implementation of a farm plan involves tracking the actions taken to execute the strategies outlined in the plan. This process is critical for smallholder farmers, especially in low-income countries, where resources are limited, and timely, efficient actions can significantly impact overall productivity. By focusing on monitoring activities rather than just outcomes, farmers and stakeholders can ensure that tasks are completed as planned, identify potential issues early and make adjustments to improve execution.

Effective monitoring provides immediate feedback on whether the farm plan is on track, fostering accountability, minimizing inefficiencies and ensuring that the farm operates optimally to meet its goals. This approach helps mitigate risks related to delays, resource misallocation or errors in execution.

The procedure for monitoring farm plan implementation should focus on the systematic tracking of actions, milestones and resources. Box 14.7. gives a possible procedure for monitoring a farm plan which a farmer could use to monitor its implementation. The level of application will vary and, if possible, the farmer should do this formally using the records discussed earlier. However, this detailed approach may be beyond the current capacity of the farmer to carry out on their own; in which case, the extension worker may need to assist the farmer. The aim, however, is to build the capacity of the farmer to be methodical and systematic about implementing their farm business plan.

Box 14.7. Procedure for monitoring a farm plan.

1. **Use a checklist**
   - Create a simple checklist of planned activities, including timelines and expected outcomes.
   - Regularly update the checklist to track task completion.
2. **Maintain basic records**
   - Keep a straightforward farm journal or log to note what was done, any deviations and reasons for those changes. For example, 'Fertilized Field A; irrigation delayed two days due to equipment repair'.
3. **Conduct regular field visits**
   - Inspect progress in person to ensure tasks align with the plan. For example, verify planting accuracy or irrigation efficiency during visits.
4. **Track resources**
   - Monitor the use of seeds, fertilizers and other inputs against the plan to avoid overuse or shortages.
5. **Get feedback**
   - Collect feedback from workers or other stakeholders to understand challenges and successes.
6. **Comparative analysis against budgets**
   - Compare planned activities and resource use against the plan's budget to identify discrepancies and assess efficiency.

14.8 Monitoring Farm Plans through Reflection and Collaboration

Effective farm monitoring goes beyond simply tracking tasks or collecting data. It involves creating opportunities for reflection, discussion and collaborative problem-solving. Regular meetings and reviews bring together key stakeholders – such as the farmer, farm family, employees and extension workers – to assess progress, address challenges and make necessary adjustments.

One effective way to facilitate this process is through farmer learning groups (FLGs), guided by extension workers. FLGs are small, farmer-led groups that foster knowledge sharing, problem solving and skill development. These groups provide a structured setting where farmers can collectively evaluate their progress, exchange experiences and explore solutions to common challenges.

For smallholder farmers, FLGs promote accountability and shared learning, ensuring that efforts remain aligned with the farm plan's objectives. Regular group discussions help farmers review activities, refine strategies and keep their farm businesses on track.

By incorporating structured monitoring processes, farmers can systematically track the implementation of their farm plans. This proactive approach not only helps limit setbacks but also enhances efficiency. Ultimately, a well-monitored plan lays the foundation for meaningful evaluation, ensuring that results reflect well-executed actions.

14.9 Evaluating the Performance of the Farm Plan

Evaluating the performance of the farm plan is a critical step in understanding the effectiveness of the strategies implemented to achieve desired outcomes. As noted earlier, while monitoring focuses on tracking actions during implementation, evaluation provides a retrospective analysis of whether those actions translated into results, offering insights into the plan's success or areas needing improvement. Obviously, there is a large overlap with evaluating the plan and evaluating the performance of the plan. Thus, the aim here is to offer a methodical approach to organizing and conducting the evaluation using all the various tools explained in this and earlier chapters.

For smallholder farmers performance evaluation is essential for optimizing resource use, improving productivity and ensuring long-term sustainability. By systematically assessing the outcomes against the goals outlined in the plan, farmers can identify what worked, what did not and why. This process informs future decision making (e.g. the next round of planning), enhances accountability and supports learning, enabling farmers to adapt to challenges and refine strategies for better results.

While farmers can assess their own farm plans, a more structured and effective evaluation process takes place when conducted collectively within the FLGs mentioned before. With the guidance and support of an agricultural extension worker, FLGs provide a collaborative setting where farmers can systematically review their farm plans, analyse results and refine strategies.

Evaluating farm performance in a group setting offers several advantages:

- **Shared learning**: farmers gain insights from each other's experiences, learning from both successes and challenges.
- **Objective assessment**: a group setting encourages a more balanced and critical review of farm operations, reducing individual bias.
- **Problem-solving support**: collective discussions help identify practical solutions to common issues.
- **Accountability and motivation**: engaging with peers fosters commitment to improving farm practices.

The steps in Box 14.8. outline a structured approach to farm plan evaluation, ensuring that assessments are comprehensive, actionable and contribute to ongoing improvements in farm management. This could be done by an individual farmer or a group of farmers – both with the help of an extension worker.

Evaluating the performance of a farm plan is not merely an exercise in measuring success but a vital learning process that helps farmers and agricultural advisers understand how to optimize efforts. For smallholder farmers this evaluation ensures that the hard work and resources invested yield maximum benefits. By systematically analysing outcomes, identifying gaps and applying lessons learned, farmers can continuously refine their plans to achieve greater efficiency, resilience and sustainability in the face of challenges.

14.10 Integrating Monitoring, Evaluation and Learning with Planning

Farm management is a continuous process of planning, decision making and learning from past experiences. As noted, monitoring and evaluation are central to this process, providing valuable feedback to guide decision making and improve farm operations over time. Recognizing that farm management is dynamic allows farmers to adjust their strategies based on new information and changing conditions. Integrating monitoring and evaluation into the planning process leads to more informed decisions and increases the chances of achieving desired outcomes.

Learning from past experiences is key to successful farm management. Monitoring and evaluation help farmers identify what works and what does not, ensuring that mistakes are not repeated and successful practices are reinforced. This ongoing learning process improves the farm's ability to adapt and enhances overall productivity. This continuous learning process also strengthens the farmer's ability to adapt to changing circumstances.

To be effective, monitoring and evaluation should align with the farm's long-term goals. Clear objectives and measurable indicators help track progress and ensure that activities provide useful insights. A strong learning culture – where farmers and workers actively seek new knowledge and share experiences – further supports continuous improvement.

By integrating monitoring and evaluation with planning and decision making, and creating an environment of continuous learning, farmers can create a flexible and responsive management system that enhances performance and supports long-term success.

Box 14.8. Steps to evaluate the performance of a farm plan.

How to check if your farm plan is working:

1. **Ask yourself, what do I want to check?**
   - What part of the farm plan do I want to evaluate?
   - Am I looking at my harvest, costs, income or how well I used my resources?
2. **Use simple ways to measure success**
   - Compare what you planned with what occurred.
   - Look at things like:
     - How much crop you harvested per hectare.
     - The cost of seeds, fertilizer and labour, compared with your earnings.
     - Whether your family had enough food and income.
3. **Compare plans with reality**
   - Look at your notes, records and assess your memory of the season.
   - Did you produce as much as you expected?
   - Did you spend more or less than planned?
   - What worked well? What went wrong?
4. **Find out why things went differently than planned**
   - Was the weather worse or better than expected?
   - Did pests, diseases or poor soil affect production?
   - Were inputs like fertilizer or labour too expensive or unavailable?
   - Did you follow your plan, or did unexpected problems force you to make changes?
5. **Talk to others for feedback**
   - Ask your family and workers what they think worked or did not work.
   - If possible, talk to neighbours or an extension worker for advice.
6. **Check if your spending made sense**
   - Did the money you spent on seeds, fertilizer and labour give you a good return?
   - Could you have spent less and still produced well?
   - What would you change next time?
7. **Think of the bigger picture**
   - Did the plan improve the farm's soil, water and long-term sustainability?
   - Did it help your family financially and provide enough food?
   - Did the plan help you make progress towards achieving the farm goals?
8. **Write down what you learned**
   - Keep simple notes on what worked well and what you would do differently next time. This will help you make better decisions in the future.
9. **Make a better plan for next time**
   - Adjust your farm plan based on what you learned.
   - Find ways to reduce costs, improve productivity and handle risks better.
10. **Share what you learned**
    - Talk to family members, neighbours, farmer groups and/or extension workers about what worked for you. Learning from each other helps everyone improve.

14.11 Advisory Support for Monitoring and Evaluation

Effective monitoring and evaluation require more than just record-keeping; it depends on a farmer's ability to interpret data, assess progress and adjust plans accordingly. Many smallholder farmers, however, have limited time, resources or technical capacity to track and analyse their farm performance in a systematic way. Advisory services play a crucial role in supporting farmers throughout this process. Their assistance can make the difference between record-keeping as a passive activity and monitoring and evaluation as active tools for improving farm operations and outcomes.

A key role of advisory services is helping farmers establish simple and effective monitoring systems. Given that most smallholder farmers cannot maintain detailed records, extension workers can assist in identifying key performance indicators that align with the farm's specific goals. For a rice farmer aiming to improve yields, these indicators might include planting dates, seed variety used, fertilizer application and final harvest weight per hectare. For a goat farmer, key indicators might focus on growth rates, feed conversion, kidding frequency, mortality rates and market weights at sale. By focusing on a small set of relevant indicators, advisors can help ensure that farmers are tracking useful information without creating an undue burden.

Beyond helping farmers establish what to monitor, advisory services also provide crucial support in interpreting data and assessing progress. Many farmers record information but struggle to extract meaningful insights from it. Regular on-farm visits by extension workers can help farmers compare their recorded data with their expected outcomes and adjust their practices accordingly. If a farmer's actual rice yield is lower than expected, an extension worker might analyse whether the cause lies in soil fertility, input quality, human error or water management. If a farmer's production costs are higher than planned, an extension worker might review expenditure records to identify inefficiencies. This process of analysing farm data ensures that record-keeping is not just an administrative exercise but a foundation for decision making and improvement.

Advisory services also play a role in facilitating group learning, which is particularly valuable for smallholder farmers who may lack access to formal training. Peer-to-peer learning groups allow farmers to compare progress, discuss challenges and share strategies for improvement. Extension workers can support this by organizing regular meetings where farmers can review their monitoring records, discuss trends and benchmark their performance against others in similar conditions. Demonstration farms provide another learning opportunity, allowing farmers to observe best practices in farm monitoring and evaluation in a practical setting.

Evaluating a farm plan goes beyond simply checking records; it requires an understanding of why certain strategies succeeded or failed. Advisory services can support this by guiding farmers through simple cost-benefit analyses and helping them assess whether changes in production methods, input use or market strategies have led to measurable improvements. This process ensures that farmers can refine their plans for the next production cycle based on real data rather than assumptions. In addition, extension workers can provide market insights, helping farmers align their production with demand and make better financial and marketing decisions.

For many smallholders, especially those with limited literacy, advisory services can also introduce practical, low-technological solutions to simplify record-keeping and evaluation. In cases where written records are impractical, alternatives such as tally marks for sales and input use, colour-coded logs to indicate yield levels or voice-recorded notes on mobile phones can help farmers track key information without needing to write extensive records. Where available, digital farm management tools can offer additional support, enabling farmers to store, analyse and share their data more effectively.

By integrating monitoring and evaluation into their broader support services, advisory service providers can help smallholder farmers not only track progress but also use the information to refine their strategies and improve profitability. Through regular guidance, hands-on support and accessible learning approaches, they ensure that monitoring and evaluation are active tools that strengthen farm decision making rather than a passive exercise in record-keeping.

14.12 Conclusion

ME&L are essential components of effective farm management, playing a critical role in ensuring profitability and sustainability. By systematically tracking and assessing farm activities, farmers gain valuable insights into their costs and returns, enabling them to make informed decisions and optimize their operations. Record-keeping is foundational, as maintaining accurate and organized records allows farmers to capture and analyse data related to financials, production, inventory, labour and compliance.

Benchmarking enhances this framework by allowing farmers to compare their performance against industry standards or similar farms. This comparison can reveal areas where a farm has a competitive advantage, such as specialized production techniques or efficient resource management. By identifying these strengths, farmers can leverage them to differentiate their products, enhance market positioning or explore new market segments. Benchmarking involves identifying key performance measures, collecting comparable data, evaluating performance against successful farms and implementing best practices. This systematic approach helps farmers continuously improve and adapt their strategies.

Balancing data accuracy and practicality is crucial for effective monitoring, evaluation and benchmarking. While precise data is essential, recognizing the practical challenges farmers face in data collection is also important. Collecting relevant data that captures the essence of farm performance without overwhelming farmers is key. The iterative nature of planning, decision making and learning is fundamental in farm management. Monitoring and evaluation, coupled with benchmarking, serve as feedback mechanisms that inform and guide the planning process, ensuring that future decisions are grounded in evidence and lessons learned.

Advisory services play a vital role in these processes. Farmers benefit from periodic technical assistance and mentoring from agricultural experts or advisors, providing additional expertise and guidance. These services complement the monitoring process, enabling farmers to capitalize on strengths and address challenges more effectively. In conclusion, monitoring, evaluation and benchmarking, supported by robust recordkeeping and advisory services, enable farmers to thrive profitability, improve decision making, and foster a culture of continuous learning and improvement. This approach allows farmers to navigate the complexities of modern agriculture with confidence and achieve sustainable success.

References

Candler, W. and Sargent, D. (1962) Farm standards and the theory of production economies. *Journal of Agricultural Economics* Available at(2), 283–290.

Chwawa, T. (2019) Integration of information and communication technology (ICT) in the development of a recordkeeping database of smallholder sheep farming systems in the Western Cape Province of South Africa. Thesis. Stellenbosch University, South Africa. Available at: https://scholar.sun.ac.za/bitstreams/6c2b150b-90b0-408c-a977-6653b0ab178c/download (accessed 19 March 2025).

Manteaw, S.W., Akpotosu, B.W., Folitse, B.Y. and Mahama, S. (2021) Assessing farm records-keeping behavior among small-scale pineapple farmers in the Nsawam Adoagyiri Municipality, Ghana. *Ghana Journal of Agricultural Science* 56(2), 34–45. DOI: 10.4314/gjas.v56i2.4.

Prajapati, M.R., Vahoniya, D.R. and Lad, Y.A. (2015) A study on the status of farm recordkeeping practices among farmers in anand taluka. *International Journal of Business and General Management* 4–6, 35–42.

Resti, Y., Reynoso, G.G., Probst, L., Indriasari, S., Mindara, G.P. *et al.* (2024) A review of on-farm recording tools for smallholder dairy farming in developing countries. *Tropical Animal Health and Production*. DOI: 10.1007/s11250-024-04024-9.

Upton, M. (1973) *Farm Management in Africa*. Oxford University Press, London.

Additional Reading

Collinson, M. (1983) Farm Management in Peasant Agriculture. Westview Press, Boulder.

Deomampo, N. (2003) Experiences with farm recordkeeping in Asia with particular reference to the philippines, workshop on sustainable farm household information systems. Workshop on Sustainable Farm Household Information Systems, FAO, Rome, Italy. (FAO unpublished).

Dillon, J.L. and J.B. Hardaker (1993) Farm Management Research for Small Farmer Development. FAO Farm Systems Management Series No. 6, Food and Agriculture Organisation of the United Nations, Rome.

Kahan D. (2010) Farm Business Analysis Using Benchmarking. Farm Management Extension Guide No. 4, FAO.

Makeham, J.P. and L.R. Malcolm (1986) The Economics of Tropical Farm Management. Cambridge University Press.

Malcolm B., Makeham J. and Wright V. (2012) The Farming Game. Cambridge University Press.

Minae, S. (2002) FAO workshop "local level sustainable information systems to fight hunger and poverty". FAO unpublished.

Upton, M. (1987) African Farm Management. Cambridge University Press.

Part 4: Farm Management Advisory Service Support

This part of the book highlights the significance of agricultural extension, knowledge transfer and capacity building in enabling farmers to adopt effective management practices. Various advisory support models – provided by government agencies, NGOs, and the private sector – are examined, with recommendations to enhance the accessibility and effectiveness of farm management advisory services. This part also advocates for strengthening the advisory system by investing in the capacity development of agricultural advisors in farm business management. Additionally, it emphasizes the need to engage young people in addressing the challenges and opportunities of food and agriculture in the 21st century.

15 The Farm and Agribusiness Management Advisory System

Extension and advisory services have long supported farmers of all scales in navigating the ever-changing agricultural landscape. They have played a pivotal role in agricultural development for food and nutrition security, food sovereignty and economic stability. Over the decades, their aim has evolved but remains focused on helping farmers achieve their farming and livelihood goals. This is accomplished by building farmers' capacity to manage their enterprises, navigate the social, environmental and legislative contexts in which they operate, and sustain their own life-long learning to adapt and benefit from changing systems. At its core, extension and advisory services are about empowering farmers to make better and more informed decisions.

Originally focused on transferring technical knowledge from researchers to farmers, extension and advisory services have evolved significantly over the past 40 years. They are no longer confined to primary production, one-way communication or viewing farmers as end-users of externally developed technologies. While knowledge transfer remains a core function, extension now encompasses the entire agricultural value chain, emphasizing multi-directional dialogue and recognizing farmers as active contributors to agri-food systems. As one philosopher noted over a century ago, 'the farmer is the first active agent in human society'. Over time, extension and advisory services have come to honour this perspective.

Within these services is the critical branch of farm management extension which helps farmers navigate the complexities of modern agriculture. Historically rooted in public-sector initiatives, farm management extension has expanded to include agribusiness advisory support, drawing on economics and marketing to guide decision making. This chapter traces the development of farm management extension as a specialized category of agricultural advice.

According to the global forum for rural advisory services (GFRAS), over 80 countries provide agricultural extension services, employing more than one million agents, globally (Swanson and Davis, 2014). This underscores the global relevance of extension in equipping farmers with the skills to manage their operations as businesses, enhancing profitability, competitiveness and sustainability. However, only a small percentage of farmers – typically those already profit-minded and market-oriented – currently benefit from farm management advice. Yet, there is growing interest among smallholder farmers transitioning to commercial approaches, recognizing the need to improve profitability. These emerging farmers require tailored support to navigate market-oriented systems effectively.

This and the subsequent chapters are dedicated to exploring this vital function within the broader framework of extension and advisory services. The evolution of farm management extension reflects broader shifts in agricultural advisory systems. Since the 1990s, there has been a transition from public-sector, supply-led models to pluralistic, demand-responsive frameworks involving private-sector actors, NGOs and innovative initiatives like agri-clinics and agribusiness service centres. While private-sector involvement offers opportunities, it is not a panacea for small-scale family farm development. Case studies highlight both the potential and limitations of private-sector engagement, emphasizing the need for diverse extension systems that cater to varied clienteles.

Public-sector providers remain essential, particularly for smaller-scale, less commercial

 Farm Management in a Changing World: Improving Smallholder Livelihoods in Low-income Countries (Kahan and Worth)
DOI: 10.1079/9781836990536.0015

farmers, while fostering collaboration among stakeholders. Regulatory oversight may be needed to align private-sector activities with development goals. By understanding the strengths and weaknesses of different models, this chapter underscores the importance of tailored, effective advisory services to support smallholder farmers in the developing world.

Ultimately, extension systems – like the farmers they serve – must adapt to the evolving agricultural landscape, embracing diversity in providers and approaches to meet the needs of farmers at all scales. This chapter highlights the critical role of farm management extension in empowering farmers to make informed decisions, ensuring the sustainability and competitiveness of their enterprises in a rapidly changing world.

15.1 What Is Farm Management Extension?

While operating in the same context and with similar objectives with respect to farmers, farm management extension represents a specialized area within the broader field of agricultural extension. General agricultural extension focuses primarily on improving agricultural practices such as crop production, pest management, and livestock care – often by providing technical information, demonstrations, and skills training. It also plays an important role in enhancing farmers' capacity to make better-informed decisions within the extension worker's area of focus. Farm management extension helps farmers analyse, interpret and make informed decisions. As a specialized form of agricultural extension its aim is to improve decision-making processes and maximize the utilization of scarce resources. As should be the case with all forms of extension, farm management extension encourages farmers to think critically about their choices and learn from them. Farm management extension helps farmers to not only address technical aspects but also to incorporate economic considerations, ensuring that they can optimize their farm operations both in terms of production and profits. This makes farm management extension an invaluable complement to general extension, particularly where farmers face complex trade-offs and need to balance multiple goals under conditions of risk and uncertainty.

As we previously noted in Part 3 of this book, farm management extension services have at their disposal a diverse array of tools and methods to enhance farmers' decision-making abilities and optimize farm operations. As a reminder, these tools include, among others, budgeting, record-keeping, cash flow analysis and farm business planning. In this digital age, electronic technology has become a cornerstone of farm management extension. This is initially covered in Chapter 10, and further discussed in Chapter 16. Farm management software, mobile applications and online platforms facilitate efficient data management, analysis and dissemination of knowledge. Ensuring these tools are relevant and useful for smallholders in low-income countries is challenging. These tools should assist farmers in organizing and analysing farm data, enabling well-informed decisions, pinpointing areas for improvement, and monitoring the performance of their farming enterprises and overall farm business. In this context, the role of farm management extension becomes even more important.

The farm management tools and techniques developed in Western countries were designed for commercial farmers who have more resources and the skills to collect and analyse farm and market information for informed decision making. While highly effective in developed countries, these practices face unique challenges when applied to the multitude of smallholder farmers in developing regions with diverse farming environments. Smallholders often operate on smaller plots of land, have limited access to resources, and possess lower levels of education and technical skills. This diversity and multiplicity present significant hurdles in adopting and effectively using advanced farm management tools. Smallholders often lack skills to gather essential farm and market data, complicating their decision making. Efforts should still be made to help these farmers, as much as possible within their circumstances, to adopt the basic economic principles behind the techniques – even if they cannot use all the farm management tools available to commercial farmers.

The importance of farm management in low-income countries, particularly among smallholder farmers, is often undervalued. Traditional agricultural systems can generally be regarded as well managed, and one of the founding fathers of agricultural economic development argued that smallholder farmers were rational and efficient in their use of scarce resources – implying that farm management extension might not significantly improve productivity (Schultz, 1964). However, as technologies evolve and market dynamics grow in importance, smallholder farmers increasingly require the skills to adapt their farms, allocate resources more effectively, and pay greater attention to both the non-production aspects of farm management and the challenges posed by climate change.

As we discussed in previous chapters, even farmers with limited education make rational decisions, often internalizing processes such as budgeting and planning. While these may not be formally documented, the concept of profit is generally well understood. The key issue is to strengthen this capacity. Systematic data presentation can support these farmers in making better decisions regarding changes in their farm enterprise combinations, production methods, post-harvest strategies, input and output marketing and financing. Clearly showing expected profits and livelihood options enables more informed and strategic decision making.

As farmers become more market-oriented, advisory support also needs to evolve. Small family farmers increasingly require assistance in managing and marketing their products, improving farm efficiency and profitability, and in this way optimizing their farm income. Even if overall farm profitability is satisfactory, it remains necessary to examine individual farm enterprises for cost-reducing and income-increasing opportunities. When profit levels are low, the entire farm business must be scrutinized for ways to enhance profitability and income. This involves examining yields, output prices, input levels for individual enterprises, marketing strategies and the intensity of the entire farming system. And this must be seen within a longer-term context to ensure sustainability.

The challenge for extension lies in ensuring the relevance and utility of the farm management tools for smallholder farmers in low-income countries. While these tools are designed to be adaptable to various farming systems, scalable to different farm sizes, and accessible to farmers at different levels of technical proficiency, their effectiveness depends on appropriate adaptation and contextualization. Public and private advisory services must find efficient ways to support the large number of smallholder family farmers, helping them leverage these tools to organize and analyse farm data, make informed decisions, identify areas for improvement and monitor their farming enterprises' performance.

In low-income countries, given the scarcity of public funds, there is a need to use scarce farm management skills sparingly with a focus on impact. In some countries, more attention is given to farm management analysis where the results are used to influence policy change and contribute to the design of extension programmes. In Latin America, in particular, farm management advice has been used to design land reform measures as part of settlement-cum-land consolidation schemes. Diagnosis of this kind also has the potential to provide a useful input into the type of extension messages that extension services need to communicate and the kind of training they may need to offer. Under land reform schemes, extension workers often have the task of providing the technical and economic background information necessary to allocate resources more efficiently to maximize farm income.

Farm management extension services also provide an added framework for the activities of general extension workers. Farm management offers a perspective outside technical subject areas, considers all aspects of farming and their interrelationships, and is positioned to draw up priorities for action. Improved use of farm management techniques may allow agricultural extension workers to identify areas where new technologies would be particularly beneficial to farmers so that they can inform research workers of farmers' needs or facilitate local innovation. In this way, farm management advice can benefit a large number of smaller farmers indirectly.

The style of consulting or the way that extension services work typically involves helping farmers learn how to analyse, interpret and define appropriate managerial actions for themselves and, alternatively, doing some or all of this work to support smallholder development. Extension workers have the important functions of information gathering, interpretation and dissemination, and act as conduits feeding information to farmers and the wider rural community. In this way, farm management extension provides a combined educational and service role. But the long-term objective is to build as much as possible the capacity of the farmers to do these things themselves – perhaps collectively – with minimal outside assistance.

Given the dearth of information and data among smallholders and their limited decision-making capacity, the role of advisory service providers is vital. These providers must bridge the information and skills gap by simplifying tools and processes, offering hands-on support, and ensuring that the management practices are accessible and actionable for smallholder farmers. Efficiently supporting a large number of smallholders requires leveraging community-based approaches, such as group training sessions and peer learning networks, to maximize impact. This is further developed in Chapter 16.

As noted previously, as the agricultural landscape has become increasingly market-oriented and the provision of extension services more pluralistic, the concept of farm management extension has evolved significantly. Farm management extension, once mainly focused on helping farmers make better decisions to use resources efficiently and increase profits, has expanded to include business-oriented advice for a wider range of people involved in modern value chains – such as processors, marketers and other actors beyond the farm. These farm management and agribusiness advisory services are now offered by both public and private agencies, highlighting a collaborative, multifaceted approach to supporting modern agricultural systems. However, since the focus of this book is on developing the small family farm, this is largely the role of the public sector advisory or extension service, the main contributor to supporting small family farming.

This chapter explores the evolution of extension services and uses the terms *farm management extension* and *agribusiness advisory services* to describe a broad spectrum of support offered to farmers of all sizes and other value chain actors. This support ranges from traditional production advice to farm business management and more specialized agribusiness guidance.

Modern extension services are delivered by a wide range of agencies and organizations that work both independently and in collaboration. These services target individual farmers, value chain actors and organized farmer groups, such as cooperatives and agribusinesses. The chapter also examines the broader agribusiness landscape, which includes industrial agriculture and off-farm agricultural enterprises – businesses managed by individuals or companies not directly involved in primary production. These larger enterprises are increasingly supported by private sector advisory services that include business development advisors and agribusinesses, and generally lie beyond the scope of small family farm development.

As this book focuses on small family farmers, we will continue to use the term *farm management* (and, more specifically, *farm business management*) to reflect the emphasis on decision making and resource use at the farm level.

15.2 Early Initiatives to Develop Smallholder Family Farming

In the 1990s, the FAO promoted the farming systems development (FSD) methodology, a participatory and iterative approach to agricultural development that placed farmers and their livelihoods at the centre of attention (FAO, 1989). Rooted in the recognition of the complexity and diversity of farming systems, FSD sought to understand and address the interactions between crops, livestock, natural resources, markets and socio-economic conditions. The methodology combined economic analysis with diagnostic analysis, planning and implementation, focusing on identifying and prioritizing farmers' constraints and opportunities. Central to FSD was collaboration among farmers, extension workers and researchers, to co-develop sustainable, locally appropriate interventions aimed at enhancing productivity, income and resilience.

The diagnostic phase of FSD involved comprehensive assessments of existing farming systems through farm surveys, participatory rural appraisals (PRA), and stakeholder consultations to identify biophysical and socio-economic constraints, resource-use patterns and farm household goals. Based on this understanding, the next phase of planning involved the design of tailored interventions to address the needs and capacities of farmers, such as introducing improved crop varieties, promoting sustainable water management practices or diversifying income sources. The final phase of implementation involved pilot testing interventions, scaling them up, and monitoring and evaluating their impact to refine strategies as needed. A key element of FSD was the delineation of recommendation domains to ensure interventions were relevant to specific farming system contexts, thereby ensuring their scaling-up.

FSD proved particularly valuable where smallholder farmers operate in resource-constrained and highly variable environments. The approach enabled the identification of context-specific solutions aligned with farmers' priorities, such as improving food security, adapting to climate variability and enhancing market access. For example, FSD facilitated integrated crop-livestock systems in sub-Saharan Africa, improved soil fertility management in Southeast Asia and supported irrigation development in arid regions. Its participatory nature ensured that interventions were farmer-driven, increasing their adoption and long-term sustainability.

FSD also served as a bridge between agricultural research and practical solutions for farmers, integrating external research outputs into development initiatives. The iterative nature of the methodology allowed for continuous learning and adaptation, a critical feature in dynamic agricultural systems influenced by climate change, market fluctuations and evolving socio-economic conditions.

Despite its strengths, FSD faced limitations. Its participatory approach, while a cornerstone of its success, was time-intensive and resource-demanding, often delaying the implementation of urgently needed interventions. The approach required skilled facilitators and interdisciplinary teams, which were not always available in resource-constrained settings. Additionally, FSD primarily focused on the farm level, often overlooking broader systemic issues such as policy constraints, market dynamics and institutional weaknesses that profoundly impact farming systems and farmer decision making.

Over time, FAO's approach to agricultural development evolved significantly, transitioning from the farming systems development (FSD) methodology to a stronger emphasis on farm management training and extension services for smallholder farmers with a focus on business. While the FSD approach concentrated on understanding and optimizing the complex interactions within farming systems, it became clear that achieving long-term sustainability required a more business-oriented emphasis.

15.3 Status of Farm Management Extension Globally

In the early 21st century, FAO held regional consultations with agricultural extension agencies to review their structure, functions and the role of farm management within their services. These consultations, accompanied by regional studies and workshops, were conducted in Africa, Asia and Latin America. Farm management extension during that period was an integral part of public sector service delivery, managed within government agricultural ministries and departments. These services aimed to provide advice to all small-scale farmers, addressing challenges with significant public value, such as food security, soil conservation and poverty alleviation. These were traditionally public sector funded models of national agricultural extension service delivery. These initiatives played a crucial role in spreading 'green revolution' type technologies. However, these national systems frequently faced budget constraints, which hindered their ability to innovate and respond effectively to the emerging challenges of market liberalization. Centralized management frequently led to inefficiencies, hindering the ability to adapt to emerging needs or integrate new agricultural technologies. Over the past two decades, due to budget constraints, public agricultural extension has increasingly turned to the private sector to fill the gaps in service delivery. This shift has led to a broader extension system where public and private actors, often supported by NGOs, work together to provide more diverse and inclusive agricultural services. This evolution is especially clear in farm management, where effective support requires coordination among multiple actors to improve market access, financial management and resource use.

In regions such as Africa and Asia, the public sector has, historically, been the cornerstone of extension services. These services were centralized, often limiting direct feedback from farmers, which is crucial for refining and improving support mechanisms. For example, the green revolution in Asia was significantly propelled by robust public extension-research linkages, illustrating the pivotal role of government-driven initiatives in agricultural advancement. However, this model was also criticized for being too bureaucratic, slow to adapt to changing agricultural needs and for creating environmental, social and economic problems – especially the marginalization of the smallholder farmers it was meant to support.

In Africa, farm management extension varies across regions, shaped by historical and linguistic differences. English-speaking countries have generally maintained stronger ties to traditional public extension systems than Francophone countries. However, with the growth of commercial agriculture, there is increasing recognition of the importance of farm management and a shift towards strengthening support in this area.

In parts of Eastern and Southern Africa, where agriculture became more commercialized earlier, farm management was already institutionalized by the 1970s. In West Africa, many countries began to focus on processing primary products (such as starch and oil), promoting export crops (like pineapple, groundnuts, cocoa and shea nuts), and developing large-scale marketing systems (for cotton and cocoa) or niche production (such as fruit juice and organic sesame). Policymakers came to understand that farmers would only remain engaged if they saw clear benefits and understood their options. As a result, farm management services became more important, and new providers – including farmers' organizations, marketing boards, outgrower schemes and NGOs – emerged to offer relevant advice and support.

In Asia, the green revolution and the World Bank's training and visit (T&V) system in the 1990s helped build strong links between extension and research, supported mainly by the public sector. Since then, extension services have undergone major changes, with countries like India and the Philippines moving towards decentralized systems involving public, NGO and private actors. This shift reflects a broader move towards more pluralistic advisory services and the need for efficient, cost-effective public sector support. (See Table 15.1.)

Over the past two decades, various international donors have played a critical role in strengthening farm management capacities in low- and middle-income countries. These efforts have supported a wide range of interventions – from capacity building for extension workers and farmers to the development of farm business schools, and improvements in agricultural marketing and data systems.

Table 15.2 presents a selection of donor-supported farm management projects implemented across Africa, Asia and the Caribbean. These projects reflect the growing recognition of farm management as a vital component of agricultural development, aiming to enhance productivity, improve decision making and support the integration of smallholder farmers into more competitive and sustainable agricultural systems. The projects were implemented by bilateral donors, international organizations such as FAO and multilateral financing institutions, often in collaboration with national governments.

In French speaking West Africa, in particular, attention was also given to farm management supported by the French Agricultural Research Centre for International Development (CIRAD). Many countries in the region focused on cash crops and value-added activities. Policymakers saw that farmers would keep producing only if they understood the benefits and had clear choices. Farm management service provision became perceived as new, innovative and of increasing importance. Of particular significance is the approach known as management advice for family farms (MAFF) implemented in Senegal, Burkina Faso, Cameroon, Chad, Mali, Benin and the Ivory Coast. The objective of MAFF was to improve farm profits and address smallholder household food security needs. The content of extension was a combination of technical and farm management disciplines (see Box 15.1.).

Table 15.1. Models of extension in Asia. From FAO, 2002, reproduced from unpublished material.

Country	Level of organization	Public sector performance	NGO	Private sector	Provision of farm management advice
India	Provincial/ state	Well organized Top heavy	High	High	Low
Pakistan	Provincial/ state	Well organized Top heavy High overheads	Low	Low	Nil
Bangladesh	National	Many organizational changes Narrow commodity focus	High	Moderate	Low
Sri Lanka	National	Lack of coherent policy Sectoral structure Weak coordination between government departments	Low	Low	Low
Thailand	National	Increases in number of extension staff at provisional and district level. Ratio of extension worker to farm family 1:1000	Low	High	Moderate
Vietnam	National	'Top down' orientation Rigid and bureaucratic Farming systems approach followed	Low	Low	Nil
Malaysia	National	Well organized Commodity-based extension	Low	High	High

In contrast, the Pacific region experienced since the early 2000s, a marked deterioration in the effectiveness of extension services with farm management heavily impacted by the overall decline in agency effectiveness and the shift out of agriculture towards the service sector (McGregor, 2002).

In Latin America, despite a strong tradition of centralized, supply-based extension systems modelled after the USA, recent decades have seen a shift towards programmes that combine production focus with broader support services. This included a special emphasis on farm management or agro-enterprise development emphasizing the importance of business management. This brought the introduction of business management principles into farm management, especially during periods of agrarian reform which created diverse agricultural enterprises requiring nuanced management strategies (Berdegue, 2005, unpublished material).

Overall, these regional comparisons and the mapping of extension services highlight a global trend: there is growing recognition of the need for continued farm management extension as smallholder agriculture evolves. This calls for strengthening and integrating farm management into all types of extension services, whether centralized or decentralized. The shift towards more pluralistic service delivery reflects a global understanding that extension must be more responsive and tailored to local needs to support modern farming and agribusiness practices effectively.

Despite these transformations, the integration of farm and agribusiness management into

Table 15.2. Some donor supported farm management projects.

Name of project	Donor	Country	Duration
Management advice for family farmers, West Africa (Mali, Burkina, Cote D'Ivoire, Cameroon, Benin)	CIRAD	West Africa	1998–2006
Capacity building farm business management for farmers and extension officers	FAO	Botswana	2002–2004
Farm demonstration and management	GIZ	Kenya	2002–2005
Agricultural support project (ASP)	SIDA	Zambia	2003–2008
Capacity building for extension workers and farmers through farm business management schools	FAO	Kenya	2004–2006
Development and promotion of location-specific integrated high-yielding rice-based technologies	JICA	Philippines	2004–2009
Strengthening farm management survey and data analysis in support of agricultural policy formulation and advisory services	FAO	Nigeria	2006–2008
Capacity Building in farm planning and management for extension workers and farmers	FAO	Malawi	2006–2008
Capacity Building in farm planning and management for extension workers in Zambia	FAO	Zambia	2006–2008
Capacity building in agricultural marketing and market management	FAO	Nepal	2008–2009
Agricultural statistics and farm management	FAO	Barbados	2008–2010
Advisory service capacity development in support of food security in Tanzania	FAO	Tanzania	2009–2012
Strengthening the agricultural extension system through agro-enterprise development	FAO	Sri Lanka	2010–2013
Management advice for family farmers	CIRAD	Myanmar	2011–2016
Farmer business school (FBS) programme	Government of Malawi	Malawi	2011–2015
Capacity building of small farmers in entrepreneurship development and market access	FAO	Philippines	2011–2013
Promotion of agricultural livelihoods and employment through investment in land reclamation and water reservoirs - farm business school component	FAO	Lebanon	2017–2022
Digital agricultural supply chain platform (DASCP) under SAAMBAT	IFAD, EIB	Cambodia	2020–2025
CAMAgriMarket App for farmer market access	MAFF and Korea agency	Cambodia	2020–Present
Livestock value chain development	FAO	Azerbaijan	2021–2024
Promoting rural development in Uganda (PRUDEV II)	GIZ	Uganda	2023–2026
Diversified resilient agriculture for improved food and nutrition security (RAINS) (FFS-FBS)	Global agriculture and food security programme (GAFSP)	Bangladesh	2023–2026

Box 15.1. The MAFF strategy. From Faure, 2001.

- Groups of 15–20 farmers meet at fortnightly intervals, and sessions include field visits and demonstrations.
- Sessions are facilitated by extension advisors with support from a MAFF coordination unit, which provides regular training, information, monitoring and evaluation.
- Enables the farmer and his family to analyse their own situation, to look ahead, to make choices, to monitor their activities and to evaluate the results.
- Takes into account technical, economic, financial and social aspects of farming.
- Involves a capacity-building process for men and/or women farmers covering agricultural production, organization of labour, management of cash flows, etc.
- A learning process that includes training, record-keeping, exchange of practices and utilization of farmers' expertise
- Provides tools for improved decision making on technical and business aspects of production, calculation of gross margins, cash flow management, etc.
- Assumes farmers' ability in calculation and writing.
- Aims at developing farmer-driven delivery services, with strong participation of farmer organizations.

public and private advisory systems remains uneven and underfunded. The wide range of skills needed to deliver effective farm business advice has not yet been fully incorporated into mainstream extension systems. There is an ongoing need for trained professionals – both public and private – who can connect agricultural practices with business skills, helping farmers not only boost productivity but also compete in markets and increase their profits.

15.4 The Performance and Impact of Farm Management Extension

As implied previously, farm management extension programmes are intended to contribute to agricultural development by (i) increasing farm income through guiding resource allocation and channelling farm operations towards farms that have secure markets; (ii) ensuring optimum allocation of scarce resources so that food supply needs are met and income can be enhanced; and (iii) utilizing scarce family labour more productively by better planning of farm operations and diversifying farm enterprises into higher value activities. There is a direct relationship between farm management extension and increased management skills, profitability and income. Farmers who participate in farm management extension programmes are expected to make changes to their farming practices to improve profitability. Extension and training programmes can drive change by sharing new knowledge and skills, connecting farmers with extension workers and encouraging interaction with fellow farmers.

The global farm management projects discussed in this section use evidence from indicators such as income growth, adoption of improved practices, and stronger market linkages to show the benefits of farm management advisory services. These examples consistently highlight the potential for transformative impact. For instance, programmes that combine farm business training, record-keeping, and market analysis have shown clear gains in farm profitability and productivity. A key example is the approach developed since the early 2000s in Senegal, Burkina Faso, Cameroon, Chad, Mali, Benin, and the Ivory Coast by CIRAD under the management advice for family farms (MAFF) programme. In Benin, for example, MAFF was successfully scaled up to reach nearly 20,000 farmers, with trained farmer-facilitators offering tailored local support. The programme led to significant changes in farming techniques, management practices and decision making. Farmers adopted new technologies, improved cash flow and storage management, and took better advantage of market opportunities. In Myanmar, MAFF was adapted to local conditions through collaboration between the French

NGO GRET and local partners. This included developing participatory tools and training farmer-facilitators. The programme reached over 580 families across 111 villages, leading to improvements on individual farms and collective gains through group learning and knowledge sharing. Its impact extended beyond participants, influencing broader community practices and building leadership within farmer organizations. These initiatives highlight the importance of tailoring extension services to the specific needs and contexts of different farming communities.

In Kenya, farm management extension projects have encouraged a shift towards higher-value, income-generating activities. Farmer learning groups established through farmer field schools have enabled economies of scale, enhanced market access and boosts to incomes. A 5-year GTZ project highlighted a transformation in extension worker attitudes, prioritizing financial and economic analysis alongside technical recommendations. This approach facilitated diversification, double cropping and value-added activities, moving many farmers away from subsistence maize production (Maingi, 2000, unpublished material; Kahan, 2007a).

Similarly, in Nigeria, several farm management projects implemented over the same period enhanced financial and business management skills among smallholder farmers, contributing significantly to increased agricultural productivity. Specifically, financial literacy training and business management education empowered farmers to manage their resources effectively, invest strategically in agricultural inputs and achieve higher profits (Falola *et al.*, 2023).

In Malawi, recent training programmes in financial literacy, cost management and marketing helped farmers understand production costs and develop enterprise budgets. These efforts improved productivity and resilience, even with limited access to financial services and technology (Wiafe and Yona, 2016). The projects also secured market access, improved resource use and increased incomes, with farm business analysis showing clear efficiency gains from simple management changes. In Uganda, an entrepreneurship training programme for soybean farmers led to significant income growth. Trained farmers earned about 380,000 Ugandan shillings more per year than those who were not trained (Van Campenhout *et al.*, 2017). Skills like record-keeping, budgeting and credit management helped farmers use resources more effectively, reduce waste and boost profits.

Pilot projects in the Philippines in the early 2000s showed the value of including farm management techniques in extension services. A JICA-funded model project introduced farm record-keeping using group study and on-farm demonstrations, leading to increased farm income in the provinces of Bataan, Antique, and Cavite. These activities gave farmers practical tools to improve profits, while ongoing support and structured guidance helped sustain better farm management practices (De Torres, 2001, unpublished material).

The use of technology in farm management extension has shown strong potential. In Mexico and Colombia, combining AgTech tools with business training helped transform smallholder farming. Farmers using farm management software and digital marketplaces saw productivity rise by 45% (Salman and Bernal, 2023), showing how technology can modernize traditional advisory services and boost competitiveness.

In Tanzania, agronomic and business training led by Mount Meru Millers and RLDC significantly improved sunflower production by combining technical knowledge with financial and marketing skills (Ngaleya and Kuzilwa, 2016).

Participatory approaches, like FAO's farmer field and business school in Bangladesh, have strengthened household food security by offering hands-on training in sustainable practices. In Malawi, an IFPRI study found that farm business schools (FBS) gave smallholders valuable skills in gross margin analysis, accounting, marketing and decision making. Farmers who joined the programme earned about US$20 more annually than those who did not (Chilemba and Ragasa, 2018). While the income gain was modest – mainly due to limited access to capital – it showed that better financial and business management skills can improve farm profitability even in resource-constrained settings.

In Kenya, the One Acre Fund has greatly improved smallholder farmers' productivity and incomes by offering a full support package that includes quality inputs, training and market access. Farmers in the programme have seen yield increases of 50–100%, showing the strong impact of tailored agricultural support (One Acre Fund, 2023).

Farmers who also receive formal training in farm business management are more likely to adopt better practices, leading to sustainable profits. These training programmes create change by sharing knowledge, connecting farmers with extension workers and peers, and encouraging collaborative learning. The results are clear: farmers gain practical skills that improve both short-term income and long-term resilience. For example, in Eastern Ethiopia, farmers who took part in business skills training at farmers' training centres earned significantly more – around Birr 9,557–10,388 (about US$350) annually – than those who did not (Wordofa and Sassi, 2018). These income gains were linked to better productivity and stronger market participation.

Despite these impacts, knowledge adoption alone does not automatically translate into productivity or income gains. There are often many other factors such as the specific biophysical conditions of the farm, supporting infrastructure, market and financial access and general government policies. But perhaps most directly, the capacity constraints of extension workers. Smallholder farmers, in remote areas often face difficulties accessing high-quality advisory services. Many other factors play a role, including farm conditions, infrastructure, access to markets and finance, and government policies. One of the biggest challenges is the limited capacity of extension workers, especially in remote areas where farmers often struggle to access quality advice.

Programmes like MAFF have shown that customized approaches, such as training illiterate farmers and building local leadership, can help overcome some of these challenges. Still, many farmers are hesitant to adopt new practices when information is limited, as the perceived risks remain too high.

Social and educational barriers also affect adoption. Low literacy can make tasks like record-keeping and financial planning difficult. However, younger, better-educated family members often support these tasks, and these barriers are expected to ease over time with generational change. To overcome these constraints, farm management support should be part of a broader, integrated approach that includes investment in infrastructure, education and value chain development. Single, stand-alone services are often less effective than a coordinated package of support – a key lesson from successful initiatives.

In summary, farm business management extension equips smallholders with essential financial and decision-making skills. These skills help farmers manage risks, use resources more efficiently and build resilient farm businesses that can respond to market changes and climate pressures. As agriculture continues to change, farm management extension must adopt innovative approaches and adapt to different farming contexts. Moving forward, collaboration among public, private and NGO partners will be essential to ensure all farmers – especially those in marginalized communities – can access the support they need for lasting impact.

15.5 Growing Momentum: The Expanded Role of the Private Sector, NGOs and Cooperatives

The move to privatize national extension services has been encouraged by multilateral and bilateral donors, mainly due to limited public funding. As a result, there has been a shift towards more private funding for agricultural extension. Common alternative funding methods include voucher systems, subcontracting services to non-public organizations, and commodity levies that involve farmers in sharing costs. Other approaches include setting up extension funds and using outgrower systems.

Agricultural advisory services have grown to include active roles for NGOs, farmer organizations and cooperatives to work alongside public and private actors. These models are discussed in more detail in a forthcoming section of this chapter. By supporting a more diverse advisory system, NGOs and farmer-led groups help make services more inclusive, fair, and better suited to the needs of smallholder farmers.

15.5.1 Private sector agricultural extension

Private sector extension services are not new in low-income countries. They have roots in the colonial period, especially in Africa and Asia, where they supported export crops like palm oil in Malaysia and cotton in Mali. These models combined both extension and research. In Mali, for example, smallholder farmers benefited from a self-funded cotton research and extension system, showing how the private sector has long been involved in supporting specific commodities.

Private sector extension services do more than just replace public ones. They come in different forms (see Box 15.2.). A common model is when private input suppliers give farmers technical advice when they buy products. These companies often have advisers who offer farm management support related to the inputs provided, similar to contract farming (as explained in Chapter 8). Some efforts also involve working with traders and buyers to share market information. In this way, private extension services are often linked to other business activities, helping improve product quality for buyers and building stronger partnerships between farmers and suppliers (see Box 15.3.).

Private sector extension services, while providing a valuable complement to public advisory systems, face significant structural and operational challenges that limit their effectiveness and inclusivity. Adapting these models to meet the broader and diverse needs of modern agriculture has proven difficult. One major obstacle is the high cost of delivering services to remote rural areas. Without subsidies or external funding, providing advisory support in these locations is often economically unviable. Furthermore, the initial investment required to establish private advisory services – including infrastructure, training and logistics – is often prohibitively high, deterring new entrants and limiting competition. As a result, private extension providers tend to prioritize accessible and profitable regions, leaving small family farmers and resource-poor smallholders underserved.

Research from Pakistan underscores these challenges. A study by Davidson *et al.* (2001) revealed that private advisory service delivery in Pakistan predominantly targeted wealthier farmers in profitable regions. This approach aggravates disparities among farmers. These findings highlight systemic inequities in the

Box 15.2. Privatization approaches.

- **Extension contract system:** in this model, an extension worker or company provides inputs and advice to farmers or farmer groups. After the harvest, the farmer pays back the input costs and gives the worker or company a percentage of the profits above a set target. If the harvest falls short due to poor advice or delayed inputs, the payment is reduced accordingly.
- **Contract farming:** in this arrangement, an agribusiness firm supplies the farmer with inputs, technology and production support. In return, the farmer agrees to sell the produce only to the firm at a premium price, meeting set quality standards. This setup gives the farmer support throughout the production and marketing process, while the firm secures high-quality produce directly, avoiding middlemen.
- **Agribusiness service consultancy:** the Agribusiness Service Consultancy model offers specialized, market-focused advice to agribusinesses and farmers along the value chain. Provided by private firms, NGOs or public agencies, these services cover areas like business planning, financial management, marketing and value chain integration. The goal is to help clients boost productivity, profits and competitiveness. These consultancies often work on a fee-for-service basis or with donor or government support, delivering tailored, high-quality guidance that promotes innovation and sustainable agricultural growth.
- **Public extension via private delivery:** government agencies grade and certify agricultural consultancy firms, which then bid competitively to offer extension services at various levels (state, district, village). The service costs are shared between the government and the farmers. If farmers are dissatisfied with the service, they have the option to replace the consultancy firm with another.
- **Service for vouchers:** instead of receiving direct public extension services, farmers are given vouchers based on factors like land size, type of enterprise and their specific information needs. These vouchers can be used to access services from any approved agricultural consultancy – public or private. Once the voucher period ends, farmers must pay the full cost of services themselves. The vouchers are then passed on to the next group of priority farmers, gradually building farmer independence and reducing reliance on public support.

Box 15.3. Case study: India's agriclinics and agribusiness centres (ACABC) scheme. From MANAGE, 2008.

The AgriClinics and AgriBusiness centres (ACABC) scheme, promoted by the Indian government since 2002, is an innovative model aimed at delivering agricultural extension services to smallholder farmers through technically trained agricultural graduates, known as 'agripreneurs'. These agripreneurs operate at the village level, providing critical advisory services through AgriClinics, which are linked to agribusiness centres that supply inputs and farm equipment. The services offered by AgriClinics encompass a broad range of agricultural needs, including advice on new technologies, laboratory tests for soil, water and input quality, and plant protection services. In some regions, such as Tamil Nadu, AgriClinics are owned by cooperative banks and managed by agricultural graduates, while in other cases, graduates run their clinics independently. Between 2002–2008, the scheme facilitated employment for 4,152 graduates, who served 2.3 million farmers across 126,000 villages, with a significant proportion of the ventures being self-financed.

prioritization of service delivery, which disproportionately benefits already advantaged groups.

Market failures exacerbate these challenges, particularly for smallholder farmers operating in subsistence-based agricultural systems. Many of these farmers cannot afford the fees associated with private advisory services, and a lack of awareness about their potential benefits diminishes demand. This issue is particularly acute in specialized areas like farm management, where the advantages of advice – such as financial planning or production optimization – may take years to materialize. Consequently, private service providers struggle to sustain operations in low-demand regions, perpetuating a cycle of low adoption and inadequate service provision.

Private sector service providers also face internal challenges. Many lack the technical skills and resources needed to offer full advisory support, especially in areas like farm management and agribusiness. Contract farming, which connects farmers to markets through agreements with private firms, often struggles with issues like side-selling, where farmers sell to others instead of their contract partners. This weakens trust and makes these models less sustainable. Other problems, such as high costs, poor infrastructure and remote locations, also limit the reach of private services. Many providers invest little in staff training, resulting in a shortage of skilled advisers and a 'brain drain' as experienced staff leave for better jobs.

Uganda's National Agricultural Advisory Services (NAADS) is an example of a private sector model supported with public funds. It aimed to help farmer groups manage funds and hire private extension providers. While promising in theory, NAADS faced problems like uneven service quality, weak oversight and financial barriers for small-scale farmers. Many farmers could not or would not pay for services, leading to low participation and mixed results. Operational inefficiencies and limited accountability also held the programme back. These experiences show that integrating private sector models into extension systems is complex and must be supported by strong public oversight and clear rules to ensure equity and effectiveness.

Building a sustainable and inclusive agricultural advisory system requires a balanced approach that brings together private sector innovation, public sector oversight, and support from NGOs and farmer organizations. Strong partnerships, reliable funding and ongoing capacity building are essential to ensure that all family farmers – regardless of their size or income – can access quality advisory services. This integrated model can help close existing gaps, boost productivity and support fair and inclusive growth across the agricultural sector (see Boxes 15.4. and 15.5.).

15.5.2 NGOs, farmer organizations and cooperatives

NGOs, farmer organizations and cooperatives play a vital role in delivering business management advisory services to farmers, especially in areas where public and private services are limited or unavailable. These groups help smallholder and marginalized farmers by promoting

Box 15.4. Example of public extension via private delivery. From IFAD, 2022.

In Bangladesh, this approach has been employed under projects such as, promoting agricultural commercialization and enterprises (PACE), where private service providers worked alongside public institutions to deliver extension services. Farmers were able to choose between multiple providers, with costs partially subsidized by the government. The government's role was to ensure that service quality met minimum standards and that providers remained accountable to farmers' needs. This type of contract-based extension has been effective in some cases, particularly when integrating ICT tools like mobile apps for advisory services, which were also tested in the project.

Box 15.5. Key learning from private sector provision of farm management services.

- Private sector models may prioritize wealthier or more commercial farmers, leaving smallholders underserved.
- Extension activities may also focus on product sales rather than balanced advisory services, which can lead to biased recommendations.
- Smallholder farmers with limited purchasing power might not benefit from these services, unless supported by partnerships or subsidies to bridge affordability gaps.
- Private sector providers often require skills development.
- Problems of access to rural areas – delivery of services is expensive.
- High start-up costs of service provision.
- Problems of market failure.
- Low demand for services by farmers.
- Low awareness among farmers that services are on offer.
- Inability of small farmers to pay for services.
- Lack of intermediaries to organize farmers, etc.
- Problems of contract farming – side selling.
- Lack of capacity of private service suppliers to provide advice.
- Market distorting practices of public/ donor agencies.
- Frequent requirement of public sector investment to create a demand for private services.
- Frequent requirement by the private sector for support from the public sector:
    - Creates an enabling environment for private sector investment.
    - Provides capacity building support.
    - Addresses the public sector component (e.g. organizing farmers into groups/cooperatives).
- In principle, training programmes should be cost shared.
- Knowledge services tend to be embedded in private sector goods.
- Farmer groups and cooperatives are well placed to charge for advisory services.

sustainable practices, building capacity and improving economic outcomes.

NGOs often focus on reaching farmers who are underserved by traditional extension systems. They use participatory methods to strengthen skills in business management, entrepreneurship, market access and financial literacy. NGOs also support public and private advisory efforts by organizing smallholder farmers into producer groups or cooperatives, helping them connect to markets.

The positive impact of NGO-led projects is clear in many settings. For example, in Ethiopia, the NGO Fair Planet helped smallholder farmers triple their food production by providing quality vegetable seeds and training (Fair Planet, 2023). In Tanzania's Njombe District, farm management training improved productivity and reduced poverty. However, the study noted that advisory support alone was not enough – issues like land access, markets and support services also needed attention for lasting results.

Farmer organizations and cooperatives also serve as private advisory providers and platforms for collective action. By sharing resources and knowledge, and helping farmers access inputs and markets, they strengthen members' business skills and improve their position in the value chain. Cooperatives play a dual role: they offer technical and business advice while also acting as aggregators to increase farmers' bargaining power.

These organizations can complement the work of NGOs by promoting collective decision making and building social capital, as discussed in Chapter 8. They also offer services like training, credit access and marketing support, helping smallholders overcome market barriers. In Zambia, cooperatives under the agricultural support programme delivered strong benefits. In East Africa, the East African Farmers' Federation (EAFF) supports member cooperatives through training in financial literacy, business planning and market access. EAFF connects farmers to buyers, processors and exporters, helping them secure better prices and improve profitability.

Despite their contributions, the success of NGOs and cooperatives has been uneven. NGOs often follow a welfare focused model, which can conflict with market-based approaches. Many struggle to sustain their services when donor funding ends or to become commercially viable without disrupting existing private providers. They also face capacity gaps, especially in delivering market-oriented advice. Similarly, cooperatives face challenges with governance, financing and business skills, and often require external support to grow and improve services.

To maximize the impact of farm business management advisory services, a coordinated approach is needed. This includes ongoing capacity building, supportive policies and a balance between commercial and welfare-based strategies. Multi-stakeholder platforms involving NGOs, farmer organizations, cooperatives and government agencies can improve coordination, ensure inclusivity and strengthen the overall system. This is discussed further in Chapter 16.

15.6 Lessons Learned

Providing effective advisory services to farmers is not without its challenges. In this section, some of the key encounters faced by farm management advisory and extension support and the importance of addressing them for successful implementation, are discussed.

15.6.1 Reaffirming the role of public sector extension services

Funding shortages have increasingly led governments to rely on private-sector financing of extension. As previously pointed out, this shift has not always resulted in a stronger engagement of the private and NGO sectors in providing extension services in an effective and sustainable manner. The question arises among the donor community, policymakers and practitioners: what should be the role of government in this situation? Should government continue to provide services directly, or should it solely be there to facilitate their provision? The answers lie in the importance of government involvement, particularly in addressing public goods, targeting poor and vulnerable smallholder family farmers, and addressing challenges related to climate change. However, the issue of where government extension services best fit, both geographically and thematically, remains. To address the issue effectively, governments may play the role of providing oversight, regulation and coordinating the broadened range of service providers whilst offering advisory services of a public goods characteristic to smallholder family farmers. This would involve partnering with the private sector and NGOs to leverage their resources, expertise and outreach capabilities (see Box 15.6.).

15.6.2 Lack of planning and supervision in public-sector extension programmes

One of the biggest challenges in delivering advisory services, including farm management, is the lack of proper planning and supervision. Many extension programmes are launched without a clear roadmap, leading to fragmented and uncoordinated efforts that often fail to meet farmers' needs. Without thorough planning, programme goals and methods remain unclear, resulting in wasted time, resources and effort. Extension workers – both frontline and specialists – may struggle to prioritize tasks, allocate resources effectively or monitor progress. This can lead to inconsistencies in how farm management advice is delivered across regions or among different service providers.

Box 15.6. The role of government in farm management extension service delivery.

- Organizing platforms for public and private actors to exchange information on bottlenecks and opportunities.
- Facilitate private sector investment in rural infrastructure and communications.
- Provide incentives for sustainable use of production resources.
- Invest in establishing private sector service provision especially through providing capital and credit support.
- Provide subsidies or co-financing for the development of high-risk interventions.
- Establish and enforce a commercial code that ensures fair and expeditious judicial processes for resolving contract disputes.
- Ensure the availability of production and price information to facilitate market activity and to monitor market progress.
- Assemble a body of knowledge on effective extension service provision through supporting research and ensuring that records of performance are kept and maintained.
- Build capacity of extension workers and farmers through training and extension programmes.
- Provide public goods for private sector advisory service delivery.
- Provide 'best-practice' and ethics frameworks for conducting extension whether public, private or NGO based.

A lack of supervision further affects accountability and the quality of service. Without regular monitoring, it becomes difficult to evaluate the performance and impact of extension efforts. This weak oversight limits opportunities for learning and improvement, both for extension staff and the farmers they serve. To overcome these issues, farm management extension programmes must incorporate strong planning and supervision mechanisms. This includes setting clear goals, identifying target beneficiaries and outlining specific activities and strategies. Well-defined guidelines for programme design, implementation and evaluation are essential to improving the relevance and effectiveness of advisory services.

15.6.3 The need for efficient use of manpower resources

One of the biggest challenges facing agricultural advisory systems is the shortage of front-line extension workers, which limits the ability to scale up farm management advice. Given these staffing constraints, it is essential to carefully plan, assign roles strategically and adopt innovative methods that extend the reach and effectiveness of existing personnel.

Technology offers practical tools to support this effort. Mobile applications, online training platforms and digital communication channels allow extension workers to deliver training, share information and engage with farmers more efficiently. These tools help overcome barriers of distance and time, enabling services to reach a larger number of farmers and provide more timely support. These innovations and their implications will be discussed further in Chapter 16.

Beyond technology, collaboration is also critical. By partnering with farmer organizations, community-based groups, NGOs and private companies, extension services can pool resources, share expertise and use wider networks to broaden their reach. This pluralistic and collaborative approach improves the use of limited manpower while contributing to a more integrated and effective strategy for supporting smallholder farmers.

While farm management advisory services are commonly included in national extension systems, they continue to face staffing challenges. In public sector systems especially, it is important to make better use of existing personnel. Extension services encompass multiple roles, including front-line extension workers, subject matter specialists and farm management experts. To be effective, these roles must be clearly defined and well-coordinated. Front-line workers are essential for identifying farmer needs and linking them to appropriate support, while farm management specialists, often based at district or regional levels, provide vital business guidance to support front-line staff.

As agriculture becomes increasingly market-oriented, the roles of extension staff must evolve. Field-level advisers are shifting from being mainly technical experts to acting as facilitators of learning and brokers of information (Suleiman, 2012). This transition requires retraining staff in participatory teaching methods, problem-solving and the organization of producer groups. Extension workers must also be prepared to connect farmers with markets and value chains, where coordination among stakeholders and contract agreements are common.

In this new context, farm management specialists and front-line staff must act as facilitators and brokers – helping local actors build trust, collaborate and manage shared resources sustainably. Extension workers may also play a role in resolving disputes that emerge from more complex market relationships. However, such skills are still rare in many advisory services. Strengthening these capacities will be essential, and Chapter 17 explores how to address these needs, in greater detail.

15.6.4 Upgrading staff skills and competencies to support farmers

An FAO assessment of farm management training programmes in 2007 revealed key weaknesses in extension staff skills and in the training programmes meant to address them. Training was often fragmented and lacked a long-term vision for building capacity. Many programmes were too theoretical, offering limited practical application, and left participants unprepared for real-world challenges. A lack of qualified trainers, limited follow-up and few refresher courses reduced the long-term impact of training efforts. Additionally, materials were often generic and poorly suited to the specific needs of diverse farming communities, including those with low literacy or numeracy. Budget constraints, dependence on donor funding and weak institutional capacity further limited the ability to deliver effective, coordinated and relevant training to farmers and extension workers (Kahan, 2007).

Improving the skills of agricultural training staff is essential for successful, market-oriented farm management advisory services. While the need for technically skilled trainers is widely recognized, challenges remain in ensuring both the quality of trainers and the funding required for proper training. The lack of skilled trainers and limited resources weaken the extension system's ability to function as an effective training institution. To overcome this, significant investment is needed to strengthen training frameworks and make them more practical, demand-driven and business-oriented – treating trainees as clients and ensuring training meets their evolving needs (see Box 15.7.).

Extension materials

In many low-income countries, quality farm management materials are limited within extension systems. There is a general lack of training content tailored to the specific needs of small family farmers and extension workers. Where materials do exist, their quality is often inadequate, particularly in terms of didactic value and relevance to local agro-ecological and socio-economic conditions. This gap includes a shortage of basic resources such as pamphlets, brochures and simple guides on farm management and marketing, typically developed by specialist advisers.

Materials must be customized for different categories of learners and adapted to specific farming systems and agro-ecological zones. There is also a pressing need for more innovative resources that address literacy and numeracy challenges through participatory approaches and varied forms of communication.

Such materials could include:

- Awareness-raising leaflets on farm management;
- Farm recording and accountancy forms;
- Self-learning and teaching guides on basic farm management techniques for farmers;

Box 15.7. Good practices in developing training programmes.

1. Training programmes should be based on what farmers need and want, and should built on the capacities they already have. While it is important to listen to their requests, farmers may not always be aware what training they need or that could open new opportunities for them. Showing them the financial and other benefits of the training can help create interest. When farmers see value in the training and have reliable income, they may be willing to share the cost. As their skills grow, more advanced business management topics can be introduced.
2. The content of training programmes must be tailor-made or customized. They should be relevant to situation and requirements of both the target recipients and their level of literacy and ability to assimilate the information provided as well as the different agro-ecological conditions of farmers.
3. Training design should explore innovative ways of building local capacities. Counselling, learning by doing and success case replication should be considered ways of promoting developing entrepreneurship skills.
4. Training of extension personnel and farmers should be practical, continuous, regular, participatory and closely monitored. Priority should be given to short courses provided in manageable doses with practical timing. The programme should include assessment and follow-up on application in the real world on what has been learned.
5. Training should be cost-effective and aimed at maximizing outreach. This can be achieved by utilizing local training capacity, organizing farmers into extension groups to reduce costs and identifying, training and deploying potential trainers from among extension staff and farmers.

- Market information sheets tailored to specific commodities, farmer categories and regions;
- Advanced training modules for farm management specialists;
- Adaptations of existing published materials or regional/global training programmes;
- Booklets and brochures focusing on strategic (long-term) farm management;
- Audio-visual materials such as instructional videos, animations or podcasts; and
- Interactive digital tools, including mobile apps and e-learning modules.

Extension workers require a wide range of materials, from simple information tools to more advanced training packages, all of which should be tailored to local conditions and learner profiles. In addition to content development, many extension offices lack the basic hardware needed to support training delivery and materials production. This includes visual aids, data projectors and equipment for producing training content – such as video cameras, recorders and video editing tools – as well as appropriate computer software. Building the capacity to produce high-quality extension materials requires not only access to such equipment but also investment in human resource development and technical support. Adequate budgetary provisions must be made to develop this capacity and to promote awareness of the business opportunities that improved extension services can offer both to extension workers and to farmers.

Backstopping services

In addition, greater attention needs to be given to the technical back-up services that are essential to supporting extension service delivery, including training trainers, testing new technologies and developing training materials. An important area of back-up service is the support given by 'core trainers' initially involved in implementing training of trainer programmes. Besides providing initial training, they have a role to ensure that follow-up mentoring and coaching advice is available. However, these services often rely heavily on public sector support, highlighting the importance of sustainable funding mechanisms to ensure scalability and continuity. For the market-oriented services to become and remain attractive and useful to all the clients it is essential that the direct service providers have continuous access to backup services that help to build, maintain and update their capacity. This issue will be elaborated on in Chapter 16.

Ultimately, strengthening agricultural extension services requires a comprehensive approach. This includes not only improving training and developing better materials but also providing strong institutional support to help navigate policy challenges and take advantage of market opportunities.

In addition, extension workers at all levels need training in the use of emerging digital agricultural technologies, especially those related to farm management. They should also be equipped with the necessary tools, such as smartphones and other digital devices, to apply these technologies effectively in the field.

15.6.5 Technical and management skills development: the need for synergies

While farm management skills – particularly in business and marketing – are essential, they are not sufficient on their own to fully transform smallholder family farming. To sustainably enhance productivity, profitability and incomes, smallholders need an integrated approach that combines technical agronomic expertise with financial literacy and business management capabilities.

Historically, training and extension efforts have focused primarily on technical production and the transfer of technologies. While this remains important, farm management is a vital complement. However, neither technical training nor business management training alone can adequately address the wide range of challenges smallholder farmers face.

Recognizing these realities, there is increasing acknowledgement that integrated support – bringing together agronomic and business expertise – holds greater potential for advancing smallholder agriculture. Collaboration between technical specialists and farm management advisors ensures that farmers receive well-rounded support, covering both production and economic concerns.

Extension systems that deliver integrated training packages allow farmers to adopt new technologies more effectively while also managing the financial and market risks that come with innovation. Building synergies between technical know-how and business skills is therefore essential for achieving sustainable agricultural development, strengthening livelihoods and empowering smallholder farmers economically.

15.6.6 Strengthening collaboration between researchers, extension workers and farmers

Weak links between research institutions, extension services and farmers are a major barrier to agricultural innovation. When researchers work in isolation, focusing mainly on theory, their findings often do not reach farmers in a practical way. At the same time, extension workers may lack the skills, support or funding to turn research into useful advice for farmers. As a result, research can become outdated or irrelevant to farmers' real needs, and adoption of new practices remains low.

This gap is largely due to a lack of strong communication and feedback between researchers, extension staff and farmers. Without regular exchange, research efforts risk missing the actual challenges faced on the ground. Farm management extension specialists can play a key role in closing this gap. They help analyse on-farm trials, make sure results are practical and economically sound and support the design of extension programmes suited to different farming situations.

To improve collaboration, structured feedback loops are needed. Regular dialogue between farmers, extension staff and researchers helps ensure that research is based on real-world issues and that extension messages stay relevant. Farm management specialists can lead these efforts by translating field experiences into valuable insights for research and extension planning.

In many sub-Saharan African countries, platforms for collaboration – known as innovation platforms – have been created to bring together researchers, extension workers, farmers and private sector actors like input dealers, traders and financial service providers. These platforms encourage information sharing, joint problem-solving and action towards shared goals, helping to promote effective agricultural innovation (Nederlof and Wongtschowski, 2011). Examples are provided in Box 15.8.

Experience shows that traditional top-down approaches often fail because they do not involve farmers enough. Participatory research, where farmers manage on-farm trials and demonstrations, leads to higher adoption and greater impact, and contributes significantly to building farmer innovation capacity. Farmer-to-farmer learning networks also help spread new technologies, especially when backed by accurate, research-based information and farm management support.

To be effective, recommendations must be adapted to local conditions and farmer preferences, offering flexible options instead of one-size-fits-all solutions. Information and communication technologies (ICT) tools can further improve communication, allowing quick and relevant feedback between researchers, extension workers and farmers.

Overall, joint research and extension efforts – built on strong funding, participation and feedback – can make innovations more practical and widely adopted. By combining technical and farm management expertise in a collaborative and inclusive way, these approaches deliver more effective and farmer-focused results.

15.6.7 Ensuring independence from political agendas and vested interests

Political interference in extension services, prioritizing government agendas over farmer needs, is common. There are also several instances where vested interests have influenced the design and implementation of public sector agricultural extension services. These influences often result in the extension services serving the interests of political groups and agribusinesses rather than the farmers' actual needs.

Box 15.8. Research-based innovation platforms.

Case 1 – maize and legumes, Nigeria
The innovation platform brought together farmers, researchers, capacity-building organizations, national extension services, the private sector and local government. Together, participants set up training programmes and joint experiments and supported farmers' organizations. The platform triggered the development of an apex farmers' organization to allow direct negotiations between farmers and private companies. Achievements of the innovation platform included: improved maize–legume production systems; facilitation of mutual learning processes among platform members; participatory experimentation with farmers; inter-organizational coordination to support change processes; development of an apex farmers' organization and new networks; and training of lead farmers in pilot villages to disseminate new practices (Nederlof *et al.*, 2011).

Case 2 – a research-supported platform
A government-initiated, inter-ministerial innovation platform in Mozambique collaborated on with civil society and private sector actors to develop a policy for sustainable biofuels. Between 2008 and 2012, researchers from Wageningen University supported the platform by making an inventory of biofuel activity ties in the country (traditional research), facilitating and documenting platform meetings (knowledge management) and fundraising and lobbying to create an enabling environment for collaboration. Supporting different stakeholder groups throughout the policy process was challenging, as stakeholders had conflicting needs and interests (Schut *et al.*, 2011).

Case 3 – a research-initiated platform
The Nile Basin development challenge used innovation platforms in a 2 year (2011–2013) project to improve rainwater management in Ethiopia. Researchers played multiple roles: establishing the platforms, facilitating and managing them, obtaining funding for pilot interventions, building the capacity of members and processing documentation. Because innovation platforms were new to their members, building trust took time. The power dynamics and representation of stakeholders was also a challenge; it was necessary to get the right balance to prevent powerful individuals from hijacking the research agenda (CGIAR, 2013).

To prevent this, public extension services must remain independent and focused on helping farmers make informed, independent decisions about their farm businesses. This can be achieved through greater transparency, more diverse and pluralistic extension systems, continuous training for extension workers and strong regulatory oversight. Most importantly, strengthening farmers' ability to speak up and influence services is the best way to protect the independence and effectiveness of extension.

15.6.8 Monitoring, evaluating and learning (ME&L)

Farm management programmes often lack well-developed systems for monitoring, evaluation and learning (ME&L). While most public advisory services have basic monitoring systems focused on tracking activities, inputs and outputs – such as the number of farm visits or demonstration plots – they rarely assess the performance or impact of farm management interventions. Measuring performance is easier than assessing impact, as it avoids the complex task of linking results directly to advisory services. Impact assessment is more challenging because it depends on farmer behaviour, which is influenced by many external factors. In systems with multiple advisory providers, it becomes even harder to track results, as no single provider may take full responsibility or receive full credit.

To improve effectiveness, ME&L systems must support continuous learning and adaptation by capturing lessons from experience in delivering farm and agribusiness services. A strong ME&L system should be inclusive, involving all stakeholders and covering both public and private service providers. It should help improve operations, ensure accountability and provide useful feedback for policymakers and planners, particularly on funding decisions. Special attention should be given to tracking training outcomes and changes in farmer skills and competencies. Importantly, ME&L must focus not just on collecting data but also on learning from it. Real improvement comes when results are used to guide changes and strengthen future advisory efforts.

15.6.9 Funding shortages in support of farm management extension

The growing demand for funds to promote and scale up farm business management advisory support in the rural areas has placed a strain on public sector budgets. Donor funding has often been sought to fill the gap, although public sector support is not widely regarded as an attractive priority. Due to limited funding, finding ways to reduce public-sector spending is increasingly necessary. As we noted previously, to ensure that efficient use is made of scarce financing for farm management extension, collaborative programmes between the public and private sectors need to be more readily encouraged. The underlying premise is that both the public and private sectors have their own unique characteristics and strengths which when combined offer superior advisory support resulting in a more efficient and effective means of delivery. Collaboration between the government and the private sector can take place, as we have seen, through various mechanisms such as service contracts, management contracts, leasing arrangements and joint ventures.

15.7 Shifting Perspectives: The Agricultural Advisory Service System

In Chapter 1, we examined the major trends that have reshaped agricultural extension services in recent decades and acknowledged the increasing complexity of the agricultural advisory system. This evolution underscores a shift towards market orientation, demand-responsiveness, and the growing diversity of service providers, including public and private sectors, NGOs and farmer organizations. Market orientation emphasizes the need for farmers, particularly small family farms, to align their production with evolving market demands, focusing on profitability, competitiveness and sustainability. Demand-responsive advisory systems prioritize tailored support that directly addresses farmers' specific needs, fostering innovation and improving market access.

This shift acknowledges that modern agriculture connects production, markets and the environment. Value chain development has become central, requiring advisory services to go beyond production challenges and address broader market demands and quality standards. As seen throughout this book, effective advisory services integrate technical, economic and business guidance, while facilitating market linkages. Mixed systems that combine public and private efforts have proven particularly effective. Public sector extension prioritizes inclusivity, sustainability and equity, ensuring marginalized farmers are supported, while the private sector drives innovation, efficiency and market alignment. Together, these elements create an adaptive system that meets farmers' diverse and complex needs while promoting long-term sustainability.

The evolution of agricultural extension has also broadened the training curriculum for extension workers. Beyond primary production, training now encompasses information management, financial literacy, marketing, natural resource management, farm management and agribusiness. Advisory services must address not only production-related issues but also economic and market challenges. Farm management, with a strong market focus, equips farmers with the knowledge and skills to make informed decisions. Extension workers, just as they may specialize in technical areas like poultry production, can develop expertise in farm business management and agribusiness, enabling them to support farmers in navigating value chains and optimizing resources. This reinforces the rationale for this book.

The agricultural advisory system has shifted from viewing the farm solely as a production unit to recognizing it as part of a broader market-driven and sustainable value chain. Moreover, addressing the unique challenges faced by women and youth in agriculture is essential to ensuring equity and inclusion. These developments reflect the continuing evolution of advisory services, driven by increasingly complex value chains and the need for market-oriented, multi-actor approaches to support small family farm innovation and sustainability.

The agricultural advisory system, shown in Fig. 15.1, outlines the range of services needed to support small family farmers in improving their livelihoods and highlights the different organizations involved in providing these services. These services help smallholders manage production, market and environmental risks in a changing agricultural environment. Advisory support now goes beyond just transferring knowledge and technology – it includes areas such as research, marketing, business management, post-harvest handling, value addition, food safety, mechanization and finance. While public research and extension services once led this work, today's advisory systems take a broader approach that includes market access and a wider range of actors. In both public and private systems, advice is often delivered along with agricultural inputs.

The figure also illustrates efforts to improve service access for both men and women farmers and their organizations, from both the user (demand) and provider (supply) sides. It shows the importance of involving a mix of service providers – public agencies, private companies, producer organizations and civil society –working together in a coordinated system. The system now also includes social and institutional innovations that help connect stakeholders, improve market access, raise competitiveness and build strong networks among producers, processors, traders and service providers. This approach supports collaboration among public and private sectors, civil society and farmer groups, as discussed in Chapter 1. It also stresses the need to build skills in marketing, business and post-production to increase productivity, improve food security and incomes, create jobs for the youth and promote gender equality.

15.7.1 Agricultural advisory services for market integration

One of the most progressive approaches emerging from the evolution of extension is value chain extension, which integrates the principles of value chain analysis and agribusiness development. This model not only provides technical and business support to all actors in the agricultural value chain but also empowers farmers to claim a more equitable share in markets and value-adding processes.

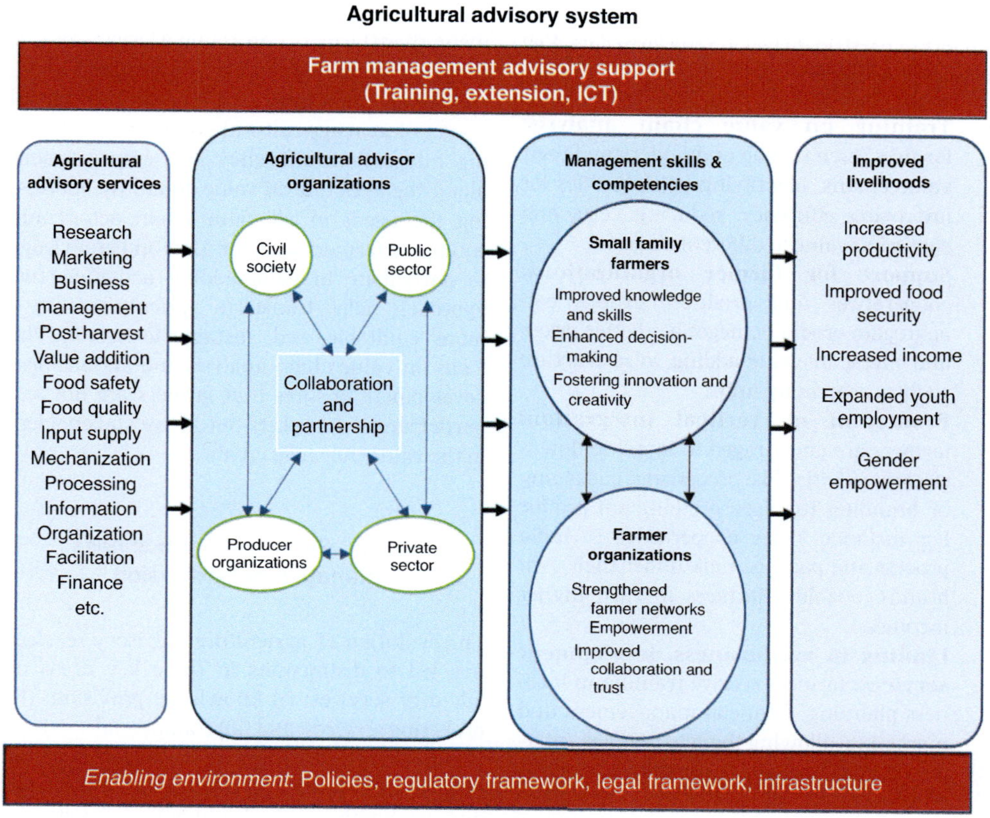

Fig. 15.1. The agricultural advisory system.

Extension across agricultural value chain actors

Traditional extension models focus mainly on farmers as producers, offering advice on crop cultivation and livestock management. Whilst the value chain landscape is much wider, this does not mean that public sector agricultural advisory or extension services need to dramatically shift their focus away from the farm. Private sector led agribusiness support, in most cases, more effectively cater to the needs and advisory demands of off-farm agro-enterprises on a cost recovery basis. In general, advisory services to support this wide range of actors aim to:

- enhance coordination between value chain participants, improving efficiency and reducing post-harvest losses;
- provide technical training on product standards, certifications and quality control to meet market demands; and
- facilitate technology transfer to processors and exporters, enabling them to handle higher volumes or process commodities into high-value products.

Some of these aims can be met through the public sector. For example, in Kenya's horticulture sector, value chain extension services target both farmers and exporters, focusing on compliance with global quality standards like Global GAP. Extension agents work with exporters to train farmers on good agricultural practices, ensuring product quality while creating a reliable supply chain.

Empowering farmers for market integration

While value chain extension helps farmers gain market access, a transformative approach ensures that farmers also secure a stake in value-adding processes. This involves equipping farmers with the skills, tools and institutional support to participate beyond the farm gate. Key strategies include:

- **Training on value chain analysis:** farmers learn to map and understand their value chains, identifying opportunities for improving efficiency, reducing costs and capturing value at different stages.
- **Support for farmer organizations:** cooperatives and producer groups can aggregate products, negotiate better prices and invest in value-adding infrastructure such as processing units.
- **Promotion of vertical integration:** farmers are encouraged to invest in downstream activities like processing, packaging or branding to capture additional profits. For instance, dairy cooperatives in India process and package milk under their own brands, enabling farmers to earn higher incomes.
- **Linking to agribusiness development services:** farmers receive training in business planning, financial management and marketing, allowing them to manage enterprises and navigate competitive markets effectively.

Enabling farmers' participation in value addition

Value chain extension goes beyond providing market information to actively involving farmers in value addition. Some examples include:

- Contract farming models with profit-sharing arrangements, where farmers receive a share of the profits from processed products.
- Promotion of niche markets, such as organic or fair-trade certifications, enabling farmers to earn premium prices.
- Support for agripreneurship, where farmers develop enterprises that focus on processing or specialized production, such as producing spices or high-value crops for export markets.

In Colombia, for instance, coffee farmers organized into cooperatives have moved beyond production to control processing, branding and even retailing. With extension services focused on marketing strategies and quality enhancement, these farmers now retain a larger share of the profits from their coffee.

Value chain extension represents a crucial progression in agricultural extension, emphasizing integration, efficiency and empowerment along the agricultural value chain. By addressing the needs of all value chain actors and equipping farmers with the tools and knowledge to participate in value-adding activities, this approach helps transform agriculture into a more equitable and sustainable sector. The focus on value chain analysis and agribusiness development ensures that farmers are not just participants in markets but active stakeholders in the value they help create.

15.7.2 Farm and agribusiness management extension

The evolution of agricultural advisory services has led to distinctions in three key areas of advisory services: (i) knowledge provision; (ii) brokering services; and (iii) commercial services. These distinctions recognize the diverse needs of farmers and value chain actors in a dynamic and competitive agricultural environment supported by both public extension services and private agribusiness advisers.

Knowledge provision focuses on equipping farmers with the technical know-how to improve the quality, quantity and timing of production. This includes advice on agronomic practices, pest and disease management, soil fertility and irrigation techniques, ensuring that farmers can optimize productivity while maintaining sustainability. Beyond production, there is also a growing emphasis on know-how related to economics, business management and markets. Farmers need guidance on cost management, enterprise budgeting and market analysis to remain competitive. Furthermore, knowledge provision extends to enabling value chain actors to meet market or value chain quality requirements, such as adhering to food safety standards or achieving certification for organic or fair-trade products. As discussed, ICTs play a crucial role in this domain by delivering real-time technical, marketing and business information to farmers, enabling informed decision making.

Brokering services emphasizes the facilitation of relationships and linkages among the various actors in the agricultural value chain. As agriculture becomes more market-driven, the role of advisory services has expanded to include facilitating and accompanying changes in value chain management. This involves helping farmers form or strengthen producer groups, cooperatives and other value chain actor associations. These groups can then negotiate better terms with buyers, access bulk purchasing discounts for inputs and engage in collective marketing. Advisory services also play a critical role in building the capacity of these groups to navigate the complexities of modern value chains. Moreover, brokers often act as intermediaries, creating linkages between farmers, processors, distributors and retailers, ensuring smoother coordination and mutual benefits across the chain.

Commercial services focus on directly supporting the economic activities of farmers and agribusinesses. This includes providing tailored technical and business advisory services that align with specific market demands. For example, advisors may help farmers adopt production practices that meet export market requirements or assist processors in ensuring consistent supply quality from their farmer networks. Commercial services also involve facilitating access to financial services, such as loans or insurance, which are critical for investment in new technologies or scaling up operations. Furthermore, advisory services often incorporate market linkages, connecting farmers with buyers, processors and even export markets, as elaborated on in Chapter 8.

This broader understanding of agricultural advisory services reflects the diverse and interlinked challenges facing modern agriculture. By distinguishing between knowledge, brokering and commercial services, advisory systems can better address the holistic needs of farmers and other value chain actors. Integrating these three components ensures that advisory services not only enhance productivity and profitability but also contribute to building resilient, sustainable and inclusive agricultural systems.

The broad range of services is given in Box 15.9.

Another development in the evolution of extension is the trend towards the provision of combined commercial and advisory services. There now appear to be many more cases of service providers – public, private and NGO – delivering advisory services to farmers in combination with tangible goods and finance. Advisory services by themselves are often inadequate to promote commercial farming among smallholder farmers. Commercialization of farming requires the availability of good genetic material, an assured market, reliable market infrastructure and a diversity of support services – among them the commercial sale of produce and finance. For example, the ATMA model in India requires a market assessment, organizing farmers into producer groups, developing high-quality planting material, post-harvest handling, processing the raw material into a value-added product and selling it in the market. This suggests a need to provide commercial services combined with extension support, institution building and marketing. The sales of inputs and final produce often lie at the heart of private sector service provision.

Other examples similarly suggest that independent stand-alone agricultural advisory service providers are likely to be less effective than those providing a package of services (e.g. inputs and markets), particularly through 'interlocking-market' linked service models. The private sector-led case studies described in this and previous chapters, highlight situations where contract farming and agribusiness linkages are established. In most cases it is the private sector (commercial bodies) that initiates the linkage and provides input supplies, technology and finance to outgrowers. In some cases, formal contracts are drawn up. However, the existence of contracts does not automatically guarantee the development of successful linkages. As noted in Chapter 8, informal trust relationships are often more important. When farmers are well informed and have other choices, and service suppliers or traders are competing to purchase outputs, the interlocking markets are more likely to benefit them.

15.7.3 The demand for farm business management advisory services

Farm and agribusiness stakeholders along agricultural value chains include a diverse range of decision makers and service providers. To

Box 15.9. Farm and agribusiness management advisory services. Adapted from Chipeta *et al.*, 2008.

- Technical expertise aimed at generating value by increasing the volume and quality of production and the timing of the supply of raw materials.
- Economics, marketing and business management expertise (e.g. farm enterprise analysis, marketing information and business planning, etc.).
- Post-production expertise aimed at creating value along the value chain through improved post-harvest handling, packaging, storage and distribution, while meeting food safety and quality requirements.
- Support in strengthening producer and other value chain stakeholder groups through improved collective marketing, business management, financial management, leadership, negotiation skills and linkages with research institutions for innovation.
- Support in facilitating value chain development and strengthening through improved coordination of production, negotiation of contracts, brand development, linking producers to buyers as well as providing advice on legal, regulatory and certification issues.
- Facilitating institutional changes – forming producer organizations, clusters, networks and linkages among different actors along value chains (e.g. convening multi-stakeholder forums to understand market opportunities and constraints along value chains, develop contractual and trust relations). ICT has a particular role to play as an instrument information among smallholder farmers and value chain actors more broadly and at scale.

deliver effective farm management training and support, extension and advisory services must engage these stakeholders within the broader agricultural framework. This requires rethinking traditional extension approaches and fostering collaboration between public and private sectors. Public extension services continue to focus on farmers, while private sector advisors increasingly support non-farm agribusinesses along the value chain.

Clients of advisory services span the entire value chain, including input providers, farmers, producer organizations, processors, traders, exporters and retailers. Each actor requires tailored business advice, best provided through collaborative efforts. For instance, input suppliers need insights into market trends, processors require support in quality control and logistics, and financial service providers benefit from advisory services to assess market prospects. Private-sector consultants often play a key role in these areas. Competitive value chains depend on expertise, trust and sustainable relationships among all stakeholders.

While smallholder family farmers remain the most critical recipients of advisory support, their needs vary greatly based on factors like farm size, resource quality, access to markets and technical skills. A farmer with poor market access faces challenges distinct from one with better linkages and technology. Effective advisory services must be context-specific, considering regional, cultural and socio-economic differences to ensure relevance and impact. Table 15.3 shows the potential range of farm management stakeholders.

The following are the major characteristics of farm level stakeholders, elaborated on more fully in Chapter 2, that can be supported with farm and agribusiness management advisory support.

Subsistence farmers

As described in Chapter 2, subsistence farmers typically focus on food production strategies that ensure sufficient staple crops and household self-sufficiency. The demand for farm management advice among these farmers is generally limited due to several challenges in commercializing their operations. These challenges include low levels of education, a tendency to be risk-averse when making substantial changes to their farming systems, limited knowledge, and a lack of the technical and management skills needed to utilize relevant information effectively. Additionally, in some countries, factors such as the small size and poor quality of land and water resources further hinder their ability to engage in competitive farming. Given these constraints, subsistence-oriented farmers are often less inclined to adapt their farming systems and pursue commercialization. As a result, they require targeted support in modern farm management to improve productivity and enhance their integration into market systems.

Small market-oriented family farmers

As discussed previously, an increasing number of small family farmers are shifting towards market-oriented production, driven by improved market opportunities and expanded support services. As they build their asset base and adopt new, environmentally suitable production processes, many are transitioning towards profit-oriented farming. With 20–30% of their farm produce already being sold in local and regional markets, this group has significant potential for broad-scale impact in agricultural transformation. As these farmers diversify production, integrate into higher-value markets and explore value-added processing, the demand for farm management advice becomes increasingly critical. Effective decision making in areas such as financial planning, resource allocation, risk management and market engagement is essential to ensure sustainable growth. Those with more abundant land and labour resources have greater potential to specialize in high-value crops, livestock or agribusiness ventures, further increasing their need for specialized advisory services in market linkages, investment planning and business management. Strengthening

Table 15.3. Stakeholder categories in farm management.

Category	Stakeholders	
Farmers	Urban vegetable farmers Fruit growers Poultry farmers Cattle raisers Short cycle animal owners	Export crop growers Small-scale and large-scale stable crop producers Specialized crop producers Large-scale farmers
Special groups, often marginalized and very vulnerable	Female-headed farm households Young farmers Farmers holding undeveloped land	Households affected by HIV/AIDS Immigrants People with disabilities
Farmers' organizations	Small farmers' associations platforms	Fisher and livestock associations Chambers of agriculture
NGOs	Field-level extension agents Subject matter specialists	Community associations and societies Local financing structures
Private sector	Commercial dealers (inputs, value-adding, markets) Industrial producers Small-scale producers	Financial institutions supporting agricultural products Large-scale producers
Agro-industrial stakeholders	Government marketing and export boards	Multinational corporations
Public sector	Policymakers Decision makers in service-providing institutions Decision makers in governmental and parastatal agencies Rural schools	Service providing agents (extension agents) Development projects and programmes National and international agricultural research institutes Teachers in colleges and universities Media
Donors	Governments as donors NGOs as donors	Private sector donors Development partners as donors

access to farm management training, extension support and financial literacy programmes, will be key to maximizing their productivity and long-term success in commercial farming.

Medium-scale farmers

Medium-scale farmers typically cultivate between 2–10 ha (World Bank, 2009; FAO, 2020). This category of farmers represents a crucial segment with a high demand for farm management extension services. Their greater resource endowment, often including access to water, markets and credit, enables them to engage more actively in commercial farming. Compared to smallholder farmers, they tend to be less risk-averse, more willing to adopt new technologies, and more likely to invest in high-value crops or livestock enterprises. Given their relatively higher levels of education, often with primary and secondary schooling, they are more aware of modern agricultural practices, market trends and financial management strategies. This category of farmers is well-positioned to benefit from both public and private farm management advisory services. Their engagement with input suppliers, extension workers and buyers makes them more accessible for extension outreach, whether through government programmes, private-sector advisory services or producer associations. As progressive and market-oriented farmers, they often play leadership roles within their communities and are active members of producer groups and cooperatives, increasing their access to inputs, services and markets. Importantly, many medium-scale farmers have the financial capacity to pay for professional farm management services, particularly in areas such as business planning, investment analysis, risk management and financial literacy. Strengthening both public and private sector extension to offer tailored advisory services will be essential to enhancing their productivity, profitability and long-term sustainability in commercial agriculture.

Commercial farmers

Commercial farmers in low-income countries, typically those with landholdings exceeding 10 ha, operate more capital-intensive farming systems and are less reliant on public extension services (FAO, 2020). Given their higher levels of education, business acumen and progressive approach, they are more likely to seek specialized agricultural advice from private-sector advisory service providers. Their demand for farm management extension support is driven by the need for strategic business planning, investment analysis, risk management, market intelligence and precision agriculture solutions. Unlike small- and medium-scale farmers, commercial farmers are less dependent on government extension services, except when accessing cutting-edge research, new seed varieties or emerging technologies. Instead, they are more influenced by fellow entrepreneurial farmers and industry networks. Many rely on private-sector agronomists, financial advisors and supply chain specialists to optimize their operations. They often have direct links to agricultural research institutions, universities and transnational agribusiness firms, enabling them to stay ahead of technological advancements and market trends.

With greater financial capacity, commercial farmers are willing and able to pay for specialized advisory services that provide high-value insights into market trends, export requirements, digital farming technologies and sustainable intensification practices. As agribusiness becomes increasingly integrated with multinational supply chains, commercial farmers are receiving production inputs, technical expertise and management support directly from private-sector firms, reinforcing the shift towards fee-based farm management consultancy services. Strengthening private-sector advisory networks will be essential in meeting the evolving demands of commercial farmers and ensuring agriculture remains competitive, efficient and resilient in the global market.

Rural women

As previously noted, rural women play a critical role in agricultural development, yet their demand for extension services is often overlooked or underserved. Traditionally, women have been more engaged in food crop cultivation, while men have focused on higher-value crops and livestock. However, changing market dynamics are creating new opportunities for women to participate in commercial farming and agribusiness, increasing their need for specialized farm management advice. As more women engage in high-value crop production and agribusiness ventures, their demand for market-oriented extension services is growing. Given the increasing entrepreneurial role of women in agriculture, farm management extension services must also address gender-specific challenges, such as access to credit, land ownership issues and labour constraints. Incorporating financial literacy, business planning and cooperative management training into extension programmes, will empower women to scale up their farming enterprises. A more inclusive, women-friendly extension system will not only improve agricultural productivity and household incomes but also contribute to more sustainable and equitable agricultural development.

The youth

The youth (especially the rural youth) faces challenges of high migration rates to urban areas, necessitating attention from national governments. The younger generation is seeking new employment opportunities and ways of making a decent and reasonable living. To mitigate this trend, farm management can play a significant role in preparing youths for modern agriculture. Emphasizing that farming can be a profitable business venture is essential to attract young people to the sector. Extension programmes can cater to youth-specific needs by offering entrepreneurship training, mentorship programmes and access to financing. Youths need to develop business acumen and the skills to treat farming as a business. This is particularly important as they inherit the land and farming operations of their elderly parents.

Farmer organizations

Farmer organizations play a crucial role in both demanding and supplying farm management advisory services. As small-scale farmers integrate into modern markets and develop links to agro-enterprises, these organizations have the potential to provide business management and market access support. By pooling resources, organizing joint input purchases and consolidating output, they strengthen bargaining power and profitability, increasing demand for specialized farm management advisory services in areas such as finance, production planning and risk management.

As they develop managerial capacity, farmer organizations can evolve into apex institutions that both seek and deliver advisory services. Larger cooperatives often employ agronomists, financial advisors and marketing specialists, enabling them to serve as intermediaries linking farmers to government extension services, private consultants and financial institutions. These organizations also advocate for policy changes, trade regulations and infrastructure investment, further supporting agricultural commercialization.

Well-governed and financially stable farmer organizations can contract private farm and agribusiness consultants to provide specialized advisory support on business planning, investment analysis and value chain integration. Larger organizations may employ in-house agricultural advisors, some of whom would require training in farm business management to deliver tailored support.

Despite their potential, many farmer organizations, however, often face capacity constraints such as weak leadership, low business skills and limited market intelligence. Strengthening these organizations through training, mentorship and institutional development will be key to enhancing their ability to demand and deliver farm management advisory services, fostering a decentralized, market-driven and sustainable extension system.

Rural entrepreneurs

Rural entrepreneurs, as we noted previously, include a diverse range of value chain stakeholders with varying capital endowments and entrepreneurial potential. They include small-scale traders, input and equipment dealers, agro-processors and retailers such as supermarket chains. In order to ensure value chain competitiveness, there is often a need and demand for information and skills development programmes among this varied range of stakeholders. It is difficult to know exactly what form of agribusiness support each entrepreneurial category will demand but in general this includes a better understanding of the concept of value chains, market price information, basic business, finance and marketing skills and contracts – negotiation and compliance. All these stakeholders should also have an understanding of linkages among actors and the importance of setting up partnerships as a strategy to boost competitiveness.

Rural entrepreneurs require business management support to better understand the changing agri-food system and the integrated system of relations and interactions with different actors with specific value chains. Traders need to have better knowledge of quality and safety issues as well as management information systems. Input dealers need to understand more clearly the importance of quality and reliability of inputs and materials as well as technical recommendations for their application and use. Agro-processors often require support in business management – simple procedures to plan, monitor and control the business. Other issues such as production scheduling, finance, inventory management, quality management and staff management are also useful to know. Agro-processors also need a better understanding of contractual relations and the development of mutual trust. As mentioned previously, support to off-farm rural entrepreneurs is more effectively supported by private sector business development service providers where fee-for-service payments are more affordable and likely to be transacted.

15.7.4 A farm management decision framework

In Chapter 4, the authors looked at the livelihoods framework for small family farms. In Chapter 5, the framework was linked with the farm management cycle from the perspective of the farmer. In this chapter (Fig. 15.2), these have been adapted into a farm management decision framework from the perspective of a farm management advisor.

The figure provides a holistic view of the elements and interactions necessary for effective decision making. Policymakers, farm management advisors and extension workers may apply it in detail and may use it to support farmers in decision making. It is preferable that planning is carried out with the farmer and the farm family, but it may sometimes be necessary to develop a plan independently which can be discussed with the farmer.

The planning process starts with the assets and the influencing context, policies and institutions – asking relevant questions to gather information needed to diagnose the family farm business. Some of these questions are as follows.

- What assets does the farmer have, or have access to?
- What condition are these assets in?
- How are the assets affected by climate change?
- How is the farm affected by globalization?
- What economic conditions are relevant?
- What new technologies are available?
- What policies influence access to, and use of, these assets? What is their impact?
- What institutions influence access to, or use of, these assets? What is their impact?

Next, the advisor can start to diagnose the farm by exploring some critical farm management challenges. Some relevant areas to diagnose are:

- Previous farm performance.
- Weather variations.
- Changes in the market.
- Level of competition.
- Linkages to the value chain.
- Changes in family labour.
- Input supply issues.
- Demands on financial capital.

It is important to look also at the value chain to determine who are the actors, what additional advisory and support services are available and what influence do they have on strategic options. How connected is the farmer to the value chain?

Then planning can begin. Start with their goals. What is the farm family trying to achieve? Once this is clearly understood and diagnosed, the advisor can consider what livelihood strategies the farm family could adopt. To do this, the advisor can follow the same systematic decision-making process suggested for farmers, which is discussed in Section 5.5.2 in Chapter 5. In the process of doing this, they consider the key decisions regarding:

- what to produce;
- how to produce it;

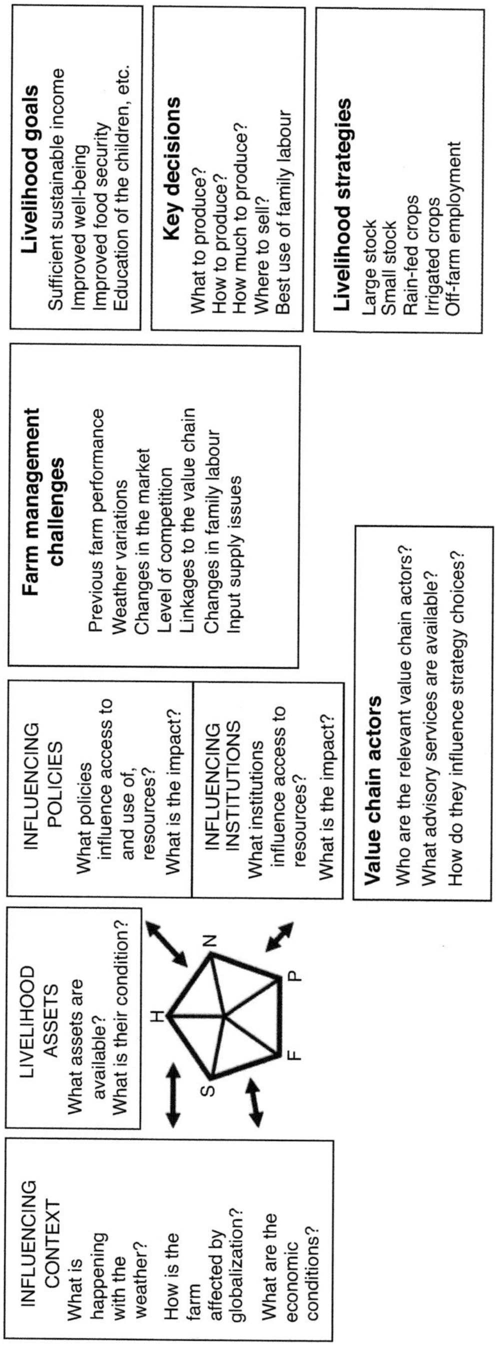

Fig. 15.2. The farm management decision framework.

- how much to produce; and
- where to sell it.

The best use of family labour must also be considered.

Answers to these questions will guide the family in choosing their livelihood strategies. In general, as shown in Fig. 15.2, these strategies will fall into one or more of the following:

- Large stock.
- Small stock.
- Rain-fed crops.
- Irrigated crops.
- Off-farm income.

Box 15.10. provides a practical example of a farm management decision framework.

15.7.5 Empowering farmers through informed decision making

Central to the evolution discussed in this book, is the understanding that sustainable agricultural development hinges on the ability of farmers to make informed, autonomous decisions about their farming enterprises. In support of this, FAO began promoting approaches that prioritized farmer empowerment, emphasizing participatory learning and decision-making processes. This led to the development of the Farmer Field School (FFS) approach, which focused on equipping farmers with the skills to analyse and adapt their practices based on local conditions and challenges. Building on this philosophy, the Farm Business School (FBS) was introduced to address the need for farmers to view their farms as businesses, teaching them how to plan, invest and market their products effectively.

The FFS and FBS models share a core philosophy of empowering farmers through experiential learning and peer-to-peer knowledge exchange. While FFS emphasizes improving technical knowledge on crop and livestock production through observation and experimentation, FBS complements this by focusing on economic and managerial aspects of farming. Farmers learn practical business skills such as budgeting, record-keeping, financial analysis and market assessment. This holistic approach recognizes that smallholder success requires not only improved production techniques but also the ability to navigate market dynamics and maximize returns.

This evolution in FAO's approach reflects a broader shift towards integrating technical, economic and social dimensions of agricultural development. By combining the insights of FFS with the business-oriented focus of FBS and their participatory learning methods, FAO has developed over decades a comprehensive framework for supporting smallholder farmers. This approach aligns with the global push towards inclusive and market-oriented agricultural systems, emphasizing the need for farmers to be proactive participants in their own development.

15.8 Challenges and Opportunities

While significant progress has been made in the evolution of extension and advisory services – particularly with the growing importance of farm business extension – many challenges and opportunities still remain.

15.8.1 Need for a conducive enabling environment to promote private sector 'business services'

A lack of supportive policies and regulations for private sector-led business services remains a major challenge to agricultural development. Without clear incentives and guidelines, businesses are less likely to invest in providing advisory services to farmers, limiting access to important support for rural communities. This also raises concerns about the quality and reliability of private advisory services. To address this, it is essential to create a policy environment that encourages private sector participation, ensures service quality and supports innovation. Measures can include offering incentives, simplifying regulations and setting clear standards to promote sustainable agricultural growth and improve rural livelihoods.

15.8.2 Leveraging private sector funds

The growing shortage of public funding for extension and advisory services has made it

Box 15.10. A practical example of a farm management decision framework.

A practical example of a framework is a decision-making grid designed to support helping a farmer choose between alternative production livelihood strategies. In this case, the advisor has identified as four options: (i) crop rotation; (ii) monoculture; (iii) organic production; and (iv) intensive. There are five criteria: (i) profitability; (ii) risk; (iii) sustainability; (iv) alignment with livelihood goals; and (v) alignment with personal values. Each criterion can be ranked between 1–10, with 1 being the lowest and 10 being the highest.

Option	Profitability	Risk	Sustainability	Livelihoods alignment	Personal values	Total
Crop rotation	8	3	9	8	8	36
Monoculture	9	6	4	6	8	33
Organic production	7	4	10	9	9	39
Intensive production	10	7	3	9	4	33

In this simple exercise:

- **Crop rotation** scores high in sustainability and aligns well with personal values, but has moderate profitability and low risk.
- **Monoculture** offers high profitability but carries higher risks and lower sustainability.
- **Organic farming** has the highest overall score and has the highest sustainability and alignment with the farmer's personal value, has moderate risk, but has the lowest profitability.
- **Intensive farming** has the highest profitability but scores low in sustainability and personal values, with high risk.

Which option should the farmer choose? The apparent choice is organic production because it has the highest overall score and appears to balance the criteria. However, the farmer may look at this more holistically and decide to use a combination of organic production and crop rotation. The decision grid set these two options as mutually exclusive, but the process of using a structured approach allowed the farmer to see the value of both and the possibility of integrating them. This will need the support of an advisor.

necessary to seek alternative sources, including private sector support. Involving the private sector can expand the reach and long-term sustainability of advisory services, helping to meet the varied needs of farmers. Private actors bring valuable skills in areas such as market access, supply chains and financial literacy – key components of farm business management training. Since their success depends on farmer productivity and profitability, agribusinesses have a strong interest in supporting effective advisory services.

The move towards a pluralistic extension system opens the door to innovative financing models that blend public and private investment. Private sector contributions can support digital advisory tools, input financing and specialized training. Other revenue-generating options – such as commercial use of surplus state-owned land, lab services, certification, consultancy and selling agricultural publications – can also help fund extension services more sustainably.

However, these opportunities come with challenges. Private funding is often driven by profit, which may shift focus towards large-scale or commercial farmers, leaving smallholders behind. Services may become too market-oriented, prioritizing short-term gains over long-term sustainability. There are also concerns about fairness, especially whether poor farmers can afford or access privately funded services. To ensure equity and effectiveness, it is essential to maintain transparency, accountability and strategic reinvestment of funds. While private funding can support public efforts, extension services must stay farmer-centred and aligned with broader agricultural development goals.

15.8.3 Effective coordination

Effective coordination among advisory service providers and commercial agricultural enterprises is essential for delivering integrated support to smallholder farmers. This means combining physical inputs with valuable advice and training. However, coordination is often weak, especially at the local level – within villages or districts – where multiple providers operate in isolation. This can lead to overlapping efforts, wasted resources and mixed messages reaching farmers.

As mentioned earlier, better coordination between agricultural research and extension services is also vital. To close the gap, systems are needed to turn research findings into practical, farmer-friendly solutions and deliver them through extension networks. Stronger links between research institutions and extension services help ensure that farmers receive timely, relevant and evidence-based recommendations to solve real challenges.

Higher education institutions, such as departments of agricultural economics or agribusiness management, can also support farm management advisory services. They do this by building the capacity of future extension professionals and aligning academic curricula with the needs of both farmers and advisory services – public and private. These departments also contribute through farm surveys and profitability analysis, which offer valuable benchmarks for farm business decision making. Strengthening partnerships between these institutions and extension services means updating academic content to match current field realities, offering practical experiences, and promoting collaborative research and outreach. This ensures the advisory system is prepared to meet new challenges and serve farmers effectively.

Lastly, vertical coordination between local, district, provincial and national governments remains a challenge. Misaligned policies, unclear roles and poor communication between levels often lead to inefficiencies and delays. Local governments, despite being closest to farmers, may find it hard to align their actions with higher-level strategies. Top-down planning often overlooks local needs, reducing the effectiveness of services. To address this, better communication and coordination across all administrative levels is needed, along with mechanisms for shared planning, mutual accountability and integrating local priorities into broader strategies. Platforms for dialogue and knowledge-sharing can help bridge both horizontal and vertical gaps, making advisory services more cohesive, responsive and farmer-focused.

15.8.4 Capacity to pay for advisory services

Many farmers particularly in low-income settings have considerable difficulty paying up front for any form of service provision. The ability of farmers to pay for farm management services depends largely on the profitability of the farm enterprise and the nature of the service. Knowledge services such as farm management and market information are intangible, and the added value is invisible. It is, consequently, unlikely to expect that training, technical assistance or information services for farmers will ever achieve total cost recovery through the payment of user fees. The issue is complicated, however, as the ability to pay also depends on the different types of farmers (subsistence, emerging commercial and fully commercial) demanding the service. Although there is recognition by some farmers, particularly those more commercially minded, of the importance of the payment principle, there is no consensus on what an appropriate cost of training and extension should be and what proportion can be expected to be covered by farmers. Farm management training and extension are likely to continue to remain a subsidized public good among small farmers until direct benefits can be realized and efficient delivery ensured. However, there are mechanisms by which these extension services can be subsumed.

A common strategy for funding extension services is to include their cost in income-generating activities through interlinked services. For example, farm business management advice can be offered through commercial relationships between large and small-scale farmers, often on a contract basis. In contract farming, larger agribusinesses form agreements with smaller farmers to create a mutually beneficial, long-term partnership. These arrangements not only provide farmers with economic security but also serve as a channel for introducing new technologies, ensuring timely input supply and embedding advisory costs into the price of produce sold.

In addition, businesses involved in buying farm products or supplying inputs may cover the cost of advisory services through small price mark-ups. The idea behind these strategies is to continue offering services that may not be immediately profitable in order to build or maintain a loyal client base for other more profitable services (see Box 15.11.).

Another way to recover costs and ensure long-term impact is to organize small and vulnerable farmers into groups, associations, clusters or networks. This approach creates economies of scale and lowers the cost of providing training and advisory services. It is especially useful for reaching underserved groups, such as women, ethnic minorities and farmers in remote areas with weak market access. Field experience shows that tailored extension services for farmer groups can be both effective and viable.

In some West African countries, contributions from farmer members and their organizations cover about 5–10% of extension system costs – and, in rare cases, up to half (Faure and Kleene, 2002). The share farmers can contribute depends on the type of farm and the kind of advice provided. While farmers' financial contributions are generally small and based on what they can afford, they should still be encouraged as a step towards full cost recovery.

In areas with weak markets, a gradual approach to cost recovery can be used, with subsidies reduced over time. This allows farmers to adjust to paying for services that were once free. In some cases, especially for farm management training, demand for the service must first be created before farmers are willing to pay. Just because farmers use a free service does not mean they are ready to pay for it. Experience shows that cost recovery methods must be tailored to each situation, and extension services need to be assessed within their specific local context.

15.8.5 Enhancing farm management extension services for women farmers

The potential of women farmers in 'farming as a business' and entrepreneurship is often overlooked, and the support they receive remains inadequate. Farm management extension services must be tailored to meet the specific needs of women – both as individual entrepreneurs and as members of groups. As previously discussed, women juggle household and farming responsibilities, often leaving little time for business activities. To help bridge this gap, farm management units within extension services should be supported to establish women's bureaux that focus specifically on gender-responsive services. Countries like Malawi, Zambia and Kenya have already implemented such structures, highlighting the important link between gender-focused support and farm business development.

To better serve women farmers, both the content and delivery methods of extension services must be adapted. Many women prefer learning in group settings – such as self-help or producer groups – which support knowledge sharing, improve market access and allow for resource pooling. Extension programmes should adopt gender-responsive strategies, such as flexible meeting times, women-led training sessions and greater recruitment of female extension officers. In addition, technology-based tools like mobile apps, community radio and digital platforms can provide farm management advice

Box 15.11. Embedded services.

- Generating income from a viable service to cover the costs of extension services. That is, cross subsidizing using income from more profitable services.
- Facilitating business links between the farm business and rural finance. In this case, the costs of farm management services are embedded in the conditions of loan repayment (piggy backing off microfinance).
- Supplying farm business management advice to larger, more commercial farming ventures that can afford to pay the 'real rates', and using the profits to supply the service to smaller farmers. In other words, cross subsidizing from larger to smaller clients.
- Obtaining income for extension services through the development of commercial operations that cover real costs (interlocking linkages).

in ways that fit women's schedules and mobility limitations. These approaches can make extension services more inclusive and responsive to women's realities (see Box 15.12.).

Finally, integrating gender-focused farm management into agricultural policies can help address structural inequalities. Ensuring women have equal access to land, credit, advisory services and markets, not only empowers them economically but also strengthens food systems and rural livelihoods. Supporting women's active participation in agribusiness and value chains can drive broader agricultural transformation and sustainable rural development.

15.8.6 Engaging the youth in agriculture

Young people are vital to the future of small family farms, yet their engagement in agriculture is increasingly uncertain. While youth living in rural areas contribute significantly to food production, household income and rural livelihoods, structural barriers prevent them from fully participating in the sector. Many perceive agriculture as a low-status, labour-intensive occupation with limited financial returns, leading to rural-to-urban migration and declining interest in farming. This trend threatens the long-term sustainability of smallholder agriculture and food security, particularly as the average age of farmers continues to rise.

In sub-Saharan Africa, where over 60% of the population is under 25, youth disengagement poses a serious challenge. Land fragmentation, limited financing and poor market access make agriculture an unattractive career choice, driving many young people towards cities in search of better opportunities. Asia faces similar issues, with rapid urbanization and industrial expansion pulling youth away from farming. Many prefer higher wages and modern working conditions in urban centres, leaving rural areas with an aging workforce and few successors to sustain agricultural production. In Latin America, rural youth migration is fuelled by land tenure insecurity, weak agricultural extension services and the growing impacts of climate change. However, initiatives promoting agribusiness training, digital farming tools and sustainable agricultural practices present opportunities to reintegrate young people into farming.

To reverse this trend, farm management advisory services must integrate youth-specific training, digital tools and agribusiness mentorship, through programmes targeted towards youth employment in agriculture. Strengthening youth access to extension services, fostering entrepreneurship and incorporating digital advisory platforms can make agriculture a more attractive and viable career choice. For a more detailed discussion on youth engagement in agriculture, refer to Section 17.4, Chapter 17.

15.8.7 More effective use of ICT

The role of ICT in farm management, as outlined in Chapter 10 is essential. Farm management specialists and extension workers need to use ICT tools to give farmers important information on production, marketing and agribusiness. This requires investment in computers, mobile

Box 15.12. Designing training programmes for women.

- Adapt programmes to women's needs and skills.
- Allow sufficient time to enable women to acquire new skills and adjust schedules to fit women's existing workloads.
- Provide training in agricultural and other productive activities, not just home and family welfare topics.
- Emphasize activities for which there is an actual income generation potential.
- Ensure the involvement and full participation of women from poorer and less educated backgrounds.
- Use trainers who are not only technically competent and up to date, but who empathize with the needs and aspirations of rural women.
- Provide practical field experience in the use of innovations.
- Shift more resources to village-based training rather than residential training.

devices and software for both service providers and farmers. Tools like the mobile phone and radio can also improve access to technical and business information.

The importance of ICT in farm management, as discussed in Chapter 10, cannot be overstated. There is a pressing need for farm management specialists and front-line extension workers to leverage ICT to provide farmers with crucial information on production, marketing and agribusiness. This entails investment in computer hardware, electronic devices and software programmes, for both advisory service providers and farmers. The integration of mobile phones and radio can further enhance access to technical and business information for farmers. An integrated strategy encompassing a full spectrum of ICT tools should be employed in a coordinated manner to further the delivery of technical and business information to farmers. However, expanding ICT coverage and infrastructure often requires collaboration between the public and private sectors, with a focus on capacity building for extension workers, farmers and rural entrepreneurs. Facilitators with ICT expertise and credibility must be engaged to support local communities in adopting and utilizing ICT effectively. Ultimately, the successful diffusion of ICT relies on strong political commitment and financial support at the policy level.

15.8.8 Promote public-private collaboration and mixed systems

Different goals and priorities among stakeholders – such as government agencies, NGOs, private companies and farmer organizations – make it difficult to deliver coordinated and effective farm management advisory services. Each group works under its own mandate and funding priorities, which often leads to duplication, gaps in support and missed opportunities for collaboration. Without a unified approach for planning, sharing resources and delivering services, efforts remain fragmented. This can result in farmers receiving mixed messages and extension programmes failing to fully meet their business and market needs.

Despite the challenges, collaboration among different stakeholders offers a valuable opportunity to improve farm advisory services by using their combined strengths. Different actors bring unique capabilities: government agencies can provide policy and regulatory support, the private sector can bring in market- based solutions and funding, while NGOs and farmer organizations help with knowledge-sharing and capacity building. Public-private partnerships (PPPs) are especially effective in bringing together resources, skills and networks to expand and strengthen advisory services. These partnerships can increase funding, improve outreach, and build more sustainable service models that go beyond government limitations.

Successful examples include India's Small Farmers Agribusiness Consortium (SFAC), which supports agribusiness ventures and agri-clinics, allowing agricultural graduates to offer expert advice. This model blends public support with private sector involvement in agribusiness. Another example is IFAD's '4Ps' (Public-private-producer partnerships), introduced in countries like India, Bangladesh and the Philippines. These partnerships combine government policy support with private sector inputs such as technology, capital and market access, helping farmers access business services, financial tools and farm management training. These cases demonstrate how well-structured PPPs can strengthen extension systems and improve farm advisory services. Another example of a public-private partnership from South Africa is given in Box 15.13.

To make the most of these opportunities, better coordination and alignment among stakeholders is essential. Creating joint platforms for dialogue, knowledge sharing and planning can help develop more integrated, farmer-focused advisory services. Setting shared goals and clear accountability can also ensure that public–private partnerships stay inclusive, transparent and aligned with national agricultural goals. While different stakeholder interests can create challenges, well-structured collaborations offer a strong way to close service gaps, improve advisory support and help farmers move towards more profitable, market-oriented farming.

15.9 Implications for Effective Farm Management Advisory Service Delivery

Delivering farm and agribusiness advisory services in a pluralistic, market-oriented system requires clear strategies to overcome challenges and ensure coordination among diverse stakeholders.

Box 15.13. Public-private partnership providing extension to sugarcane farmers. From Heerden and Naiker, 2024.

In South Africa, the collaboration between public and private sectors in agricultural extension is exemplified by the work of the South African sugarcane research institute (SASRI), which provides extension services aimed at improving sugarcane farming practices. Through this public–private partnership, SASRI conducts participatory research with farmers, allowing for farmer-driven trials to test and refine agricultural practices. One example is a collaboration between SASRI specialists and sugarcane farmers in the Midlands south region, where a chemical ripening process was tested on sugarcane varieties. This partnership resulted in better cane quality, improved recoverable value (RV) yields and higher financial returns for farmers. The positive outcomes of these trials led to widespread adoption of chemical ripening techniques, expanding the practice from 648 ha in 2018, to over 4500 ha by 2021.

These participatory research efforts not only provided practical, on-the-ground insights into farming practices but also fostered a broader adoption of improved sugarcane management. The programme integrated feedback from both researchers and farmers, ensuring the practices were economically viable and suited to local conditions. While challenges like inconsistent adoption due to external factors arose, the ongoing collaboration and open communication between farmers and researchers helped overcome these barriers.

This case underscores how public–private partnerships can effectively enhance the delivery of agricultural extension services, blending research, technical expertise and real-world farmer input, to improve farm productivity and sustainability.

Public institutions, private companies, NGOs and farmer organizations all have a role to play in providing inclusive, effective support to farmers and agribusinesses. This section outlines key strategies and activities that can strengthen the delivery of these advisory services.

15.9.1 Reorganizing the extension service for better advisory service delivery

To improve farm planning, supervision and farm business management support, many public extension services – both at national and local levels – need to be reorganized. A key step is creating a dedicated unit for farm and agribusiness advisory services. This unit should be central to extension operations and staffed with well-trained specialists in farm business management, marketing and agribusiness. These professionals should have both technical and communication skills to work effectively with frontline extension workers and farmers. Placing these experts at central, regional or district levels helps ensure their services are accessible and their knowledge is shared throughout the system.

Both centralized and decentralized extension models have benefits and drawbacks. A centralized system offers consistency in training, messaging and resource use, but it may not respond well to local needs. Decentralized models allow regional or district offices to design services based on local priorities, encouraging trust and better community engagement. A hybrid model that combines national coordination with local implementation can offer the best of both approaches – maintaining standards while allowing flexibility.

Outsourcing farm management advisory services to other public agencies, NGOs or private organizations, can improve delivery by tapping into specific expertise, such as financial literacy or market linkages. These partnerships can create more integrated, multidisciplinary support for farmers. However, success depends on strong coordination, clear roles and proper monitoring to ensure accountability and service quality.

Strong administrative and supervisory systems are also essential. Clear responsibilities at all levels help prevent overlap and ensure efficient service delivery. Monitoring and evaluation systems are key to measuring outcomes, adoption rates and farmer satisfaction. Regular data collection supports evidence-based decision making and helps managers adjust strategies and allocate resources more effectively.

Capacity building is another priority. Extension workers need ongoing training, up-to-date technical knowledge and opportunities to collaborate with research institutions and the private sector. A systems approach – coordinating across disciplines and value chain actors – is also important to tackle the complex challenges in today's agriculture.

In the end, reorganizing extension services isn't just an administrative task – it is a strategic step towards making farm advisory services more relevant, efficient and impactful. With better structures, strong partnerships and flexible planning, extension systems can better support farmers and promote sustainable agricultural development.

15.9.2 Strategic planning and implementation

A practical strategy for using both public and private advisory service providers is to focus their efforts on key areas based on location and need. As discussed earlier, public farm management services should target small-scale farmers in difficult environments with poor market access – areas where the private sector is less likely to invest. In contrast, private providers often focus on regions with good infrastructure and better-off farmers who can afford to pay for their services.

Public extension services are especially suited to helping smallholder farmers at the lower end of value chains, where food security and basic farming support are priorities. Private providers, on the other hand, are better equipped to deliver specialized technical, financial and marketing advice to farmers higher up in the value chain. Strong coordination between public and private actors is essential to ensure farmers receive full and well-rounded support.

This targeted approach allows advisory services to use limited resources more effectively. Rather than trying to cover all areas equally, it focuses on where support is most needed, leading to greater impact in those specific regions.

15.9.3 Campaigns to strengthen public engagement

Public extension services can improve their reach and impact by organizing short-term, well-planned advisory campaigns. These campaigns help raise awareness about available farm management services and encourage smallholder farmers to get involved. A mix of outreach methods – like mobile advisory clinics, booths at agricultural fairs, printed materials and social media – can be used to connect with farmers. Having a strong presence both in the community and online allows extension workers to respond quickly to farmers' questions and concerns.

India's Krishi Vigyan Kendra (KVK) programme is a good example of this approach. Run by the Indian Council of Agricultural Research (ICAR), KVKs deliver hands-on support to farmers to improve productivity and sustainability. Campaigns often begin with community meetings, where experts provide advice on crop management, pest control and sustainable practices. These sessions also include training on farm business management, helping farmers understand profitability and better use of resources. The campaigns are interactive and spread out over several months, with follow-up training and on-site demonstrations. By training lead farmers and community extension workers, KVK ensures ongoing peer support and long-term adoption of practices like crop diversification, input use and market access. These efforts show how public extension services, combined with local knowledge and active farmer participation, can boost farm profitability in a focused and time-bound way.

15.9.4 Implementing differential strategies

The conventional method of giving the same technical advice to all farmers using a single approach no longer works in today's diverse farming environment. It is being replaced by customized, client-focused strategies that address the specific needs of different groups – such as subsistence farmers, market-oriented farmers, commercial farmers, women, youths and marginalized communities. These tailored services make extension efforts more relevant, inclusive and fair, by recognizing that farmers have different needs and challenges.

Differentiated strategies are designed to meet the unique requirements of each group. For example, subsistence farmers may need help with food security and diversifying income, while commercial farmers may need advice on market access, value chains and advanced technologies. Women and youths often face extra barriers in accessing finance, training and markets, so they need specialized support. These targeted, gender-sensitive approaches improve the impact of advisory services and promote fairness across the farming sector.

By applying these strategies, extension systems can better meet the needs of all farmers, helping to boost agricultural growth while also supporting social inclusion. This leads to more sustainable and competitive farming systems.

15.9.5 Bundling services

In market-oriented farming, farmers rarely need advice on just one issue. Instead, they need bundled services that combine advisory support with access to inputs, markets and finance. For example, marketing advice is closely linked to production planning, financial management and input access. A farmer planning to invest in a new crop needs not only technical guidance but also credit and reliable inputs to put that advice into action. Without this integration, advisory services alone often fall short, as farmers may lack the means to follow through.

Farmers are increasingly looking for combined services to tackle their interconnected challenges. For instance, successful marketing depends on access to finance, which helps farmers invest in materials, equipment and improvements. Likewise, quality seeds and fertilizers are essential for applying technical advice effectively. Bundled services allow farmers to receive guidance, inputs and financial support through one streamlined system, improving efficiency and outcomes.

This integrated approach also creates opportunities for cost recovery. Advisory fees can be included in input sales or financial products like loans, easing the cost burden on farmers while ensuring the sustainability of services. Financial institutions, agribusinesses and cooperatives can help deliver these bundled services, supporting farmers in moving towards more profitable, market-driven agriculture.

By linking advisory services with finance, inputs and markets, providers can deliver complete solutions that address multiple farmer needs at once. This avoids duplication, increases the value of services and supports sustainable farming. Ultimately, bundled services lead to better productivity, reduced risk and stronger market links – contributing to rural development and economic growth.

15.9.6 Establishing and strengthening multi-stakeholder platforms

To improve coordination among advisory service providers, practical mechanisms are needed at different levels. Multi-stakeholder platforms are especially useful for this, as they bring together a range of actors – farmers, agribusinesses, government agencies, NGOs and financial institutions – to align goals, share resources and plan joint strategies. These platforms help improve communication, reduce duplication of efforts and ensure farmers and agribusinesses receive well-rounded support.

For example, platforms can connect credit providers with farm business advisory services, helping farmers access both the guidance and financing needed to take advantage of market opportunities. By linking public, private and non-profit actors, these platforms strengthen value chains, encourage innovation and improve the overall delivery of services in the agricultural sector.

15.10 Conclusion

This chapter has provided an overview of the emerging specialization of farm management extension within the broader and ever-evolving extension and advisory service system. Over recent decades, extension as a profession and a service has undergone significant transformation. It has shifted from a linear, technology-transfer model focused primarily on production to a more inclusive and participatory approach that engages farmers as active partners in learning. This evolution reflects an acknowledgment of the increasing complexity of farming systems, regardless of scale.

Modern farmers, even at small scale, face challenges that extend beyond production. They manage information, procure inputs, oversee financial operations, market their produce, maintain infrastructure and navigate their roles within the broader agricultural value chain. To succeed, they require targeted training, support and advice in farm business management – essential components for running their farms as sustainable and profitable enterprises.

Supporting other actors within the value chain through agribusiness management services is equally important. These efforts are crucial to strengthen the competitiveness of small-scale family farmers and facilitate their integration into modern agricultural value chains. Public and private sector advisory services must therefore broaden their scope in addressing the broad farm and agribusiness environment and collaborate effectively to meet the diverse and growing needs of these stakeholders.

The increasing complexity of the farming system underscores the importance of a pluralistic advisory approach involving public, private and NGO sectors. While this approach has proven essential for advancing smallholder farmers, it is not without challenges. Key issues include the need for better coordination, adequate funding, alignment of objectives and targeted support for marginalized groups such as women and youths. Additionally, the adoption of ICTs and digital agricultural technologies remains critical for driving innovation and efficiency across the sector.

To address these challenges, extension and advisory services must strengthen their capacity through specialized training in farm business management. Tailored projects, programmes and skilled advisors are essential for delivering targeted support that meets the specific needs of farmers and value chain actors, while fostering sustainable and inclusive growth within the sector. These themes are further explored in Chapter 16.

References

CGIAR (2013) CGIAR challenge program on water and food (CPWF). Communities & Collections Website. Available at: https://hdl.handle.net/10568/2983 (accessed 30 April 2025).

Chilemba, J. and Ragasa, C. (2018) The impact of a farmer business school programme on incomes of smallholder farmers: Insights from Central Malawi. International Food Policy Research Institute (IFPRI), Washington, DC. Available at: https://massp.ifpri.info/files/2018/04/MaSSP_WP_23_BGSChilemba_Final.pdf (accessed March 2025).

Chipeta, S., Christoplos, I. and Katz, E. (2008) Common framework on market-oriented agricultural advisory services. Neuchatel Group. Available at: https://cgspace.cgiar.org/items/15723521-2178-42a4-aba1-2a0382237c62 (accessed March 2025).

Davidson, P., Ahmad, M. and Ali, T. (2001) Dilemmas of agricultural extension in Pakistan: Food for thought. Agren Network Paper Nn. 116 July 2001. ODI, London, UK.

Fair Planet (2023) How one NGO is lifting farmers out of poverty. Seed World, London, UK. Available at: https://www.seedworld.com/europe/2023/05/16/how-one-ngo-is-lifting-farmers-out-of-poverty/ (accessed 2 March 2025).

Falola, A., Olowogbon, T.S. and Mukaila, R. (2023) Financial literacy of rural farming households in Kwara state, Nigeria: A guide for financial inclusion. *Journal of Rural and Community Development*.

FAO (2020) World Food and Agriculture – Statistical Yearbook 2020. FAO, Rome, Italy. Available at: https://www.fao.org/3/cb1329en/CB1329EN.pdf (accessed 2 March 2025).

FAO (1989) Guidelines on the Conduct of a Farming Systems Development Training Course. FAO, Rome, Italy

Faure, G. (2001) What is the future for management advice for family farms in West Africa. CIRAD, Montpellier, France. Available at: https://publications.cirad.fr/une_notice.php?dk=521823 (accessed March 2025).

Faure, G. and Kleene, P. (2002) Management Advice for Family Farms in West Africa: Role of Producers' Organisations in the Delivery of Sustainable Agricultural Extension Services. CIRAD, Montpellier, France. Available at: https://agritrop.cirad.fr/521837/ (accessed March 2025).

IFAD (2022) Lessons Learned from Supporting Pluralistic Extension Services in Asia and Africa. IFAD, Rome, Italy. Available at: https://www.ifad.org/en/w/publications/lessons-learned-from-supporting-pluralistic-extension-services-in-asia-and-africa (accessed 2 March 2025).

Kahan, D. (2007) Farm management extension services: A review of global experience. Occasional Paper 19, Rome, Italy. Available at: https://openknowledge.fao.org/server/api/core/bitstreams/b1e70a40-dc93-4e80-b888-536a35a41ee0/content/a1494e.htm (accessed 2 March 2025).

MANAGE (2008) Annual report 2007-2008, National Institute of Extension Management. Hydrabad, India. Available at: https://www.manage.gov.in/publications/annualreports/annualreports.asp (accessed 13 March 2025).

McGregor, A. (2002) The Role of Farm Management in Agricultural Extension in the Pacific Islands. SAPA Discussion Paper 1. FAO Sub-Regional Office for the Pacific. Apia.

Nederlof, E.S., Wongtschowski, M. and Van der Lee, F. (2011) Putting Heads Together: Agricultural innovation Platforms in Practice, Royal Tropical Institute – KIT Publishers. Available at: https://bibalex.org/baifa/en/resources/document/417494 (accessed 16 November 2025)

Ngaleya, G. and Kuzilwa, J. (2016) Entrepreneurial training amongst smallholder farmers in Tanzania: An assessment of effect on farm productivity. *Rural Planning Journal* 18(1). Available at: https://repository.irdp.ac.tz/handle/123456789/320 (accessed 13 March 2025).

One Acre Fund (2023) Annual Report. One Acre Fund, London, UK. Available at: https://oneacrefund.org/about-us/reports/annual-report (accessed 2 March 2025).

Salman, F. and Bernal, A. (2023) Cultivating change: How can sustainable AgTech empower smallholder farmers in Latin America? Available at: https://dalberg.com/our-ideas/cultivating-change-how-can-sustainable-agtech-empower-smallholder-farmers-in-latin-america/

Schultz, T. (1964) *Transforming Traditional Agriculture*. Yale University Press, New Haven.

Schut, M., van Leewis, C., Paassen, A. and Lerner, A. (2011) Knowledge and innovation management in the policy debate on biofuel sustainability in mozambique: What roles for researchers? *KM4Dev Journal* 7, 45–64.

Swanson, B. and Davis, K. (2014) Status of agricultural extension and rural advisory services worldwide, GFRAS. Available at: https://cgspace.cgiar.org/items/5d99bfa5-388f-495b-ae66-73693cdf2296 (accessed March 2025).

Van Campenhout, B., Bizimungu, E. and Smart, J. (2017) Impact pathways of a participatory local governance initiative in Uganda: A qualitative exploration. IFPRI Discussion Paper Series 01688. Available at: https://cgspace.cgiar.org/items/5a184ecc-8754-4d0f-908d-c79f6dd05420 (accessed 21 March 2025).

Van Heerden, R. and Naiker, K. (2024) Enhancing sugarcane through farmer-driven research. SASRI, Mt Edgecombe, South Africa. Available at: https://sasri.org.za/article/enhancing-sugarcane-quality-through-farmer-driven-research/ (accessed 2 March 2025).

Wiafe, T. and Yona, L. (2016) Farm business management skills – a missing link for smallholder farmers: A case of Malingunde, Malawi. *European Journal of Business and Innovation Research*. Available at: https://eajournals.org/ejbir/vol-4-issue-3-june-2016/farm-business-management-skills-missing-link-smallholder-farmers-case-malingunde-malawi/ (accessed 18 March 2025).

Wordofa, M.G. and Sassi, M. (2018) Impact of farmers' training centres on household income in eastern Ethiopia: Evidence from propensity score matching. *Social Sciences* 7(1), 4. Available at: https://www.mdpi.com/2076-0760/7/1/4 (accessed 21 March 2025).

World Bank (2009) World Development Report: Reshaping Economic Geography. World Bank. World Bank, Washington D.C., USA. Available at: https://openknowledge.worldbank.org/entities/publication/58557d74-baf0-5f97-a255-00482909810a (accessed March 2025).

Additional Reading

Adetarami, O., Olagunju, O., Oyebamiji, B.A., Odeyemi, A.A., and Johnson, S.B. (2022). *Profitability and utilisation of Farmer Business School extension approach on smallholder cocoa farmers in Nigeria*. Sarhad Journal of Agriculture, 38(3), 851–861. DOI: 10.17582/journal.sja/2022/38.3.851.861.

Collinson, M.P. (1972) *Farm Management in Peasant Agriculture*. Praeger, New York.

Collinson, M.P. (1974) *The Role of the Agricultural Economist in Regional Planning, in Planning Agriculture in Low Income Countries*. University of Reading Development Study No. 14.

IFAD (2019) *Creating Opportunities for Rural Youth: 2019 Rural Development Report*. IFAD, Rome, Italy. https://www.ifad.org/ruraldevelopmentreport/ (accessed 02 March 25)

Dillon, J. and Hardaker, B. (1993) *Farm management research for small farmer development*. Farm Systems Management Series. No. 6., FAO, Rome.

Davis, K., Nkonya, E., Kato, E., Mekonnen, D.A., Odendo, M., Miiro, R., and Nkuba, J. (2012). *Impact of Farmer Field Schools on Agricultural Productivity and Poverty in East Africa*. World Development, 40(2), 402–413. DOI: 10.1016/j.worlddev.2011.05.019.

Dorward, A.R. (1991) *Integrated decision rules as farm management tools in small-holder agriculture in Malawi. Journal of Agricultural Economics*, 42 (2).

Dorward. A.R. (1986). *Farm Management and Extension in Smallholder Agriculture. Part 1 – the role of farm management in extension*. Agricultural Administration 22.

Dorward, A.R. (1986) *Farm Management and Extension in Smallholder Agriculture. Part 2 – appropriate farm management skills and their use in extension*. Agricultural Administration 22.

FAO (2011) *The State of Food and Agriculture 2010–2011: Women in Agriculture - Closing the gender gap for development*. FAO, Rome, Italy. https://www.fao.org/4/i2050e/i2050e00.htm (accessed 02 March 25)

Feder G., Birner, R., and Anderson, J. (2011) *The private sector's role in agricultural extension systems: potential and limitations. Journal of Agribusiness in Developing and Emerging Economies, 1(1)*.

Fleming, E. and Hardaker, B. (2005) *The Relevance and Usefulness of Farm Business Management in Enhancing Small Farmer Income in the South Pacific*. Rome, FAO.

Foster, P. (1987) *The contribution of education to development*. In G. Psacharopoulos (Ed), Economics of Education: Research and Studies, Pergamon Press, Oxford, pp. 93–100.

Hardaker, J.B. (1994) *A review of some farm management research methods for small farm development in less developed countries. Journal of Agricultural Economics* 30.

Hansra B.S. and K. Vijayaragavan edit. (2003) *Agribusiness and Extension Management*. Concept Publishing Company, Delhi, India.

IFPRI, (2006) *From Best Practice to Best Fit: A Framework of Analyzing Pluralistic Agricultural Advisory Services Worldwide*. Washington D.C., IFPRI.

Kahan D. (2007b) *Business services in support of farm enterprise development: a review of relevant experiences, Occasional Paper 18*. FAO, Rome, Italy. DOI: 10.13140/RG.2.1.1662.0642 (accessed 02 March 25)

Kilpatrick, S. (1997) *Education and training: Impacts on farm management practice*. Centre for Research and Learning in Regional Australia.

Kunze, D. (2003) *Farm Management in Extension in Africa*. FAO, RAF Publication, Ghana.

Maingi, G. (2000) Impact of the GTZ farm management support project, Unpublished conference paper presented to the third African farm management congress.

Nakano, Y., Tsusaka, T.W., Aida, T., and Pede, V.O. (2018). *Is farmer-to-farmer extension effective? The impact of training on technology adoption and rice farming productivity in Tanzania*. World Development, 105, 336–351. DOI: 10.1016/j.worlddev.2017.12.013.

Rehman, T. and Dorward, A. (1984) *Farm Management Techniques and Their Relevance to Administration: Research and Extension in Agricultural Development: Part 1 – their evolution and use in developed countries*. Agricultural Administration 15.

Rehman T. and Dorward A. (1984) *Farm Management Techniques and Their Relevance to Administration: Research and Extension in Agricultural Development: Part 2 – An Appraisal of their potential in less developed countries*. Agricultural Administration 15.

Sayinzoga, A., Bulte, E., and Lensink, R. (2016). *Financial Literacy and Financial Behaviour: Experimental Evidence from Rural Rwanda. Economic Journal*, 126(594), 1571–1599. DOI: 10.1111/ecoj.12217.

Sulaiman, R. and Davis, K. (2012) The new extensionist: Roles, strategies, and capacities to strengthen extension and advisory services. *Journal of International Agriculture* 12(3), 6–18. DOI: 10.5191/jiaee.2014.21301.

Waddington, H., Snilstveit, B., Hombrados, J.G., Vojtkova, M., Anderson, J., and White, H. (2014). *Farmer field schools for improving farming practices and farmer outcomes: A systematic review*. Campbell Systematic Reviews, 10(1), 1–335. DOI: 10.4073/csr.2014.6.

Wongtschowski, M., Belt, J., Heemskerk, W. and Kahan, D. (2013) The Business of Agricultural Business Services: Working with Smallholders in Africa. Royal Tropical Institute (KIT), Food and Agriculture Organisation of the United Nations (FAO), and Agri-ProFocus. Available at: https://www.kit.nl/wp-content/uploads/2018/08/2080_the_business_of_agricultural_business_services.pdf)

World Bank. (2005) *Directions in Development Agricultural Growth for the Poor: An Agenda for Development*. World Bank., Washington, D.C.

16 Farm and Agribusiness Management: A Specialization

This chapter explores the role of specialist farm management and agribusiness advisors in managing the challenges of modern agriculture. These experts connect farmers, policymakers, researchers and value chain actors by offering key support in financial management, business planning, market analysis and investment strategies. In the public sector, they help smallholder farmers boost productivity and access markets. In the private sector, they offer tailored services to agribusinesses and cooperatives. Their work supports knowledge sharing, efficient use of resources and adoption of best practices, helping make agriculture more resilient and profitable. This specialization also creates many career opportunities within agriculture and related support services.

The chapter also looks at the training needed for farm and agribusiness advisors to do their jobs well. They need knowledge in economics, business and agriculture, as well as skills in data analysis, policy and adult education. Their training often covers diagnostic tools, budgeting, market analysis and project planning, preparing them to solve problems in a variety of settings. The chapter also stresses the value of training and collaboration among stakeholders – especially with farmers – to support innovation, build capacity and encourage sustainable farming.

Finally, the chapter highlights how public and private advisors must work together to meet the varied needs of farmers and agribusinesses. Farm management extension workers play a key role in shaping policy, designing extension programmes, and connecting small farmers to markets. Using simple budgeting tools, they help farmers plan better, manage resources, and choose the right crops and livestock based on changing markets and climate. The chapter outlines the tools and strategies these specialists use to communicate effectively with farmers, policymakers, programme managers and other value chain actors – especially in the era of big data, ICT and social media.

16.1 Farm and Agribusiness Management Advisers

To recap the discussion so far, the farm and agribusiness management advisory system operates across both public and private sectors, each playing a key role in improving agricultural productivity, efficiency and profitability. Public sector extension workers, typically employed by government agencies or agricultural institutions, focus on broad goals such as promoting sustainability, supporting small family farmers, ensuring food security and helping farmers access new markets. They work at various levels – from national policy to on-the-ground service delivery – providing technical advice and connecting farmers to government programmes.

Private sector advisors, in contrast, offer specialized services aimed at commercial farms, agribusinesses, cooperatives and other value chain actors. Their focus is on financial management, business planning, market analysis and supply chain development. These advisors are often employed by agribusiness companies, consulting firms, NGOs or work independently, and provide market-driven, client-focused solutions.

DOI: 10.1079/9781836990536.0016

They complement public sector services by delivering targeted support that responds quickly to changing market demands. This reflects the pluralistic nature of today's advisory system.

As discussed in the previous chapter, collaboration between public and private advisors is essential for building a strong, well-rounded support system for agriculture. Working together, they enable better knowledge sharing, efficient resource use and innovation across the value chain. Since this book focuses on small family farms, special attention is given to the role of public sector extension in equipping farmers with the tools and strategies needed to compete in the market. Farm management specialists, positioned throughout the advisory system, are central to building resilience and promoting sustainability. This pluralistic model ensures the diverse needs of small family farmers are met effectively.

Figure 16.1 illustrates the diversified functions and roles of advisory services in general, emphasizing the complexity and broad scope of tasks required to support agricultural development. Central to this are the qualities of agricultural advisors, which underpin their ability to perform key functions such as training, advice, information provision, process facilitation, dissemination, facilitation of linkages, mediation and other support activities.

Within the public sector, a distinction exists between subject matter specialists and front-line advisors. Subject matter specialists are experts in agricultural topics such as crop production or livestock management. They possess in-depth knowledge and technical expertise, contributing to the design of advisory materials, research and policy recommendations. These specialists provide technical support to front-line extension workers, who are in daily contact with farmers. Among the specialists is the farm management specialist whose specialization spans investigation, planning, marketing, training and extension, along with other related functions, which are elaborated on in this chapter.

As depicted in Fig. 16.2, the farm management specialist, supported by front-line agricultural extension workers, needs to cater to the diverse needs of smallholder farmers, as well as other clients and stakeholders within the broader agricultural advisory system that are necessary to support smallholder farm development. This shows how complex the advisory system is, since it must handle different goals, roles, topics and actors. A holistic and flexible approach is needed to deliver extension services

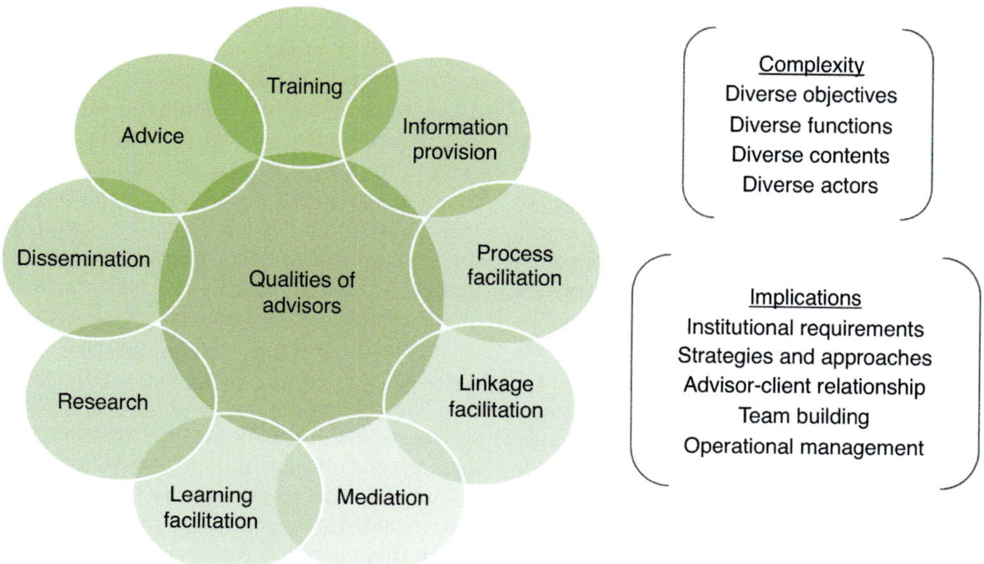

Fig. 16.1. Diversified functions and roles of advisory services. Adapted from Blum, 2008.

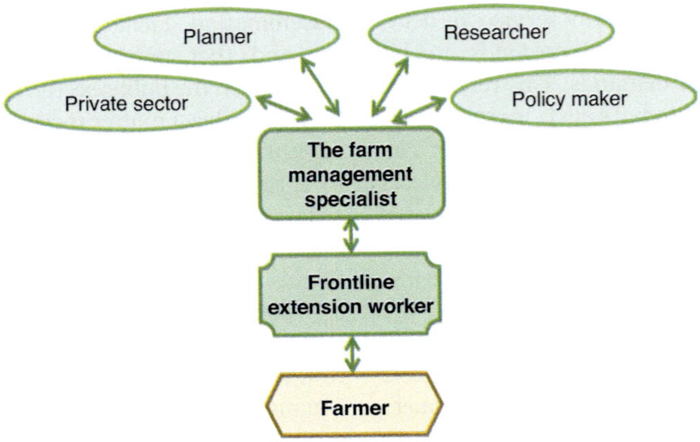

Fig. 16.2. The central role of the farm management specialist in public sector extension services.

that can meet the diverse challenges of modern agriculture.

Farm management specialists are traditionally trained in economics and possess expertise in business and marketing. They play a crucial role in agricultural extension, planning, programme design and policy analysis by bridging gaps between farmers, researchers, policymakers and the private sector. With access to farm management data, and skills in farm investigation and planning, they are well-positioned to influence policy formulation, project design and the development of national strategies. Their expertise is particularly valuable in these areas. Within public sector extension services, farm management specialists collaborate with other subject matter experts, provide guidance to frontline extension workers and support planners and policymakers. They also serve as leaders in addressing farmers' needs across agriculture and related sectors while motivating frontline extension workers in the public, private and NGO domains.

In contrast, frontline extension workers directly assist farmers with practical advice. Their role is often wide in that they assist farmers with every aspect of their farming livelihoods. Often frontline extension workers draw on the expertise of subject matter specialists, which they translate into actionable advice for farmers. One important function is that of intermediaries where they connect farmers with research, technology and market information. They facilitate knowledge exchange and learning and promote the adoption of innovative practices. By helping farmers access, interpret and adapt information, they ensure that solutions are tailored to local conditions. Their direct engagement also helps identify the capacities and needs of farmers, drive knowledge-sharing and foster the implementation of new practices and technologies. Above all, their aim is to build the capacity of farmers to make effective decisions about their farming livelihoods.

Extension workers also support farm management specialists by gathering and analysing data on farm inputs, production levels and costs to assess farm business performance. This collaboration is critical in identifying inefficiencies, reducing high-cost inputs and uncovering opportunities for increased profitability. A common approach involves farm management specialists training extension workers, who, in turn, train farmers either directly or through farmer learning groups. As core trainers, farm management specialists support both technical subject matter specialists and frontline advisers.

The placement of farm management specialists within public sector extension services depends on institutional needs, available resources and personnel capacity. In some cases, they are based at the central level but, more commonly, they operate at regional or district levels, depending on the degree of decentralization in the system. Regardless of their placement, they function at the intersection of planning, research and extension, ensuring that farm management information informs all aspects of service delivery to farmers.

16.2 Farm Management Specialist: Overview of Roles and Responsibilities

As elaborated on in Chapter 9, understanding economics is essential for smallholder farmers to manage their farms as businesses, and for extension workers to support them effectively. Economics provides the foundation for farm (business) management, enabling the application of tools and techniques to make sound decisions. Without an economics perspective, farmers risk making poor choices that hinder profitability and sustainability. This approach addresses critical questions such as selecting appropriate technologies, optimizing farm enterprise combinations, choosing profitable markets and managing the risks in farming.

While good farm management advice can come from various educational disciplines, understanding and applying basic economic principles is essential for all extension workers to be effective in their roles. Those who use economic thinking will be better at addressing complex issues and providing solutions that consider the overall economic performance of the farm to be run as a business. This means looking at the big picture, not just isolated parts. Therefore, mastering economic reasoning is key for farm management specialists who want to give comprehensive and effective advice, helping farms run efficiently and profitably.

In today's dynamic agricultural landscape, farm management specialists perform multiple roles, including:

- Influencing policy decisions to support the agricultural sector.
- Providing advisory support to farmers for improved decision making.
- Facilitating linkages between farmers, input suppliers and markets.
- Optimizing resource use by diagnosing problems identified through extension activities.
- Compiling and analysing farm enterprise data to generate actionable extension information.
- Advising on market opportunities and providing timely marketing and business insights.
- Resolving marketing challenges by fostering connections between farmers and buyers.
- Contributing to public projects and programs with expertise in business management.
- Enhancing communication between farmers and public sector actors to strengthen collaboration (see Box 16.1.).

These specialists require a wide-ranging skill set that includes finance, marketing, cost-benefit analysis, law, contracts and traditional management expertise. Additionally, they should possess the ability to work with front-line agricultural extension workers to organize farmers into groups, associations and cooperatives, with a strong understanding of contract farming to establish effective market linkages.

The farm management specialist is a critical resource for small family farmers transitioning towards market-oriented and commercial farming. They serve as sources of knowledge, facilitators of market linkages, advisory workers, innovators and organizers, empowering farmers to achieve greater profitability and sustainability.

In conclusion, the placement of farm management specialists within public sector extension services ensures optimal support for frontline workers, other subject matter specialists and agricultural stakeholders, amplifying their impact on agricultural development.

16.2.1 Knowledge, skills and qualities

To be effective in supporting farmers with technical expertise, analytical insights and practical guidance, farm management specialists must possess a strong foundation in both knowledge and skills, enabling them to address the diverse challenges farmers face. Their work requires not only technical and financial expertise but also the ability to communicate, educate and build

Box 16.1. Role of the farm business management specialist. From Kahan, 2007.

Source of knowledge and information:

- Provide information enabling farmers to make informed decisions.
- Communicate information to front-line agricultural advisers.

Facilitator of market-linkages:

- Know the agents and the opportunities.
- Gather and disseminate market information.
- Bring farmers and commercial agents together.
- Assist with the formulation of fair contracts.

Business adviser:

- Identify and respond to problems and opportunities enabling farmers to expand their options and increase their profitability.

Facilitator of innovation:

- Support farmers to engage with researchers and other stakeholders to foster innovation along the value chains.
- Facilitate the interaction among value chain and other stakeholders to collaborate on developing innovative responses to problems and opportunities.

Farmer organizer:

- Help farmers form their own organizations to achieve economies of scale, create efficiencies in buying inputs and marketing produce.
- Mobilize farmers to work together to tackle problems and seize opportunities by themselves.

Facilitator of finance and investment:

- Support farmers, farmer organizations and producer organizations in financial literacy training, loan application support and linking to investors.

trust with farming communities. This section explores the essential knowledge requirements, skills and personal qualities that define a successful farm management specialist.

Knowledge requirements

To fulfil their roles effectively, farm management specialists must possess specific qualities essential for the dissemination of knowledge and skills.

- **Technical knowledge:** they need a broad exposure to the technical aspects of their work. They need a good understanding of key elements within the different farming systems, whether it is crop production, livestock management, vegetable production or mixed farming.
- **Analytical and diagnostic skills:** farm management specialists must analyse situations effectively, identify problems, spot opportunities and suggest practical solutions. For instance, when faced with a farmer experiencing low income or profitability, specialists should diagnose the issue, pinpoint its cause and recommend appropriate solutions.
- **Investigation:** proficiency in locating, interpreting and conveying information is also crucial. They should know where to find relevant data, how to analyse it and how to present it effectively. For example, if a farmer needs help deciding whether to enter a new market, the farm management specialist should collect relevant market information and present it clearly and usefully to support the farmer's decision making.

- **Management and planning:** farm management specialists must possess skills in management, finance and marketing to assist farmers effectively. They should be able to assess the financial status of farm businesses, assist in cash flow budgeting, negotiate with financial institutions, facilitate decision making and provide information on government assistance schemes.
- **Policy awareness:** familiarity with government legislation and institutional policies impacting agriculture and rural development is also essential. Farm management specialists should understand regulations concerning agricultural product movement, pesticide and herbicide use, labour laws and tax legislation.
- **Education:** since advisory support involves education, these specialists should be well-versed in adult education approaches, group dynamics and techniques for fostering farmer participation. They should also be aware of specialized education and training programmes tailored for adults with limited formal education, including literacy and numeracy training.

Skills and qualities

The effectiveness of a farm management specialist hinges on a diverse set of skills and qualities essential for delivering valuable support to farmers and agribusiness.

- **Professional competence:** a proficient farm management specialist possesses a solid understanding of farming techniques, economics and specialized knowledge in farm business management. They are adept at applying concepts, tools and techniques of farm management to provide practical advice to farmers and explore alternative solutions.
- **Trustworthiness:** integrity and reliability are paramount for a farm management specialist. They must uphold ethical standards and demonstrate trustworthiness in all interactions with farmers.
- **Discretion, tact and confidence:** respecting farmers' privacy, exercising tact to avoid upsetting or embarrassing them, and offering assistance without being intrusive are key attributes. Farm management specialists should challenge farmers' perspectives constructively without criticism.
- **Initiative:** The ability to work independently, take proactive steps and offer guidance without constant supervision is crucial for success in this role.
- **Teaching ability and understanding:** beyond advising, farm management specialists must empower farmers to make informed decisions through sound economic reasoning. They recognize the importance of addressing non-business factors and provide support to alleviate farmers' uncertainties and fears associated with change.
- **Sense of responsibility:** prudence and accountability are essential traits. Farm management specialists strike a balance between taking responsibility for their actions and acknowledging that farmers have the ultimate authority over farm management decisions.
- **Team spirit:** collaboration with local agricultural advisors and fostering a sense of teamwork among farmer groups are vital for effective support delivery.
- **Organizational ability:** proficiency in planning, implementing and managing extension activities is crucial. Farm management specialists should be capable of coordinating extension efforts and overseeing administrative activities efficiently.
- **Communication and facilitation skills:** effective communication through various channels and the ability to facilitate learning and skill acquisition are both essential for engaging with farmers and conveying information well.

16.2.2 Diagnosis and investigation

The specialist in farm management has the following five key roles within diagnosis and investigation, aimed at providing strategic guidance to policymakers and programme managers, as well as support to extension workers and farmers.

1. Providing policymakers with guidance and support from farm-level studies conducted in the area.

2. Aiding research staff in research stations and institutions, and extension service providers, by prioritizing technical areas of research concentration.
3. Preparing area profiles based on data and information collected on the biophysical and socio-economic background of the area being considered.
4. Analysing market opportunities for high potential agricultural products and supporting private sector investments.
5. Providing extension services and (through them) frontline agricultural advisers and farmers with information on the profitability and competitiveness of farm enterprises.

Farm management specialists collect, analyse and present farm-level and market data to a diverse range of users. They play a critical role in investigating priority issues, leading or facilitating teams to conduct market appraisals, value chain analyses and farm profitability assessments. Additionally, they have a responsibility to train frontline extension workers in farm business management, guiding them in collecting and analysing farmer-level data to identify areas for improvement and devise strategies to optimize the resource use of farm businesses.

Agricultural extension workers, as facilitators and frontline advisers, must collaborate closely with farmers to understand the farmers' goals, capacities, challenges and constraints. They also require the skills to guide farmers in conducting financial assessments, analysing costs and returns, evaluating market opportunities, and implementing effective management practices. These competencies enable extension advisers to support farmers in making informed decisions that enhance farm performance and sustainability. The role of the farm management specialist cannot be understated in this support function, as they equip extension workers with the knowledge and tools needed to provide tailored recommendations that address specific challenges and to build farmer capacity to face those challenges and improve overall farm management.

16.2.3 Planning, policy support and project appraisal

As we noted in Chapter 13, farm planning allows farmers to determine the optimum combination of farm production and farm resources. It is the process that farmers use to decide how to better run their farms in the short- and long-term. Farm planning is closely related to farm investigation and analysis which is an invaluable source of information for making sound planning decisions that farmers need to take. The farm management specialist, as part of the extension service, is responsible for designing and conducting farm planning activities that support extension. To do so effectively, the specialist needs the information and skills necessary to develop relevant and useful plans.

The intention of farm planning as we have seen is to help farmers move progressively to higher levels of productivity and income along the road best suited to their specific situations and the conditions that surround them, and to sustainably achieve their livelihood goals. It identifies different routes that may be followed and assists farmers to find the one that fits their situation best.

Looking ahead to consider new opportunities and changing conditions is not natural for traditional smallholder farmers. They tend to look to the past for guidance. Farm planning teaches farmers to look to the future and run their farm as a business. Farm planning support is essential in this process as well as to inform programme managers and policy makers.

Investment planning

Farm management specialists play a crucial role in investment planning at both the project and farm level, ensuring that resources are allocated efficiently to maximize returns. In the preparation of agricultural investment projects, they contribute to feasibility studies, cost-benefit analyses and financial appraisals, helping to design projects that are both economically viable and aligned with broader agricultural development objectives. At the farm level, they support farmers in making informed investment decisions by assessing the profitability, risks and long-term sustainability of different investment options.

To fulfil this role effectively, farm management specialists must have a strong understanding of key investment appraisal concepts and the ability to apply financial analysis techniques. This includes skills in discounting project inflows (benefits) and outflows (costs) to assess net present value, internal rate of return, and other financial indicators that guide investment decision making, as discussed in detail in Chapter 13. By equipping both policymakers and farmers with sound financial assessments, farm management specialists ensure that investments – whether at the national, regional or farm level – are well-structured, financially sustainable, and capable of delivering long-term benefits.

Policy analysis

Farm analysis, along with planning, also has an important contribution to make at policy level. Farm analysis is instrumental in evaluating agricultural policy options, addressing constraints posed by limited knowledge of farm resources and production practices. At policy level, farm-level studies aim to elucidate farmer behaviour concerning input utilization, market preferences, and responses to price fluctuations and technological advancements. These studies also look at issues like market access, input availability, credit accessibility and other factors influencing farm performance. Policy analysts in government offices often need farm-level data that are essential inputs for national and sectoral studies. Although farm management specialists may not be directly responsible for policy analyses, they can contribute by providing this micro level information (FAO, 1994).

Research and extension

The synergy between research, extension and farmers is pivotal for effective technology innovation and transfer. Farm management specialists, as key figures in extension, have a dual responsibility: advising farmers and supporting research efforts. They provide invaluable guidance to farmers and extension workers, assisting in aligning new ideas with farm requirements and integrating them into farm plans. Specialists facilitate the flow of information between farmers, extension workers and researchers, contributing significantly to knowledge discovery and technology dissemination.

Extension priorities

Farm management analysis and planning also serve as the backbone for setting extension priorities, guiding the design of tailored extension programmes. By grouping smallholder farmers based on similar social and economic factors, farm management specialists can tailor extension advice to better match each group's specific needs and circumstances. While detailed farm plans provide useful information, the broader lessons learned from these analyses are even more important for guiding extension work and supporting agricultural development.

16.2.4 Farm planning for smallholder farmers

Farm planning in developed countries has traditionally involved conducting planning exercises through programme planning and linear programming on individual commercial farms. These were discussed in detail in Chapter 13. Advanced planning methods like linear programming and other mathematical techniques have been criticized for their limited use at the farm level in low-income countries. These methods often rely on computers, are difficult to understand and feel impersonal. In these countries, most farmers are smallholders working small plots of land, and reliable farm management data is often scarce and costly to collect. As a result, using such detailed planning methods for individual small farms is often too expensive, inefficient and impractical.

When applying farm management practices to small family farms, a different approach is required. Farm management specialists can either conduct detailed analyses and demonstrations using selected case study farms or develop farm models that represent the 'typical' farms commonly found in the area.

Selecting case study farms

Farm management specialists can select one or more farms as case studies, choosing them for their potential to highlight key farm management challenges and practices. While the goal is to reflect the diversity of local farms, the selection is usually purposive rather than random to allow for practical demonstrations. These case study farms are used to analyse and illustrate planning methods, with the results shared with other farmers in the area to promote wider adoption of better management practices.

Identifying a 'typical' farm

Where farms share similar characteristics, a single farm may be selected as representative. However, in areas where farms vary significantly, they must first be grouped into more homogeneous categories. Each group can then be used to identify a model farm that best represents the most common farming system in the area. Studying these typical farms provides insights applicable to a broader range of farms, allowing for the development of general recommendations. This approach is particularly useful for economic analyses, policy development and farm management planning. The methodology for identifying representative model farms is outlined in Box 16.2.

The results from the 'typical' or case study farm will not apply specifically to any one actual farm. The intention in modelling is to generate results which should reasonably or 'on average' apply to the farms they represent. The plans can then be submitted to farmers in whole or in part so that they may choose the one they think is best suited to their farms and goals.

Planning for climate change

As discussed in Chapter 7, agricultural advisory services play a vital role in supporting farmers to adapt to climate change. In this chapter, the focus is on the specific contributions of farm management specialists in climate change adaptation planning.

One of their key responsibilities is to assess, in collaboration with frontline extension workers, the vulnerabilities of smallholder farmers to climate-related risks such as unpredictable weather patterns, droughts, floods and pest outbreaks. Based on this assessment, farm management specialists work with extension workers and farmers to develop tailored farm plans that minimize these risks while maximizing resource efficiency and profitability. This may include selecting climate-resilient crops and livestock breeds, optimizing water and soil management strategies, and diversifying income sources to reduce reliance on a single enterprise.

A critical component of climate adaptation planning is financial analysis, where farm management specialists provide expertise in evaluating the economic feasibility of climate-smart practices. They assist farmers in assessing costs, potential returns and risks associated with climate adaptation measures. This may involve designing cropping plans that account for rainfall variability, preparing partial budgets to ensure financial feasibility and integrating climate resilience into farm business planning. By providing tools for budgeting, risk assessment and cost-benefit analysis, specialists empower farmers to make informed decisions that balance short-term needs with long-term sustainability.

Farm management specialists also facilitate the incorporation of innovative technologies into farm plans to address climate challenges. This includes introducing mobile applications that provide real-time weather forecasts, recommending digital tools for tracking resource use and productivity, and promoting precision agriculture techniques. These technologies enhance farmers' capacity to respond to climate variability and improve overall decision making.

Beyond direct farm planning, specialists play an important role in helping frontline extension workers and farmers build the skills to manage climate risks on their own. This includes training farmers to analyse data, assess risks and adjust their practices as conditions change. By promoting an adaptive mindset and providing practical tools, specialists help make climate resilience a regular part of farm management, not just a reaction to problems.

Additionally, farm management specialists encourage forward-looking strategies such as scenario planning, where farmers assess potential climate impacts on production and income under different conditions. They also promote diversification strategies – such as integrating multiple income streams or exploring insurance options – to enhance resilience. Through financial planning and strategic adaptation, they help farmers maintain economic viability while improving overall farm sustainability in the face of climate uncertainties.

By integrating climate change considerations into farm management, these specialists play a pivotal role in ensuring that smallholder farmers are prepared for both current and future challenges. Their expertise in farm economics and planning enables them to guide farmers in making informed, sustainable decisions that enhance productivity, profitability and resilience.

One approach to planning for climate change is the participatory integrated climate services for agriculture (PICSA) process. The

Box 16.2. Process for identifying 'typical' representative model farms.

1. **Identify homogeneous areas:** for example, by type of enterprise, topography, climate, soils, and economic and social conditions. This activity could be conducted by the farm management specialist when developing an area profile.
2. **Choose sample areas:** further classify the farms identified in Step 1 according to similar agro-ecology, topography and soil type. Representative areas could be found for each main farm type categories. Each of the areas chosen could then be studied so that the farm management specialist has figures on the total number of farms and their land holding size.
3. **Select the most typical farm size groups:** the next step is to identify the typical farms. This requires conducting a survey and listing the number of farms and their sizes for each of the sample areas identified in Step 2. Farms that are exceptionally large or small or have characteristics that vary significantly from the rest would be dropped. This stage results in a model size group for each of the sample areas.
4. **Select the most representative farms:** this stage involves taking stratified samples of the farms within the size groups as identified in Step 3. The farms could be sorted into three different categories of land size: small, medium and large. A random sample of farms from each category would be selected, with one chosen to represent each size group from each enterprise type. This figure could be reduced further to identify the most common farm in each selected type of farming area.

PICSA approach empowers smallholder farmers by equipping them with accurate, location-specific climate and weather information, alongside practical options for crops, livestock and livelihoods. PICSA supports farmers in making informed and evaluating their viability in the context of changing climate conditions. Developed with a focus on sub-Saharan Africa, where small-scale, rain-fed farming predominates, PICSA combines historical climate data, seasonal forecasts and participatory planning techniques, to enable farmers to identify strategies that enhance resilience and productivity.

Figure 16.3 provides a simplified version of the PICSA approach. Each step is designed to build farmers' understanding and decision-making capacity. These steps guide facilitators and farmer groups through activities such as mapping current practices, analysing historical and forecasted climate data, and exploring various farming and livelihood options. The approach culminates in farmers making tailored decisions for their individual circumstances, supported by participatory tools like seasonal calendars and budgets. PICSA not only fosters local adaptation but also encourages a collaborative learning process that strengthens long-term sustainability and climate resilience. Not only does this approach aim to develop responses to climate change but it also ensures that the capacity of farmers to address climate change is strengthened (Dorward *et al.*, 2015).

16.2.5 Farm planning in extension and research

The relationship between research and extension is deeply interconnected. Research provides the foundation for extension services, while extension brings relevance to research by ensuring its practical application. Without the feedback of information from farmers and extension workers, research risks becoming disconnected from real-world agricultural needs. Collaboration among research, extension and farmers is essential for effective technology innovation, development and transfer.

This relationship is highly collaborative. While research provides valuable insights and tools, farmers are increasingly taking on the role of innovators and researchers themselves, adapting and developing new practices to remain competitive in the modern agricultural landscape. They are often experimenting and adapting practices to suit their unique contexts. Extension serves as a bridge, facilitating the exchange of knowledge and ensuring that research is informed by and responsive to farmers' real-world experiences. Effective technology innovation, development and transfer require a participatory approach that values the contributions of farmers alongside those of researchers and extension workers, fostering mutual learning and co-creation of solutions.

Agricultural development requires a delicate balance between top-down and bottom-up approaches. Top-down strategies, characterized by centralized decision making and dissemination of innovations, can ensure consistency, scalability and access to advanced research. However, when applied in isolation, they risk alienating local communities and overlooking the nuanced realities of smallholder farmers. On the other hand, bottom-up approaches empower farmers by placing their knowledge, experiences and needs at the forefront of decision-making processes. While these methods promote local ownership and relevance, they may struggle to achieve broader impact without the structured support of institutional frameworks.

The challenge lies in avoiding the extremes of either approach by fostering a dynamic, two-way exchange of knowledge and resources. A balanced approach values farmers not just as beneficiaries but as innovators and co-creators of solutions, working collaboratively with researchers and extension services. By integrating the insights and capacities of all stakeholders, this synergy ensures that interventions are both grounded in local realities and informed by cutting-edge research, creating a pathway for sustainable and inclusive agricultural development.

Farm management specialists play a key role in this dynamic by introducing farmers and extension workers to new ideas and helping them evaluate whether these innovations address their needs and fit their farm plans. Farm management specialists can also play a key role in assisting farmers test and refine their own innovations to determine their viability and sustainability as well as their potential for scalability and sharing with other farmers. Specialists act as advisors, counsellors and guides, but the ultimate responsibility for decisions and their outcomes rests with the farmers. By actively participating in the engagement between researchers, extension workers and farmers, farm management specialists provide invaluable contributions to the innovation process. Their investigations not only guide extension programmes but also generate critical feedback for research (both formal and at farmer level), contributing to the discovery of new knowledge.

Farm planning in extension

Farm management analysis and planning are essential tools for setting priorities in extension work. As previously mentioned, farm modelling allows specialists to classify small farms into homogeneous groups, enabling extension services to provide targeted advice to farms with similar characteristics. However, it is essential to avoid rigid, standardized solutions and instead ensure flexibility to accommodate the unique

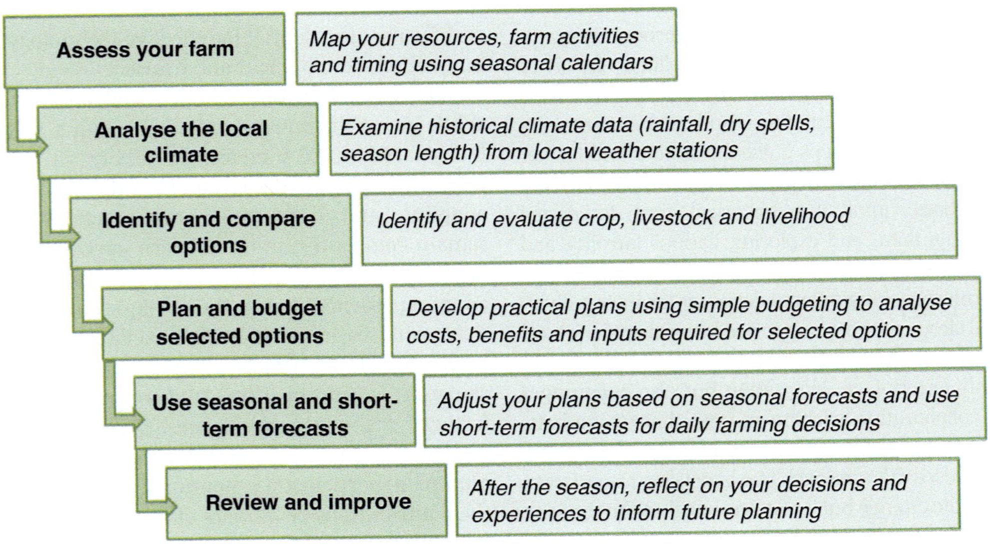

Fig. 16.3. A participatory approach to planning for climate change.

conditions and needs of individual farms. Instead, investigations should involve farmers to ensure that plans are practical and aligned with their needs, resource conditions and limitations. Collaborative farm planning, involving a team of specialists, researchers, extension workers and farmers, is key to developing realistic and implementable strategies. While this approach has its place, it must be tempered with the growing advances in agricultural extension.

Farm planning in extension is increasingly focused on the farmer as the starting point. This means planning begins with understanding the farmer's goals, abilities, resources and vision for the future. While land factors like soil fertility, water and climate are important, they should support, not override, the farmer's priorities. By putting the farmer at the centre of decision making, plans can be better tailored to their unique situation, supporting both sustainability and farmer empowerment.

To make this approach work, specialists, researchers and extension workers must understand both the farmer's context and the land's potential. Farmers should be grouped based on both personal and land characteristics. Although this may take more time, it leads to more meaningful and lasting results.

When following this expanded approach, once farm planning investigations are completed, the findings must be shared with other farmers and farmer groups. Disseminating these results strategically requires phased extension programmes. Initially, a few key areas can be selected for focused efforts, involving a small number of interested farmers. As these early participants gain experience and achieve results, they can assist their peers in exploring and adopting or adapting farm planning practices, as well as gain new ideas and insights from newly engaged farmers. Over time, this approach can be scaled to involve more farmers, promoting skill development and system improvements to enhance profitability. The key to success is building genuine learning partnerships among farmers, specialists, researchers and extension workers.

Farm planning in research

Farm management provides analytical tools to help research organizations prioritize their efforts and align them with the needs of farmers. These tools can be used to identify and analyse constraints, evaluate the suitability of technologies before conducting location-specific research and plan, monitor and evaluate on-farm trials and demonstrations.

By guiding on-farm trials and pilot projects in representative farming areas, farm management specialists can assess the profitability and adoption potential of new technologies. Aggregating data from individual farms allows for an evaluation of the broader impact of proposed innovations within a region. These insights help research institutes allocate human and financial resources effectively, ensuring that their programmes address critical farmer needs.

Farm management methods, such as enterprise budgeting and farm planning, also play a vital role in setting research priorities. Evaluating technologies and farming strategies using these tools ensures that research efforts focus on solutions with the greatest potential impact. Findings from farm-level investigations can inform decisions about resource allocation and shape research programmes to meet the needs of farmers more effectively.

By integrating farm management into both extension and research, specialists can ensure that agricultural innovation is practical, relevant and beneficial to farmers. This synergy between research, extension and farmers is crucial for driving sustainable agricultural development and improving livelihoods.

16.2.6 Linking farmers to markets and finance

Farm management specialists also play an important role as facilitators, linking farmers to markets and financial services. In today's pluralistic advisory systems, this role is increasingly being performed by private sector actors — including agribusinesses, agricultural business consultants, producer organizations, commodity associations and commercial banks — that provide business support to their members and broker linkages to markets and finance. However, public sector farm management specialists continue to have a critical role. They help organize farmers to engage effectively with buyers, build the business skills of individual farmers and farmer group leaders, and often support NGOs and private sector actors by creating an enabling

Box 16.3. Role of agricultural advisory workers in facilitating group development.

- Support groups in developing a growth policy and strategy that leverages their energetic, entrepreneurial approach while safeguarding their focus and business direction.
- Promote connections between groups and external organizations or individuals, enabling them to benefit from collective strength while maintaining individual creativity and initiative.
- Provide training to group members on effective group management and dynamics, ensuring they maximize both collective efforts and personal contributions.
- Guide groups in implementing, monitoring and evaluating their business plans to ensure sustainable progress and measurable outcomes.

environment that attracts investment and participation in agriculture (see Box 16.3.).

Farmers often struggle to participate in higher-value markets due to the high costs of marketing individually. Farmer organizations strengthen farmers' bargaining power, coordinate production and improve access to credit and support services. By working collectively, farmers can supply larger volumes, eliminate intermediaries and protect their interests more effectively. In many low-income countries, NGOs or private sector actors are often preferred to organize farmers. However, where such actors are unavailable, public extension services take on this role. Front-line extension workers often require support from farm management specialists to facilitate group formation, develop organizational capacities and improve farmers' marketing and business skills.

The level of facilitation depends on the group's starting point. Newly established or weak farmers, organizations often need strong external support to build capacity, strengthen democratic and accountable structures, and link with networks and service providers. Farm management specialists can play a key role in guiding this process, providing knowledge about markets, technology and training opportunities. Strong organizations, by contrast, may require minimal support, focusing instead on expanding their networks to amplify their influence at higher levels.

As primary farmer groups become more self-reliant, extension workers can shift their focus to forming second-tier organizations. These larger groups, composed of multiple primary organizations, enhance bargaining power for input procurement and product marketing. Operating at this level requires advanced managerial and entrepreneurial capabilities, which will evolve alongside the organizations themselves. Such structures support long-term sustainability, bolster the reputation of primary organizations and strengthen the position of farmers within value chains.

Additionally, there is a clear role for specialists trained in farm business management, to further train NGOs, private sector enterprises and farmer organizations and cooperatives, to equip them with the skills and competencies to facilitate market linkages. This also involves the development of long-term business relationships with market outlets, strengthening producer–buyer linkages, facilitating linkages with input suppliers, financial institutions and market outlets and helping to negotiate contracts. Links can be achieved both informally by building trust and formally through contracts, as discussed in detail in Chapter 8.

Farm management specialists, consequently, have a role in providing farmers and buyers with technical support in marketing, post-harvest handling and business management. They can also support small family farmers with advice in investment opportunities as a way of ensuring consistency of produce to buyers whilst reducing the risks that they face. However, to support farmers, farm management specialists need to be well informed of the marketing outlets. They should make regular studies of these outlets including issues of access, prices, quantity and quality standards, and packaging requirements. They should also know which of the outlets are reliable and trustworthy. They should know about the transport and handling options and costs relevant to each of the outlets. And they should keep abreast of contract possibilities as they present themselves in the marketplace. The tools and methods to assist them in these functions were described in Part 3 of this book.

16.2.7 Developing human resources

Access to technology and markets alone will not make a farm successful or profitable. At the centre of the enterprise is the farmer. Thus, the development of the farmer is critical to success. One of the key functions in extension is human development. Awakening in farmers the consciousness and fostering the ability to understand and make choices about the operation of their farms is the primary concern of extension. As we stated repeatedly, in the book, extension needs to build the capacity of farmers to manage their farming business, manage the context in which their business operates and manage their life-long self-learning. However, as important as this is, it is often the case that extension workers do not have knowledge and skills in capacity building. This section will provide some insight into this important aspect of extension.

Training is a shared task between farm management specialists and extension workers. To make the best use of their expertise, specialists need to work closely with front-line extension workers, who may have limited formal training in farm management and marketing. This collaboration involves creating clear frameworks for extension activities, designing training and advisory programmes, and supplying both extension workers and farmers with accurate and useful farm management information.

Specifically, farm management specialists have a role to train extension workers, farmers and leaders and members of farmer organizations and cooperatives.

Training extension workers

As noted throughout this book, extension services have traditionally focused on the technical aspects of agriculture, addressing day-to-day production problems faced by farmers. For farm management competencies and skills to be mainstreamed within extension services, concerted capacity building training programmes in farm management with a focus on business need to be designed. These skills are requisites to support market-oriented farming and are also weak amongst private service providers and NGO staff. Staff are often young and lack the practical experience and specialized expertise to provide effective and credible management advice. For farm management extension to be effective, training is essential at all levels; from front-line extension workers (public, private, NGO sectors) to subject matter specialists to policymakers and programme managers. Special guidance is needed for each of these categories of extension staff through tailor-made programmes. This is a key role for the farm management specialist.

In public sector extension services, extension workers and subject matter specialists often have varying levels of academic qualifications, leading to gaps in skills and knowledge. As mentioned in the previous chapter, budget cuts have reduced funding for training, leaving many extension staff undertrained – especially in farm management. Shortages of adequately trained extension staff limit the effectiveness of the advice offered. Capacity and skills in farm business management are also weak among the private and NGO agricultural advisers.

Training farmers

Training farmers in farm management skills and competencies is essential for enhancing their ability to select new farm enterprises and appropriate technologies as well as generally to manage their farm businesses. By providing the necessary knowledge and skills, training programmes can help farmers develop their capacity to define problems and seek solutions independently. This empowerment is particularly crucial for women farmers, who play an active role in many aspects of farming.

Training in farmer organizations and cooperatives

Leadership and members of farmer organizations and cooperatives greatly benefit from business management training. These entities often act as intermediaries between individual farmers and larger market players, and their effectiveness can significantly influence the success of the entire value chain. Training leaders and members of these organizations in business management and marketing can enhance their capacity to manage collective resources, negotiate better prices and provide support services to their members. This, in turn, leads to more robust and resilient farmer organizations at both primary and secondary levels. By fostering strong leadership and management skills within farmer organizations and cooperatives, stakeholders at all levels of the value chain can work more collaboratively and effectively, driving improvements in productivity, profitability and sustainability across the agricultural sector.

It should be recognized that training programmes can most effectively be designed though collaborative endeavours and partnerships with specialized stakeholder institutions to ensure that the content of training programmes remains relevant with the latest agricultural practices, technologies and methodologies. Partnerships with universities, research institutions and the private sector allow for knowledge sharing, access to resources and coordinated capacity-building efforts.

16.2.8 Communicating information

Effective communication is one of the core responsibilities of agricultural extension workers at all levels of an extension organization. More specifically, the role of farm management specialists involves translating complex data and analyses into actionable insights for all stakeholders. These include policymakers, extension workers and farmers. The success of this communication depends on tailoring the content, format and delivery to suit the needs of each audience. Farm management specialists must also ensure that information is timely, accurate and accessible, enabling stakeholders to make informed decisions.

Communicating to policymakers and other decision makers

Communicating farm management information to policymakers and other decision makers must be effective so that the findings of farm management investigations and diagnoses can be easily understood and appropriate actions taken. Farm management specialists should present findings in a concise, actionable and practical manner. Recommendations should be prioritized, and in some cases categorized, into immediate and longer-term actions.

Information should be communicated in forms that suit the preferences and capacities of decision makers. Verbal presentations, such as face-to-face briefings, are particularly effective as they encourage two-way communication and immediate feedback. Written reports, on the other hand, offer greater detail and serve as

Box 16.4. Business management training in the value chain beyond the farmers.

For the purposes of this book, the authors have focused primarily on farmer-level farm management extension. They have also occasionally referred to agribusiness management advice which focuses on business management skills for the overall improvement of agricultural value chains. Beyond farmers, there is a significant demand for business management training among representatives of marketing organizations, processing companies, traders, consumers and input suppliers. These stakeholders play essential roles in ensuring that agricultural products reach the market efficiently and profitably. Those specialized in agribusiness management are well-positioned to provide this support due to their comprehensive understanding of the agricultural sector and its economic dynamics. By extending training to these groups, farm management specialists can help create more cohesive and well-functioning value chains, where each participant is equipped with the skills to optimize their operations and contribute to the sector's growth.

Naturally, there will be some overlap between farm business management and agribusiness management because farmers may also have stakes in other post-farm activities in the value chain such as contracting or initial value adding such as packing. In such cases, it is useful for farm management extension specialists to collaborate with agribusiness management specialists. Also, as smallholder farmers become more market-oriented, they will need additional knowledge and skills to strengthen their engagement with the post-harvest value chain to ensure they are getting a fair deal and, where possible, to capture some of the downstream value-adding profits.

reference materials, while graphical tools like charts, diagrams and photographs enhance clarity and understanding. Farm management specialists should carefully balance these methods, ensuring the credibility and accessibility of the information.

To maximize impact, communication should be timely, aligned with decision-making schedules and linked to ongoing policy or programmatic priorities. Accurate and evidence-backed information increases its credibility, while involving stakeholders in reviewing findings ensures alignment and buy-in for proposed actions.

Communicating to extension workers

Effective communication between farm management specialists and extension workers bridges the gap between research findings and their practical application on farms. This process ensures that extension workers can translate insights into relevant, actionable advice for farmers.

To achieve this, farm management specialists must tailor their communication to the context and challenges faced by extension workers. Clear, simple, and practical recommendations that align with local agricultural conditions and farmer needs are essential. Recommendations should focus on solutions that farmers can feasibly implement within their operational and resource constraints. Additionally, timeliness is crucial; information should align with the agricultural calendar to influence critical decisions effectively, such as planting schedules or market planning.

Farm management specialists should actively support extension workers by fostering collaboration, providing training and offering feedback mechanisms. Empowering extension workers with a solid understanding of farm management principles enhances their ability to effectively engage with farmers, boosting confidence and fostering trust.

Institutional challenges, such as limited resources or bureaucratic delays, can hinder communication. Farm management specialists must work to overcome these barriers by ensuring consistent communication channels, leveraging digital tools where possible, and building strong linkages between research and field operations.

Communicating to farmers

The ultimate goal of farm management communication is to equip farmers with the knowledge needed to make informed decisions. While, ideally, farmers develop the capacity to access and analyse information independently, many smallholders continue to rely on the guidance of extension workers and specialists to navigate complex agricultural challenges.

Effective communication with farmers should be relevant, clear and actionable. Information must directly address farmers' key concerns, such as optimizing production, improving profitability or adapting to market demands. Messaging should be simple and concise, ensuring that recommendations are easily understood and applied.

Extension workers play a crucial role in this process by facilitating knowledge exchange through workshops, field days and study tours, as examples. These activities encourage peer learning, enabling farmers to share experiences, adopt best practices and collaborate on innovative solutions. By fostering a continuous learning environment, extension workers help build farmers' confidence in making strategic decisions.

Timeliness is also essential. Information must align with the agricultural calendar to effectively support key decision making. Whether advising on planting schedules, input selection or market opportunities, timely communication ensures that farmers can act when it matters most.

By tailoring communication strategies to the needs and capacities of farmers, extension workers and policymakers, farm management specialists help bridge knowledge gaps, foster collaboration and promote sustainable agricultural development.

16.3 Areas of Support: Applications and Methods

As we noted previously, farm management specialists have several specialized roles: diagnosis, planning, monitoring and evaluation, supporting the organization of farmers into farmer groups, associations and cooperatives;

assisting smallholder farmers to access finance and markets; supporting partnerships and networking; and conducting training programmes for front line extension workers and farmers. In this section, we will look how they carry out the various tasks of diagnosing and planning, linking farmers to markets, training and communicating.

16.3.1 Support to diagnosis studies and planning

As noted previously, farm management specialists are involved in diagnosing, planning and monitoring the small family farm business. They carry out these functions for all categories of their clients: farmers, extension workers, policymakers and programme managers. To conduct this role effectively, the specialist will need to select and apply the most appropriate tools discussed in previous chapters. Constraints and opportunities analysis, enterprise budgets, gross margins and benchmarking are examples of tools that are used in farm diagnosis and monitoring. Budgeting, programme planning and farm investment appraisal are common tools used for planning. They are described and elaborated on in Chapters 12 and 13.

The effectiveness and, therefore, the choice of farm management tools depends on several factors. First, the technique must be valid, meaning it must be based on reliable data and appropriate assumptions. The application of a tool that uses inaccurate data will not provide useful results. Second, the extension worker and farmer using the technique need the capability to understand and apply it. The application of the tools varies in complexity and, if they require advanced knowledge or resources that are not readily available, they are unlikely to be helpful. Third, the results must be usable, meaning the information must be clear, relevant and available at the right time to support smallholder farmers.

Choosing the right farm management technique often depends on the context. Some techniques, such as linear programming, require high-quality, detailed data, which can be expensive to collect – as noted previously. These might work well for large-scale farms where the benefits of precision outweigh the costs. However, for smallholder farmers, simpler, less data-intensive methods are usually more practical. A sophisticated farm planning model is unlikely to be realistic for a farmer with limited resources or technical training, but a simpler tool like a checklist or a quick evaluation method can provide valuable insights.

Anderson and Hardaker (1979) argued that many of the traditional farm management techniques used for the analysis and planning of larger commercial farms are overly complex and reliant on extensive data collection, often failing to address the practical challenges faced by smallholder farmers. They advocated for simpler methods that require less detailed analysis but still provide valuable insights. One such approach is rapid rural appraisal (RRA), which allows for the efficient collection of essential information without the need for large-scale data sets, helping to identify the main problems farmers face and enabling targeted solutions. This idea of simplifying farm management remains relevant, particularly for smallholder farmers who are unlikely to have access to sophisticated tools or expertise.

Concerns were raised again in the late 1980s by Dorward about the complexity and data demands being applied to smallholder farming (Dorward, 1986). It was observed that some of the traditional management techniques described in the earlier chapters, were too complex and data-intensive for smallholder farmers and extension workers. This complexity hindered the development of practical farm management skills and led to calls for a reorientation of the discipline. Simplifying problems, breaking them into manageable sub-problems and focusing on fewer goals were recommended approaches. Simple decision-making tools such as gross margins, partial budgets or straightforward decision rules were singled out as appropriate and relevant for front-line extension staff to help smallholders address fundamental questions:

1. Should they grow a specific crop?
2. How much should they grow?
3. How should they manage it?

These concerns remain valid today. However, with the expertise of farm management specialists and recent technological advancements – particularly in the use of spreadsheets and the accessibility of farm management software – many of these challenges can be mitigated. As previously discussed, the emergence of digital tools and mobile applications in recent years has significantly transformed farm management practices, making decision making more efficient and data driven. These technologies provide farmers with user-friendly interfaces that offer real-time information, such as weather updates, market prices and financial tracking, without the need for extensive data analysis. For instance, mobile apps like AgriEdge and FarmLogs enable farmers to monitor crop health, manage resources efficiently and make informed decisions, thereby enhancing productivity and profitability. The need for simplification continues to drive innovations in farm management. Today, tools and practices have emerged that align with this thinking, focusing on reducing the data demands and technical barriers for smallholder farmers.

Additionally, participatory rural appraisal (PRA) has evolved to incorporate technology, facilitating farmer involvement in decision making and data collection. This integration allows for the efficient gathering of essential information, helping to identify the main problems farmers face and enabling targeted solutions. Moreover, emerging technologies such as remote sensing and drones offer low-cost, accessible ways for farmers to gather crucial information, making it easier to manage their farms without complex technical knowledge. These developments highlight the ongoing shift towards more accessible, practical farm management solutions that still provide actionable insights for farmers. However, farmers still require the skills and competencies to apply these new innovations to address their farm problems, and this reinforces the role of the farm management adviser.

Use of computer spreadsheets for data management and analysis

As highlighted in Chapter 10, electronic spreadsheets are a valuable tool for managing and analysing farm-level data. Programmes such as Microsoft Excel or Google Sheets offer accessible platforms for calculating budgets, tracking financial data and conducting basic analyses – requiring only a modest level of computer literacy. Their capacity for 'what if' analysis enables users to simulate different scenarios, compare the outcomes of proposed changes, and assess the technical, financial and economic feasibility of investment projects. Tools, such as break-even analysis and sensitivity testing, help investigate potential risks and guide decision making.

While spreadsheets are well-suited for evaluating a limited set of alternatives, they do not optimize outcomes. More advanced analytical tools, such as Linear Programming (discussed in Chapter 13), may be needed for complex decision making and creating comprehensive farm plans. However, even basic spreadsheet models – when thoughtfully designed – can substantially enhance the quality of farm-level decision making. See Box 16.6. for examples of spreadsheet applications.

Although smallholder farmers are increasingly encouraged to use spreadsheets to support farm and business management, many will need guidance – especially in the initial stages. Extension workers, as well as farm and agribusiness management specialists from both public and private sectors, play a key role in helping farmers utilize spreadsheets effectively. These professionals can offer technical assistance in setting up spreadsheets, entering data correctly and applying appropriate functions and formulas for analysis.

Farm and agribusiness specialists can assist in designing customized templates that capture essential data for sound decision making. They help interpret financial indicators, analyse production trends and assess the profitability of farming practices – transforming raw data into meaningful insights.

Beyond technical assistance, specialists can provide training programmes and capacity-building initiatives to enhance the ability of farmers to use spreadsheets independently. By introducing farmers to spreadsheet functions, data management techniques and analytical tools, they help build the confidence and skills necessary to adopt spreadsheets as everyday management tools. Regular feedback and review of farmers' spreadsheets further reinforces learning and ensures the accuracy and utility of the data being collected and analysed.

This continuous support – delivered through collaboration between farm and agribusiness management specialists, service

Box 16.5. Simplified applications.

1. **Simple, practical approaches:** organizations like the University of Minnesota Extension have developed tools such as partial budgets that help farmers make financial decisions without extensive data collection. These tools enable farmers to evaluate trade-offs, such as whether switching crops or investing in new equipment will yield benefits.
2. **Rapid information gathering:** methods like Rapid Rural Appraisal (RRA) are still widely used. They rely on qualitative assessments and minimal data collection to identify key constraints and opportunities for farmers.
3. **Technology and software:** advances in technology have introduced user-friendly farm management platforms like Krisiyukta, which cater specifically to smallholder farmers. These tools provide simple interfaces and decision-making aids, allowing farmers to manage their operations more effectively with minimal technical knowledge.
4. **Accessible educational resources:** institutions like Colorado State University Extension have created decision tools using Microsoft Excel. These allow farmers to input basic data and receive actionable insights, making them highly suitable for smallholders with limited literacy or resources.

providers and farmers – has the potential to significantly improve the productivity, profitability and resilience of smallholder farm operations.

Use of AI in farm and agribusiness management

As noted in Chapter 10, the integration of artificial intelligence (AI) into farm management represents a transformative advancement, particularly for agricultural advisors and farm management specialists. AI equips advisors with powerful tools to deliver customized recommendations and precise, data-driven insights, enhancing their ability to support small family farmers in critical decision-making processes. By employing AI-based analytics, farm management specialists, in particular, have a notable role. The technology has the potential to assess productivity more accurately, predict market and environmental fluctuations and develop strategies to improve resource efficiency and farm profitability. An example of the use of AI in farm management is given in Box 16.7. However, to make AI more readily accessible to farm and agribusiness management specialists, they must be trained to use AI technologies more effectively to interpret complex data outputs and translate these into practical guidance for farmers. Additionally, it is important to develop localized, farmer-centric AI applications that align closely with the specific needs and realities of smallholder agriculture. Creating partnerships between private sector technology providers, agricultural extension services and farm management specialists can further accelerate the adoption and effective use of AI tools. With these measures, widespread adoption of AI-supported farm management can drive innovation, increase resilience and enhance overall agricultural development within farming communities.

16.3.2 Designing extension programmes

One of the important contributions farm management specialists can make to address the challenges faced by farming communities, is to support extension workers prioritize and design extension programmes. This is most effectively done when advisors work as a part of a team comprising extension staff, subject matter specialists and other stakeholders. In making their contribution, farm management specialists need to employ a systematic, data-driven approach to identify priority areas for intervention, optimize resource use and deliver solutions tailored to the specific needs of farmers.

Participatory and inclusive planning

While extension planning is primarily the responsibility of the extension service, one of the essential elements of designing extension programmes is ensuring that the planning process is participatory and inclusive. As stated

Box 16.6. Spreadsheet applications by farm management specialists.

1. **Farm-level profitability analysis:** spreadsheets enable farm management specialists to organize and analyse financial data to guide farmers in improving profitability. By creating detailed income and expense sheets, specialists can help farmers track input costs, labour expenses, equipment use, and revenues from crop and livestock production. This allows calculation of key financial indicators such as profit margins, return on investment and break-even points. Comparing different farming practices or crop options within spreadsheets makes it easier to identify the most profitable strategies.
2. **Farm planning and group modelling:** for planning purposes, spreadsheets serve as flexible tools for modelling different farming systems. Specialists can develop comprehensive plans that include crop rotation schedules, labour management strategies and resource allocation. When working with groups of farmers, these models can be aggregated to develop a collective strategy, revealing synergies and opportunities for scaling up. Spreadsheets also allow easy updating as conditions evolve, helping to maintain relevance and improve long-term planning.
3. **Policy advisory support:** farm management specialists with spreadsheet expertise also contribute to policy formulation by synthesizing agricultural data into meaningful models. Using data on yields, costs and farm structures, they can develop projections that inform decisions on subsidies, land use regulations or sustainability incentives. Cost-benefit analyses can be carried out to evaluate the implications of policy alternatives, helping policymakers understand their economic and environmental trade-offs. This data-driven approach enhances the development of practical, evidence-based agricultural policies.

Box 16.7. Integrating AI in agricultural extension services for smallholder farmers in Nigeria. From Ibrahim, 2023.

In Nigeria, integrating AI into agricultural extension services has shown promise in enhancing smallholder farming practices. Through a comprehensive literature review, data analysis and case studies, the current state of AI applications in agriculture were assessed and the gaps and challenges highlighted. The study reveals that AI technologies offer innovative solutions for knowledge dissemination, decision support and resource management in agriculture, contributing to increased yields, reduced environmental impact and improved livelihoods for farmers. Extension agents trained to use AI tools, delivered insights on optimal planting times, pest and disease management and efficient resource use during field visits and via mobile platforms. However, challenges related to data quality, accessibility, and ethical and policy considerations needed to be addressed. The study demonstrates the transformative potential of AI in improving farm management for smallholder farmers in low-income countries. While infrastructural and educational challenges remain, the project highlights the importance of capacity building and ongoing support to fully leverage AI's benefits in agricultural extension services.

above, extension programmes must be developed collaboratively through teams that also include farmers. Farmers are key stakeholders in this process, and their active participation ensures that the programme aligns with their needs and aspirations.

All the members of the planning team, including the farm management specialists, must ensure that the programmes are inclusive and engage women, youths and marginalized communities, where relevant. Inclusivity fosters trust and strengthens the likelihood of widespread adoption of new ideas and practices. Additionally, local leaders and development partners should be involved in the planning process to support implementation and promote long-term success.

The planning process

PRE-PLANNING. As with farm planning, extension programme design starts with the farmer. It begins by identifying the target audience. Groupings of farmers should be based on common characteristics of farmers and of production potential. One approach to identifying farmers is set out in Table 16.1. It describes key factors (measured on a continuum) framing the characteristics of farmers including motivation and aspiration, innovation, land, primary purpose, self-reliance, production paradigm and technology paradigm. Understanding the farmers at this level positions extension to determine its engagement strategy. It helps avoid assumptions about the farmers.

Effective extension programme design begins with understanding the farmers it aims to support. By identifying and grouping farmers based on key characteristics, extension services can develop targeted strategies that align with their needs, resources and aspirations.

The first step is to engage with farmers through discussions, observations and meetings to understand their motivations, production systems and resource use. Farmers are assessed using key characteristics, such as their level of market engagement, production methods and risk tolerance. For example, some farmers may be primarily subsistence-focused, while others actively seek new technologies or market opportunities. At this stage, farmers are not yet grouped, but their individual characteristics are documented along a continuum rather than fitting into rigid categories.

The second step is to analyse the findings and create groupings by common patterns. Once data is gathered, the third step is to identify patterns and form logical groupings based on shared characteristics. Instead of forcing farmers into predefined categories, the analysis determines natural groupings based on how they align across multiple characteristics. For example, if a group of farmers shares a strong interest in market-oriented production but has limited access to credit, they may form a distinct category requiring specific extension support.

Finally, once farmer groupings have been established based on shared characteristics, the fourth step is to assemble the planning team, which may include extension workers, farm management specialists and researchers. Their role is to translate these insights into practical extension strategies that respond to the distinct needs of each group.

For example, if one group consists of subsistence-oriented farmers with limited access to markets, the extension approach may focus on improving productivity and household food security. If another group is market-driven but constrained by access to finance, the focus may shift to enterprise development, financial planning and market linkages. The planning team ensures that extension messages, training methods and support services are designed to address the specific realities of different farmer groups, making interventions more relevant and effective.

This farmer identification process should be repeated regularly, such as bi-annually, and updated using insights from extension worker reports. The results serve as a foundation for extension programming, ensuring that interventions align with farmer realities. Farm management specialists use this understanding to identify key enterprises, assess potential financial impacts and set advisory priorities. While precise financial calculations may not always be needed, broad profitability estimates help guide decisions on which enterprises or practices to prioritize. These findings are then taken into planning sessions with farmers and extension workers.

EXTENSION PROGRAMMES. Typically, extension programmes will cover three extension imperatives: (i) long-term farmer capacity-building programmes; (ii) policy-driven and urgent response programmes; and (iii) innovation-focused programmes. Each planning process starts with the step: team formation and alignment. The aim of this step is to assemble a team of stakeholders with technical and local knowledge. This is critical to ensure establishing a participatory, collaborative process. In this first stage, shared goals are defined and developed, and roles are clearly delineated to ensure balanced participation.

In each of the following planning processes, the support of the farm management specialist is essential to address the economic aspects of the extension programme design.

PLANNING LONG-TERM FARMER CAPACITY-BUILDING PROGRAMMES. This aspect of the planning process is for developing extension priorities

Table 16.1. Factors framing the characteristics of farmers.

Factor	Nature of continuum	Key question framing the continuum	Typical questions to answer
Motivation and aspiration	Status quo – intentional change	Is the farmer's aim to maintain their status quo or are they looking to change things?	Does the farmer prefer to experiment with different technologies or to be given advice about how to produce something? What changes have they made to their farm which they decided to do on their own, not because they were advised to do so?
Innovation	Farmer as an innovator – Farmer as an end-user	Does the farmer see themselves as an innovator or end-user of technology?	Have they ever developed a new way of producing crop/livestock? Give an example Where did they learn to farm the way they do?
Land	As a function of technology used in the context of economic viability	How economically viable is the current land (e.g. fertility, size, condition) given the current technology mix?	What is the size of the land they currently farm? What is the condition of the land? What is it suitable for? What forms of technology do they use? (e.g. tractors, hand weeding, mechanical harvesting, etc.) (Use this information to determine the gross margins for their enterprises; and then rework them using different technologies.)
Primary purpose	Household food consumption – market-orientated/ profit driven	Is the farmer primarily producing food for their home or producing for profit?	Why do they produce what they produce? What is their main reason for farming? How much of what they produce do they consume/use at home? Why?
Self-reliance	Dependent on external support – self-reliant	Is the farmer dependent on external support or are they self-reliant?	How do they decide what, when and how much to produce? How do they decide on the technology they uses?
Production paradigm	Low external input (including organic) – high external input	Does the farmer use inputs from their or others' farm (e.g. organic) or do they purchase inputs?	Why do they produce the way they do? Why do they use the production system they currently use? What do they use for fertilizer/ pesticides/herbicides animal medicines? Why? Where do they get it?
Technology paradigm	Labour intensive /'traditional' technologies – capital intensive mechanization	Is the farmer more inclined towards labour intensive/ more traditional technologies or towards mechanization technologies?	For what farming operations do they use labour? Why? For what farming operations do they use machinery and equipment? Why? What tools do they use for which farming operations?

over a longer timeframe, such as a year or more. It places farmers at the centre, ensuring that their aspirations, challenges and capacities directly shape the programme. The aim is to enhance the capacity of farmers to manage their farm enterprises and overall farming systems, navigate the broader contexts in which they operate, and cultivate self-directed, lifelong learning. After the planning team has been established, the planning process typically follows the following steps.

1. **Engagement and dialogue:** participatory tools such as focus groups, community meetings and farmer-led diagnostic workshops are used to identify the most pressing needs and opportunities. A key element of this process is facilitating open discussions that allow farmers to articulate their goals and the barriers they face. Farm management specialists, extension workers and other subject matter specialists also contribute insights to provide a broader perspective on needs and opportunities.
2. **Priority setting:** farmers and extension workers collaborate to prioritize issues based on their significance, potential for improvement and alignment with community values. This process may also incorporate input from non-farmer stakeholders, who can offer additional insights and a wider perspective.
3. **Programme design:** detailed action plans are developed through a co-creation process, outlining specific interventions, timelines, roles and responsibilities. Farmers provide input in the development of budgets, resource allocation and monitoring mechanisms, ensuring that the programme remains relevant and feasible.
4. **Monitoring, evaluation and learning:** systems are established for regular feedback from farmers and other stakeholders, allowing for continuous adjustments and refinements as the programme progresses.
5. **Iterative implementation:** flexibility is built into the programme to allow for real-time adjustments based on feedback from farmers and other stakeholders. As new challenges and opportunities arise, the programme can be adapted to remain effective and responsive to the evolving needs of the farming community.

By following this structured approach, farmer capacity-building programmes can remain dynamic, responsive and aligned with farmers' evolving needs. A well-planned programme not only strengthens individual farm enterprises but also contributes to broader agricultural development by fostering adaptability, resilience and long-term self-reliance among farming communities.

POLICY-DRIVEN, RESPONSE PLANNING. This process responds to critical or time-sensitive issues, such as climate change adaptation, pest outbreaks or new regulations. It retains participatory elements but balances them with the need for urgency. After the planning team has been established, the planning process typically follows the following steps.

1. **Stakeholder consultation:** farmers, policymakers, researchers and extension workers participate in structured consultations to gather diverse perspectives on the policy or issue. Methods such as focus groups, surveys and rapid assessments, help streamline input collection. The process ensures that minority and outlier voices are actively considered alongside majority opinions, capturing a broad spectrum of experiences and needs.
2. **Presenting and tailoring interventions:** while policy objectives or external drivers (e.g. climate action goals) provide the overall framework, farmers contribute to shaping interventions to fit local conditions. Their involvement helps ensure that proposed solutions are practical and feasible, with considerations for adoption and adaptation requirements.
3. **Strategic programme design:** targeted action plans are developed to focus on high-impact, implementable interventions. These may include deploying climate-smart practices, disseminating critical information or ensuring access to necessary resources. The aim is to align policy implementation with realistic, time-sensitive extension efforts.

4. **Pilots and scalability:** where feasible, pilot programmes are established in key areas to test and refine approaches before broader implementation. Effective record-keeping and documentation are essential for tracking results, refining strategies and facilitating the wider dissemination of successful interventions.
5. **Adaptive rollout and continuous learning:** as the programme is implemented, mechanisms are established for tracking progress, gathering feedback and refining strategies. Lessons from early implementation phases inform necessary adjustments, ensuring that interventions remain relevant and effective. Ongoing engagement with farmers and other stakeholders helps to identify emerging challenges or unintended consequences, allowing for timely course corrections and improved scalability.

This process balances urgency with participation, ensuring that solutions are responsive to both immediate needs and local contexts. A well-executed response plan strengthens resilience, enabling farming communities to navigate challenges while remaining actively engaged in shaping sustainable solutions.

PLANNING FOR INNOVATION-FOCUSED EXTENSION PROGRAMMES. This process acknowledges the important contribution of researchers and research institutions towards developing new technologies and other innovations. At the same time, it embraces and gives exposure to local innovations. This process involves a multidisciplinary team of extension workers, farm business management specialists, researchers and smallholder farmers in co-developing a 1–2 year extension plan to introduce and explore innovations in production and business management. It is outlined here as follows:

1. **Innovation scoping:** the process begins with scouting innovations from researchers, agribusinesses and farmer-led practices. These innovations are evaluated for relevance, feasibility and impact. Existing data and information from other extension processes may also be utilized. Ultimately, the aim is to establish an inventory of plausible innovations that can be shared with farmers.
2. **Programme design:** clear objectives are established, and innovations are prioritized for trials. Plans are developed for on-farm and off-farm trials in production and business areas. Capacity-building activities are designed to equip farmers with technical and managerial skills. Resources are allocated and funding or partnerships are identified to support the programme.
3. **Knowledge exchange and dissemination:** team-led demonstrations and field days are organized to share trial results. Platforms for business networking and innovation sharing are created. Accessible communication channels, such as newsletters, radio or social media, are utilized to disseminate information widely.
4. **Monitoring, evaluation and learning:** participatory monitoring frameworks with measurable indicators are developed. Regular feedback loops are established to facilitate adjustments and learning. The process remains flexible to adapt based on trial outcomes and stakeholder feedback.
5. **Scaling and sustainability:** pathways for scaling successful innovations are identified. Farmer champions are supported to lead peer-to-peer dissemination efforts. Innovations are embedded in existing structures or linked to ongoing extension support to ensure long-term sustainability.

Box 16.8. lists some of the factors to consider when selecting technologies for adoption and scaling and Table 16.2 provides an example of how these factors can be applied methodically and systematically.

16.3.3 Accessing markets and finance

Farm management specialists can help farmers secure market access and financial resources, both of which are essential for sustainable farm businesses. By facilitating connections with markets, they enable farmers to identify opportunities, strengthen value chain participation and negotiate fairer terms. Similarly, access to finance remains a significant challenge for smallholder farmers. Specialists help bridge this gap by supporting financial literacy, advocating

Box 16.8. Factors to consider when identifying and selecting innovations to disseminate.

1. **Identify limiting factors** that impact the profitability of an enterprise, such as crops or livestock. This involves listing all the factors affecting profitability and focusing on those with significant potential for improvement while discarding less relevant ones. By narrowing the scope, farm management specialists can prioritize the most impactful areas for intervention.
2. **Estimate financial gains** by comparing the performance of average farmers to a practical target level already achieved by above-average farmers in the same area. The financial potential for improvement is calculated based on this difference, providing a clear indication of the possible benefits of adopting better practices.
3. **Determine the scope of impact** by assessing the percentage of producers who are not achieving the desired practice. Since advisory services are directed towards farmers rather than specific land areas or livestock populations, understanding the extent of the target audience helps refine the focus of the programme.
4. **Identify barriers to adoption** to ensure successful implementation. This involves analysing the reasons behind farmers' underperformance, such as lack of education, insufficient capital, traditional practices or non-profit motives. Understanding these barriers allows specialists to design strategies that address these challenges effectively.
5. **Evaluate advisory effectiveness** by estimating the likely impact of extension efforts on each identified factor. This is done using an effectiveness index ranging from 0.1 to 1, taking into account past results, the influence of other advisory services, and existing barriers to delivery. This evaluation ensures that extension efforts are targeted and realistic.
6. **Setting priorities** based on this evaluation, by combining the estimated financial gains with the advisory effectiveness index to calculate weighted financial returns. This process identifies areas with the highest potential returns, enabling the programme to focus on interventions that will have the greatest impact.
7. **Consider intangible benefits** such as food safety, environmental sustainability and community development. Although these effects are often difficult to quantify, they play a vital role in the overall success and sustainability of the programme.
8. **Identify subject areas for extension**, selecting technologies or practices that are easy to implement and align with farmers' existing systems. For more complex changes that may require longer timelines, strategies should be developed to provide ongoing support and resources to farmers.
9. **Engage farmers** in the planning process by discussing proposed actions to ensure they align with farmers' needs and priorities. By clarifying the reasons for each action and organizing priorities based on short-, medium- and long-term benefits, the programme becomes more relevant and practical for the farming community.

for inclusive credit policies and linking farmers with financial institutions. Strengthening both market access and financial inclusion empowers farmers to invest in their enterprises, increase profitability and build resilience.

Access to markets

As mentioned frequently in this chapter and in Part 3, the farm management specialist has a key role to play in facilitating connections with markets. The first step is often to assess the existing market conditions and identify potential opportunities that align with the farmers' products. The specialist can analyse local and regional markets for demand, pricing and distribution channels. By conducting market research, they can help farmers understand what products are in high demand, the best times to sell and the price fluctuations they may encounter. In many cases, the specialist may focus on value addition – such as processing raw agricultural products or creating niche products – that can enhance the marketability of the farmers' goods.

Similarly, here, there is a definitive role for front-line extension workers to support farmers to link to markets. The farm management specialist needs to work with this cadre of

Table 16.2. Application of factors set out in Box 16.8.

1	2	3	4	5	6	7	8	9	10
Factors limiting profitability	Desirable practices	Method of assessing financial benefit	Estimated financial gain	Percentage of farmers affected by problem	Barriers to desirable practices	Project effectiveness of extension (range 0.1–1.0)	Weighted financial gain (col 4 and 7)	Intangible benefits and costs	Areas of advisory activity
Soil type suitability	Cultivate the crop on better soils and good cultivations	The gross margin from utilizing better soils is high at $20/ha. It has been estimated that there is 100 ha of this crop on poor soils	$2000	5%	1. Limited cash flow to shift production 2. How does the change fit in with other crops?	0.3	$600	Benefits: better and more efficient application of inputs Costs: additional costs of cultivations	Activities aimed at showing importance of: 1. Supplying organic matter 2. Good timing of cultivations 3. Improved drainage
Quality of seed	1. Use of high-quality seed 2. Drought resistant	Yield increased by 10%	$6	10%	Poor storage	0.7	$4.2	Costs: cost of seed is higher than current practice	1. Activities aimed at seed multiplication on farms where suitable 2. Earlier delivery of seed 3. Closer liaison between growers and seed producers
Weed control	1. Weed free seed bed 2. Spray with herbicide or inter-row cultivation	7% of cultivated area of the crop is affected by showing a potential increase of 5%	$450	20%	1. Cost 2. Tradition 3. Lack of knowledge of value of herbicides 4. Abuse of herbicides	0.5	$225	Costs: additional application of labour	1. Inform farmers of value of herbicides 2. Importance of good soil structures to get best results 3. Timing of herbicides 4. Serious effect on yield of weed competition for moisture, nutrients, etc.

extension workers to facilitate this process. In this respect they have a responsibility to train extension workers on market access strategies, enabling them to guide farmers in negotiating with buyers, understanding pricing strategies and organizing collective marketing efforts. Extension workers can also help farmers with post-harvest handling, storage techniques and packaging, which can significantly increase the shelf life and quality of the products, making them more appealing to potential buyers.

Farm and agribusiness management specialists also play a vital role independently and often together to strengthen the position of farmers within modern value chains. They are expected as part of their duties to facilitate connections between farmers and buyers by identifying potential trading partners and arranging interactions that lead to formal market linkages. By enhancing the farm's integration into the value chain, farm management specialists ensure that farmers can negotiate better terms, secure consistent markets and participate more equitably in the distribution of value. In doing this, collaboration with other stakeholders is vital. In many cases, the specialist should work with local government bodies, agricultural cooperatives, NGOs and private sector actors to build a network of market linkages. This could include introducing farmers to buyers, retailers or even export companies who are looking for specific crops. Additionally, the specialist can help connect farmers to digital platforms that allow them to sell directly to consumers or other market actors, expanding their reach beyond local markets. Building infrastructure for transportation, storage and communication is also crucial, and often requires collaboration with local governments or NGOs that can facilitate the necessary investments.

One of the key technical elements in this process is for farm management specialists to guide farmers in understanding and negotiating contract terms that ensure fair prices and secure supply agreements. Specialists may need to train farmers on how to read and interpret contracts, as well as how to negotiate terms that protect their interests, ensuring that the contract is clear and mutually beneficial. In some cases, the specialist might also help draft sample contracts or connect farmers with legal advisors who can offer support in formalizing agreements. The specialist may also need to train farmers in record-keeping, cost analysis and risk management. The latter is particularly important since not all risks can be mitigated through contracts. Farmers and buyers must understand and address shared risks to maintain stable market relationships.

In addition to direct negotiations, market linkage agreements often involve establishing relationships with key intermediaries such as aggregators, processors or cooperatives. The specialist can facilitate these relationships by explaining how these intermediaries operate and how smallholder farmers can work with them to ensure consistent access to markets. The specialist would need to guide farmers in establishing agreements with these intermediaries that ensure fair treatment, including transparent pricing, payment schedules and dispute resolution mechanisms.

Another crucial technical aspect is negotiation. A farm business management specialist would assist farmers in developing negotiation skills to ensure they can secure the best possible terms with buyers, wholesalers or retailers. These negotiations may involve not only price but also payment terms, delivery schedules and any quality guarantees that can impact the final deal. For example, the specialist could help farmers understand how to negotiate for better payment terms or discuss volume discounts or incentives for consistent supply.

Lastly, quality standards and certifications are another technical aspect that cannot be overlooked. When entering formal markets, especially for export or higher-value markets, farmers must meet specific product quality standards and certifications (e.g. organic, fair trade). The farm management specialist can work with farmers to ensure they understand these requirements, including the production processes, post-harvest handling and necessary documentation for certification. This can involve practical training in quality control, sorting and grading of produce to meet the required standards. Such knowledge is key to ensuring that small family farmers can successfully enter and remain competitive in higher-value markets.

Access to finance

As we are aware, smallholder farmers need access to finance to grow their businesses – a common challenge. Limited financial options prevent many from purchasing inputs or investing in new enterprises, restricting their ability to seize opportunities and develop entrepreneurial behaviour. Women, subsistence farmers and youths face even greater difficulties in accessing capital. While informal credit sources are often relied upon, market-oriented farmers increasingly require access to formal credit options. Special attention needs to be given by extension workers to addressing the financial access barriers faced by women, who often encounter greater challenges due to societal norms, limited mobility, illiteracy and a lack of collateral, such as land ownership.

Farm management specialists can play a vital role in helping these men and women farmers to navigate and secure financial resources. They can support women by advocating for more flexible collateral requirements and ensuring that credit policies accommodate their needs. For instance, alternatives to land as collateral can be examined and practices like requiring a husband's signature for loans should be dropped. Specialists with extension workers can also help women improve their financial literacy and numeracy skills, which may be prerequisites for accessing loans and participating in entrepreneurial training. By facilitating linkages to financial institutions and negotiating fair terms, farm management specialists can help remove traditional barriers and empower farmers, especially women, to access the resources they need to grow their businesses.

16.3.4 Facilitating partnerships and networking

Farm management specialists play a pivotal role in fostering partnerships and networks that broaden the support needed for small family farmers. By creating and supporting connections among stakeholders such as farmers, extension workers, service providers, training institutions, research organizations and value chain actors, they ensure collaboration, knowledge sharing and coordinated efforts that improve the impact and sustainability of the range of advisory services offered.

To build and maintain effective networks, farm management specialists can actively organize regular communications among stakeholders, creating platforms for agricultural advisors to exchange experiences, learn about best practices and stay informed about new innovations. This may involve hosting meetings, virtual forums and workshops, developing directories of stakeholders for efficient communication and forming advisory committees with representatives from various sectors to facilitate dialogue and cooperation (see Box 16.10.).

Strengthening linkages among stakeholders is another critical aspect of their work. Specialists connect farmers with institutions and organizations that can transfer knowledge and build capacity. These efforts include coordinating joint training sessions with research institutions, linking farmers to input suppliers for access to better technologies and facilitating connections with financial institutions to improve access to credit.

Collaborating with research institutions

Collaborating with research institutions, such as university agricultural economics departments, is a key strategy for farm management specialists. By leveraging academic expertise and research capabilities, they can integrate cutting-edge knowledge into advisory services. This collaboration includes organizing farm surveys to collect profitability data, conducting comparative farm performance analyses to establish benchmarks (as discussed in Chapter 14), and using these insights to support farmers and extension workers in making informed decisions and identifying areas for improvement (see Box 16.9.).

Sustaining coordination and collaboration

To sustain coordination and collaboration, farm management specialists can employ various strategies, including encouraging frequent communication among partners, organizing workshops and field trips and facilitating partnerships or joint ventures along the value chain. Formal and informal learning platforms, such as business clubs and associations, foster continuous learning and peer-to-peer knowledge exchange.

16.3.5 Supporting the development of human capital – training and mentoring

As noted previously, farm management specialists are involved in the training of farmers,

Box 16.9. Farm business management support.

- Encourage frequent communication among potential and existing partners to stay abreast of the partners' changing views regarding objectives and expected benefits.
- Encourage frequent meetings and information exchanges including inter-group exchange visits, meetings, workshops, joint training and local radio.
- Organize special workshops and field trips to maintain interests and collaboration.
- Encourage the formation of partnerships or joint ventures in initiatives such as sharing machinery and farm investments including the following: joint ventures which can be set up between farmers, cooperatives and agribusiness companies anywhere along the value chain including the input, production, grading and processing, and marketing stages.

Box 16.10. Partnerships with university agricultural economics departments.

Through farm surveys aimed at calculating the profitability of crop and livestock enterprises, data can be collected to conduct comparative farm performance analyses. This methodology involves gathering performance data from a large group of farms and selecting the most profitable one-third as benchmarks. These benchmarks represent the average performance measures of the most successful farms, providing valuable insights into optimal farm management practices and economic viability.

The comparative farm profitability analyses generated through this partnership serve as valuable tools for farmers, extension workers and policymakers alike. Farmers gain insights into the profitability of their enterprises compared to industry standards, enabling them to identify areas for improvement and make informed management decisions. Extension workers can tailor their advisory services based on the benchmark data, offering targeted support to enhance farm profitability.

extension workers and farmer organizations and cooperatives. This section will discuss how this can be done in practice.

Training extension workers

Farm management advice differs from general agricultural extension by focusing on helping farmers analyse, interpret and define their managerial actions. Instead of offering ready-made solutions, farm management specialists in business guide farmers in decision-making processes. This requires close coordination between specialists and front-line extension workers, who often lack formal training in farm management and marketing. Specialized training is essential to equip extension workers with the skills and knowledge necessary to provide effective support and guidance to farmers. Examples of training curricula in farm business management for advisory service providers – public or private – are given in the Appendix.

Farm management training must be an explicit and integral component of capacity-building efforts within these agricultural extension services at all levels. Frontline extension workers, subject matter specialists and extension directors all require a solid foundation in farm management to effectively support farmers in making informed business decisions, optimizing resource use and improving farm profitability.

To ensure sustained impact, farm management training should be embedded within a country's extension training programme through a structured, multi-stage approach. This includes pre-service training for new recruits, induction training for newly appointed extension staff, in-service training to continuously enhance skills and remedial training to address specific gaps in knowledge and practice (see Box 16.11.). By systematically incorporating farm management into extension capacity-building programmes, extension services can strengthen their advisory role and enhance the effectiveness of farm planning and farmer decision making.

Training farmers

As discussed, farm management specialists are crucial in fostering business management and entrepreneurial learning among farmers.

Their role involves motivating farmers, providing ongoing encouragement and creating an environment conducive to learning. Advisory workers must often adapt their approach, shifting from simply offering advice to actively training farmers when critical skills or knowledge gaps are identified. Facilitating learning processes and raising awareness about the importance of entrepreneurial and management skills are integral to this work.

Training in farm management is fundamental to empowering farmers to select new enterprises, adopt appropriate technologies and manage their farm businesses effectively. Well-designed training programmes equip farmers with the skills to independently identify problems and seek solutions. This is especially critical for women farmers, who play a significant role in agriculture. Initiatives such as the farm business school (FBS) and participatory farm management have proven successful in building farmers' skills and competencies.

Efforts to adapt farm management tools for illiterate or semi-literate farmers highlight the importance of accessibility in training programs. For instance, initiatives like those by the University of Reading, UK, have developed tools that use symbols instead of numbers and words, making them more accessible to farmers with limited literacy skills. These initiatives underscore the need for practical, responsive and inclusive training programmes that address the unique challenges faced by smallholder farmers. By focusing on these aspects, farm management specialists can support more effective and sustainable farm management practices.

One method of fostering learning among farmers is through participatory approaches that involve them directly in the management process. Participatory farm management offers an alternative to conventional methods, making decision-making tools more accessible, particularly for smallholder farmers in unpredictable environments. By engaging in participatory processes, farmers develop problem-solving skills and gain practical experience in managing farm enterprises under real conditions (see Box 16.12. and Fig. 16.4.).

There are various methods for training farmers.

STUDY OR LEARNING GROUPS. Study groups emphasize farmer-to-farmer learning, and enable members to glean insights from successful peers, whether within or beyond the group. Collaboratively, they visit each other's farms to exchange insights on new methods or technologies and may also engage extension staff for specialized knowledge sessions. Learning groups and networks are integral to fostering successful and profitable farm businesses, comprising individuals closely associated with the farmer's operations. Whether meeting formally or communicating informally, these groups convene to deliberate on challenges and opportunities. While encouragement is beneficial, such groups ideally evolve organically among farmer-entrepreneurs, suppliers and stakeholders. They serve as platforms for sharing experiences and addressing emerging issues, aligning well with the learning-by-doing approach.

STUDY TOURS. Farmers observe 'good entrepreneurial practices' and learn from the practices of others and are often taken on visits to markets, input suppliers and buyers to make contacts and develop networks. Farmers also visit farmers in other areas to exchange experiences and see new techniques and innovations. Such visits can transform a farmer's view on production and marketing. These visits, however, should be followed up with periodic evaluations to assess their impact and the ultimate application of new practices and techniques adopted.

TALKS AND SEMINARS. Presentations from farm and agribusiness practitioners can be a valuable addition to both classroom-based and field-based training, enriching the learning experience for farmers. Incorporating guest sessions led by experienced practitioners allow farmers to gain practical insights, real-world examples and industry perspectives that reinforce training content. These visits can cover a range of relevant topics, including entrepreneurship, successful businesses initiatives, new value adding opportunities and new technologies and practices. By engaging directly with professionals who have hands-on experience, farmers can better understand business strategies, market dynamics and innovative solutions, making the

Box 16.11. Categories of extension service training in farm management.

Pre-service training: professional training received before extension workers are appointed to the extension service. As there are usually few professionals at undergraduate or diploma levels, it is sometimes necessary to employ graduates of secondary vocational schools. Such graduates will not have the breadth or depth of training obtainable in college or university. As it is essential that front-line extension workers are better trained than farmers they work with, what may be lacking in pre-service academic training must be made up to the extent possible in induction and in-service training.

Induction or oriented training: providing a general understanding of the organization and the role of extension workers within it. All extension workers need special training to prepare them for their specific jobs and orient them towards what farm management extension is: its principles, objectives and methods of teaching and extension support. Induction training should also develop an attitude of personal dedication to the service of farmers. This kind of training supplements whatever preservice training the new personnel may have had and gives it a context for the extension worker's new job.

In-service training: on-the-job training and scheduled training while away from the job. On-the-job training can be ad hoc or structured with planned learning outcomes scheduled around the work of the trainees. Farm management specialists are well positioned to offer such training. (Trained extension workers can conduct similar training for farmers.) There are different forms of in-service training, including refresher courses, remedial training, mini-training programmes, practical workshops, study tours, seminars and field trips.

Training for promotion: pathway for advancement. After serving for a period in the extension service, many extension workers feel the need for additional academic training to improve performance in their present positions or to qualify for promotion. Those extension workers aspiring to become subject matter specialists could take courses leading to graduate degrees in farm management or marketing.

Note on 'remedial training': remedial training is aimed at correcting or improving knowledge and skills in a particular area – in this case farm management and marketing. Such programmes assume that extension workers on the programme would have been exposed to farm management as a discipline during their formal education. Remedial training is intended to complement their technical skills with training in farm management. Remedial training could be conducted as induction or in-service training programmes.

Box 16.12. The case for participatory farm management methods for smallholder farmers. From Dorward *et al.*, 2007.

The following four main limitations of conventional farm management methods were identified as ones that participatory methods could address.

1. Conventional farm management methods generally work on the premise that profit maximization or increasing worth are the main objectives of users. In reality, small-scale farmers operating in harsh and unpredictable environments frequently have other objectives such as to reduce risk through improving food security. Resources other than cash are also important in decision making.
2. Conventional farm management methods focus on the final output (e.g. profit at the end of the production period) and do not take into account changes with time during a production period or season. Farmers often take important decisions during a season, depending on conditions at that particular point (e.g. when it is evident how good the rains have been, how much labour they have or how healthy a crop looks).
3. Conventional farm methods are relatively complex and difficult to use, particularly for non- or semi-literate farmers. They are, therefore, not easily used by the majority of the world's farmers, either on their own or with advisers.
4. Conventional methods often require a limited amount of equipment to use, e.g. personal calculators or even computers. Even the use of pen and paper can be inappropriate with non- or semi-literate farmers.

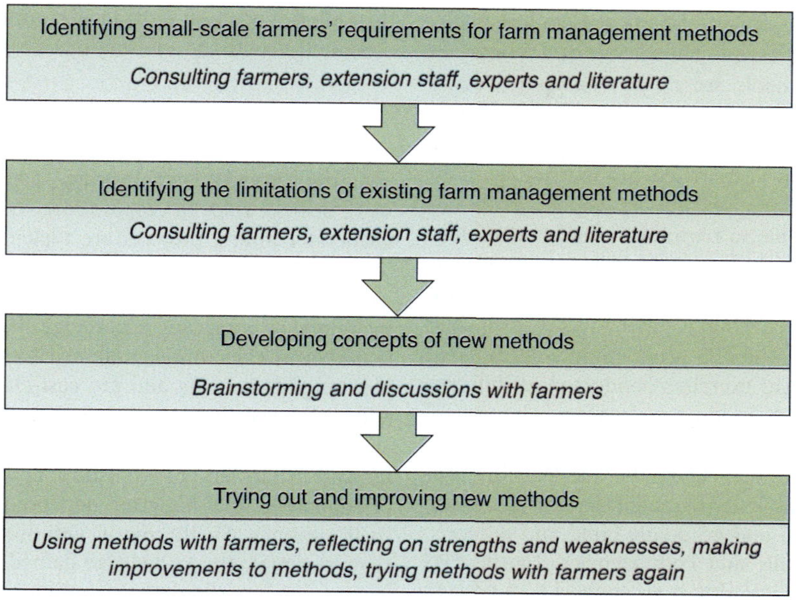

Fig. 16.4. Sequence of activities in the development of participatory farm management methods.

training more practical and applicable to their own farming operations.

WRITTEN INFORMATION. Fact sheets can be prepared and distributed to farmers. In addition to providing specific technical and marketing information, they can also identify potential stakeholder partners who can provide advice and support.

SUCCESS CASE REPLICATION. Success case replication serves as a strategy for disseminating sustainable and profitable farming practices. By identifying successful farmers or agribusinesses that have achieved noteworthy success, their methods and approaches can be replicated by others. This approach emphasizes knowledge sharing and encourages farmers to adopt proven strategies and techniques, facilitating the spread of innovations and fostering agricultural productivity and income growth.

The method hinges on farmer-to-farmer learning, leveraging innovative entrepreneurs who identify new practices and effectively tap into market niches, which may be overlooked locally. This creates opportunities to promote and support the replication of these experiences for the benefit of other farmers. Successful case replication involves original innovators training aspiring entrepreneurial farmers and sharing invaluable information and insights.

Operating as a 'horizontal' extension mechanism, the approach relies on one farmer instructing others. To mitigate competition among farmers, the focus often shifts to markets where there is untapped demand for the new product, ensuring that the knowledge and experience shared by the entrepreneur benefit both the trainer and the trainees (Orsini, 2000).

FARM BUSINESS SCHOOLS. The Farm Business School (FBS) is a curriculum-based extension approach designed to build farmers' entrepreneurial capacity. References to FBS project experiences are made in Chapter 15. The FBS has gained recognition as a valuable approach for supporting the capacity development of farmers, providing comprehensive training programmes that equip them with essential business management skills. By enhancing their knowledge in areas such as financial management, marketing strategies and business planning, farmers are empowered to make informed decisions and effectively manage their farm enterprises.

Learning takes place in the context of the participants' farming businesses and takes place through schools set up at community level. Farmers work in small groups at their own pace and at an agreed time and duration. The FBS aims to help farmers learn how to make their farming enterprises and operations profitable and to be able to respond to market demands. It helps farmers learn and increase knowledge, change attitudes and develop the skills needed for profitable farming – while working on their own farms. The FBS programme is farm-based with practical exercises conducted at different times of the production season. The school is organized around marketable enterprises selected by the participants. The learning programme is designed to match the activities of the farm season and gradually build and reinforce farmers' skills and competencies. In the FBS programme learning is recognized as a process that requires time for skills to be developed. The knowledge and skills learned are made to match the knowledge and skills needed in the various stages of development of the farm business.

FARMER FIELD SCHOOLS. Similar to FBS, farmer field schools (FFS) offer hands-on, participatory learning experiences that often include farm management and marketing-related modules as part of their training programmes. These sessions help farmers develop business acumen, improve decision making and adopt market-oriented production strategies. Thus, through both FBS and FFS, farmers gain the tools and techniques needed to enhance productivity, optimize resource utilization and improve profitability.

As an outcome of the seasonal farmer field and business school programmes high-performing farmers are often identified to lead a further round of farmer-to-farmer training providing farmer-run advisory services. The 'high performing' farmers are identified by front-line extension workers as having the communication skills and motivation to be in the position to organize farmers into learning groups to continue scaling-up the training programme through peer-to-peer exchange of knowledge, thereby creating a snowball effect. The impact of this process depends on the careful selection of farmers. Downward accountability is ensured through self-evaluation by local committees, in conjunction with the advisory service providers.

Training programmes vs training projects

Training projects and training programmes serve distinct roles in empowering smallholder farmers. Training projects are typically short-term initiatives focused on achieving specific goals, such as introducing a new agricultural technique or improving knowledge about a particular crop. These projects often deliver immediate, tangible benefits and are cost-effective for addressing specific issues. However, their short duration and narrow focus can limit long-term impact, as farmers may not have enough time or support to fully integrate new practices into their routines. Furthermore, without ongoing reinforcement, the knowledge gained may not be sustained, particularly in communities with limited access to resources.

Post-training support

After completing farm management training programmes, farmers require ongoing support that aligns closely with the training content. This continuity reinforces learning, fosters coherence and enhances the overall effectiveness of the programme.

Entrepreneurial skills are most effectively developed through hands-on practice. Learning occurs through a cycle of planning, acting and reflecting, where farmers are exposed to new ideas, experiment with them and evaluate the outcomes. This process is equally relevant whether farmers are addressing challenges or pursuing opportunities.

Although the process is straightforward, it often proves challenging in practice, as it demands consistent guidance and support. Farmers may need help applying the concepts and tools they learned during training to their specific circumstances. While training equips farmers with essential tools and competencies, participants may initially lack the confidence to apply these skills independently. True learning only takes place when farmers put their knowledge into practice, leading to changes in behaviour. Extension workers can play a crucial role in this process by providing ongoing support, facilitating specialized expertise and ensuring that farmers can implement what they have learned effectively. Their primary aim is to improve farmers' performance post-training by encouraging reflection, guiding implementation and helping farmers evaluate and enhance their actions.

An experienced extension worker can also serve as a mentor, supporting farmer learning by teaching, facilitating knowledge exchange, challenging assumptions, steering decisions and encouraging reflection. However, the responsibility for decision making ultimately rests with the farmer.

Group-based activities are often the most effective approach for post-training support. Bringing together farmers with similar or related businesses into farmer learning groups as discussed before, fosters shared learning, collaboration and peer support. These group settings are especially beneficial when combined with technical guidance from experts, particularly when participants share common challenges, opportunities or interests and are willing to collaborate and exchange information about their farm performance. Building trust and confidence among group members may require the extension worker's active involvement.

In addition to targeted, on-farm support, extension workers can provide broader guidance to help farmers adapt their practices, reflect on their progress and apply their learning in ways that improve overall outcomes. This integrated approach ensures that training translates into practical, meaningful improvements in farm management and business performance.

COACHING AND MENTORING. Farm management specialists play a critical role beyond their training and advisory responsibilities by coaching and mentoring both front-line extension workers and farmers. This involves providing regular support to ensure that the knowledge and skills gained through formal training are effectively applied and further developed. Coaching focuses on observing how training outcomes are implemented in practice, helping extension workers and farmers build their capacity in knowledge, skills and behaviour.

Through coaching and mentoring, the capacities of extension workers and farmers are genuinely enhanced, maximizing opportunities for growth. While formal training equips them with the necessary tools, true learning occurs when individuals apply what they have learned in real-world settings, leading to sustained changes in their thinking and behaviour. Coaching bridges this gap, elevating performance by encouraging reflection and fostering continuous improvement.

One of the key strengths of coaching is its flexibility. Unlike formal training, it can be applied on an as-needed basis, addressing specific challenges or opportunities as they arise. It also enables targeted, practical support for longer-term development, helping extension workers and farmers overcome weaknesses and achieve their full potential in a cost-effective manner.

Coaching and training work hand in hand. For example, during a coaching session with an extension worker, a farm management specialist may identify gaps in knowledge or skills that limit the worker's ability to support farmers effectively. In such cases, the coaching process can pause, and additional training can be organized to address these gaps. This complementary relationship ensures that both extension workers and farmers receive the support they need to succeed in applying farm business management principles and improving their overall performance (see Box 16.13.).

COUNSELLING. Business counselling is another important role of farm management specialists, though it is typically reserved for specific situations (see Box 16.14.). This type of counselling requires personalized, one-to-one interaction, along with a foundation of trust and open communication. It can be particularly valuable in addressing specific challenges faced by extension workers or farmers. In smallholder agriculture, where advice is primarily delivered through front-line extension workers, the opportunity for direct interaction between farm management specialists and farmers, is limited. However, in larger-scale farming systems, direct contact between specialists and farmers allows for deeper engagement and greater understanding of the farmer's perspective. While business counselling is currently uncommon in small-scale farming systems in low-income countries,

Box 16.13. Stages of coaching. From Thorpe and Clifford, 2003.

1. clarify the overall need for, and goals of, the coaching programme;
2. agree on specific development needs;
3. formulate a detailed plan for the coaching programme;
4. conduct a task or activity from the plan;
5. review activities and planning for improved performance; and
6. exit from the coaching relationship.

Box 16.14. Counselling, extension, consulting and mentoring.

Counselling involves using specialized skills and knowledge to assist clients in identifying their needs and determining potential courses of action. The process is client-centred, with the individual taking ownership of their issues and maintaining full responsibility for the decisions made and their outcomes. The counsellor's role is to support and guide rather than take control.

Extension is a collaborative learning process where advisors and clients share knowledge and experiences to address practical challenges. While clients are encouraged to take responsibility and help themselves, the advisor has a more active role in influencing decisions. However, the ultimate responsibility for decisions and their consequences rests with the client.

Consulting is a professional service where a consultant provides expert advice based on specialized knowledge to help clients address specific issues. This approach assumes that the consultant has a higher level of expertise in the subject matter than the client and is better equipped to handle the challenges at hand.

Mentoring refers to an ongoing relationship between a knowledgeable mentor and a client. This relationship evolves over time and covers a broad range of topics as the client's business develops. Mentoring often involves guiding the client through decision-making processes and supporting the implementation of those decisions in practical settings.

its potential is increasing as farmers adopt a more commercial mindset and, in some cases, as the scale of farming operations grows.

Farm management specialists, both public and private, can also extend business and marketing counselling to rural entrepreneurs and enterprises within agricultural value chains. The goal of such counselling is to help these businesses adapt to economic changes and develop effective strategies to navigate them. This requires specialists to establish a professional and confidential relationship with entrepreneurs, allowing for open discussions of critical and sometimes sensitive issues. Since every business has unique needs, counselling must be tailored to address specific challenges and goals. Specialists can help entrepreneurs rethink their perspectives, behaviours and responses to challenges, guiding them towards more effective solutions. However, due to the limited scale of some ventures, the opportunity to provide this kind of one-to-one counselling remains relatively rare.

16.3.6 Communicating information

Farm business management specialists can employ a variety of methods to communicate effectively with smallholder farmers. One of the most practical and widely used approaches is through group training sessions such as the farmer field and business schools described above. These sessions provide an interactive platform where farmers can gather, discuss and learn about farm management principles. By fostering a group learning environment, farmers are able to share experiences and challenges, while the specialist can tailor content to address their specific needs. These sessions often include hands-on demonstrations, such as preparing simple farm budgets or calculating input costs, which make the learning process more accessible and directly applicable to the farmers' daily practices.

Using mass media and big data in communication with farmers

Extension workers and farm management specialists have many options for communicating with farmers. Choices depend on the capacity of the farmers, the availability of equipment and infrastructure, and the recurring challenge of connectivity. And this is an area where collaboration between public, private and NGO sector service providers will produce better results. By combining their efforts, these sectors enhance the dissemination of critical agricultural knowledge, improve market access and foster resilience among smallholder farmers through the integration of ICTs, mass media and social media into everyday farming activities.

Farmers and agribusiness stakeholders could greatly benefit from access to information, such as the prices and availability of inputs like seeds, fertilizers, chemicals, credit, irrigation tools and other equipment. Providing this information, alongside technical guidance on recommended seeds and inputs, would support both farmers and agribusinesses in improving productivity and market access. Combining this approach with internet-based tools creates opportunities to develop entrepreneurship training programmes for small producers and rural entrepreneurs.

As discussed in Chapter 10, big data and mass media play an increasingly vital role in farm and agribusiness management, particularly in analysis and planning. Big data enables the collection and processing of vast amounts of information, including weather patterns, soil conditions, market prices and crop performance. By analysing these datasets, farm management specialists can identify trends, predict outcomes and support farmers in making informed decisions. For example, they can forecast market demand, optimize planting schedules based on climate conditions and provide targeted recommendations for farm planning.

Mass media platforms, such as television, radio and community broadcasts, complement these efforts by disseminating critical agricultural information to wide audiences, including remote farming communities. Seasonal weather updates, market price fluctuations and policy changes can be communicated efficiently, ensuring farmers stay informed. Additionally, mobile apps and SMS systems – powered by big data analytics – deliver real-time, localized advice directly to farmers, increasing the relevance and immediacy of information. These technologies also enable two-way communication, allowing farmers to provide feedback, ask questions and share experiences, fostering a participatory approach to agricultural development.

Given that many smallholder farmers may have limited literacy skills, visual tools such as infographics, charts and short instructional videos can simplify complex concepts. For instance, specialists can create easy-to-understand visuals that illustrate financial planning steps or explain input–output calculations for different crops. By combining big data insights with mass media outreach, farm management specialists can enhance decision making, improve productivity and drive innovation across the agricultural value chain.

Webinars

Webinars can also be an effective method for farm business management specialists to communicate with smallholder farmers, particularly as access to mobile phones and internet connectivity expands. While there may be challenges related to internet access and bandwidth, specialists can adapt webinars to local conditions by offering them in formats that are more accessible. For example, using audio-only webinars, or providing recorded sessions that farmers can access at their convenience via mobile phones, can help overcome some of the infrastructure limitations. In rural areas with limited internet access, webinars can be conducted in partnership with local institutions like agricultural cooperatives or community centres that have better connectivity, allowing farmers to attend in a group setting. This method can significantly extend the reach of training, especially for areas where it may not be feasible for a specialist to visit in person.

The interactive nature of webinars is another benefit, as it allows farm business management specialists to engage directly with farmers, answer their questions and address specific concerns in real-time. This engagement can be enhanced by including features like live polls, chat functions and Q&A sessions, which help foster a two-way communication channel. Additionally, webinars can be used to bring in experts from other regions or countries to share successful case studies or innovative practices, providing farmers with new perspectives and ideas that they may not have encountered otherwise. For example, a webinar on sustainable agricultural practices could feature farmers from other countries who have successfully implemented similar techniques in their own contexts, offering practical insights that are relevant to the local farming community.

While webinars offer great potential, their success hinges on ensuring that farmers have the necessary skills and access to technology. In regions with low levels of digital literacy, it may be beneficial for specialists to conduct preliminary workshops on using digital tools before launching webinars. Furthermore, a combination of online and offline methods may be most effective – where webinars serve as an additional tool, complementing face-to-face learning, group discussions and mobile technology. This blended approach can maximize learning opportunities and ensure that farmers are fully supported as they adopt new farm business management practices.

WhatsApp, Facebook and Instagram

As discussed briefly in Chapter 10, social media platforms like WhatsApp, Facebook and Instagram have become pivotal tools in agricultural extension and farm business management, especially for smallholder family farmers in low-income countries. These platforms facilitate the dissemination of vital information, enable real-time communication and foster community building among farmers, extension agents and agricultural experts.

Social media offers numerous advantages for extension to reach smallholder farmers. Its capacity for rapid information dissemination allows farmers to access best practices, market trends, weather forecasts and pest control measures. Instructional videos, live discussions and downloadable materials, enhance farming techniques and decision making. Additionally, platforms like Facebook groups foster community building by connecting farmers facing similar challenges. These online communities enable peer-to-peer learning and shared experiences, often leading to innovative solutions that benefit the wider farming community. Social media also provides a cost-effective communication channel, reducing the need for physical travel and enabling extension services to reach remote areas where traditional methods may be impractical.

16.4 Challenges and Opportunities in Implementing Farm Management Extension Programmes

The successful implementation of farm management extension programmes is crucial for enhancing smallholder productivity, profitability and sustainability. However, challenges persist, such as resource constraints, farmer resistance and understanding smallholder decision making. Opportunities exist to improve their impact through innovation, tailored advisory services and practical tools. These aspects are explored next.

16.4.1 Limited budgetary and other resources

Farm management programmes often operate in settings with limited financial, human and technological resources, especially in rural and low-income areas. These constraints can reduce the reach and impact of extension services. Budget shortages often mean there is not enough funding for field visits, training or the development of learning materials. At the same time, a lack of skilled staff adds pressure, as overworked advisors struggle to support the many and varied needs of smallholder farmers. Technological challenges – such as poor access to tools and weak connectivity – also make it harder to share timely and useful information.

To address these challenges, farm management specialists should partner with local NGOs, cooperatives, and private sector actors to share resources and expand their reach, as discussed previously. Building the skills of extension workers is also key. This can be done through focused training and the use of simple, low-cost tools suited to local conditions. Technology such as mobile phones and digital platforms can help deliver training and advice more efficiently, but it is important to provide user-friendly solutions that account for limited access and digital skills. By using these strategies together, farm advisors can better support smallholder farmers, even in resource-limited areas, and help improve productivity and resilience.

16.4.2 Resistance to change among smallholder farmers

Resistance to change among farmers represents one of the most persistent and complex challenges faced by farm management specialists. The reluctance to adopt new practices is deeply rooted in cultural norms, fear of risk, limited access to information and a lack of trust in external advisors or innovations. In many farming communities, traditional methods and practices have been passed down through generations and are closely tied to the community's identity and way of life. As a result, proposed changes may be viewed not only with scepticism but also as a potential threat to the social fabric and established norms.

Fear of risk is another significant barrier that we have previously discussed. This is particularly true for smallholder farmers who often work with very limited resources. For them, trying a new crop, method or technology can be risky – one failure could seriously affect their income and food security. The lack of safety nets like insurance or guaranteed buyers makes this fear even greater, pushing farmers to stick with familiar but less productive practices. Past negative experiences with advisors, where promised results did not happen, can also create mistrust and reluctance to change. When farmers do not see clear and relatable proof of benefits, they are less likely to try something new.

To overcome this resistance, farm management specialists need to use participatory approaches that involve farmers in the decision-making process and offer solutions that match their needs. Building trust is key – this means being present, showing an understanding of local realities and helping farmers achieve small, visible improvements. Sharing success stories, setting up field demonstrations, and working with respected local farmers can provide relatable proof that change is worthwhile. Offering support such as crop insurance, input access or guaranteed markets can also reduce the risks farmers face. By combining trust-building, clear evidence and practical risk-reduction tools, advisors can help farmers become more open to innovation and long-term improvements.

16.4.3 Understanding smallholder farmer decision making

Understanding smallholder farmer decision making is a critical yet challenging aspect of farm management work. As mentioned throughout this book, the choices of farmers are shaped by a complex interplay of factors, including risk aversion, social norms and cognitive biases. These influences, while rooted in the realities of smallholder farming, can often result in sub-optimal decisions that limit productivity, sustainability and resilience. Risk aversion is a significant driver of smallholder decision making. For many small family farmers, the stakes are incredibly high, as their livelihoods depend on making informed and rational decisions. This leads to a natural tendency to avoid risks, even when potential rewards are substantial.

Social norms and peer influence play a strong role in how farmers make decisions. As noted in Chapter 5, farmers are influenced by the practices and opinions of neighbours and local leaders. This can be a barrier if farmers fear being judged for doing things differently, but it can also be an opportunity. When respected community members or early adopters try new practices, they can encourage others to follow.

Cognitive biases also affect decision making. For example, farmers may prefer sticking to old methods (status quo bias) or may be too confident in traditional practices even when better options exist. In facing complex choices, they often rely on simple mental shortcuts (heuristics), which can lead to poor decisions – like overusing inputs or ignoring the value of crop diversification. These thinking patterns are discussed in more detail in Chapters 5 and 9.

To support better decision making, farm management specialists must employ strategies that address these challenges comprehensively and empathetically. By identifying and engaging opinion leaders within the community, extension workers can encourage these individuals to participate in pilot projects or adopt new practices. The success of these early adopters serves as a compelling demonstration for others, fostering a ripple effect that drives broader acceptance.

Countering cognitive biases requires simplifying and contextualizing agricultural information. Extension workers should present data and recommendations in a clear, relatable manner, emphasizing practical applications and addressing specific challenges farmers may have. For instance, using visual aids, relatable success stories or hands-on demonstrations can make complex concepts more accessible. Additionally, guiding farmers to recognize and overcome biases, such as over-reliance on traditional practices, helps them make more informed decisions.

16.4.4 Use of analytical tools for smallholders

Many smallholder farmers transitioning towards market-oriented farming often struggle to apply the farm management tools and methods discussed in Part 3 of this book. These tools can be difficult to use effectively due to their complexity and the intensive data required. Traditional farm management methods typically involve detailed record-keeping, precise calculations and a solid understanding of economic principles – all of which can be overwhelming or impractical for small-scale farmers. Many face limitations in literacy and numeracy, alongside a lack of infrastructure and resources to support accurate data collection. As a result, decision making often relies on intuition or tradition rather than data-driven insights, which may hinder productivity and efficiency. Simplified analytical tools are, therefore, essential for subsistence farmers who have not received formal farm management training.

Farm management specialists can play a critical role in addressing this gap by helping smallholder farmers use decision support tools that are simple, practical and better aligned with their realities. Tools such as gross margin analysis, partial budgeting and straightforward decision rules provide a manageable framework for key decisions around crop selection, resource use and farm practices. For example, a simple gross margin analysis can help evaluate the profitability of different crops, while a partial budget can support decisions about adopting new inputs or practices. These tools are accessible and actionable, enabling farmers to make informed choices with minimal complexity and, in many cases, without needing direct extension support.

More educated and entrepreneurial small family farmers who participate in farm business management training are generally able to apply these tools more confidently. Tools like whole-farm budgeting, farm planning and simple investment appraisal methods can help them make strategic decisions and improve their farm profitability. For more commercially oriented small farmers who have participated in farm business management training, it is expected that they will begin to use more advanced decision support tools – such as investment analysis using discounting, business planning and risk assessment. As these farmers become more commercially focused and develop the necessary skills, they are increasingly capable of using these tools independently.

However, the application of more complex tools – such as comparative analysis, benchmarking, linear programming and programme planning – is likely to require direct support from farm management specialists, often in collaboration with extension workers. Implementing these tools typically involves collecting external data beyond the farm's boundaries, which can be costly and difficult for individual farmers to obtain. In such cases, the extension service should take responsibility for gathering and providing this data to ensure these advanced tools can be effectively used in practice.

A summary of the suitability of different tools for various farmer categories is presented in Table 16.3, along with the required level or intensity of farm management advisory support that is expected. The table should help guide extension workers and specialists in tailoring their advice to the farmer's needs and capacity.

Table 16.3. Use of analytical tools by small farmer category.

Tool/method	Level of complexity	Category of smallholder farmer	Extension support needed for tool application by farmer
Gross margin budgeting	Simple	– Small subsistence farmer (transitioning) – Small market-oriented family farmer – Small commercial farmers	Limited support needed
Break-even budgeting	Simple	– Small subsistence farmer (transitioning) – Small market-oriented family farmer – Small commercial farmers	Limited support needed
Record-keeping	Simple	– Small subsistence farmer (transitioning) – Small market-oriented family farmer – Small commercial farmers	Limited support needed
Simple investment analysis	Simple	– Small subsistence farmer (transitioning) – Small market-oriented family farmer – Small commercial farmers	Limited support needed
Risk management	Simple	– Small subsistence farmer (transitioning) – Small market-oriented family farmer – Small commercial farmers	Limited support needed
Market planning	Simple	– Small subsistence farmer (transitioning) – Small market-oriented family farmer – Small commercial farmers	Limited support needed
Partial budgeting	Simple	– Small subsistence farmer (transitioning) – Small market-oriented family farmer – Small commercial farmers	Limited support needed

Continued

Table 16.3. Continued

Tool/method	Level of complexity	Category of smallholder farmer	Extension support needed for tool application by farmer
Whole-farm budgeting	Moderate	– Small subsistence farmer (transitioning) – Small market-oriented family farmer – Small commercial farmers	Guided support needed
Cash flow (analysis/ planning)	Moderate	– Small subsistence farmer (transitioning) – Small market-oriented family farmer – Small commercial farmers	Guided support needed
Labour planning	Moderate	– Small subsistence farmer (transitioning) – Small market-oriented family farmer – Small commercial farmers	Guided support needed
Trade-off analysis	Moderate	– Small subsistence farmer (transitioning) – Small market-oriented family farmer – Small commercial farmers	Guided support needed
Discounted investment analysis	Moderate	– Small commercial farmers	Guided support needed
Market margins	Moderate	– Small commercial farmers	Guided support needed
Farm business planning	Moderate	– Small subsistence farmer (transitioning) – Small market-oriented family farmer – Small commercial farmers	Guided support needed
Participatory budgeting	Complex	– Small subsistence farmer (transitioning) – Small market-oriented family farmer – Small commercial farmers	Intensive advisory support needed

Continued

Table 16.3. Continued

Tool/method	Level of complexity	Category of smallholder farmer	Extension support needed for tool application by farmer
Resource efficiency ratios	Complex	– Small subsistence farmer (transitioning) – Small market-oriented family farmer – Small commercial farmers	Intensive advisory support needed
Problem analysis	Complex	– Small subsistence farmer (transitioning) – Small market-oriented family farmer – Small commercial farmers	Intensive advisory support needed
Farm planning: programme planning	Complex	– Small market-oriented family farmer – Small commercial farmers	Intensive advisory support needed
Linear programming	Complex	– Small commercial farmers	Intensive advisory support needed
Profit derivation network	Complex	– Small market-oriented family farmer – Small commercial farmers	Intensive advisory support needed
Benchmarking	Complex	– Small subsistence farmer (transitioning) – Small market-oriented family farmer – Small commercial farmers	Intensive advisory support needed
Comparative analysis	Complex	– Small subsistence farmer (transitioning) – Small market-oriented family farmer – Small commercial farmers	Intensive advisory support needed
Planning for risk	Complex	– Small market-oriented family farmer – Small commercial farmers	Intensive advisory support needed

The success of these tools depends on effective training and hands-on support. Extension workers must go beyond simply introducing the tools – they must demonstrate their use through workshops, on-farm visits and guided exercises. A hands-on approach builds confidence and ensures that farmers understand how to apply the tools in real situations. For instance, using actual farm data to walk through a gross margin calculation can help show how the tool supports input use or crop diversification decisions.

Local adaptation is another crucial element in making these tools effective for smallholders. Farmers operate in diverse environments with varying constraints, such as limited access to markets, unpredictable weather patterns and distinct cultural practices. Extension workers must customize analytical tools to reflect these specific conditions, ensuring that the recommendations generated are relevant and realistic. This might involve simplifying input categories for farmers with limited resources, incorporating region-specific price data, or adjusting decision rules to account for local cropping systems.

Despite these advancements, challenges remain in promoting the adoption and sustained use of simplified tools. Farmers may initially struggle to see the value of these tools, particularly if they have relied on traditional methods for generations. Extension staff must address this scepticism by demonstrating tangible benefits, such as improved profitability or reduced waste, that result from data-driven decision making. Additionally, ensuring that tools are easy to update and use over time is vital, as overly rigid or cumbersome processes can lead to disuse.

By addressing these challenges and providing ongoing support, farm management specialists together with frontline agricultural extension workers should empower smallholder farmers to adopt and benefit from these simplified analytical tools. These efforts will not only enhance the farmers' decision-making capabilities but also pave the way for more efficient resource use leading to increased productivity and sustained profitability.

16.4.5 Experiential learning methods

Extension workers, supported by farm management specialists, can significantly enhance the impact of their efforts by promoting farmer-to-farmer learning. These peer exchanges enable farmers to share practical experiences and insights directly, making new practices more relatable and encouraging broader adoption.

As noted earlier, participatory approaches such as FFS, FBS and farmer learning groups, are powerful tools for strengthening farmer education, promoting skill adoption and empowering rural communities. These approaches emphasize hands-on, field-based learning, where farmers engage directly in implementing and observing new practices and technologies. By seeing results firsthand, farmers gain confidence in the value of innovations, which supports long-term adoption. These group-based methods also encourage dialogue and knowledge exchange among participants, fostering a collaborative environment where farmers collectively solve problems and learn from one another's experiences. This not only strengthens community cohesion but also builds local capacity for innovation and resilience.

A key strength of these participatory methods is their season-long engagement, which allows learning to unfold gradually and contextually across the crop cycle. Ongoing workshops and follow-up sessions reinforce understanding, support adaptation to changing conditions and help farmers assess progress. For extension advisors, these approaches offer a valuable platform to engage more meaningfully with farmers, providing tailored advice and building lasting trust. By integrating experiential learning into extension systems, advisors can shift from top-down knowledge delivery to inclusive, participatory processes that empower farmers to improve productivity, enhance sustainability and strengthen their livelihoods.

16.4.6 Use of digital communication

Many of the communication and information-sharing tools promoted in modern extension programmes, including social media and other digital platforms, rely heavily on digital access. However, limited internet connectivity, lack of affordable devices and low digital literacy among small-scale farmers pose significant challenges for extension workers. These barriers hinder the effective dissemination of knowledge and limit farmers' ability to engage with extension services, access real-time information and participate in online networks.

Addressing these challenges requires targeting the interventions that are referred to in earlier chapters: improving the rural internet infrastructure, providing affordable digital tools and offering tailored digital literacy training. Without these efforts, the potential of digital tools to enhance farmer engagement, share best practices and foster collaboration, remains underutilized. Bridging the digital divide is essential to ensure that all farmers, regardless of their location or resources, can benefit from the advancements in agricultural communication and contribute to the resilience and productivity of their communities.

In the meantime, advisory services can adopt a blended approach, combining digital tools with traditional methods such as radio broadcasts, printed materials and face-to-face interactions to ensure inclusivity. Community-based intermediaries, such as farmer cooperatives or local champions, can also play a vital role in bridging the gap by relaying digital content to those without direct access.

16.5 Conclusion

Farm management specialists are pivotal in addressing the evolving needs of small family farm businesses and theory integration into modern value chains. Their role extends beyond traditional advisory functions in management, marketing and business, encompassing capacity building, innovation facilitation and fostering connections between value chain stakeholders. By bridging the gap between research, policy and practical farm management, these specialists enable farmers and agricultural entrepreneurs to enhance productivity, profitability and income.

A key area of focus is the collaboration of public and private sector advisory service providers. Public sector farm management specialists often address foundational needs, providing broad-based support and training to frontline extension workers and through them to smallholders and emerging on-farm agricultural enterprises. In contrast, private sector advisors offer tailored, specialized business management services that address specific value chain demands, including market access, financial planning and supply chain optimization. The collaboration between these sectors ensures a cohesive, scalable approach to advisory services, leveraging the strengths of each.

Capacity development lies at the heart of creating a sustainable advisory system. Participatory training models equip farmers with critical business and management skills, enabling them to make informed decisions and transition towards market-oriented farming. Equally, continuous in-service training for extension staff in farm management ensures that advisory support remains relevant and impactful. These efforts create a strong foundation for innovation and the adoption of sustainable practices.

Market linkage facilitation is another critical function. Farm management specialists also have a role in acting as connectors, enabling farmers to access reliable markets, negotiate fair contracts and adopt practices that align with quality and packaging standards. This role not only optimizes returns for farmers but also strengthens the overall efficiency of the agricultural value chains.

In conclusion, farm business management specialists are instrumental in transforming agriculture into a sustainable, market-oriented sector. Their ability to adapt to changing contexts, embrace technology and foster collaboration is essential for ensuring the success of small family farmers and agricultural entrepreneurs.

References

Anderson, J.R. and Hardaker, J.B. (1979) Economic analysis in design of new technologies for small farmers. In: *Planning Agricultural Research: A Sourcebook*. International Service for National Agricultural Research, The Hague, Netherlands, pp. 57–70.

Blum, M.L. (2008) *Trends, frameworks and requirements for a comprehensive approach to advisory services in Africa*, FAO/ADB 6th Thematic Meeting, Rome, Italy. 8 January.

Dorward, A.R. (1986) Farm management and extension in smallholder agriculture. Part 2 – appropriate farm management skills and their use in extension. *Agricultural Administration* 22. Available at: https://www.sciencedirect.com/science/article/abs/pii/0309586X86900750

Dorward, P., Shepherd, D. and Galpin, M. (2007) The development and role of novel farm management methods for use by small-scale farmers in developing countries. *Journal of Farm Management* 13(2).

Dorward, P., Clarkson, G. and Stern, R. (2015) *Participatory integrated climate services for agriculture (PICSA): Field manual*. Walker Institute, University of Reading.

FAO (1994) In: Upton, M. and Dixon, J.M. (eds) *Methods of Micro-Level Analysis for Agricultural Programmes and Policies: A Guideline for Policy Analysis*.

Ibrahim, U. (2023) Artificial intelligence in agricultural extension for sustainable development. *International Journal of Applied and Scientific Research (IJASR)* 1(3), 259–268. DOI: 10.59890/ijasr.v1i3.740.

Kahan, D. (2007) The role of the farm management advisor in extension. In: *Farm Management Extension Guide 6*. FAO.

Orsini, J.B. (2000) *Success Case Replication: A Manual for Increasing Farmer Household Income by Mobilising Successful Farmers and Groups to Train Their Peers*. ESCAP/FAO, Bangkok, Thailand.

Thorpe, S. and Clifford, J. (2003) *The Coaching Handbook*. Kogan Page Publishers.

Additional Reading

Brown, M. (1979) From Farm Income Analysis to Agricultural Project Analysis, World Bank Staff Occasional Papers. Farm Budgets.

Collinson, M. (1983) Farm Management in Peasant Agriculture. Westview Press, Boulder

Dorward, P.T., Shepherd D.D. and Wolmer, W.L. (1997) Developing Farm Management Type Methods for Participatory Needs Assessment. Agricultural Systems, Vol. 55 No.2.

FAO (1989) Guidelines on the Conduct of a Farming Systems Development Training Course.

Rehman, T. and Dorward, A. (1984) Farm management techniques and their relevance to administration, research and extension in agricultural development: Part 1- Their evolution and use in developed countries; Part 2 An appraisal of their potential in LDCs. Agricultural Administration, 15, 177 190, 239–254.

Rehman, T., and Dorward, A. (1984) Farm management techniques and their relevance to administration, research and extension in agricultural development: Part 2—An appraisal of their potential in less developed countries. Agricultural Administration, 15(4), 239–253.

17 Farm Business Management Support for Small Family Farm Development: Future Challenges and Opportunities

This book has explored how farm business management can strengthen smallholder family farms, ensuring their productivity, profitability and sustainability. It began by emphasizing the crucial role that small-scale family farms play in contributing directly to local and national economies, and indirectly to the global economy, particularly in enhancing food security. It has discussed the fragility of the agriculture sector and the agri-food system in general. In many ways, small-scale family farmers are trapped in a system that does not effectively address the persistent challenges they face, such as limited access to land, capital and infrastructure, as well as weak market linkages. Competing against large-scale, industrial farms in an increasingly volatile agricultural landscape presents a formidable task. Now, in the face of climate change, their mere survival is at stake. However, the vast and rapid changes taking place in the agricultural landscape are not only a threat but also an opportunity. The threat is clear: it is difficult to compete with large-scale, industrial agriculture. They have little leverage in a complex, global system that is highly competitive. The opportunity is equally clear: if these farmers can be assisted to embrace and apply the principles and practices of farm business management, they not only have the chance to survive, but to greatly strengthen their livelihood systems and to make them more sustainable and profitable. Being so important to the overall agri-food system, it is essential that this category of farmers be given the kind of support that will enable them to take advantage of this opportunity.

If they can adopt effective farm business management practices, they not only stand a chance of survival but can position themselves as competitive and resilient agricultural entrepreneurs. Through capacity-building initiatives, the integration of new technologies and enhanced access to financial and market systems, these farmers can make informed decisions that strengthen their livelihoods. (See Fig. 17.1.)

This chapter examines the challenges and opportunities shaping the future of smallholder farming. It emphasizes the importance of supporting business-oriented farming, equipping farmers with financial and management skills, and fostering an environment where advisory services, digital tools and market integration facilitate the transition from subsistence farming to commercially viable enterprises.

This final chapter gives special attention to youths and their role in the future of smallholder family farm agriculture. It seeks to inspire those of you who are youths venturing into this field of service, as well as policymakers and those currently working in the field, into action. Youth involvement is pivotal to driving innovation, enhancing market access and securing the economic viability of small family farms. This chapter highlights the vast opportunities for young women and men to participate as entrepreneurs in 'farming as a business' or to support smallholder farmers by offering a wide range of goods and services. These emerging opportunities demonstrate how youth engagement can shape the future of modern farming.

DOI: 10.1079/9781836990536.0017

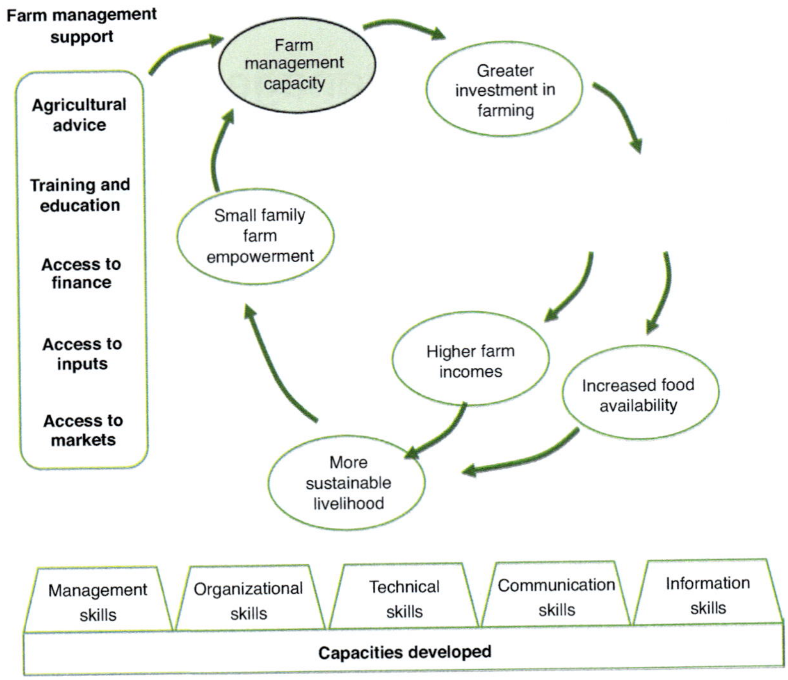

Fig. 17.1. The empowering cycle of farm management impacts.

17.1 Empowering Small Family Famers in a Changing Agricultural Landscape

Small family farmers face significant barriers in adapting to rapid changes in agriculture, climate, markets and technology. Empowering smallholders demands addressing their unique challenges, leveraging emerging opportunities and enhancing their participation in modern agricultural value chains. The following sections explore how targeted support, collective action and entrepreneurial capacity-building can strengthen the resilience and competitiveness of small family farms in today's dynamic agricultural context.

17.1.1 The evolving agricultural landscape: challenges and opportunities

Small family farmers in low-income countries face a unique and evolving set of challenges that hinder their ability to achieve sustainable livelihoods. These challenges include limited access to critical resources such as quality seeds, fertilizers and finance, as well as inadequate infrastructure, such as roads and storage facilities, which restrict their market access and competitiveness. Additionally, many small-scale farmers lack the technical knowledge and skills necessary to adopt modern farming practices that could enhance productivity, profitability and long-term sustainability. Climate change further exacerbates these difficulties through unpredictable weather patterns, increased incidences of pests and diseases and the rising frequency of extreme weather events, making farming even more precarious.

As discussed in Chapter 2, small family farms employ diverse livelihood strategies based on their resource base, market access and household priorities. These strategies can be broadly grouped into three categories. *Subsistence-oriented small family farms* primarily produce for household consumption, generating little or no surplus for sale. Operating in marginal environments with limited access to markets and financial resources, these farms rely on diversified income sources such as off-farm employment, remittances from family members and informal safety nets within rural communities. Given their economic vulnerability, these households prioritize stability over risk-taking, often engaging in low-input, traditional farming methods that emphasize food security rather than commercial viability.

In contrast, *market-oriented small family farms* are better integrated into local and national markets and have access to essential infrastructure and financial services that enable them to produce surpluses for sale. Their livelihood strategies focus on improving productivity through the adoption of modern inputs, diversification into high-value crops and participation in cooperative models that enhance market access. With targeted support in financial management, risk mitigation and business planning, many of these farms have the potential to transition towards *fully commercial small family farms*, which are deeply embedded in competitive value chains. These commercially driven farms operate with a clear profit-oriented approach, often specializing in specific commodities and leveraging economies of scale to maximize returns. Their strategies involve investment in advanced technologies, contract farming and supply chain partnerships, ensuring long-term growth and sustainability.

Significant opportunities exist for market-oriented small family farmers with the necessary management skills and willingness to embrace innovation. Advances in digital technology are transforming how farmers access knowledge, respond to market signals and adapt to climate variability. Farmer organizations and cooperatives are playing an increasingly vital role in overcoming market barriers by enabling smallholders to pool resources, negotiate better prices and benefit from economies of scale. Strengthening entrepreneurial and business skills among this category of farmers is crucial to improving financial decision making, resource management and overall profitability. By developing robust farm management competencies, market-oriented small family farmers can build more resilient, sustainable farming systems capable of thriving in today's competitive agricultural landscape.

17.1.2 Strengthening the position of smallholder farmers in modern value chains

For smallholder farmers to enhance their economic viability, integration into value chains is crucial. Many face challenges such as low production volumes, weak market linkages and price volatility, making financial planning and investment decisions difficult. Limited access to reliable buyers and fair pricing mechanisms often compels them to sell at unfavourable rates, jeopardizing their long-term sustainability. Additionally, land fragmentation, a widespread issue in many low-income countries, further reduces farming efficiency and limits opportunities for economies of scale, making it harder for smallholders to compete in modern agricultural markets.

By working collectively, farmers can secure contracts with larger buyers, reducing transaction costs and improving market stability. Shared infrastructure, such as storage facilities and transport services, further enhances efficiency and reduces post-harvest losses. Contract farming arrangements offer an additional layer of security by guaranteeing buyers for produce before the planting season, minimizing risks associated with market fluctuations. Engaging in value-added activities, such as food processing and packaging, allows farmers to move beyond selling raw commodities, thereby increasing their earning potential and creating new revenue streams.

Land fragmentation, a common challenge in many low-income countries, limits the efficiency, productivity and sustainability of smallholder farming. Cooperative approaches can address this issue by consolidating smallholder resources and facilitating access to mechanized services and modern farming technologies. Shared-use systems, such as tractor rental schemes and cooperative-owned machinery, enhance productivity by reducing individual farmers' capital costs. By fostering collaboration, farmer organizations and cooperatives play a pivotal role in strengthening smallholders' access to inputs, markets and finance, thereby improving efficiency and promoting long-term sustainability.

Beyond securing access to markets, smallholder farmers can also become active participants and stakeholders within the value chain itself. Again, by organizing into cooperatives or farmer-owned enterprises, they can collectively invest in processing facilities, distribution networks and input supply businesses, capturing more value from their produce. Smallholders can also engage as shareholders in agribusiness ventures, ensuring a more equitable distribution of profits and decision-making power. Strengthening their role as co-owners and value chain actors not only improves their economic resilience but also fosters a more inclusive and sustainable agricultural system.

17.1.3 Supporting small-scale family farmers as entrepreneurs

Agricultural advisory and extension services play a pivotal role in enabling small-scale family farmers to evolve as successful entrepreneurs. These farmers, already known for their adaptability and resourcefulness, often experiment with innovative ways to manage their farms, adopt new crops and livestock breeds, and leverage technologies to boost productivity and profitability. As discussed in the previous chapter, farm management specialists can transform entrepreneurial traits – such as market orientation, calculated risk-taking and a focus on profitability – into comprehensive skills that align with modern agricultural value chains and market demands. They achieve this by providing farmers with essential business and marketing knowledge, thereby enhancing farm profitability and competitiveness. This process often involves reinforcing management skills and developing detailed farm business plans. By integrating these entrepreneurial traits with structured management practices, specialists can help farmers move from subsistence-focused decisions to market-oriented strategies that balance immediate needs with long-term sustainability.

Farmer entrepreneurship requires a shift in perspective, where agricultural practices are seen as business opportunities rather than solely as a means of survival. Extension services can guide this transformation by emphasizing forward-thinking approaches, encouraging farmers to identify and seize new market opportunities. This involves helping farmers envision possibilities beyond their current operations, take calculated risks and innovate in response to market and environmental changes. Such support ensures that farmers are equipped to navigate dynamic and interconnected agricultural value chains, where relationships with suppliers, traders, processors and transporters significantly impact success. Extension services, moreover, can advocate for improvements in infrastructure and facilitate solutions and can work with policymakers to create a more conducive environment, ensuring that legal and regulatory frameworks support smallholder businesses.

Limited access to finance is another critical challenge, as many small-scale farmers struggle to invest in modern inputs and equipment. Advisory services can facilitate access to affordable credit by connecting farmers with financial institutions or designing collective financing mechanisms through farmer organizations and cooperatives. Extension workers can also guide farmers in preparing business plans and financial proposals to improve their ability to secure funding.

Social and cultural barriers, including resistance to change and gender inequality, further complicate the entrepreneurial journey. Extension services can address these issues by promoting inclusivity, providing targeted support for women farmers and fostering a culture that values innovation. Training programmes tailored to the unique needs of smallholder farmers can empower them with skills in financial management, marketing and strategic planning, enabling them to operate more effectively as entrepreneurs.

Agricultural advisory services must also integrate digital tools and technology to enhance decision-making and management capabilities. By introducing mobile applications, farm management software and digital platforms, extension workers can provide farmers with real-time access to weather forecasts, market prices and pest management advice. Training programmes focused on building digital literacy and data management skills ensure farmers can fully utilize these tools to improve efficiency and profitability.

Developing small-scale family farmers into successful entrepreneurs necessitates a coordinated and sustained effort. Agricultural extension services play a pivotal role in this transformation by addressing individual, organizational and systemic challenges to build a supportive ecosystem. Through capacity-building initiatives, policy advocacy and improved access to resources, these services empower farmers to realize their entrepreneurial potential, thereby driving rural development and sustainable agricultural growth.

Farm business management specialists and extension workers are instrumental in strengthening market access, enhancing financial management and fostering collaboration among smallholders. By supporting the development of collective value-adding enterprises and facilitating ownership stakes in key value chain functions, they integrate small-scale family farmers into the broader agricultural economy. This integration not only boosts individual farm profitability but also contributes to the resilience and sustainability of the agricultural sector as a whole.

17.2 Risk Management for Sustainable Farming

Risk management is a fundamental strategy to attain the goal of sustainable agricultural development. Effective risk management enables smallholder farmers to engage proactively with the uncertainties that increasingly challenge agricultural production and livelihoods. Given the complexity and interconnectedness of agricultural risks, an integrated approach – encompassing targeted extension support, strategic use of resources and adaptive practices – is essential for fostering resilience, productivity and long-term sustainability of farm systems. The following subsections explore practical strategies and considerations to effectively support farmers in navigating both general agricultural risks and specific challenges posed by climate change.

17.2.1 Supporting farmers to manage agricultural risks

The importance of managing risk has been a central theme throughout this book. Risk is an inherent part of farm management. Farmers must learn to deal with risk in an integrated way and, thus, extension and advisory services, and their personnel, need to be well versed in addressing key risks.

Assisting farmers in managing risk

Extension workers and farm management specialists also have a vital role to play in assisting farmers in making decisions regarding risk management. To be able to provide advice and support will require a sound and practical understanding of the risks that farmers are likely to face, and the range of risk management strategies open to them. It will require a good understanding of the sources and types of risk and some of the 'good practices' of risk management described earlier in this book. With these skills, extension workers and farm management specialists should be able to provide farmers with the information they will need to better assess the risks that they face and their likely consequences. The workers and specialists should be in the position to help identify the most appropriate risk management options, assess their benefits and costs and the impact of the risk management strategies on the farm business. In order to develop appropriate strategies, it is essential to consider them in the context of the farmer's location, the unique goals of the farm family and the range of decision-making options open to them. This will require a good understanding of the farmer and their surrounding environment.

Risk management options

As discussed in Chapter 6, the main sources of risk are production and technical risk, marketing/price risk, financial risk, institutional risk, and human and personal risk. For each of these types of risks there are a range of possible management strategies. For example, production risk management includes adopting risk reducing strategies like drought resistant varieties and irrigation, diversification or climate-smart strategies. Marketing risk management includes options around sales and pricing strategies and value-adding.

However, risks do not occur in isolation, nor in predictable patterns or combinations. Similarly, risk management strategies cannot be selected or applied discretely; they must be considered collectively. And very often, they must be bundled. For instance, a maize farmer in a drought-prone region could purchase weather-based insurance bundled with a loan to buy drought-resistant maize seeds. Alongside this, the farmer would receive training on using efficient water management techniques, such as drip irrigation or mulching, to conserve moisture. If the rainfall is below a certain threshold, triggering the insurance policy, the farmer would receive a payout to cover their financial losses. This reduces their dependence on external aid, helps them repay the loan and builds long-term resilience.

For small-scale vegetable farmers, bundling risk management strategies to address marketing and price risks is essential to ensure stable income and reduce vulnerabilities to market fluctuations. A well-designed bundle might include contract farming agreements, access to market information systems and training in value addition and diversification. For instance, the vegetable farmer growing cabbages and peppers could sign a contract with a local school to supply vegetables for its school feeding programme. Through a mobile market information platform, the farmer could also identify surplus produce buyers during peak harvest periods. Simultaneously, training in value addition would allow the farmer to process excess peppers into chili sauce for local sales, further stabilizing income. The combination of a guaranteed contract, market intelligence and diversified income streams ensures that the farmer is less vulnerable to price drops or oversupply in local markets.

Risk management is itself a risk

Extension workers and farm management specialists will appreciate that assessing risk and choosing tools for managing risk is not certain. Risk management operates with unknown variables, from beginning to end. There is uncertainty about many things such as prices, weather, labour, disease and pests. Additionally, the strategies themselves are, at best, only estimates about how to cope with the possible range of risks.

The more experienced the farmer, the extension worker and farm management specialist are in assessing and responding to risks in farming, the better they will be at making reasonable risk management decisions. However, it is crucial to recognize that, even with extensive experience, no set of risk management tools and strategies can fully eliminate all risks inherent in agriculture. Therefore, while risk management aims to mitigate potential negative impacts, it cannot guarantee complete protection against all uncertainties. Risk aversion is, after all, risky.

Collaborative networks and partnerships

Here too, collaborative networks and partnerships play a vital role in enhancing risk management for smallholder farmers. By joining farmer organizations and cooperatives, farmers can pool resources, access shared infrastructure facilities and transportation and, as noted previously, collectively negotiate better terms for inputs and market access. These networks also foster knowledge-sharing, enabling farmers to exchange experiences and learn from one another's successful risk management practices. Partnerships with financial institutions, research organizations and local governments further strengthen farmers' ability to manage risk by improving access to credit, insurance and climate-resilient technologies. For example, cooperatives can negotiate group insurance policies or secure better prices for bulk purchases of inputs like seeds and fertilizers. Additionally, connecting smallholder farmers to broader value chains through partnerships with private sector actors can mitigate marketing risks and create new opportunities for income generation. These collaborative efforts not only reduce individual vulnerability but also enhance collective resilience, ensuring that farmers can navigate risks more effectively while fostering sustainable agricultural development.

17.2.2 Adaptive risk management for climate change

As discussed at length in Chapter 7, climate change poses a significant threat to agricultural productivity and food security, particularly for smallholder farmers in low-income countries. As weather patterns become more unpredictable, with increased frequency and severity of extreme events such as droughts, floods and storms, the need for robust farm management support becomes ever more critical. Effective farm management support can help farmers anticipate and respond to these climatic changes by providing them with the tools, knowledge and strategies necessary to mitigate adverse impacts. This support includes access to accurate weather forecasts, climate-smart agricultural practices and resilient crop varieties that can withstand extreme conditions.

Informed decision making and strategic planning are essential for managing the myriad production and market-related risks exacerbated by climate change. Smallholder farmers must navigate fluctuating weather conditions, pest and disease outbreaks and volatile market prices, all of which can jeopardize their livelihoods. Farm management support can equip farmers with the skills to conduct risk assessments, develop contingency plans and make data-driven decisions. For instance, by using predictive models and risk management tools, farmers can optimize planting schedules, diversify their crops and implement integrated pest management strategies. Additionally, access to market information and price trends can help farmers make better decisions about when to sell their produce and how to negotiate better prices, thereby reducing market-related risks.

Adaptive management practices are key to enhancing the resilience and sustainability of smallholder farms in the face of climate change. These practices involve continuous monitoring, learning and adjusting strategies based on observed outcomes and changing conditions. For example, conservation agriculture techniques, such as minimum tillage, crop rotation and the use of cover crops, can improve soil health, increase water retention and reduce erosion, making farms more resilient to extreme weather events. Agroforestry, which integrates trees and shrubs into agricultural landscapes, can also provide multiple benefits. Similarly, regenerative agriculture can help farmers adapt to climate change by focusing on practices that enhance soil health, increase biodiversity and improve ecosystem resilience.

Institutional support and effective partnerships also in this sphere, play a critical role in helping smallholder farmers adapt to climate change. Governments, research institutions, NGOs and private sector actors must collaborate to develop and disseminate climate-resilient technologies, create enabling policies and provide financial support mechanisms. Public-private partnerships can facilitate the adoption of innovative solutions, such as renewable energy systems for agriculture (e.g. solar-powered irrigation) or affordable precision agriculture technologies that cater to the needs of small-scale farmers. These partnerships can also support the establishment of micro-finance schemes or index-based insurance products that provide financial safety nets against climate related shocks.

At the local level, community-based initiatives are essential for fostering resilience. These initiatives promote knowledge sharing and resource pooling, enabling smallholder farmers to collectively respond to climate risks. For example, farmer cooperatives can provide access to shared resources such as storage facilities, drought-tolerant seeds or water harvesting systems. Institutional support, coupled with strong partnerships, bridges the gap between farm-level adaptation measures and systemic efforts to build climate-resilient agricultural systems. By aligning efforts across multiple stakeholders, the agricultural sector can achieve sustainable and inclusive climate change adaptation.

17.3 Leveraging Technology for Better Decision Making

As discussed in previous chapters, the agricultural sector is evolving with the use of innovative tools that have the potential to improve farm management decision making among farmers. Mobile applications, spreadsheets, financial planning tools and AI are examples of decision aids that provide practical management solutions for modern farming. Agricultural advisers and farm management specialists can leverage these tools to translate data into actionable advice. Additionally, mobile applications, referred to previously, have emerged as essential aids for smallholder farmers to receive real-time updates on market prices, weather conditions and agronomic and agribusiness data. Furthermore, many of these applications can now offer interactive advisory support, allowing farmers to consult specialists remotely and receive customized recommendations based on their specific farming conditions. Digital extension services further enhance accessibility by delivering training content via SMS, voice messaging and online platforms, ensuring that even those in remote areas benefit from expert guidance. However, the impact of these new technologies hinges on the provision of appropriate advisory support.

Advisory service providers, with expertise in farm and agribusiness management, again play a critical role in bridging the gap between technology and practical application. By equipping farmers with essential digital skills, they help ensure that technological advancements translate into tangible productivity gains. Training programmes focused on digital literacy and data management are essential and need to be tailored to farmers' educational backgrounds and local contexts, making these tools accessible and relevant to their daily operations.

Participatory approaches that include farmers in technology research and development are critical in ensuring the design of these technologies is tailored to their specific needs. Incorporating feedback mechanisms allows for continuous improvement, making tools more user-friendly and effective. This sense of ownership increases farmers' willingness to adopt new technologies and apply them in their farm management practices. Agricultural advisory services are essential to provide the requisite training and extension support to small family farmers with the motivation and capacity for entrepreneurship. As noted before, the role of extension services needs to extend beyond training to include providing ongoing support and facilitating access to the broader range of technological solutions.

Thus, the integration of digital tools into farm management has a huge potential to transform smallholder agriculture, enabling farmers to make better decisions, improve productivity and enhance sustainability. With effective support from advisory services and targeted training programmes, farmers can leverage this technology to strengthen their businesses, improve rural livelihoods and contribute to broader economic development.

17.4 Youth in Agriculture: The Key to Sustainable Transformation

Ensuring the long-term sustainability and competitiveness of small-scale family farming requires not only enhanced business management practices but also the active involvement of the next generation. Youth engagement plays a crucial role in fostering innovation, expanding market opportunities and strengthening agricultural value chains. By empowering young people in agriculture, challenges such as poverty, unemployment and inequality can be addressed, securing the sector's future.

The following sections examine the opportunities available to youths who are well-trained in agriculture and related fields, as well as the potential roles they can assume. To fully harness the potential of skilled youth in agriculture, policymakers must create an enabling environment. This involves improving access to land, credit, education and markets – essential elements for young agripreneurs. Establishing targeted programmes and policies that provide these resources can significantly enhance youth participation in agriculture. By fostering professionalism among young agriculturalists, their role in advancing the sector can become both impactful and sustainable.

17.4.1 Opportunities for the next generation in agriculture

Agriculture is undergoing rapid transformation, and the youth has the potential to lead the way. As global demand for food rises and climate challenges intensify, the sector needs fresh thinking, new technologies and bold entrepreneurship. Young people bring energy, adaptability and digital expertise – qualities that can reshape agriculture into a more dynamic, profitable and sustainable field.

Opportunities extend far beyond traditional farming. Across the sector, young professionals are developing agri-tech startups, building digital tools that connect farmers to markets and pioneering climate-smart farming techniques. Careers in agribusiness management, farm advisory services, research and policy development are expanding, offering diverse pathways for those eager to contribute. With the right skills, training and support, young people can move beyond being workers in agriculture to becoming decision makers, innovators and business leaders.

Beyond securing jobs, young professionals have the potential to become key players within agricultural value chains. By forming cooperatives or investing in processing, logistics and input supply businesses, they can capture more value from their produce and strengthen rural economies. Owning a stake in the value chain – rather than simply supplying it – allows smallholder farmers and agripreneurs to build more resilient and profitable enterprises.

Equipping young people with access to finance, business training and strategic partnerships can transform agriculture from a sector of necessity into one of opportunity. Leveraging their proficiency with digital tools, young professionals can drive the adoption of precision agriculture, mobile finance solutions and data-driven decision making in farming. With the right investments in education, access to finance and professional networks, the next generation can modernize agriculture and transform it into a thriving industry. The challenge lies not in the availability of opportunities within agriculture, but in providing young people with the support needed to seize them.

17.4.2 Engaging the next generation in agriculture

For those working in agricultural development, the challenge is clear: how can the sector attract and retain young talent? With an aging farming population and growing food security concerns, ensuring the active participation of young professionals is not just beneficial – it is essential. The future of agriculture depends on whether it can be positioned as a sector that offers real opportunity, career growth and economic security.

Agriculture must be presented as more than just a livelihood – it must be seen as an industry of innovation, investment and influence. Development practitioners, extension workers and policymakers have a critical role in shifting this perception. If young people are to engage, they must see agriculture as a space where they can prosper financially, apply cutting-edge skills and contribute meaningfully. Enhancing entrepreneurship training, broadening access to financial services and promoting agribusiness incubation and mentorship, are essential actions to make the agriculture sector a more attractive career path. Additionally, offering competitive compensation and benefits can significantly boost the sector's appeal.

Young people need more than employment; they need opportunities to shape, and take ownership of, the agricultural sector. Supporting youth-led agribusinesses, fostering cooperative ownership models and ensuring young farmers have a stake in value chains, can drive long-term economic stability. Policymakers and development practitioners must move beyond simply supporting youth in farming – they should position them as key value chain actors, potential investors and an integral partner in decision-making processes. Encouraging youth self-employment in farming and agribusiness ventures will empower youth as owners of agriculture's future.

Technology adoption is equally critical. Young professionals are well-positioned to integrate digital solutions into farming – from climate monitoring tools to financial planning apps. The use of AI, in particular, offers educated youth opportunities to act as intermediaries between technology and farmers. It not only helps address the issue of digital illiteracy among farmers but also creates additional employment opportunities for the youth in rural areas. Ensuring that these intermediaries are adequately trained and receive ongoing support is crucial to the success of such initiatives. Technology alone is not enough – policymakers and development practitioners must ensure that these tools are accessible, scalable and integrated effectively into extension services. Training in digital agriculture, data analysis and business intelligence can provide young agripreneurs with the competitive edge they need.

The agricultural sector's future will be defined by the investments made today in the youth. Governments, NGOs, educational institutions and private sector actors must collaborate to create a clear path for young professionals – one that includes training, access to capital and leadership opportunities. By making these commitments now, agriculture can secure not only its workforce but its long-term resilience, profitability and sustainability.

17.4.3 Opportunities for youths as farm and agribusiness management specialists

The agricultural sector offers vast opportunities for young people, not only as entrepreneurs but also as providers of advisory services that support smallholder farmers. As highlighted throughout this book, farm and agribusiness management specialists play a crucial role in helping smallholders access markets, integrate into modern value chains and adopt innovative practices. These roles create employment in both public and private sectors, positioning young professionals as farm business management specialists or agribusiness consultants that contribute to the transformation of agriculture.

Farm business management specialists diagnose key challenges faced by smallholders and develop tailored solutions to enhance productivity, profitability and resilience. Their expertise in farming systems, market dynamics and sustainability enables them to facilitate access to credit, quality inputs and advisory support. They also strengthen connections between farmers, cooperatives and buyers, improving smallholders' competitiveness in modern value chains.

Young professionals are well-positioned to bring innovation and efficiency to farm advisory services, thanks to their adaptability and familiarity with modern technologies. They can offer mechanization services – such as tractor rentals, planting and harvesting support, and post-harvest processing – as well as engage in agro-input supply, financial services and digital extension. These services equip farmers with the tools and knowledge needed to boost productivity and profitability.

Beyond traditional advisory roles, young professionals also facilitate access to mobile applications that provide market information, precision agriculture tools and cooperative business models that promote efficiency and sustainability. More recently, they have begun to harness the potential of AI. By leveraging the skills of these young service providers, small family farmers can access AI-driven insights that would otherwise be beyond their reach, enabling more informed decision making and improved farm management.

Employment opportunities in this field span multiple sectors. In the public sector, young professionals can work as extension agents, supporting farmers through government-led rural development and modernization programmes. In the private sector, they can work as independent consultants or join agribusiness firms, helping to commercialize agriculture through tailored solutions for farmers. Expertise in financial management, strategic planning, marketing and participatory approaches enables young professionals to strengthen smallholder farmers' competitiveness in evolving markets.

To bridge the digital divide, farm management specialists can collaborate with private sector intermediaries to translate AI-generated insights into practical, accessible advice for farmers. These intermediaries – often young and educated individuals from local communities – interpret complex data and deliver actionable recommendations. This approach not only increases the accessibility of AI-driven tools but also fosters rural employment by creating new opportunities for youths. With proper training, these intermediaries become essential links between advanced AI systems and small-scale farmers, ensuring that technology-driven solutions are effectively integrated into local farming practices.

By equipping young professionals with the necessary skills, technology and business acumen, the agricultural sector can harness their potential to drive innovation, enhance productivity and promote sustainability. Their engagement in farm and agribusiness advisory services can transform smallholder farming, making it more profitable, resilient and prepared for future challenges.

Encouraging youth employment in farm management and agribusiness advisory services not only addresses the aging farming population but also fosters a dynamic and innovative agricultural sector. Supporting the success of small family farms while creating meaningful career opportunities for young professionals establishes farm and agribusiness advisory services as a promising and essential profession for the future.

17.5 Capacity Development for Scaling of Farm Management Field Programmes

Chapter 15, identified some of the key challenges faced by farm management advisory and extension support and the importance of addressing them for successful implementation. They included:

- Lack of clarity and recognition of the importance of public sector extension services.
- Lack of planning and supervision in public-sector extension programmes.
- The need for efficient use of manpower resources.
- Redefining extension worker roles.
- Upgrading staff skills and competencies to support farmers.
- Strengthening linkages between research and extension.
- Ensuring independence from political agendas and vested interests.
- The need for monitoring, evaluating and learning.
- Funding shortages in support of farm management extension.

Most of these challenges are related to capacity development at the level of the individual extension worker and farm management specialist as well as at the institutional level. As discussed in Chapter 15, the record of farm management extension and training delivered through agricultural extension services has been mixed, with many initiatives funded by donors, such as FAO, CIRAD, SIDA and USAID, often limited to short-term project durations. However, capacity development must be seen as a long-term investment to be taken to scale and ensure sustainable outcomes. A strategic, holistic approach is essential for enabling smallholder farmers, extension workers and youths to tackle modern agricultural challenges and seize entrepreneurial opportunities effectively. This approach must integrate three interconnected dimensions: (i) the enabling environment; (ii) organizational systems; and (iii) individual skills.

The enabling environment encompasses the broader social, economic and institutional systems that underpin agricultural advisory services. Supportive policies and investments are crucial for fostering access to markets, financial services and innovative technologies. For farm management extension workers, this includes professional development opportunities and access to resources, while for youths, it involves addressing barriers such as inadequate infrastructure and limited financing to promote agro-entrepreneurship.

The organizational dimension focuses on strengthening the institutions – public, private and civil society – that deliver training and advisory services. These organizations must broaden their scope beyond technical support to include farm business management and entrepreneurship training. Investing in staff capacity ensures that extension workers can effectively bridge the gap between farmers and specialized resources, acting as trainers, facilitators and advisors to promote sustainable farming practices.

At the core of capacity development is the individual dimension, which targets the skills, knowledge and attitudes of farmers, extension workers and youths. Training programmes for smallholder farmers should emphasize experiential learning, encouraging them to explore innovative practices and take ownership of their development. Extension workers, traditionally focused on technical agriculture, need specialized training in business management and entrepreneurial development to better support farmers. Similarly, empowering the youth with entrepreneurial skills and technical expertise positions them as drivers of innovation and change in the agricultural sector.

Integrating these dimensions ensures that capacity development is not a standalone activity, but a coordinated, continuous process aligned with the needs of all stakeholders. When policies, organizations and individuals work together, the agricultural sector can foster entrepreneurship, build resilient farming systems and contribute to sustainable rural development. This holistic strategy equips smallholder farmers, extension workers and youths to meet the evolving demands of modern agriculture while improving the livelihoods of smallholder farmers.

17.6 Conclusion

The transition to 'farming as a business' presents a significant challenge for small family farmers in the 21st century. To remain viable and competitive in a rapidly evolving agricultural landscape, smallholder farmers must move beyond traditional subsistence methods and embrace an entrepreneurial approach to agriculture. This shift requires more than just improving productivity and efficiency; it demands the development of business management skills, adoption of innovative practices and integration of technology into daily farm operations. Strengthening farm business management is essential for ensuring the long-term sustainability and resilience of smallholder farming.

As efficiency improves and factors such as specialization, technological advancements, production complexity and capital investment continue to shape the agricultural sector, farm consolidation is expected to increase. Larger, more commercially oriented smallholder farms will likely seek professional farm business management support to enhance decision making. The extension audience will also evolve, with greater participation from youths and women and a growing interest in agribusiness, climate resilience and sustainable farming practices.

With the right support mechanisms, smallholder farmers can adopt business-oriented strategies that enhance efficiency, profitability and resilience. By leveraging digital tools, strengthening value chains and improving access to financial and market systems, farmers can build sustainable enterprises better positioned to compete in modern agricultural markets. Implementing these strategies will not only transform smallholder farming but also contribute to national economic development. By empowering small scale family farmers with the skills and resources to manage their farms as businesses, these strategies can increase agricultural productivity, improve food security and boost rural incomes. As farmers become more efficient and market-oriented, the agricultural sector can contribute more substantially to national economies, driving economic growth and development – as well as maintain and enhance smallholder farmers' contribution to food security. This transformation can help countries graduate from low- to middle-income status, fostering broader socio-economic progress and reducing poverty levels.

The future of farm management holds immense promise, particularly with the active involvement of the youth. Young people bring energy, creativity and a willingness to embrace new technologies, making them key drivers of agricultural innovation and sustainability. By investing in farm business training, digital literacy and entrepreneurial development, we can equip young innovators to develop digital tools, mechanization services and advisory solutions that help smallholder farmers make better decisions and improve their operations. Their participation in agribusiness and value chain services will not only support smallholders but also create meaningful economic opportunities for youths themselves.

As smallholder farming evolves, continuous learning, strategic investment and innovative partnerships will be essential in creating an agricultural sector where farmers thrive. A stronger focus on sustainable farming practices and climate resilience will further ensure that agricultural development is both economically viable and environmentally sustainable. By adopting resource-efficient and adaptive practices, small-scale farmers will be better prepared to navigate the challenges of climate change and market volatility.

Looking ahead, the transformation of smallholder farming into a competitive and sustainable business sector depends on innovation, investment and collaboration. With the right policies, capacity-building efforts and access to resources, small family farmers can continue to play a crucial role in global food security and economic stability, securing a more resilient future for agriculture and rural communities.

In conclusion, while the transition to 'farming as a business' presents significant challenges, it also offers tremendous opportunities for small-scale family farmers and the broader agricultural sector. By implementing targeted strategies and leveraging the potential of young people, we can foster a more productive, resilient and sustainable agricultural landscape. This positive outlook on the future of farm management underscores the critical role of innovation, collaboration and continuous learning, in achieving long-term success and prosperity for these farmers and their communities. And it underscores the need for developing specialized support services in farm and agribusiness management.

Appendix: Example Training Programme in Farm Business Management for Small Family Farms

Training is a central theme throughout this book. The following table outlines a proposed training programme for farmers and extension workers, drawing on key content from seven of the technically focused chapters of this book. These chapters provide the core knowledge and skills essential for building capacity in farm business management. While the training themes, learning objective and topics have been structured as a guide, trainers should adapt the content and methods to match the existing skills and learning needs of participants.

Training theme	Learning objective	Topics covered
The essentials of management: drawn from Chapter 6		
Importance of effective farm management	Understand the role of farm management in improving productivity, profitability and sustainability	• Definition and scope of farm management • Key management functions: production, marketing, financial, labour, risk and organizational management • Relationship between management, profitability and sustainability • Impact of external factors (market trends, climate change, policies)
Production management	Develop skills for planning and managing crop and livestock production effectively	• Crop and livestock management decisions • Cropping systems, crop intensity and rotation • Production scheduling and resource allocation • Sustainable production practices and precision farming • Input procurement and enterprise policies
Marketing management	Develop the ability to identify markets, set prices and effectively sell products	• Understanding market demand and consumer trends • Marketing channels and strategies for smallholders • Price setting and forecasting • Value addition and branding • Overcoming marketing challenges (quality standards, transport, market access)
Financial management	Improve financial literacy and decision making for managing farm resources	• Importance of farm financial planning • Budgeting and cash flow management • Understanding capital structure and credit options • Debt management and investment planning • Record-keeping and financial analysis

Continued

Training theme	Learning objective	Topics covered
Organizational and information management	Strengthen the ability to structure farm operations and make data-driven decisions	• Record-keeping and inventory management • Efficient resource allocation • Time and labour management • Information gathering for better farm decisions • Role of digital tools in farm data management
Labour management	Develop strategies to optimize labour use and workforce productivity	• Family vs hired labour: planning and integration • Efficient labour allocation for peak and off-peak seasons • Managing workload, motivation and conflict resolution • Work-life balance and succession planning
Risk management	Develop the capacity to identify, assess and mitigate different types of risks	• Types of farm risks: production, market, financial, institutional and personal • Risk management strategies (crop diversification, climate adaptation, financial planning) • Developing risk mitigation plans • Case studies on managing uncertainty
Adapting to climate change: drawn from Chapter 7		
Impact of climate change on farming	Understand the direct and indirect effects of climate change on small family farms	• Rising temperatures and shifting rainfall patterns • Increased pest and disease pressures • Water scarcity and irrigation challenges • Economic risks and sustainability threats
Farm management for climate adaptation	Develop strategies for effective decision making to enhance farm resilience	• Risk management in farming • Crop diversification and income strategies • Community collaboration and shared resources • Financial planning for climate resilience
Sustainable production systems	Implement environmentally sustainable farming techniques for long-term viability	• Agroecological approaches • Soil health management and conservation • Biodiversity conservation and ecosystem services • Nutrient and pest management
Climate-smart agricultural practices	Apply climate-smart techniques to improve productivity and reduce environmental impact	• Conservation agriculture (minimum tillage, crop rotation, mulching) • Regenerative agriculture principles • Sustainable mechanization • Precision irrigation techniques
On-farm irrigation management	Enhance water use efficiency for improved crop yields and sustainability	• Types of irrigation systems (drip, sprinkler, flood) • Maximizing water use efficiency • Investment decisions in irrigation technology • Climate challenges affecting irrigation
Sustainable mechanization	Assess the role of mechanization in improving farm efficiency while ensuring sustainability	• Benefits and challenges of mechanization • Choosing appropriate farm machinery • Financial strategies for mechanization investment • Shared access models (cooperatives, rentals)
Digital agricultural technologies	Utilize digital tools to enhance farm productivity and resilience	• Precision agriculture and IoT applications • Remote sensing and AI for farming • Smart irrigation and farm management software • Digital marketplaces and e-marketing

Continued

Training theme	Learning objective	Topics covered
Market-oriented farming and adaptation	Balance profitability and climate adaptation strategies in a changing agricultural landscape	• Trade-offs in specialization vs diversification • Sustainable value chains • Certification schemes and carbon credit markets • Government and policy support
Economic concepts and principles for farm management: drawn from Chapter 9		
Foundations of farm economics	Understand key economic principles that influence farm decision making	• The role of economics in farm management • Key concepts: gross income, costs of production, profit • Opportunity cost and resource allocation • Comparative advantage and enterprise selection
Costs and profitability analysis	Applying tools to assess and improve farm profitability	• Fixed and variable costs • Gross margins and enterprise profitability • Understanding total costs and farm profit • Enterprise budgeting and cost control
Risk and uncertainty in farming	Identifying and managing risks affecting farm operations	• Types of risk: production, market, financial, institutional, personal • Managing price and climate variability • Diversification as a risk management strategy • Role of insurance and financial planning
Resource allocation and efficiency	Develop strategies to maximize farm productivity through efficient use of resources	• Optimizing land, labour and capital • The law of diminishing returns • Input-output relationships and marginal analysis • Factor-factor relationships and input substitution
Farm financial management	Strengthen financial decision making and long-term farm sustainability	• Cash flow management • Investment decisions and time value of money • Discounting and present value of future earnings • Depreciation and asset management
Market and price decision making	Enhance the capacity to navigate markets and price their products effectively	• Supply, demand and price determination • Price fluctuations and market cycles • Marketing strategies and value addition • Cost-benefit analysis for production and sales decisions
Budgeting: drawn from Chapter 11		
Introduction to budgeting in farm management	Understand the role of budgeting as a management tool for decision making and financial planning	• Definition and importance of budgeting • Budgeting as a planning aid • How budgeting helps in resource allocation and financial stability • Common budgeting challenges for smallholder farmers
Types of budgets and their uses	Learn about different types of budgets and when to use them	• Enterprise budgets: assessing profitability of individual farm enterprises • Partial budgets: evaluating the financial impact of incremental changes • Break-even budgeting: identifying the minimum yield or price needed to cover costs • Whole-farm budgeting: understanding the financial health of the entire farm

Continued

Training theme	Learning objective	Topics covered
Enterprise budgeting for farm planning	Develop the capacity to create and interpret enterprise budgets	• Components of an enterprise budget (revenues, costs, profitability measures) • How to calculate gross margins and break-even points • Comparing enterprises to determine the best farm mix • Using enterprise budgets to improve farm profitability
Break-even analysis and risk assessment	Determining the profitability threshold for their farm enterprises	• Understanding break-even price and break-even yield • How to calculate break-even points for different scenarios • Using break-even analysis to manage price and production risks • Case studies on break-even analysis in real farm situations
Partial budgeting for decision making	Evaluating the impact of small management changes on overall farm profitability	• When to use partial budgeting • Four key components: additional costs, additional income, costs saved and income forgone • Developing and interpreting a partial budget • Case studies of farm decisions using partial budgeting
Whole-farm budgeting for long-term planning	Develop skills in creating a full financial plan for farm operations	• The role of whole-farm budgeting in financial planning • Assessing available farm resources (land, labour, capital) • Integrating different enterprises into a whole-farm plan • Identifying risks and constraints in farm financial planning
Discounted cash flow budgeting and long-term investments	Introduce the concept of time value of money and its role in investment planning	• Time value of money and why future cash flows must be discounted • How to calculate net present value (NPV) for farm investments • Using discounted cash flow budgeting for long-term planning • Practical examples of investment decisions using discounted cash flow methods
Sources of budget information and participatory budgeting	Develop the capacity to access, interpret and use budget data effectively	• Sources of budgeting data (farm records, market reports, research data) • Challenges of collecting accurate farm data • Participatory budgeting as a tool for farmer-driven financial planning • Conducting participatory budgeting exercises with farmer groups
Diagnosis: drawn from Chapter 12		
Introduction to farm business diagnosis	Understand the importance of diagnosing farm business performance	• Role of diagnosis in farm management key diagnostic methods

Continued

Training theme	Learning objective	Topics covered
Farm management methods and decision making	Learn systematic approaches to farm business decision making	• The farm management cycle (diagnosis, planning, implementing, monitoring, evaluation) • Decision-making tools
Financial performance analysis	Develop skills in assessing financial health	• Gross margin analysis • Whole-farm income analysis
Market analysis and market margins	Identify market opportunities and profitability factors	• Understanding market trends • Market segmentation • Calculating market margins
Constraints and opportunities analysis	Assess farm business challenges and areas for improvement	• Identifying farm constraints • Conducting opportunity analysis
Efficiency measures in farm business	Improve farm productivity and cost efficiency	• Technical vs economic efficiency • Yield and resource efficiency measures
Trade-off analysis for decision making	Understand cost-benefit analysis for farm decisions	• Balancing productivity, cost and sustainability • Opportunity cost evaluation
Cash flow analysis for farm households	Manage farm finances effectively to ensure liquidity	• Identifying cash inflows and outflows • Developing cash flow projections
Risk and sensitivity analysis	Prepare for uncertainties in farming	• Types of risks in farming • Conducting sensitivity analysis
Planning: drawn from Chapter 13		
Introduction to farm planning	Understand the role of planning in farm management	• Importance of planning • Benefits for sustainability and profitability
The planning process	Learn step-by-step methods for structured farm planning	• Goal setting (SMART goals) • SWOT analysis • Action planning, implementation and monitoring
Strategic planning	Develop long-term strategies for farm success	• Mission, vision and objectives • Market trends and SWOT analysis • Aligning operations with long-term goals
Market planning	Align farm production with market demand	• Market research and customer needs • Pricing and selling strategies • Distribution channels and customer relationships
Farm planning	Optimize resource allocation for efficient farming	• Enterprise selection and diversification • Crop rotation and land use • Budgeting and profitability analysis
Labour planning	Efficiently manage farm labour resources	• Assessing labour needs • Gender considerations in labour roles • Labour scheduling and peak season strategies
Cash flow planning	Maintain financial stability and liquidity	• Estimating income and expenses • Managing financial risks and uncertainties • Tools for cash flow projections

Continued

Training theme	Learning objective	Topics covered
Investment planning	Evaluate investment opportunities for farm growth	• Assessing investment risks and returns • Payback period and rate of return methods • Financial decision making
Whole-farm planning	Integrate all farm planning aspects into a cohesive strategy	• Inventory of resources • Enterprise combination and gross margin analysis • Long-term sustainability planning
Monitoring and evaluation: drawn from Chapter 14		
Introduction to monitoring and evaluation (M&E)	Understand the role of M&E in farm management	• Definition and importance of M&E • How M&E improves decision making and profitability
The farm management cycle	Learn how M&E fits into the farm business cycle	• Stages of farm management (implementation, monitoring, evaluation) • Link between planning, monitoring and continuous improvement
Monitoring costs and returns	Track financial performance for profitability	• Cost and return tracking methods • Identifying unnecessary expenses and cost-saving measures
Key components of M&E systems	Implement an effective M&E system	• Data collection methods • Performance indicators (profitability, resource use, efficiency)
Record-keeping for smallholder farmers	Develop simple and effective farm record-keeping habits	• Types of farm records (production, inventory, labour, sales, etc.) • Digital vs manual record-keeping methods
Comparative analysis and benchmarking	Use benchmarking to assess farm performance	• Comparative analysis techniques • Identifying strengths and weaknesses through benchmarking
Monitoring the implementation of a farm plan	Ensure that farm activities align with planned goals	• Creating checklists for tracking activities • Regular field visits and resource tracking
Evaluating the performance of the farm plan	Assess whether planned strategies are effective	• Performance assessment techniques • Identifying gaps and areas for improvement
Integrating M&E into farm decision making	Use M&E insights to adapt farm strategies	• The role of continuous learning in farm management • Making data-driven decisions for long-term success

Index

Printed and bound by CPI Group (UK) Ltd, Croydon, CR0 4YY
13/09/2026
14961084-0005